D1691944

Jürn W. P. Schmelzer and
Ivan S. Gutzow
Glasses and the Glass Transition

Related Titles

Slezov, V. V.

Kinetics of First-order Phase Transitions

2009
ISBN 978-3-527-40775-0

Le Bourhis, E.

Glass
Mechanics and Technology

2008
ISBN 978-3-527-31549-9

Sinton, C. W.

Raw Materials for Glass and Ceramics
Sources, Processes, and Quality Control

2006
ISBN 978-0-471-47942-0

Drummond, C. H. (ed.)

65th Conference on Glass Problems
A Collection of Papers, Volume 26, Number 1

2006
ISBN 978-1-57498-238-1

Skripov, V. P., Faizullin, M. Z.

Crystal-Liquid-Gas Phase Transitions and Thermodynamic Similarity

2006
ISBN 978-3-527-40576-3

Schmelzer, J. W. P.

Nucleation Theory and Applications

2005
ISBN 978-3-527-40469-8

Jürn W. P. Schmelzer and Ivan S. Gutzow

Glasses and the Glass Transition

With the collaboration of

Oleg V. Mazurin
Alexander I. Priven
Snejana V. Todorova
Boris P. Petroff

WILEY-VCH

WILEY-VCH Verlag GmbH & Co. KGaA

The Authors

Dr. habil. Jürn W.P. Schmelzer
Institut für Physik
Universität Rostock
Rostock, Germany

Prof. Ivan S. Gutzow
Bulgarian Academy of Sciences
Institute of Physical Chemistry
Sofia, Bulgaria

Prof. Oleg V. Mazurin
Thermex
St. Petersburg, Russian Federation

Dr. Snejana V. Todorova
Bulgarian Academy of Sciences
Geophysical Institute
Sofia, Bulgarien

Dr. Boris B. Petroff
Bulgarian Academy of Sciences
Institute of Solid State Physics
Sofia, Bulgarien

Prof. Alexander I. Priven
Thermex
St. Petersburg, Russian Federation

Cover

The cover-picture shows a sample of diopside glass heat-treated at 870°C for about one hour. In the bulk of the sample, one can observe a spherulite of a wollastonite-like phase having a composition similar to that of the diopside glass. The crystalline dendrites below appeared on the surface of a crack (by courtesy of Prof. Vladimir M. Fokin, St. Petersburg, Russia).

All books published by **Wiley-VCH** are carefully produced. Nevertheless, authors, editors, and publisher do not warrant the information contained in these books, including this book, to be free of errors. Readers are advised to keep in mind that statements, data, illustrations, procedural details or other items may inadvertently be inaccurate.

Library of Congress Card No.: applied for

British Library Cataloguing-in-Publication Data:
A catalogue record for this book is available from the British Library.

Bibliographic information published by the Deutsche Nationalbibliothek
The Deutsche Nationalbibliothek lists this publication in the Deutsche Nationalbibliografie; detailed bibliographic data are available on the Internet at http://dnb.d-nb.de.

© 2011 WILEY-VCH Verlag GmbH & Co. KGaA, Boschstr. 12, 69469 Weinheim, Germany

All rights reserved (including those of translation into other languages). No part of this book may be reproduced in any form – by photoprinting, microfilm, or any other means – nor transmitted or translated into a machine language without written permission from the publishers. Registered names, trademarks, etc. used in this book, even when not specifically marked as such, are not to be considered unprotected by law.

Typesetting le-tex publishing services GmbH, Leipzig
Printing and Binding betz-druck GmbH, Darmstadt

Cover Design Grafik-Design Schulz, Fußgönheim

Printed in the Federal Republic of Germany
Printed on acid-free paper

ISBN Print 978-3-527-40968-6

ISBN ePub 978-3-527-63654-9
ISBN ePDF 978-3-527-63655-6
ISBN Mobi 978-3-527-63656-3

Foreword

First I would like to stress that the main authors of this monograph – Jürn W.P. Schmelzer and Ivan Gutzow – are renowned experts on properties of glasses, relaxation and phase formation processes in glasses that include glass transition, liquid–liquid phase separation, crystal nucleation, crystal growth and overall crystallization processes. In the present book, their attention is concentrated on the description of glasses and glass transition. In analyzing this circle of problems, of special relevance is their strong background on thermodynamics; they always bring their research projects into a solid thermodynamic framework. Their new monograph *Glasses and the Glass Transition* is no exception; it could be also be named, for instance, *Thermodynamics of the Vitreous State*. In this book, they review, organize and summarize – within a historical perspective and discussing alternative approaches – the results of their own publications on different thermodynamic aspects of the vitreous state performed after the publication of their book, *The Vitreous State: Thermodynamics, Structure, Rheology, and Crystallization* published by Springer in 1995.

After the introduction, in Chapter 2, Schmelzer and Gutzow disclose their ideas on the nature of glasses through an overview of the basic laws of classical thermodynamics, the description of nonequilibrium states, phase transitions, crystallization, viscosity of glass-forming systems, thermodynamic properties of glass-forming melts, glass transition, and on the overall thermodynamic nature of the glassy state. Most of the presented concepts are discussed within a thermodynamic perspective.

In Chapter 3, they present and discuss in detail a "generic theory of vitrification of liquids" and explain the application of thermodynamics of irreversible processes to vitrification. Then, the authors review relaxation of glass-forming melts, define the glass transition, comment on the entropy at very low temperatures, the Kauzmann paradox, the Prigogine–Defay ratio – including several quantitative estimates of this parameter – the concept of fictive temperature as a structural order parameter, the viscosity and relaxation time at T_g, and finally frozen-in thermodynamic fluctuations. Once more, all these important characteristics of the vitreous state are discussed within thermodynamic insights. These concepts are developed in more detail in the application to relaxation and the pressure dependence of the viscosity in Chapter 4, and an analysis of systems with a glass-like behavior in Chapter 5.

In order to apply thermodynamic methods, in general, and to glasses and glass transition, in particular, the thermodynamic properties of the respective systems, such as the equations of state, must be known. In order to give an overview on the present state in this direction, the book also comprises two special chapters: Chapter 6, authored by Oleg Mazurin on the collection and analysis of glass property data; and Chapter 7, written by Alexander Priven, discusses the available models to correlate certain glass properties to their chemical composition.

Mazurin was the head of the famous Laboratory of Glass Properties of the Grebenshchikov Institute of Silicate Chemistry of the Russian Academy of Sciences in the former USSR, for almost 40 years. Then he retired and dedicated his efforts to his present passion: collection and critical analysis of glass property data. He worked intensively and published numerous papers on many types of glass properties, and perhaps most importantly, from the early stages he started to collect property data from his own group and from published literature to finally mastermind the assembly of an impressive and most useful glass database – SciGlass – which currently contains the properties of more than 350 000 glass compositions. In my opinion this database is a must in the library of any active glass research group in the world. I am particularly proud to say that my students and I have been using the SciGlass database for several years from its very beginning. In this chapter, Mazurin discusses the power of SciGlass, which is regularly updated on an annual basis, and its multiple utilities. The author emphasizes how one can and should use SciGlass to compare the values of properties measured by any author against all the available data. He also stresses a nasty and frequent problem; that is, the poor quality of data published by some research groups.

Having at one's disposal such a comprehensive overview on existing experimental data on glass properties, the next question arises whether it is possible to theoretically predict – based on such knowledge – some properties of glass-forming systems, for which experimental data are not available. This task is reviewed in the chapter written by Priven. It is probably fair to say that Priven dedicated almost his entire career to this very important scientific and technological theme: development and testing of models to predict the compositional dependence of important glass properties, such as density, thermal expansion coefficient, refractive index, liquidus and viscosity, to their chemical composition. Priven has been involved in the arduous and complex quest of what he calls the "silver bullet." Although I never carried out any specific research on this particular subject, I have always been a keen user of Appen's model, and lately of Priven's model, to calculate glass properties from chemical composition for the development of new glasses and glass-ceramics (which always have a residual glass matrix). Priven discusses the strengths and weaknesses of several available models, for example, Winkelman and Schott's, Gelhoff and Thomas's, Gilard and Dubrul's, Huggins and Sun's, Appen's, Demkina's, Mazurin's, Fuxi's, Lakatos's and of his own model. He also discusses the numerous difficulties to develop an accurate model due to the nonexistence of data for many compositions and nonlinear effects, such as the anomalous effects of boron and alumina, which fortunately have been already solved by some of these models. At present one can use Priven's model and some others for surprisingly good predictions (say within 5–10%) of several properties of glasses containing up to 30 elements. But the "silver bullet" – to accurately predict the properties of all possible glasses with combinations of all the 80 "friendly" elements of the periodical table – is far from being found (please check the Zanotto paper on this topic published in the *Journal of Non-Crystalline Solids*, 347 (2004) 285–288).

After completion of the general task, that is, the thermodynamic description of glasses and the glass transition, the overview on existing data on glasses and the methods of prediction of glass properties, Schmelzer and Gutzow go over to a more detailed discussion of some peculiar properties of glasses and glass-like systems and their possible technological applications. Chapter 8 deals with "glasses as accumulators of free energy, of increased reactivity and as materials with unusual applications," whereas Chapter 9 is devoted to the third law of thermodynamics and its application to the vitreous state. It is usually stated that the third law is not applicable to glasses as they never reach equilibrium for temperatures tending to zero. Therefore, their entropy has to be larger than zero for $T \to 0$. The authors present a detailed historical development, describe the application of thermodynamics to nonequilibrium states, show some thermodynamic and kinetic invariants at T_g and give an extended discussion on the current controversial issue of zero-point entropy of glasses. The book is completed by an interesting analysis of the etymology of the word "glass."

Summarizing, the present book presents a thorough discussion about the nature of glass, with a rich historical background (a characteristic of Ivan Gutzow) on the most basic properties of glasses including vitrification kinetics, relaxation and glass transition. It cites more than 650 articles. This book will certainly be a very useful reference for experienced researchers as well as for post-graduate students who are interested in understanding the nature of glass and the application of the laws of thermodynamics to nonequilibrium materials such as glasses.

Edgar D. Zanotto
August 2010

Head of the Vitreous Materials Laboratory
Federal University of São Carlos, Brazil
Member of the Brazilian Academy of Sciences
Member of the World Academy of Ceramics
Fellow of the British Society of Glass Technology

Contents

Foreword *V*

Preface *XVII*

Contributors *XIX*

1 Introduction *1*
Jürn W. P. Schmelzer and Ivan S. Gutzow

2 Basic Properties and the Nature of Glasses: an Overview *9*
Ivan S. Gutzow and Jürn W. P. Schmelzer
2.1 Glasses: First Attempts at a Classification *9*
2.2 Basic Thermodynamics *14*
2.2.1 The Fundamental Laws of Classical Thermodynamics and Consequences *14*
2.2.2 Thermodynamic Evolution Criteria, Stability Conditions and the Thermodynamic Description of Nonequilibrium States *22*
2.2.3 Phases and Phase Transitions: Gibbs's Phase Rule, Ehrenfest's Classification, and the Landau Theory *26*
2.3 Crystallization, Glass Transition and Devitrification of Glass-Forming Melts: an Overview of Experimental Results *36*
2.4 The Viscosity of Glass-Forming Melts *46*
2.4.1 Temperature Dependence of the Viscosity *46*
2.4.2 Significance of Viscosity in the Glass Transition *54*
2.4.3 Molecular Properties Connected with the Viscosity *57*
2.5 Thermodynamic Properties of Glass-Forming Melts and Glasses: Overview on Experimental Results *59*
2.5.1 Heat Capacity *59*
2.5.2 Temperature Dependence of the Thermodynamic Functions: Simon's Approximation *65*
2.5.3 Further Methods of Determination of Caloric Properties of Glass-Forming Melts and Glasses *74*
2.5.4 Change of Mechanical, Optical and Electrical Properties in the Glass Transition Range *76*

| 2.6 | Thermodynamic Nature of the Glassy State *82* |
| 2.7 | Concluding Remarks *88* |

3 Generic Theory of Vitrification of Glass-Forming Melts *91*
Jürn W. P. Schmelzer and Ivan S. Gutzow

3.1	Introduction *91*
3.2	Basic Ideas and Equations of the Thermodynamics of Irreversible Processes and Application to Vitrification and Devitrification Processes *95*
3.2.1	Basic Assumptions *95*
3.2.2	General Thermodynamic Dependencies *96*
3.2.3	Application to Vitrification and Devitrification Processes *100*
3.3	Properties of Glass-Forming Melts: Basic Model Assumptions *103*
3.3.1	Kinetics of Relaxation *103*
3.3.2	Thermodynamic Properties: Generalized Equation of State *105*
3.4	Kinetics of Nonisothermal Relaxation as a Model of the Glass Transition: Change of the Thermodynamic Functions in Cyclic Cooling-Heating Processes *107*
3.4.1	Description of the Cyclic Processes under Consideration *107*
3.4.2	Temperature Dependence of the Structural Order Parameter in Cyclic Cooling and Heating Processes *108*
3.4.3	Definition of the Glass Transition Temperature via the Structural Order Parameter: the Bartenev–Ritland Equation *110*
3.4.4	Structural Order Parameter and Entropy Production *113*
3.4.5	Temperature Dependence of Thermodynamic Potentials at Vitrification *115*
3.4.5.1	Configurational Contributions to Thermodynamic Functions *115*
3.4.5.2	Some Comments on the Value of the Configurational Entropy at Low Temperatures and on the Kauzmann Paradox *121*
3.4.6	Cyclic Heating-Cooling Processes: General Results *123*
3.5	The Prigogine–Defay Ratio *125*
3.5.1	Introduction *125*
3.5.2	Derivation *127*
3.5.2.1	General Results *127*
3.5.2.2	Quantitative Estimates *133*
3.5.2.3	An Alternative Approach: Jumps of the Thermodynamic Coefficients in Vitrification *135*
3.5.3	Comparison with Experimental Data *137*
3.5.3.1	The Prigogine–Defay Ratio *137*
3.5.3.2	Change of Young's Modulus in Vitrification *140*
3.5.4	Discussion *142*
3.6	Fictive (Internal) Pressure and Fictive Temperature as Structural Order Parameters *143*
3.6.1	Brief Overview *143*

3.6.2	Model-Independent Definition of Fictive (Internal) Pressure and Fictive Temperature *146*
3.7	On the Behavior of the Viscosity and Relaxation Time at Glass Transition *149*
3.8	On the Intensity of Thermal Fluctuations in Cooling and Heating of Glass-Forming Systems *152*
3.8.1	Introduction *152*
3.8.2	Glasses as Systems with Frozen-in Thermodynamic Fluctuations: Mueller and Porai-Koshits *153*
3.8.3	Final Remarks *158*
3.9	Results and Discussion *158*

4 Generic Approach to the Viscosity and the Relaxation Behavior of Glass-Forming Melts *165*
Jürn W. P. Schmelzer

4.1	Introduction *165*
4.2	Pressure Dependence of the Viscosity *166*
4.2.1	Application of Free Volume Concepts *166*
4.2.2	A First Exception: Water *169*
4.2.3	Structural Changes of Liquids and Their Effect on the Pressure Dependence of the Viscosity *171*
4.2.4	Discussion *173*
4.3	Relaxation Laws and Structural Order Parameter Approach *174*
4.3.1	Basic Equations: Aim of the Analysis *174*
4.3.2	Analysis *175*
4.3.3	Discussion *177*

5 Thermodynamics of Amorphous Solids, Glasses, and Disordered Crystals *179*
Ivan S. Gutzow, Boris P. Petroff, Snejana V. Todorova, and Jürn W. P. Schmelzer

5.1	Introduction *179*
5.2	Experimental Evidence on Specific Heats and Change of Caloric Properties in Glasses and in Disordered Solids: Simon's Approximations *182*
5.3	Consequences of Simon's Classical Approximation: the $\Delta G(T)$ Course *194*
5.4	Change of Kinetic Properties at T_g and the Course of the Vitrification Kinetics *195*
5.5	The Frenkel–Kobeko Postulate in Terms of the Generic Phenomenological Approach and the Derivation of Kinetic and Thermodynamic Invariants *198*
5.6	Glass Transitions in Liquid Crystals and Frozen-in Orientational Modes in Crystals *208*
5.7	Spectroscopic Determination of Zero-Point Entropies in Molecular Disordered Crystals *212*

5.8	Entropy of Mixing in Disordered Crystals, in Spin Glasses and in Simple Oxide Glasses *213*
5.9	Generalized Experimental Evidence on the Caloric Properties of Typical Glass-Forming Systems *215*
5.10	General Conclusions *219*

6   Principles and Methods of Collection of Glass Property Data and Analysis of Data Reliability *223*

Oleg V. Mazurin

6.1	Introduction *223*
6.2	Principles of Data Collection and Presentation *225*
6.2.1	Main Principles of Data Collection *225*
6.2.2	Reasons to Use the Stated Principles of Data Collection *228*
6.2.3	Problems in Collecting the Largest Possible Amounts of Glass Property Data *230*
6.2.4	Main Principles of Data Presentation *231*
6.3	Analysis of Existing Data *232*
6.3.1	About the Reliability of Experimental Data *232*
6.3.2	Analysis of Data on Properties of Binary Systems *233*
6.3.2.1	General Features of the Analysis *233*
6.3.2.2	Some Factors Leading to Gross Errors *237*
6.3.2.3	Some Specific Examples of the Statistical Analysis of Experimental Data *239*
6.3.2.4	What is to Do if the Number of Sources Is Too Small? *243*
6.4	About the Reliability of the Authors of Publications *246*
6.4.1	The Moral Aspect of the Problem *246*
6.4.2	An Example of Systematically Unreliable Experimental Data *247*
6.4.3	Concluding Remarks *251*
6.5	General Conclusion *253*

7   Methods of Prediction of Glass Properties from Chemical Compositions *255*

Alexander I. Priven

7.1	Introduction: 120 Years in Search of a Silver Bullet *255*
7.2	Principle of Additivity of Glass Properties *257*
7.2.1	Simple Additive Formulae *257*
7.2.2	Additivity and Linearity *258*
7.2.3	Deviations from Linearity *259*
7.3	First Attempts of Simulation of Nonlinear Effects *260*
7.3.1	Winkelmann and Schott: Different Partial Coefficients for Different Composition Areas *260*
7.3.2	Gehlhoff and Thomas: Simulation of Small Effects *260*
7.3.3	Gilard and Dubrul: Polynomial Models *262*
7.4	Structural and Chemical Approaches *264*
7.4.1	Nonlinear Effects and Glass Structure *264*
7.4.2	Specifics of the Structural Approach to Glass Property Prediction *266*

- 7.4.3 First Trials of Application of Structural and Chemical Ideas to the Analysis of Glass Property Data *267*
- 7.4.4 Evaluation of the Contribution of Boron Oxide to Glass Properties *267*
- 7.4.4.1 Model by Huggins and Sun *268*
- 7.4.4.2 Models by Appen and Demkina *268*
- 7.4.5 Use of Other Structural Characteristics in Appen's and Demkina's Models *271*
- 7.4.6 Recalculation of the Chemical Compositions of Glasses *272*
- 7.4.7 Use of Atomic Characteristics in Glass and Melt Property Prediction Models *278*
- 7.4.8 Ab Initio and Other Direct Methods of Simulation of Glass Structure and Properties *279*
- 7.4.9 Conclusion *280*
- 7.5 Simulation of Viscosity of Oxide Glass-Forming Melts in the Twentieth Century *280*
- 7.5.1 Simulation of Viscosity as a Function of Chemical Composition and Temperature *280*
- 7.5.2 Approaches to Simulation of Concentration Dependencies of Viscosity Characteristics *282*
- 7.5.2.1 Linear Approach *282*
- 7.5.2.2 Approach of Mazurin: Summarizing of Effects *283*
- 7.5.2.3 Approach of Lakatos: Redefinition of Variables *284*
- 7.5.2.4 Polynomial Models *284*
- 7.5.3 Conclusion *285*
- 7.6 Simulation of Concentration Dependencies of Glass and Melt Properties at the Beginning of the Twenty-First Century *286*
- 7.6.1 Global Glass Property Databases as a Catalyst for Development of Glass Property Models *286*
- 7.6.2 Linear and Polynomial Models *286*
- 7.6.3 Calculation of Liquidus Temperature: Neural Network Simulation *289*
- 7.6.4 Approach of the Author *291*
- 7.6.4.1 Background *291*
- 7.6.4.2 Model *292*
- 7.6.4.3 Comparison with Previous Models *294*
- 7.6.4.4 Conclusion *296*
- 7.6.5 Fluegel: a Global Model as a Combination of Local Models *296*
- 7.6.6 Integrated Approach: Evaluation of the Most Probable Property Values and Their Errors by Using all Available Models and Large Arrays of Data *297*
- 7.7 Simulation of Concentration Dependencies of Glass Properties in Nonoxide Systems *299*
- 7.8 Summary: Which Models Were Successful in the Past? *301*
- 7.9 Instead of a Conclusion: How to Catch a Bluebird *306*

8		**Glasses as Accumulators of Free Energy and Other Unusual Applications of Glasses** *311*
		Ivan S. Gutzow and Snejana V. Todorova
8.1		Introduction *311*
8.2		Ways to Describe the Glass Transition, the Properties of Glasses and of Defect Crystals: a Recapitulation *313*
8.3		Simon's Approximation, the Thermodynamic Structural Factor, the Kinetic Fragility of Liquids and the Thermodynamic Properties of Defect Crystals *318*
8.4		The Energy, Accumulated in Glasses and Defect Crystals: Simple Geometric Estimates of Frozen-in Entropy and Enthalpy *324*
8.4.1		Enthalpy Accumulated at the Glass Transitions *324*
8.4.2		Free Energy Accumulated at the Glass Transition and in Defect Crystals *327*
8.5		Three Direct Ways to Liberate the Energy, Frozen-in in Glasses: Crystallization, Dissolution and Chemical Reactions *331*
8.5.1		Solubility of Glasses and Its Significance in Crystal Synthesis and in the Thermodynamics of Vitreous States *332*
8.5.2		The Increased Reactivity of Glasses and the Kinetics of Chemical Reactions Involving Vitreous Solids *339*
8.6		The Fourth Possibility to Release the Energy of Glass: the Glass/Crystal Galvanic Cell *340*
8.7		Thermoelectric Driving Force at Metallic Glass/Crystal Contacts: the Seebeck and the Peltier Effects *344*
8.8		Unusual Methods of Formation of Glasses in Nature and Their Technical Significance *348*
8.8.1		Introductory Remarks *348*
8.8.2		Agriglasses, Glasses as Nuclear Waste Forms and Possible Medical Applications of Dissolving Organic Glasses *350*
8.8.3		Glasses as Amorphous Battery Electrodes, as Battery Electrolytes and as Battery Membranes *352*
8.8.4		Photoeffects in Amorphous Solids and the Conductivity of Glasses *353*
8.9		Some Conclusions and a Discussion of Results and Possibilities *354*
9		**Glasses and the Third Law of Thermodynamics** *357*
		Ivan S. Gutzow and Jürn W. P. Schmelzer
9.1		Introduction *357*
9.2		A Brief Historical Recollection *360*
9.3		The Classical Thermodynamic Approach *363*
9.4		Nonequilibrium States and Classical Thermodynamic Treatment *366*
9.5		Zero-Point Entropy of Glasses and Defect Crystals: Calculations and Structural Dependence *368*
9.6		Thermodynamic and Kinetic Invariants of the Glass Transition *369*
9.7		Experimental Verification of the Existence of Frozen-in Entropies *371*

9.8	Principle of Thermodynamic Correspondence and Zero-Point Entropy Calculations *376*
9.9	A Recapitulation: the Third Principle of Thermodynamics in Nonequilibrium States *377*

10 On the Etymology of the Word "Glass" in European Languages and Some Final Remarks *379*
Ivan S. Gutzow

10.1	Introductory Remarks *379*
10.2	"Sirsu", "Shvistras", "Hyalos","Vitrum", "Glaes", "Staklo", "Cam" *380*
10.3	"Vitreous", "Glassy" and "Glasartig", "Vitro-crystalline" *382*
10.4	Glasses in Byzantium, in Western Europe, in Venice, in the Balkans and Several Other Issues *384*
10.5	Concluding Remarks *385*

References *387*

Index *407*

Preface

Nearly all authors of modern science-based books aspire to write the definitive summary of a chosen subject, a field of inquiry, or the history or future of an emerging discipline. Books such as the *Theory of Metals and Alloys* by N. Mott and H. Jones, *Introduction to Solid State Physics* by C. Kittel, and *The Nature of the Chemical Bond* by L. Pauling readily come to mind in this regard. This book, more correctly a treatise, by Jürn W.P. Schmelzer and Ivan S. Gutzow, with collaboration by Oleg V. Mazurin, Alexander I. Priven, Snejana V. Todorova, and Boris P. Petroff, rises to this level in its comprehensive summary of the field of glass science. Drawing heavily on thermodynamics, kinetic theory, and the physical sciences, virtually all aspects of the material glass including the transition of an undercooled melt to the glassy state, are summarized in an authoritative, scholarly, and convincing manner. Not only is the scope and volume of material examined for this treatise impressive, so is the way in which it was compiled. That is, the compilation was made through an exhaustive, comprehensive review and evaluation of the major literature on this subject from the present day contributions back to the beginning of the last century.

The authors were aided immeasurably by a facility to read articles as they appeared in their original language thereby helping to gain special and even essential historical insight into the value and scientific correctness of the material being reviewed. This in turn allowed a more global approach in direct support of producing an authoritative discourse on glass and the glass transition. Worthy of exceptional praise is their discussion of glass and third law of thermodynamics found in Chapter 9. It gives credence to accepting glass science as a major field in its own right and the need for nearly all theorists to understand what glass is.

The completion of this task is, in fact, a life-long labor of love for these authors, one which could only be undertaken by a handful of scientists across the globe who have shared a similar devotion to this subject throughout their professional lives. In short, they have accomplished their intended purpose in preparing this treatise: writing the definitive description of glass and the glass transition. It holds potential to be recognized as a remarkable milestone.

And what might be the expected outcomes of this Herculean task? As a university professor, I use to make sure my students understood that the penultimate essence of science is to predict – to predict facts that are known and can be measured, to predict those that are unknown, and make sure the predictions are quantitative.

That is, science must be quantified through mathematical treatment. Through the authors' mathematical description of the material of glass and its properties including the glass transition, practicing glass scientists and engineers will find this book invaluable in helping them understand and predict the behavior of glass in a wide range of settings and applications. It will no doubt be a springboard to the development of advanced and possibly even more mathematically rigorous theories of the vitreous state as new observations are recorded, and their meaning explored.

Additionally, it will be difficult to prepare a manuscript or professional presentation on glass or the glass transition without understanding or referencing this contribution to the scientific literature. In a similar way, countless students engaged in thesis work on glass will surely become familiar with this treatise, if their academic work involves fabrication, characterization, or application.

For all of these reasons, it is confidently predicted this treatise will gain international recognition – not only across the entire materials spectrum – but also in the broader fields of physics and chemistry. And for this potential, the authors and their collaborators should take a well-deserved bow.

L. David Pye *Professor of Glass Science, Emeritus, Alfred University, USA*
August 2010 *Past President of the International Commission on Glass*
Past President of the American Ceramic Society
Honorary Member of the German Society of Glass Technology
Honorary Fellow, the British Society of Glass Technology

Contributors

Oleg V. Mazurin, Thermex, St. Petersburg, Russia, mazurin@itcwin.com, Ph.D. (1953); D.Sc. (1962). Head of the Laboratory of Glass Properties (Grebenshchikov Institute of Silicate Chemistry of the Academy of Sciences of the USSR) from 1963 to 1999. Editor of the *Journal Glass Physics and Chemistry* from 1990 to 2000. Regional editor of the *Journal of Non-Crystalline Solids* from 1983 to 1990. Author and co-author of more than 400 papers, 14 books in Russian (including one hand-book in 9 volumes) and 3 books in English (including one hand-book in 5 volumes). Leading author of the scientific database for the Glass Property Information System SciGlass. Awarded by the Morey Award "for outstanding achievements in glass science" from the American Ceramic Society, Columbus (1994) and by the President's award from the International Commission on Glass (2001). Field of expertise: studies of electrical conductivity, viscosity, thermal expansion, relaxation characteristics of inorganic glasses and melts, glass transition phenomenon, phase separation in glasses, collection and analysis of published data on properties of inorganic glasses.

Alexander I. Priven, ITC Inc., St Petersburg, Russia, priven@itcwin.com, Ph.D. (1998), D.Sc. (2003). Author of methods for prediction of various physical properties of glasses and glass-forming melts in wide composition and temperature ranges. The dominant area of interest is simulation of the dependencies of physical properties of glasses and glass-forming melts on concentration, temperature and thermal history. Took part in the development of SciGlass Information System and other material property databases. Worked as a consultant for Samsung Corning Precision Glass Co. (South Korea), Corning Inc. (USA), ITC Inc. (USA).

Snejana V. Todorova, Geophysical Institute, Bulgarian Academy of Sciences, Sofia, Bulgaria, s_todorova@gbg.bg, Associate Professor in geophysics at BAS and specialist of cloud microphysics, Dr. Todorova came to the problems of glass transition from joint investigations with I. Gutzow and J. Schmelzer on the kinetics of nucleation. Of particular significance in writing the present book was also her interest in processes of planetary and cosmic formation of vitreous ice and in the possible existence of intermediate amorphous stages in vapor condensation.

Boris P. Petroff, Institute of Solid State Physics, Bulgarian Academy of Sciences, Sofia, Bulgaria, borissp@yahoo.com, As a theoretical physicist with excellent knowledge in many fields of solid state and intermediate liquid/solid forms of condensed matter and especially in liquid crystals, Boris Petroff provided for the present book the necessary links between typical glasses and other forms of systems with frozen-in defect structure.

Reflections of some of the discussions held over the course of preparation for the present monograph (Ivan S. Gutzow, Jürn W.P. Schmelzer, and Oleg V. Mazurin (from left to right))

1
Introduction

Jürn W. P. Schmelzer and Ivan S. Gutzow

The present book is devoted to the ways of formation, to the analysis of the properties and to the theoretical description of matter in a very particular state: in the vitreous state. In such a state, as it is usually assumed, amorphous solids may exist with a kinetically frozen-in, rigid structure and correspondingly frozen-in thermodynamic and mechanical properties. In technical applications and in every-day life as materials, in solid state physics, in scientific literature and also in the title of the present book these amorphous solids are called glasses. Glasses are formed in a process, which is also very particular in its physics, in its kinetics and thermodynamics, in the process of glass transition or vitrification. The reverse process is denoted as devitrification, and is frequently but not always accompanied by crystallization processes. Definitions of both these notions are given in this book, accounting for present-day results of structural analysis, statistical physics and especially of thermodynamics of glasses, since thermodynamics is the science that decides over the states of matter.

It turns out that many solids, even with a partly crystalline structure, behave in their properties and thermodynamic state-like glasses: such solids with a frozen-in defect crystalline structure (e.g., defect molecular crystals) are sometimes called glass-like, or even glassy crystals. It is shown in one of the following chapters, that highly oriented liquids, so-called liquid crystals, also form glasses, corresponding to the particular structure of the precursor liquid. Another very particular class of solids with usually crystalline structure but with glass-like magnetic properties are the spin glasses.

The process of glass transition has general, characteristic and remarkable features, which are observed not only in the vitrification of glass-forming melts and in other processes of glass formation to be discussed in details in the present book (e.g., vapor quenching in the formation of amorphous layers or electrolytic deposition of metallic alloy glasses), but also, according to some authors, in quite different events, for example, in cosmic processes. There are also many models in statistical physics, the solution of which leads to virtual systems with glass-like thermodynamic or kinetic properties; or on the contrary, to solutions of the theory developed,

Glasses and the Glass Transition, First Edition. Jürn W.P. Schmelzer and Ivan S. Gutzow.
© 2011 WILEY-VCH Verlag GmbH & Co. KGaA. Published 2011 by WILEY-VCH Verlag GmbH & Co. KGaA.

which against the expectations of their authors, are far from the behavior of real (or "common, molecular or laboratory") glasses.

Out of many possible real or imagined glassy or glass-like systems of different solids with amorphous or defect crystalline frozen-in structure, are the common glasses of every-day life, in particular the technical glasses, who give well-known and best studied examples of the process of glass transition and of the properties of vitreous states. Besides systems with inorganic composition (silicates, phosphates, halides, elemental glasses, etc.), what must also be included are the numerous representatives of organic chemistry (polymers, molecular glasses), aqueous solutions, metal alloys: practically representatives of any known structure and chemical composition. There are, as we have discussed in several publications, serious expectations, based on well-founded kinetic and structural criteria, that almost any substance could be vitrified at appropriately chosen conditions. These expectations are substantiated in our foregoing monograph [1] also devoted to the vitreous state.

However, the best known representatives of vitreous solids still remain the "common" inorganic glasses, which in their composition are oxides and especially silicates. The properties of these silicate, phosphate, borate, and so on, oxide glasses are summarized in a series of monographic reference books and the database organized by Mazurin and his collaborators [2, 3] discussed in detail in one of the chapters of the present book.

Silicate glasses belong with pottery, ceramics and bronze to the oldest materials employed by man. This early widespread application of glasses is in some respect also due to their broad distribution in nature. As an example, magmatic rocks can be mentioned, which to a large degree consists of vitreous silicates, or completely amorphous natural glasses such as obsidian or amber. It is well-known that the natural glass obsidian served as a material for the preparation of the first cutting tools of primitive men. Obsidian remained in the ancient cultures of Central America as the material for objects not only of art but also for the horrible ritual knives of the high priest of these Native American societies. Amber is most probably the first organic glass to be appreciated and used by mankind: impressive with both its beauty and for its unusual dielectric properties, known from ancient times in Greek natural philosophy.

The wide distribution of glasses in nature is not due to chance. The inner part of the Earth, characterized by very high values of pressure and temperature, is most probably itself an enormous reservoir of highly pressurized glass-like or glass-forming melts. Processes of crystallization and glass-formation connected with the eruption of volcanoes and the more or less abrupt cooling and even quenching processes of parts of this melt determine to a large degree the course of geological processes and the structure and properties of the lithosphere. Natural glasses are widespread not only on Earth but also on the Moon as it became evident from the investigation of samples of lunar rocks brought to Earth by the lunar expeditions in the mid-1970s.

Of particular interest is also the vitreous form of water. In this respect it is also worth mentioning that according to estimates made by some authors (see [4, 5]) water in the universe as a whole appears to be practically 99.9% in this vitreous form. In its vitreous form water is the main constituent of comets (of the comet "head"). At earth conditions, glassy water or aqueous solutions can be vitrified only by superfast quenching methods (described in its principle features in Chapter 2) or (as thin layers) by water vapor quenching on cold substrates (held below 120 K). On the possible role of vitreous water in the dissemination of life in the universe (the panspermian hypothesis of Hellenistic philosophy) see the considerations given below and in Chapter 8 of the present book.

The first applications of natural glasses in primitive societies for a limited number of purposes were followed by the beginning of glass production in Mesopotamia, Egypt and ancient Rome, by medieval European and Middle East glass-making and then by a long evolution to the modern glass industries and to glass science. From the point of view of the variety of properties of glasses and of the spectrum of possible applications, the significance of the vitreous state in its different forms in present-day technology and the technical importance of different glasses can hardly be estimated. The validity of this statement becomes evident if one tries to imagine for a while things surrounding us in every day life without the components made of vitreous materials. Technical glasses (like chemically resistant glasses) or optical glasses are well-known to everyone. Imagine, for example, a chemical plant, a physical laboratory, a car or a dwelling house without glasses or let us think about the importance of silicate glasses and optical glasses (with their complex compositions) in general for optical devices, and in particular, in microscopy and astronomy.

In addition to the classical oxide and particularly silicate glasses in the last decades of the twentieth century new structural and chemical classes of vitreous materials gained scientific and technical importance. They consist, as said at the beginning of this introduction, of substances of any class, or of mixtures and solutions of different substances, for which the possibility of existence in the vitreous state at these times was thought as being exotic or even impossible. One example in this respect are metallic glasses which are formed usually in processes, connected with hyper-quenching rates (e.g., in splat-cooling, to be discussed in Chapter 2) of metallic or metal-semi-metal alloy systems. Metallic glasses, first synthesized in the early 1960s, in a period of about ten to twenty years were transferred from a stage of exotic research to the stage of production and world-wide technological application. Similar examples are supplied by glassy polymers or vitreous carbon, glass-forming chalcogenide or glassy halide systems. The development of modern methods of information technology, for example, cable TV is also based on glass: on defect-free, extremely translucent glassy fibers with appropriate optical characteristics.

With the dramatic increase of the number of substances obtained in the vitreous state, the variety of properties and possible applications of glasses has also increased. Beyond the traditional applications in technology and science glassy materials are also used as substitutes for biological organs or tissues, for example, as prostheses (as vitreous carbon in heart transplants, as bio-glass ceramics in bone operations) and even in ophthalmology. Glass-forming aqueous solutions with biologically relevant compositions are used as a carrier medium for the freezing-in of biological tissues. Thus, it seems that even life can be frozen-in to a glass solving the problems of absolute anabiosis within the vitreous state. Life, frozen-in to a glass, supports the already mentioned idea of Hellenistic philosophy for the spreading of life from planet to planet. This ancient hypothesis, further developed at the beginning of the twentieth century by Arrhenius [6], was exploited not only in science fiction, but also in more or less, still fantastic proposals to guarantee immortality. The survival of animals (insects, reptilian, etc.) and of marine life at the polar regions of the Earth is also due to controlled crystallization or to the vitrification of cellular biological liquids [7–9]. Freezing-in of domestic animal sperms at liquid nitrogen temperatures to a glassy aqueous solution is for many years a known practice in present-day veterinary zoo-techniques. Porous silicate glasses are used to supply nutrient solutions to microbial populations and slowly soluble glasses containing exotic oxides are used as an ecologically compatible form of micro-element fertilization. These are only a few examples of the biological significance of vitreous materials.

Besides pure glasses, glass ceramics, like Pyroceram, Vitroceram, various sitals, that is, partially crystalline materials formed via the induced devitrification of glasses and glass-forming melts, are also gaining in importance in modern technology, architecture and in the immobilization of ecologically hazardous waste materials. A well-known example of the last mentioned application gives the vitrification of radioactive waste, originating from the nuclear fuel cycle and from nuclear weapons reprocessing. In this way dangerous radioactive materials are immobilized and stored as insoluble glasses, stable for millennia to come.

In glass-ceramic materials the transformation of the melt into the desired vitrocrystalline structure is initiated by a process of induced crystallization usually caused by the introduction of insoluble dopants (more or less active "crystallization cores") or of appropriate surfactants into the melt. As a result heterogeneous materials are formed in which the properties of both glasses and crystals are combined. In this way, an astonishing variety of new products with extreme properties and unusual possibilities of application is obtained. Classical enamels of every-day cooking ware, the mentioned glass-ceramic materials and so-called glass ceramic enamels for high temperature applications give additional examples in this sense. The physics and the physical chemistry of these materials, the kinetics and the various methods of their formation and the employment of glass ceramics, only mentioned here, is described in detail in many books and review articles and also the previous monograph of Gutzow and Schmelzer [1].

The widespread application and development of different vitreous materials and their production was connected with a thorough study of related scientific and technological aspects, resulting in the publication of a number of monographs, devoted to special classes of vitreous materials or special technological processes like the technology of silicate glasses, glassy polymers, metallic glasses, and so on. In these books specific properties of various vitreous materials are not only discussed in detail, but the attempt is also made to point out the fundamental properties and features which are common to all glasses, independent of the substance from which they are formed and the way they are produced. Latter topic is the focus of the present book. Therefore, particular attention is directed to the specification of the thermodynamic nature of any glass, regardless of its composition or other specific properties. Special glasses or the particular technologies connected with their production are discussed here only so far as it is desirable as an illustration of general statements or conclusions.

We are interested in the present book mainly in finding and describing in an appropriate way the common, general features of glasses. This refers to any of the following chapters and especially to Chapters 2, 3 and 4. There we have tried to elucidate the main, the most characteristic properties of glass-forming substances, common to all or at least to all real, physical glass-forming systems as yet known. In Chapter 2 the thermodynamically significant experimentally known properties of glasses are summarized and compared with the properties of the other forms of existence of matter. Glass transition is paralleled with phase transitions and the similarities and differences are reviewed. In Chapter 3 an attempt is made to correlate in the framework of a generic and generalized phenomenological approach both the kinetics of glass transition and its thermodynamic description. In the same chapter, following the same general approach, developed mainly by its authors in collaboration with several colleagues [1, 10–12], the thermodynamic nature of glasses is analyzed and defined. In doing so the general formulations of the thermodynamics of irreversible processes are applied and developed in a form, first proposed by De Donder [13] and Prigogine [14, 15], convenient to treat glasses as representatives of nonequilibrium systems with frozen-in structure. In Chapter 4 the same approach is used to analyze from a thermodynamic point of view the main kinetic characteristics of glasses: their rheology and viscous flow in particular. Particular emphasis is given in both chapters on the description of viscosity of nonequilibrium systems: here again a proposal by Prigogine [16] is followed in a generic approach, developed by Gutzow, Schmelzer et al. in [1, 17], as it follows from the derivations of Chapters 3 and 4. It is shown in Chapter 5 that many of the common properties of typical glasses are repeated or mimicked in a particular way by many other solids with frozen-in defect structures: even by those, which are crystalline.

A detailed analysis of the experimental results and their initial phenomenological interpretation, given in Chapter 2, leads to the conclusion that from a thermodynamic point of view glasses are frozen-in nonequilibrium systems. The detailed thermodynamic description of such states and their specific thermodynamic and kinetic properties are outlined in Chapters 3, 4 and 5. The respective discussion is based on the general postulates of thermodynamics of irreversible processes and, in particular, on the method of description of nonequilibrium states, developed by De Donder [13], Prigogine [14, 15], Glasstone, Laidler, and Eyring [18], Davies and Jones [19, 20] and many following authors as this is given in detail in Chapters 3, 4, 5, 8 and 9.

In Chapters 6 and 7, O.V. Mazurin and A.I. Priven explain and elucidate another problem of present-day glass science: which are the most reliable and convenient ways to collect and preserve, to calculate and to predict the most significant properties of single and multi-component glasses using existing experimental data, summarized in current literature and databases, as it is given the already cited series of reference monographs devoted mainly to silicate and other oxide glass-forming systems.

In Chapter 8 an attempt is made to summarize results on properties of glasses which from the standpoint of every day users are unexpected: their vapor pressure, solubility and electrochemistry. These properties not only illustrate some of the most significant features connected with the thermodynamic nature of glasses, but also give some indications of new and unexpected applications of the different substances in vitreous form. These applications include the usage of glasses as accumulators of hidden potential, of energy, and increased reaction power and (very unusual but possible, e.g., with metallic glasses) as electrochemical power sources. In this chapter also the possibilities of glassy states as a medium of frozen-in life and as promising soluble micro-fertilizers are discussed together with possible employment of soluble glasses in solving some medical problems.

Finally, in Chapter 9 the properties of glasses at extremely low temperatures are considered in detail. In doing so the authors of this chapter were mainly interested in elucidating a very general thermodynamic problem: what would be the most appropriate formulation of the third principle of thermodynamics for nonequilibrium systems, transferred into the vicinity of absolute zero of temperatures. The respective considerations require an analysis of the classical, sometimes already forgotten, formulations of this law as developed by the greatest representatives of classical thermodynamics at the beginning of the twentieth century, like Nernst, Einstein and Planck. Thus, Chapter 9 shows how glass science in its general formulations can give new visions in treating nonequilibrium systems, in general. Glasses, it turns out, are simply the best known representatives of the great class of systems in nonequilibrium. With Chapters 8 and 9 opening new horizons of applications of the vitreous states, the present book deviates from most existing monographs on glass science, treating glasses mostly in their common uses and classical ways of theoretical analysis.

The transformation of more and more substances into the frozen-in, nonequilibrium state of a glass is connected also with a substantial change in the meaning of the word "glass." Originally under the term glasses only amorphous (in the sense of nonstructured) frozen-in nonequilibrium systems were understood. At present every frozen-in nonequilibrium state (nonamorphous systems included) is denoted sometimes as a glass, for example, frozen-in crystals, crystalline materials with frozen-in magnetic disorder (spin glasses) and so on. The etymology of the word glass in its conventional use is given here in the concluding Chapter 10.

Despite its mentioned distinguishing features, the present book is in many respects a direct continuation and development of ideas formulated and exploited in several preceding books by the present authors which, to their satisfaction, were met with interest by the glass community. These publications are firstly the already mentioned collection of available data of Mazurin and colleagues on the properties of oxide glasses (and of silicate glasses in particular) [2, 3]. Secondly, the present book incorporates the concepts from the book by Gutzow and Schmelzer [1] in which the basic ideas on the phenomenology of glass transition, on simple glass models, on molecular statistics and on the crystallization of glass were summarized, corresponding to the state of knowledge at the time of its publication. In its phenomenological aspect this monograph by Gutzow and Schmelzer [1] is to a great extent based on an approximate way of treating the thermodynamics of glasses, proposed many years ago by Simon [21–23]. This fruitful but nevertheless approximate approach is brought in the mentioned monograph to its completion and many useful results are obtained in its framework. Moreover, ways are initiated there, based on the thermodynamics of irreversible processes, to develop a more general approach in the kinetics of glass transition and on the relaxation of glasses, which are developed here to a new level of understanding and application in Chapters 3 and 4. These new ideas are developed in the mentioned parts of the present book in the form of a new generic description on the kinetic processes, connected with glass formation and glass stabilization, as glass relaxation is also called. In doing so we have tried to take account of the whole development in this field of glass science, and especially in what was called in Russian literature the kinetic theory of glass transition; in fact, new thermodynamic foundations are given to this approach in both mentioned chapters of the present book. Moreover, out of it a full generic thermodynamic theory of glass transition is derived there in the frameworks of the thermodynamics of irreversible processes and the course of thermodynamic functions upon which glass transition is constructed. In treating technological problems in the already mentioned multi-volume monograph and reference book, compiled and edited by Mazurin et al. [2, 3], also opened are new, empirical and theoretical ways in predicting glass properties: by establishing connections, property vs. composition, to predict the properties of still not synthesized multi-component glasses out of their composition.

As it is seen from the above summary, the present book is directed towards the fundamental problems of glass science which are important for understanding the properties of glasses as a particular state of matter. In this discussion of the basic ideas concerning the vitreous state the historic course of their evolution is also briefly mentioned. In this discussion, a chronologically exact or comprehensive description is not attempted, but a characterization of the inner logics of the historical evolution, the interconnection of different ideas within it. This approach implies that in addition to the most fruitful concepts, which revealed themselves as real milestones in the evolution of glass science, proposals were also analyzed, which already at the time of their formulation or by the subsequent developments, were shown to be incorrect or even misleading, at least, as far as it is known today. It is the opinion of the authors that only by such an approach can a correct picture be given of the evolution of science as a struggle between different or even contradicting ideas. On the other hand, the detailed analysis of different proposals and the proof that some of them are not correct or even misleading is of an undoubted heuristic value. Such an approach can, the present authors hope, also prevent an over-enthusiasm with respect to insufficiently substantiated new or super-new hypotheses or to old, already refuted ideas presented in a modern form.

In the list of literature the interested reader may search for ideas and developments, results and interpretations, which could not be included in this volume. The present authors had to concentrate on those problems and solutions that formed the main roots of development of the knowledge of the vitreous state into a new well-founded science. As far as the present authors took an active part in this development, their results and publications are in many cases discussed in detail. In the theoretical interpretations in the present volume, as already mentioned, the phenomenological approach and the thorough comparison with experimental data is preferred: in this way, at least serious mistakes, common in the history of glass science will hopefully not be repeated. Where possible, as given in Chapters 6 and 7, a general survey of existing experimental data in glass science literature as a method of prediction and scientific prognosis and further development should be recommended. Several statistical model considerations are also introduced here in Chapters 3, 4, 8, and 9; although, from previous developments it is known that theoretical modeling can be sometimes dangerous. That is, in many cases it is not sufficiently evident, as to what extent inevitable approximations change the real picture of the systems investigated. In this sense phenomenology in many cases has been proven to be a more simple tool in glass science, especially when thermodynamics of irreversible processes is used: glasses are nonequilibrium systems and have to be treated accordingly, even classical thermodynamics may here also be misleading.

The present book represents again an attempt to take up Tammann's approach, made in his book *Der Glaszustand* [24] and repeated by two of the present authors in their preceding monograph [1]: to summarize the basic ideas of glass science, including the newest developments, remaining in the framework of only one volume. Many important topics, they initially wanted to include, could not be incorporated due to the lack of space. Other topics – new problems, theoretical approaches and complicated new attempts at a structural or general statistical description – seemed to us to be too complex to be included into one volume, directed not only to the well-established glass scientist, but also to the newcomers to a new and exponentially developing science. To simplify things, in many cases only reviews of new developments are given, in order to find a compromise. Hereby we tried to follow – hopefully successfully – the advice given by Einstein: "Everything should be done as simple as possible but not simpler."

2
Basic Properties and the Nature of Glasses: an Overview

Ivan S. Gutzow and Jürn W. P. Schmelzer

2.1
Glasses: First Attempts at a Classification

From a molecular-kinetic point of view all substances can exist in three different states of matter: gases, liquids and solids. These three states of aggregation of matter (from the Latin word: *aggrego* – to unite, to aggregate) are distinguished qualitatively with respect to the degree of interaction of the smallest structural basic units of the corresponding substances (atoms, molecules) and, consequently, with respect to the structure and mobility of the system as a whole.

Gases are characterized, in general, by a relatively low spatial density of the molecules and a relatively independent motion of the particles over distances significantly exceeding their size. The average time intervals τ_f of free motion in gases are considerably larger than the times of strong interaction (collisions, bound states) in between two or more atoms or molecules. In a first approximation the free volume in a gas is equal to the volume occupied by the system. The molecules can be treated in such an approximation as mathematical points (perfect gases). However, in more sophisticated models, volume, shape and the interaction of the molecules have to be accounted for. Gases are compressible: with a decreasing volume of the gas its pressure increases as expressed, for example, for a perfect gas, by Boyle–Mariotte's law.

Liquids have a significantly higher density than gases and a considerably reduced free volume. Thus, an independent translation of the building units of the liquid is impossible. The molecular motion in liquids and melts gets a cooperative character and the interaction between the particles determines to a large extent the properties of the system. Moreover, the compressibility is much smaller than for gases, simple liquids are practically incompressible.

According to a simple approximation due to Frenkel [25] liquids can be described in the following way. The motion of the building units of the particles in a liquid can be considered as oscillations around temporary average positions. The temporary centers of oscillations are changed after an average stay time, τ_R. The mean distance between two subsequently occupied centers of oscillation is comparable

Glasses and the Glass Transition, First Edition. Jürn W.P. Schmelzer and Ivan S. Gutzow.
© 2011 WILEY-VCH Verlag GmbH & Co. KGaA. Published 2011 by WILEY-VCH Verlag GmbH & Co. KGaA.

with the sizes of the molecules. Every displacement of the building units of the liquid requires thus a more or less distinct way of regrouping the particles and an appropriate configuration of neighboring molecules, for example, the formation of vacancies in terms of the "hole" theories of liquids to be discussed in Chapter 3. Though such a picture of the molecular motion in liquids can be considered only as a first approximation, it explains both the possibility of local order and the high mobility of the particles as a prerequisite for the viscous flow and the change of the shape of the liquids. New insights into the nature of the motion of the building units in glass-forming liquids were developed in a series of publications by Götze [26–28] at the beginning of the 1980s (see also the further work on these ideas cited in [29]).

A quantitative measure for the ability of a system to flow is its shear viscosity, η. According to Frenkel the shear viscosity, η, and the average stay time, τ_R, are directly connected. This connection, discussed further in Section 2.4.3, becomes evident by the following two equations (Frenkel [25]):

$$\tau_R = \tau_{R0} \exp\left(\frac{U_0}{k_B T}\right) \qquad (2.1)$$

and

$$\eta = \eta_0 \exp\left(\frac{U_0}{k_B T}\right). \qquad (2.2)$$

By U_0 the activation energy of the viscous flow is denoted, k_B is the Boltzmann constant and T the absolute temperature. More accurate expressions for the temperature dependence of η are given in Section 2.4.1. Nevertheless, Eqs. (2.1) and (2.2) already show in a qualitatively correct way the significant influence of temperature both on the viscosity η and on the relaxation time, τ_R.

Liquids like gases have no characteristic shape but acquire the shape of the vessel they are contained in. They are amorphous in the classical sense of the word, that is, a body without its own shape (from the Greek word *morphe*: shape; *amorph*: without shape). This classical meaning of the word amorphous is different from its modern interpretation. Today amorphous bodies are understood as condensed (i.e., liquid or solid) systems without long-range structural order being a characteristic property of crystals only.

Solids in classical molecular physics were identified initially with crystals. Their structure can be understood as a periodic repetition in space of a certain configuration of particles composing a certain elementary unit. In addition to the local (short-range) order found already in liquids, a long-range order is established in crystals, resulting in the well-known anisotropy of the properties of crystals. The motion of the atoms is, at least for a perfect crystal, an oscillation around time-independent average positions. This type of motion is connected with the absence of the ability to flow and the existence of a definite shape of crystalline solids.

The properties of gases and liquids are scalar characteristics, while the periodicity in the structure of the crystals determines their anisotropy and the vectorial nature of their properties. Liquids and solid crystals belong to the so-called condensed states of aggregation of matter. In condensed states the intermolecular forces cannot be neglected, in principle.

This classification is, of course, useful only as a first rough division between different states of aggregation of matter. It has its limitations. For example, it was shown that some gas mixtures may undergo decomposition processes, which are the result of the interaction of the particles. Liquids can be brought continuously into the gas phase (cf. van der Waals [30–32] or in a modern interpretation [33, 34]) and vice versa. Perfect, absolutely regular crystals do not exist in nature; moreover, under certain conditions crystals can also show some ability to flow, in particular, so-called plastic crystals. On the other hand, liquids are known (usually denoted as liquid crystals [35]) in which the optical properties of crystals are mimicked in a curious way (e.g., under flow). We will also come across cases where orientational disorder is frozen-in in crystals in the same way as in vitrified undercooled liquids and in other amorphous solids.

The elementary structural classifications given above employ criteria pertaining to the topological form of order (or disorder) exclusively (cf. [1]). Despite this limitation, one of the first questions discussed with the beginning of a scientific investigation of glasses was the analysis of the following problem: to which of the mentioned states of aggregation can glasses be assigned to. Experimental results indicated on one hand that glasses exhibit a practically infinite viscosity, a definite shape, and mechanical properties of solids. On the other hand, typical properties of liquids are also observed in glasses: the amorphous structure, that is, the absence of a long-range topological and orientational order, and the isotropy of its properties.

As a solution to this problem, Parks [36], Parks and Huffman [37] and Berger [38] (see also Blumberg [39]) and subsequently other authors, proposed to define the vitreous state as the fourth state of aggregation in addition to gases, liquids and (crystalline) solids. In this connection we have to mention that similar proposals have also been developed (but never accepted generally) with respect to other systems with unusual structures and properties (liquid crystals, elastomers, gels, etc.) thus introducing the fourth, fifth and further states of aggregation of matter. Already the considerable increase of the number of states of aggregation, which would follow from the acceptance of such proposals, shows that the generalization obtained with the classical division of the states of aggregation would be lost. A considerably more powerful argument against such proposals is connected with the limits of existence, stable coexistence and the possible transformations between the different states of aggregation.

Sometimes the partially or totally ionized state of matter, the plasma state, is denoted as the fourth state of aggregation (compare Arcimovich [40] and Frank-Kamenetzki [41] for its definition and description). The details of the transition of matter into the plasma state cannot be discussed here. It is only to be mentioned that it is quite different in its very nature, when compared with the transformations between the different states of aggregation – gases, liquids and crystals – discussed so far. It seems also that, in attributing the term *fourth state of matter* to the plasma state, physicists are more or less emotionally influenced by the beautiful schemes

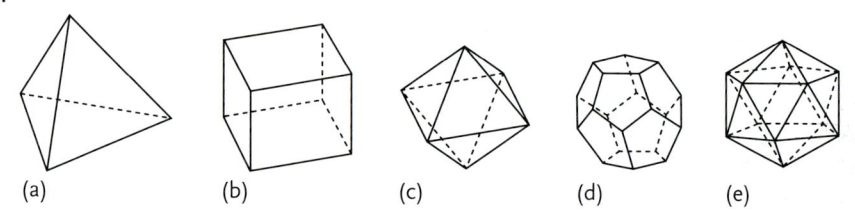

Figure 2.1 The five regular polyhedra (Platonic bodies): (a) tetrahedron; (b) cube; (c) octahedron; (d) dodecahedron; (e) icosahedron.

of ancient Greek philosophy (e.g., by Anaxagoras and, especially, by Empedocles; see, e.g., Bernal [42]) and its four elemental forces constituting the Universe: air (= gas), water (= liquid), earth (= solid), and fire (= plasma).

It must also be mentioned in connection with further discussions of the structure of glasses that in the same Hellenistic philosophical schemes the five regular (Platonic) polyhedra were also mystically introduced (Figure 2.1) as representing the four elements: tetrahedron (= fire), cube (= earth), octahedron (= air), icosahedron (= water) and the dodecahedron (because of its 12 pentagonic faces, corresponding to the zodiacal signs) as representing the Universe (or the Aristotelian fifth element: the famous *quinta essentia*). In attributing such meaning to the Platonic bodies, ancient Greek and Medieval and Post-Medieval philosophy (e.g., Kepler) have correctly chosen the icosahedron as representing liquid structures, as proven in many present-day models of liquids and glasses [1] and the cube as giving the basic features of a crystalline solid. The octahedron gives, on the contrary, in the framework of these classical schemes an idea of free movement and thus of the vapor phase.

The modern concepts concerning the division of matter into different states of aggregation and the structural characteristics of these states stem from the molecular kinetic ideas of the eighteenth century. These ideas were supplemented in the nineteenth century by a simple but unambiguous thermodynamic analysis. Thermodynamics defines the states of aggregation as thermodynamic phases and the transitions in between them as particular cases of transformation between thermodynamic phases. This approach requires first an exact definition and thorough discussion of the significant thermodynamic attributes to the notion of thermodynamic phases and to their classification, to the kinetics and thermodynamics of phase transformations, as they are given in Section 2.2.3.

If we accept the point of view that the states of aggregation are thermodynamic phases, we could call glasses an additional (e.g., a fourth) state of aggregation only, if we could prove that glasses fulfill the requirements thermodynamics connects with the definition of thermodynamic phases. However, such a proof cannot be given, as it is shown in the subsequent analysis. On the contrary, it turns out that glasses are not thermodynamic systems in the classical sense of this statement: they are in fact an example of nonequilibrium states. Thus, in order to understand the nature of glasses first a knowledge of the essentials of classical thermodynamics is necessary: we have to know what thermodynamic phases are and what glasses are not. This is the reason why in Sections 2.2.1 and 2.2.2 basic thermodynamic ideas and principles are summarized and briefly discussed in their application to our field of interest. In this way, a more correct understanding of the states of aggregation and of thermodynamic phases and of transitions taking place between them becomes possible. A more general discussion on this subject is then given in Section 2.2.3 in terms of existing phenomenological approaches, formulated in the framework of classical thermodynamics.

However, in order to analyze glass formation and the nature of glasses – as a particular physical state of nonequilibrium, of frozen-in disorder – an additional knowledge of the basic principles of thermodynamics of irreversible processes is required. Thermodynamics of irreversible processes is in some respects the continuation of classical thermodynamics into the field of nonequilibrium states and processes. It is the science describing in a thermodynamically correct way frozen-in states, their stability and the changes, connected with the processes of relaxation, leading to stable or metastable equilibrium. In many respects, this approach is a more general formulation of thermodynamics. The discussion of these topics and its application to glass formation is given in detail in Chapter 3. This approach is unfortunately little known even among scientists seriously involved in the problems of the analysis of glassy states. Such knowledge is, however, a necessity for the formulation and understanding of these problems and most of the following discussions. Thus a minimum of knowledge of the foundations of the thermodynamics of irreversible processes, at least as they are required here, is outlined in the following discussion in a simplified manner for the reader's help. A more detailed introduction into this rapidly developing branch of thermodynamic science may be found in the literature cited here, and especially in the books and monographs like [14, 15, 43–48].

Our subsequently performed discussion of existing experimental and theoretical evidence shows, as first outlined in a famous series of publications by Simon [21–23], that glasses are nonequilibrium (i.e., in classical terms: nonthermodynamic) systems and thus they cannot be described comprehensively in the framework of classical thermodynamics (or thermostatics, as this science is also denoted). However, we now also know that glasses and glass transitions give a classical example of systems and processes, which can and have to be analyzed by irreversible thermodynamics. Glasses are thus not "*a hard knock to thermodynamics*" and it is not true that "*…their …thermodynamic description is in principle not possible*" [49, 50] as it is sometimes stated even in serious journals by authors seemingly unaware of present-day irreversible thermodynamics. Glasses are on the contrary a brilliant case for illustrating the possibilities of thermodynamic analysis and of phenomenological treatment in general, however, only when thermodynamics is applied in its correspondingly enlarged formulations, appropriate for treating irreversible processes. It can be even stated that the treatment of frozen-in disorder and of nonequilibrium states, as this was demonstrated first on the example of several silicate and organic glasses (see Sections 2.5) and then of the vitreous state as a whole, determined to a great extent the development of thermodynamics of irreversible processes as a science.

The peculiar combination of properties, the nonequilibrium ("nonthermodynamic" [23, 51]) state of glasses led also in recent years in a similar way, as in Berger's [38] times, to various new (and not so new) proposals for the definition, redefinition and classification of vitrified matter. Thus, in terms of statistical physics the "nonergodic character" of the vitreous state is underlined in present day definitions of glass (see [52, 53]). In terms of the modern formalism of phase transitions even generalized symmetry considerations are applied to describe the process of vitrification. It has, however, also to be mentioned that sometimes (mostly without proper knowledge of the necessary theoretical and experimental evidence) suggestions are advanced for describing both the nature and the physical state of glasses and their properties in ways much more inappropriate then classifications in terms of the states of aggregation and the employment of classical thermodynamics only.

In the following sections of the present chapter we try to give a first, approximative approach into present-day state of the art in the understanding of the structure, thermodynamic and kinetic properties of glasses. Also several thermodynamically based definitions of glasses are advanced here as the best known examples of solids with frozen-in disorder and – as it will be shown – with frozen-in values of the thermodynamic properties. In fact, any definition of glass and of the vitreous state, in general, has to be connected with thermodynamics – classical or irreversible – because thermodynamics is the science, deciding what the nature of the states under consideration is.

2.2
Basic Thermodynamics

2.2.1
The Fundamental Laws of Classical Thermodynamics and Consequences

Classical thermodynamics is based mainly on three postulates, the three fundamental laws of thermodynamics. According to the first postulate there exists a function of state U, the so-called internal energy of the system, which depends only on the actual state of the system and not on the way the system was brought into it. The change of the internal energy can be expressed as [54, 55]:

$$dU = dQ + dA + dZ . \tag{2.3}$$

It follows that the internal energy of a thermodynamic system is changed, if energy is transferred to it from other bodies either in form of heat, dQ (microscopic form of energy transfer), by work, dA (macroscopic form of energy transfer) or by the transfer of some amount of matter, dZ. If we specify the expressions for dQ, dA and dZ, with the second law of thermodynamics in Eq. (2.4)

$$dS \geq \frac{dQ}{T} , \tag{2.4}$$

then the following relationship is obtained between another function of state, the entropy, S, the absolute temperature, T, the internal energy, U, the pressure, p, the volume, V, the chemical potentials, μ_j, and the mole numbers, n_j, of the independent molecular species (components) in the system:

$$dU \leq TdS - pdV + \sum_j \mu_j dn_j \,. \tag{2.5}$$

This equation is valid only for homogeneous macroscopic bodies. Electromagnetic fields, elastic strains or surface effects are not considered here, but are partly discussed later.

In Eqs. (2.4) and (2.5), the equality sign holds for so-called quasi-static or reversible processes. Reversible processes are defined by the criterion that a process carried out with the system can be reversed without any change to remain either in the state of the system itself or in other thermodynamic bodies. Absolutely reversible processes are, however, an idealization. They can be realized in nature in an approximate way, if the variation in the state of the system proceeds via a sequence of equilibrium states, that is, when the characteristic times of change of the external parameters are small as compared with the corresponding relaxation times of the system to the corresponding equilibrium state. Thus the definition of quasi-static or quasi-stationary processes requires the specification of the notions of thermodynamic equilibrium and equilibrium states.

Equilibrium states are defined in various ways in textbooks of thermodynamics. The obvious (and directly observable) property of equilibrium is that no macroscopically measurable change is taking place in the system during the time of observation. However, in a kinetically frozen-in system at nonequilibrium, even at considerable deviations from the equilibrium values of its thermodynamic functions, no process is taking place as well (see the subsequent discussion in Chapter 3 in the present book and De Donder's [44], Callen's [43] as well as Prigogine's and Defay's [14] books). In systems where by a stationary flow of energy or matter seemingly constant steady state values of the thermodynamic parameters are sustained, likewise no time-dependent changes can be observed. Thus, in stationary processes and in any stationary (i.e., time-independent) state quasi-equilibrium is or can be, at least, mimicked. This is the reason why we employ here in the following two more general definitions of thermodynamic equilibrium initially formulated in the framework of classical thermodynamics by Gibbs [56, 57]:

1. At equilibrium, the thermodynamic properties (thermodynamic potentials, functions and coefficients) of the system are unambiguous functions of the macroscopic external state parameters.

   At constant values of these parameters, macroscopic processes cannot take place and, thus, are not observed in equilibrium states. The macroscopic properties of a system in equilibrium are determined only by the external parameters. The thermodynamic functions of the system depend only on the actual values of the external state parameters and not on the way, these values have been established. Thus, the thermodynamic properties of equilibrium bodies

do not depend on the pre-history, and at constant external parameters they remain constant in time.

2. The state-determining thermodynamic potential of a system has in equilibrium an extremal value.

   At constant pressure ($p = $ const.) and constant temperature ($T = $ const.) and for a closed system ($dZ = 0$) the Gibbs free energy, G, is the state-determining thermodynamic potential. It has a minimum at both stable or metastable equilibrium. For isolated systems (constant volume, energy and numbers of particles), the entropy is the state determining potential, it reaches its maximal value at equilibrium.

At the lowest possible extremal value of the thermodynamic potential G, the system is in stable equilibrium. At higher values corresponding to local minima of G, metastable thermodynamic states can be realized. However (and this is not always recognized in the literature), metastable states are also thermodynamic equilibrium states (see the mechanical illustration of stable, metastable and unstable equilibrium, as depicted in Figure 2.2). States in unstable thermodynamic equilibrium, which are also determined via both definitions 1 and 2 correspond, however (at constant pressure and temperature), to a maximum or, in general, to saddle points of the Gibbs thermodynamic potential.

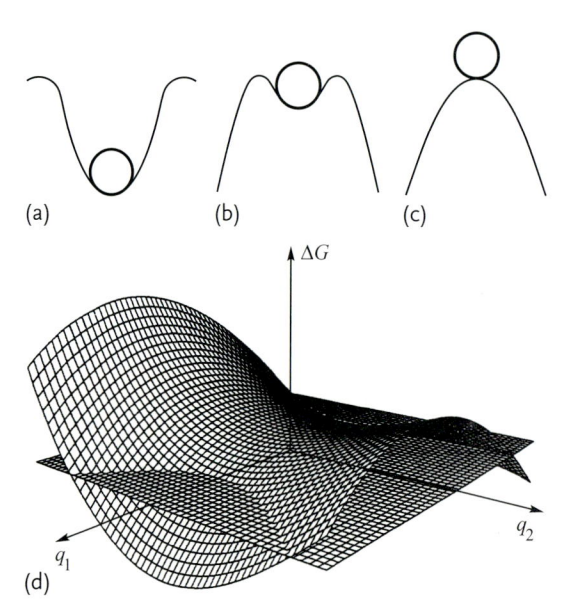

Figure 2.2 Mechanical analogy of the three forms of thermodynamic equilibrium: (a) stable; (b) metastable; (c) unstable equilibrium. In (a)–(c), the three possibilities are illustrated for a one-dimensional situation while in (d) a multi-dimensional generalization, a saddle point, is shown. Such saddle points are of particular significance in the analysis of phase formation processes (cf. [1]).

According to definitions given above, thermodynamic equilibrium states are characterized by extremum values of the thermodynamic potentials. By this reason, in nonequilibrium states and, in particular, in frozen-in systems, as shown below, an excess value of the entropy (and of any other thermodynamic function: volume, enthalpy etc.) is kinetically fixed. In contrast, in stationary quasi-equilibrium states, as shown in the thermodynamics of irreversible processes, a constant value of the entropy is exchanged per unit time with another system (e.g., with the surrounding as it is assumed in Prigogine's famous thermodynamic model of life). Both frozen-in and stationary states of a system are characterized by constant values of the thermodynamic potentials (including entropy); however, this constancy is retained in a very different way. It is obvious that a stationary process cannot be sustained in a closed system. However, frozen-in states (as, it is proven, glasses are) can be generally defined as systems with stationary (in the sense of constant) entropy values at zero entropy exchange with the surrounding. Thus, they are possible in closed systems, where at equilibrium maximal entropy values are to be expected.

As it follows from the previous discussion, the second definition of an equilibrium state in classical thermodynamics is essential in distinguishing between true equilibrium and quasi-equilibrium. In describing stationary states, constant fluxes between two equilibrium systems have to be assumed (e.g., the mentioned entropy exchange with the surrounding) as time-independent boundary conditions. In contrast, for frozen-in states the evolution to equilibrium is impeded by the vanishing of the respective kinetic mechanisms. Thus, as mentioned above, frozen-in states are nonthermodynamic systems in the classical sense of this word, and to them neither of the two above given definitions can be applied. In terms of the entropy, S (at appropriately chosen boundary conditions), equilibrium is characterized by a maximum entropy, for steady states (in the vicinity of thermodynamic equilibrium) the entropy production (dS/dt) has a minimum (as shown by Prigogine) and for frozen-in systems ($dS/dt = 0$) holds; however, the entropy is in this case (at constant energy, particle numbers, and volume) lower than the maximum possible value.

The description of the freezing-in process can be performed in terms of classical thermodynamics based on an approximation, first proposed by Simon [58, 59] in treating vitrification. According to Simon, the freezing-in process takes place abruptly at a given value of the thermodynamic state parameters, that is, when in their change some particular value, the temperature $T = T_e$ (the freezing-in temperature or the *Glaseinfriertemperatur*) is reached. This assumption and its consequences are analyzed in detail in the present chapter. This approach is extended in Chapter 3 by a more accurate theoretical description, employing the concepts of present-day nonequilibrium thermodynamics, permitting the detailed description of freezing-in phenomena as a continuous process proceeding in some temperature interval (or in an interval of other relevant thermodynamic control parameters like, e.g., pressure). In establishing the nature of glasses as nonequilibrium systems, Simon [23] first applied many years ago equilibrium definition 2, demonstrating that (in the framework of his approximation) no extremums in the course of the thermodynamic potentials are observed in dependence on temperature upon vitrification.

For quasi-stationary processes, Eqs. (2.4) and (2.5) may be written as:

$$dS = \frac{dQ}{T}, \tag{2.6}$$

$$dU = TdS - pdV + \sum_j \mu_j dn_j. \tag{2.7}$$

An integration [54] of Gibbs's fundamental Eq. (2.7) yields

$$U = TS - pV + \sum_j \mu_j n_j. \tag{2.8}$$

From the combined first and second laws of thermodynamics as given by Eq. (2.7) it follows that for equilibrium states the internal energy is unambiguously determined by the values of S, V and the set of mole or particle numbers of the different components, $\{n_j\}$. If the functional dependence $U = U(S, V, n_1, n_2, \ldots, n_k)$, the caloric equation of state, is known, all thermodynamic properties of the system can be determined by a derivation of U with respect to the corresponding independent variables, that is,

$$T = \left(\frac{\partial U}{\partial S}\right)_{V,\{n_j\}}, \quad -p = \left(\frac{\partial U}{\partial V}\right)_{S,\{n_j\}}, \quad \mu_j = \left(\frac{\partial U}{\partial n_j}\right)_{S,V,n_{i,i\neq j}}. \tag{2.9}$$

Thus it becomes evident why in analogy to classical mechanics the internal energy U (and other thermodynamic functions having at corresponding conditions analogous properties) are denoted as and fulfill the role of thermodynamic potentials.

For the analysis of the properties of different systems, including glass-forming melts, two other thermodynamic functions, the enthalpy H and the already mentioned free enthalpy G (or Gibbs's free energy) are of particular importance. This peculiar role is connected with the circumstances that condensed systems (including glass-forming melts) are usually investigated at a constant external pressure, the atmospheric pressure. The variable parameter is then the temperature of the system.

The enthalpy H is determined by

$$H = U + pV. \tag{2.10}$$

For a constant value of the external pressure, which is assumed to be equal to the pressure inside the system (mechanical equilibrium), Eqs. (2.3), (2.7) and (2.10) yield

$$dH = dQ + dZ. \tag{2.11}$$

Consequently, for closed systems ($dZ = 0$) and isobaric conditions ($p = $ const.) the energy supplied to the system in form of heat is equal to the enthalpy change.

From the definition of the heat capacity at constant pressure C_p,

$$C_p = \left(\frac{dQ}{dT}\right)_p, \tag{2.12}$$

one thus obtains directly

$$C_p = \left(\frac{\partial H}{\partial T}\right)_p, \tag{2.13}$$

or with Eq. (2.6)

$$C_p = T\left(\frac{\partial S}{\partial T}\right)_p. \tag{2.14}$$

Similarly one obtains for the heat capacity at constant volume C_V

$$C_V = \left(\frac{dQ}{dT}\right)_V, \tag{2.15}$$

$$C_V = \left(\frac{\partial U}{\partial T}\right)_V, \tag{2.16}$$

$$C_V = T\left(\frac{\partial S}{\partial T}\right)_V. \tag{2.17}$$

The Gibbs free energy G is defined by

$$G = H - TS, \tag{2.18}$$

or

$$G = U - TS + pV, \tag{2.19}$$

and with Eqs. (2.7) and (2.8)

$$dG = -S\,dT + V\,dp + \sum_j \mu_j\,dn_j, \tag{2.20}$$

$$G = \sum_j \mu_j n_j \tag{2.21}$$

is obtained.

In analogy to U, the Gibbs free energy G is also a thermodynamic potential if this quantity is known as a function of T, p and $\{n_j\}$. In this case one gets similarly to Eq. (2.9):

$$S = -\left(\frac{\partial G}{\partial T}\right)_{p,\{n_j\}}, \quad V = \left(\frac{\partial G}{\partial p}\right)_{T,\{n_j\}}, \quad \mu_j = \left(\frac{\partial G}{\partial n_j}\right)_{p,T,n_{i,i\neq j}}. \tag{2.22}$$

The application of Eqs. (2.9) and (2.22) requires the knowledge of the functions $U(S, V, \{n_j\})$ or $G(T, p, \{n_j\})$, respectively. These functions, the equations of state of the equilibrium systems, reflect the properties of the particular system and cannot be established by thermodynamics.

From the considerations given above, it is evident that the thermodynamic functions (entropy, S, volume, V, etc.) are first-order derivatives of the appropriate thermodynamic potential, the Gibbs free energy, G. Specific heats, C_p, as evident from Eqs. (2.12), (2.13) and (2.22), and all other thermodynamic coefficients (the coefficients of thermal expansion, of compressibility, etc.) are, in general, second-order derivatives of the thermodynamic potentials (e.g., of the Gibbs free energy, G). More generally speaking, thermodynamic coefficients fulfill the role of susceptibilities, describing the response a system will have to the change of external parameters. The reciprocal values of such thermodynamic coefficients like compressibility determine the values of the elastic constants in solid state mechanics and in the theory of elasticity.

In the subsequent analysis, we investigate the peculiar course of change of the thermodynamic potentials, thermodynamic functions and coefficients in different processes of change of state of a system, and especially in the glass transition, in order to draw a picture of glasses as a state and of their formation as a process. The determination of the thermodynamic potentials of a real system requires either calculations based on appropriate models given by statistical mechanics, or a distinct set of experimental measurements and their results. Following the second approach, the usual way consists in the experimental determination of the temperature dependence of the specific heats C_p at constant (i.e., normal) pressure, p. An additional enthalpy determination (e.g., the heat of the transition if phase transitions are involved) is also necessary, if this commonly employed way of experimentally based thermodynamic potential calculation has been chosen.

Once the dependence $C_p = C_p(p, T)$ is established by measurements, the thermodynamic functions H, S and G can be directly determined by (see Eqs. (2.13), (2.14) and (2.22))

$$H(p, T) = H(p, T_0) + \int_{T_0}^{T} C_p \, dT, \tag{2.23}$$

$$S(p, T) = S(p, T_0) + \int_{T_0}^{T} \frac{C_p}{T} \, dT, \tag{2.24}$$

$$G(p, T) = G(p, T_0) - S(p, T_0)(T - T_0) - \int_{T_0}^{T} dT \int_{T_0}^{T} \frac{C_p}{T} \, dT. \tag{2.25}$$

Normally, the experimenter is interested in the knowledge of C_p, H, S and G at constant (usually normal atmospheric) pressure, where the determination of both $C_p(T)$ and $H(T_0)$ is performed via calorimetric measurements.

Equations (2.23)–(2.25) hold for any arbitrary reference temperature T_0. They can be simplified based on the third law of thermodynamics. The third law of thermodynamics, established first by Nernst in 1906 (see Nernst [60] and the monograph by Bazarov [55]), reads in the formulation given by Planck (see Planck [61]) that when approaching the zero-point of the absolute scale of temperature the entropy of any system in thermodynamic equilibrium becomes a constant, we denote here by S_0, which is independent of the actual value of pressure, p, or other possible thermodynamic parameters and the state of aggregation of the substance considered. Since the entropy is defined thermodynamically, according to Eq. (2.6), only with an accuracy to an additive constant, this constant can be set, according to Planck's proposal, equal to zero. The latter assumption can be given an additional physical justification based on quantum statistical considerations (cf. Landau and Lifshitz [62, 63]). Thus, for any equilibrium system $S_0 = 0$ holds. In further analysis of the present chapter and, in particular, in Chapters 3 and 9 we will see that the third law in the sketched above form does not hold for glasses. However, as will also be shown, a generalization of the third law can be proposed which is valid both for equilibrium and nonequilibrium systems.

In a mathematical formulation, the third law for equilibrium systems can be written thus as:

$$\lim_{T \to 0} S = 0 , \qquad (2.26)$$

with the consequences

$$\lim_{T \to 0} C_p = 0 , \quad \lim_{T \to 0} C_V = 0 , \qquad (2.27)$$

$$\lim_{T \to 0} \left(\frac{\partial G}{\partial T} \right) = \lim_{T \to 0} \left(\frac{\partial H}{\partial T} \right) = 0 . \qquad (2.28)$$

The statement expressed by Eq. (2.28) is historically known as Nernst's heat theorem.

With the third law Eqs. (2.23)–(2.25) are simplified to

$$H(p, T) = H(p, 0) + \int_0^T C_p \, dT , \qquad (2.29)$$

$$S(p, T) = \int_0^T \frac{C_p}{T} \, dT , \qquad (2.30)$$

$$G(p, T) = G(p, 0) - \int_0^T dT \int_0^T \frac{C_p}{T} \, dT . \qquad (2.31)$$

Another formulation of the third law of thermodynamics can be given with Boltzmann's interpretation of the entropy S of a system as:

$$S = k_B \ln \Omega . \qquad (2.32)$$

Here Ω is the number of distinct microstates corresponding to one and the same macroscopic state. In such an approach, Planck's formulation of the third law in terms of Boltzmann statistical physics is equivalent to the statement that to the macroscopic state of the system at temperatures T approaching zero only one microstate or a relatively small number (cf. Fermi [64]) of microstates corresponds. It is shown below that for any system with frozen-in disorder, and especially for glasses, as first recognized by Einstein [65] and as it was proven by Simon [21, 59], $S_0 \gg 0$ always holds. This difference is one of the most striking deviations of the properties of vitreous substances from the behavior of systems in thermodynamic equilibrium. The intriguing history of the way this significant result has been established in science is described in detail in Chapter 9.

2.2.2
Thermodynamic Evolution Criteria, Stability Conditions and the Thermodynamic Description of Nonequilibrium States

So far we have discussed only thermodynamic dependencies, derived by classical thermodynamics for equilibrium states and for reversible processes in between them. In general, for systems in nonequilibrium the same classical approach states that the inequality

$$(dU)_{S,V,\{n_j\}} \leq 0 \tag{2.33}$$

has to hold. At fixed values of entropy, volume and at constant composition (i.e., at constant mole numbers of the different components, forming the considered body) the internal energy of a system decreases until an equilibrium state corresponding to a minimum of U is reached.

Similarly one obtains for the Gibbs free energy at fixed values of p, T and $\{n_j\}$:

$$(dG)_{p,T,\{n_j\}} \leq 0. \tag{2.34}$$

If we consider a homogeneous system and divide it artificially into two parts, specified by the subscripts (1) and (2), respectively, then the entropy, S, the volume, V, and the mole numbers of the different components, n_j, of the whole system can be written as:

$$S = S_1 + S_2, \quad V = V_1 + V_2, \quad n_j = n_{j1} + n_{j2}. \tag{2.35}$$

If S, V and n_j are kept constant, spontaneous deviations of S_1, V_1 and n_{j1} result in corresponding deviations of S_2, V_2 and n_{j2}, that is,

$$\delta S_1 + \delta S_2 = 0, \quad \delta V_1 + \delta V_2 = 0, \quad \delta n_{j1} + \delta n_{j2} = 0. \tag{2.36}$$

In the vicinity of equilibrium, which is characterized by a minimum of $U = U_1 + U_2$, such changes do not vary the total value of U. As a result one obtains from the necessary equilibrium conditions $\delta U = 0$, for example,

$$\delta U = \left(\frac{\partial U}{\partial S_1}\right) \delta S_1 = \left(\frac{\partial U_1}{\partial S_1} - \frac{\partial U_2}{\partial S_2}\right) \delta S_1 = 0 \tag{2.37}$$

and (compare Eq. (2.9))

$$T_1 = T_2 . \tag{2.38}$$

Similarly one obtains

$$p_1 = p_2 , \tag{2.39}$$

and

$$\mu_{j1} = \mu_{j2} \quad \text{for} \quad j = 1, 2, \ldots, k . \tag{2.40}$$

So, in any two parts of a system coexisting in equilibrium at planar interfaces, the values of temperature, pressure and the chemical potentials are the same. However, the above equilibrium conditions have to be modified if interfacial contributions to the thermodynamic functions have to be accounted for as is the case in the thermodynamics of nucleation phenomena (cf. [1]).

The sufficient equilibrium criterion

$$(\delta^2 U)_{S,V,n_j} > 0 \tag{2.41}$$

may be written also as (see Prigogine and Defay [14], Kubo [54]):

$$\delta S \delta T - \delta p \delta V + \sum_{j=1}^{k} \delta \mu_j \delta n_j > 0 . \tag{2.42}$$

If only the temperature of a homogeneous system is changed as a particular case we obtain

$$\left(\frac{\partial S}{\partial T}\right)(\delta T)^2 > 0 \tag{2.43}$$

and, consequently (cf. Eqs. (2.14) and (2.17), see also Prigogine and Defay [14])

$$C_p > 0 , \quad C_V > 0 . \tag{2.44}$$

Equations (2.30) and (2.44) allow us to agree with Eq. (2.32) the conclusion that the entropy S of an equilibrium system for $T > 0$ is always positive. Equation (2.34) shows further on that for fixed values of p, T and $\{n_j\}$ processes are possible in nonequilibrium systems, leading to a change in G. This conclusion implies that the values of p, T and $\{n_j\}$ do not determine completely the state of a body in nonequilibrium.

Suppose now that it is possible to introduce some additional macroscopic state variables, $\xi_1, \xi_2, \ldots, \xi_m$, which together with p, T and the overall particle or mole numbers of the different components, n_j, of the system under investigation contain the whole information about the macroscopic properties of the corresponding nonequilibrium system, determined by its momentary microscopic structure. In such cases G should be determined by an expression of the form

$$G = G(p, T, \{n_j\}, \xi_1, \xi_2, \ldots, \xi_m) . \tag{2.45}$$

Since in equilibrium G is a function of p, T and $\{n_j\}$ only, in this limiting case the additional parameters ξ_j must also be functions of p, T and $\{n_j\}$, that is, the relation

$$\xi_j^{(e)} = \xi_j^{(e)}(T, p, n_1, n_2, \ldots, n_k) \tag{2.46}$$

has to be fulfilled.

Two possible approaches can be followed in the introduction of such additional, internal structure parameters into nonequilibrium thermodynamics. The first one is mainly due to De Donder (see his *Book of Principles* [13] written in the English version in collaboration with van Rysselberghe and also [14, 43, 48, 55]). It stems historically from his treatment of chemical reactions in the framework of irreversible thermodynamics. Due to this origin, the structure parameters ξ_i are commonly denoted as De Donder's reaction coordinates. Similar dimensionless internal thermodynamic parameters were also introduced (using even the same symbol, ξ, to denote it) somewhat later by Mandelstam and Leontovich [66, 67] in developing a general theory of irreversible processes and in treating, in particular, the dissipation of sound.

Suppose we are considering, as done by De Donder, a system in which chemical changes are taking place (or more general, any type of structural reorganization, e.g., polymerization, isomerization, etc.). Then the properties of the system will change (at constant external parameters, e.g., p and T) in time, t, with changing degree of completion, $\xi(t)$, of the reactions considered. At the beginning of this process (at $t = 0$) we have $\xi(t) = 0$ and $\xi(t)$ increases with time. The reaction rate, $(d\xi/dt) = f(T, p, \xi)$, is, in general, a function of the external parameters and of the state of the system. However, the state of the system itself depends on the value of ξ or, if several independent reactions are taking place, on the degree of completion of all these reactions, $\{\xi_i(t)\}$.

Suppose, further, that we have (e.g., by a sudden temperature quench) stopped one of the reactions, initially taking place with measurable velocity at $T = T_1$ and $p = p_1$ and that we have thus frozen-in the concentration of the corresponding reagents, fixing the value of $\{\xi_i\}$. After the quench, the state of the resulting system and its thermodynamic properties will depend not only on the new values of external state parameters (T_2, p_1), but also on the values of the order parameters, $\{\xi_i(t)\}$, beforehand reached at $T = T_1, p = p_1$. Since the further reaction is hindered by kinetic reasons, the value of the structure parameter can be considered as a measure of deviation from equilibrium.

In a broader sense, the parameters ξ_i can be considered, as we have shown in [1, 68], as a generalized structure parameter of the system with normalized values, varying in the limits $0 \leq \xi_i \leq 1$ and describing its structure (e.g., at $T = T_1$) in some way, which is not necessarily to be specified in the framework of thermodynamics. After that, when we have abruptly brought the system from $T = T_1$ to $T = T_2$, it can be assumed that a fixed, frozen-in value $\xi(T_1)$ of the structural parameter contains the whole structural information, the whole previous history of the system, which is now at a temperature $T = T_2$, in nonequilibrium. The deviation from equilibrium at the new conditions can be connected with the difference

$[\xi_i^{(e)}(T_1) - \xi_i^{(e)}(T_2)]$, when we attribute to $\xi_i^{(e)}(T_2)$ (or the sum of these parameters) the meaning of the equilibrium value of ξ_i at $T = T_2$. If desired, we can specify ξ_i in terms of volume changes, entropy changes, and so on by using appropriate structural models of statistical mechanics. Examples in this respect are given in Chapter 3.

It should be mentioned that the order parameter ξ was initially introduced by Mandelstam and Leontovich [66, 67] in a similar generalized manner: it originally indicated in their treatment the time-dependent, not specified change of the structure of the systems under consideration when a sound wave passes through it. This approach led then to a new formulation of the theory of sound propagation.

If the order parameters ξ are defined in such a way that they increase with the approach to the respective equilibrium values $\xi^{(e)}$, we have:

$$\frac{\partial G}{\partial \xi} \leq 0. \tag{2.47}$$

The significance of above equation becomes evident, when it is recalled that De Donder's new thermodynamic function, the affinity, is defined (see [14, 44, 47]) by the change of ξ (for simplicity, we restrict here the considerations to the case of only one structural parameter leaving the generalization to Chapter 3) as

$$A = -\left(\frac{\partial G}{\partial \xi}\right)_{T,p} \geq 0, \tag{2.48}$$

and accordingly

$$\left(\frac{\partial A}{\partial \xi}\right)_{T,p} = -\left(\frac{\partial^2 G}{\partial \xi^2}\right)_{T,p}. \tag{2.49}$$

Provided, the system is otherwise in equilibrium, the necessary and sufficient condition for equilibrium with respect to changes of the structure parameter is then given by $A = 0$ resulting in an additional condition for thermodynamic equilibrium [14]. It will be frequently employed in the subsequent analysis, in particular, in Chapter 3. From Eqs. (2.47)–(2.49), we have then further on $(\partial A/\partial \xi) \leq 0$. The significance of the affinity function in thermodynamics, in general, and in discussing thermodynamics of glasses and processes of glass formation and relaxation, in particular, becomes fully evident in Chapter 3.

It turns out that the derivative of A with respect to parameters describing structural changes has the same significance in determining the stability of a system as the analysis of the variation of the thermodynamic potential when performed in the way indicated in the previous section (see Eq. (2.43)). In the framework of De Donder's approach we have to investigate the sign and the value of the affinity A, its nullification indicating equilibrium. Moreover, it can be shown [14] that the driving force for any reaction (more general in the above-introduced terms, of any structural change) in the system is determined by the momentary value of the affinity, A. It thus also defines, as discussed in Chapter 3 in more detail, the reaction rates in the system and the rate, $(d\xi/dt)$, of its general structural change. When, at $A \neq 0$, we observe $(d\xi/dt) = 0$, we have, according to De Donder's definition, a kinetically frozen-in nonequilibrium state. The extent of disorder and of nonequilibrium, frozen-in in the glass, is determined via its structural parameter, ξ, by the freezing-in temperature, T_e, introduced by Simon [59] in the framework of his already mentioned approximate treatment.

Similar in this respect is the notion of fictive temperature, proposed by Tool [69], which can be defined in a first approximation by the reverse function of the supposed or experimentally determined $\xi(T)$-dependence as $T_{\text{fict}} = T(\xi_g)$, where ξ_g indicates the value of the structural parameter frozen-in in the glass. In this way, the deviation of the frozen-in system from equilibrium, for example, after a temperature quench from T_1 to T_2 is performed, can be characterized (at least in such first approximation) also by the difference $(T_1 - T_2)$. Further examples of the significance of the thermodynamic structural parameter, ξ, and of fictive temperatures (and fictive pressure) in characterizing vitreous states are given in the next sections of the present chapter and in Chapter 3. The chapter also describes how the fictive temperature T_{fict} can be determined from the temperature course of various thermodynamic properties upon vitrification (see also Mazurin [70]). A general thermodynamic definition of fictive temperature and the similar concept of fictive pressure is given also in Chapter 3.

2.2.3
Phases and Phase Transitions:
Gibbs's Phase Rule, Ehrenfest's Classification, and the Landau Theory

The term thermodynamic phase (from the Greek word *phasis*: form of appearance) was introduced by J.W. Gibbs in his papers published first in the period from 1875–1878 to characterize a definite equilibrium form of appearance of a substance. In Gibbs's words the definition reads: "*In considering the different homogeneous bodies which can be formed out of any set of component substances, it will be convenient to have a term which shall refer solely to the composition and thermodynamic state of any such body without regard to its quantity or form. We may call such bodies as differ in composition or state different phases of matter considered, regarding all bodies which differ only in quantity and form as different examples of the same phase. Phases which can exist together, the dividing surfaces being plane, in an equilibrium which does not depend upon passive resistances to change we shall call coexistent,*" [56, 57].

According to this classical definition thermodynamic phases are different equilibrium forms of appearance of the same substance. Every thermodynamic phase is physically homogeneous and, in the absence of external fields, its thermodynamic parameters are the same in each part of the volume occupied by it. In other words, Gibbs's definition implies that a phase is characterized by a well-defined equation of state. Different thermodynamic phases may coexist in mutual contact in equilibrium; the coexisting phases are divided by interfacial boundaries, where the thermodynamic properties of the substance change abruptly. The transformation of a substance from one phase into the other is called a phase transition.

A definition of thermodynamic phases, similar to the above-cited original version by Gibbs, we can also find in another classical source (in Landau and Lifshitz's treatise on theoretical physics in [62, 63]): "*Suppose that the equilibrium state of a heterogeneous system is determined by two thermodynamic parameters, e.g., by its volume, V, and energy, U.*" It may be possible, it is said further on that the considered initially homogeneous equilibrium state disintegrates into two different homogeneous parts. Then "*such two different states of matter, which can exist together simultaneously in mutual contact and equilibrium, are to be called different phases of the same substance.*" The same meaning also has the definition, recommended in one of the newest physical encyclopedia [71] for the notion phase in thermodynamics: *…a distinct equilibrium state of a substance, having different physical properties from other also possible equilibrium forms (other phases) of the same substance.*

The notation thermodynamic phase is more restrictive than the much broader term state of aggregation, we introduced in one of the preceding sections. Inside a given state of aggregation a substance may exist in several different phases like, for example, the different crystalline phases of ice or the different crystalline modifications of SiO_2 (see, e.g., Tammann [72]). A thorough classical discussion of these problems may be found in the lecture course of van der Waals and Kohnstamm [32] and in Storonkin's monograph [73]. It has also to be pointed out, as first mentioned by van der Waals (see again [32]), that the given definition of thermodynamic phases, as it was initially formulated by Gibbs himself, automatically implies that thermodynamic phases are equilibrium systems in the full sense of both conditions 1 and 2 given in Section 2.2.1. This point is well recognized and stressed in the above-cited present-day definitions.

At Gibbs's times, only equilibrium systems were (and could be) anticipated and thermodynamically treated. Attempts to introduce into science notions like nonequilibrium phases (see [74]) led and lead to controversial results: as it is to be expected, a consistent thermodynamic analysis of systems constituted of nonthermodynamic bodies is, in general, not possible in the framework of classical thermodynamics.

In Gibbs's thermodynamic description of heterogeneous systems, consisting of more than one, say of r, thermodynamic phases, the thermodynamic potential, G, has to be written as a sum of the contributions of all the r phases as:

$$G = G_1 + G_2 + \ldots + G_r . \tag{2.50}$$

Interfacial contributions are neglected here as in the previous discussion. This neglect is possible as far as sufficiently large volumes of the different phases are assumed. The necessary equilibrium conditions for a stable coexistence of the different macroscopic phases read then (cf. Eqs. (2.38)–(2.40)) as:

$$T_m = T_r, \quad p_m = p_r, \quad \mu_{jm} = \mu_{jr}, \quad m = 1, 2, \ldots, r - 1 . \tag{2.51}$$

The properties of the phase m can be determined if the functional dependence

$$G_m = G_m(T_m, p_m, n_{1m}, \ldots, n_{km}) \tag{2.52}$$

is known. According to the above definition the properties of a phase do not depend on its total amount. Consequently, the state of one phase is characterized by $k + 1$ variables, for example, temperature T, pressure p and independent molar fractions $x_j, j = 1, 2, \ldots, k - 1$. The total number of variables needed for the description of the properties of an r-phase system is, therefore, $r(k + 1)$. In equilibrium, between these variables $(r - 1)(k + 2)$ independent relationships exist (compare with Eq. (2.51)). The number of independent variables or the number of degrees of freedom is, therefore, $f = r(k + 1) - (r - 1)(k + 2)$ or

$$f = k + 2 - r. \tag{2.53}$$

Equation (2.53) is the so-called Gibbs phase rule. The indicated way of derivation of Gibbs' phase rule demonstrates once more that the notation phase was introduced and applied by Gibbs to equilibrium states and can be properly used only in this context.

Experience, summarized in its historic development at the end of the present section, has shown that real systems may undergo different types of transitions. The first and remarkably successful attempt to classify possible types of phase transitions was made by Ehrenfest [75]. His analysis, applied mainly to one-component systems (or systems with constant composition), can be summarized as follows. For a one-component closed system the change of the Gibbs free energy in a reversible process is given by (cf. Eq. (2.20))

$$dG = -S dT + V dp. \tag{2.54}$$

Since for one-component systems $G = \mu n$ holds (see Eq. (2.21)), Eq. (2.54) may be rewritten as:

$$d\mu = -s dT + v dp, \tag{2.55}$$

where s and v are the molar entropy and volume of the system. Equation (2.55) yields

$$s = -\left(\frac{\partial \mu}{\partial T}\right)_p, \quad v = \left(\frac{\partial \mu}{\partial p}\right)_T. \tag{2.56}$$

The two possible dependencies $G = G(T, p = \text{constant})$ and $G = G(p, T = \text{constant})$ for two different phases are shown in Figure 2.3. Only in the point of intersection of the $G(T)$- or $G(p)$ curves are the necessary conditions for equilibrium fulfilled. For the values of p and T corresponding to the point of intersection of the respective curves representing the different phases, and denoted by T_e and p_e, a coexistence of the two different phases is possible.

For constant values of p, T and $\{n_j\}$, spontaneous processes are connected with a decrease of the Gibbs free energy (cf. Eq. (2.34)). Consequently, in Figure 2.3a,b the lower branches of the curves correspond to the thermodynamically preferred state. In a process, for example, of continuous heating, after T_e is surpassed (Figure 2.3a), a transition from phase (2) to phase (1) is to be expected from a thermodynamic

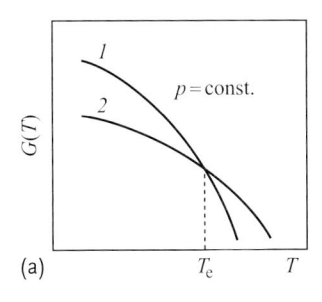

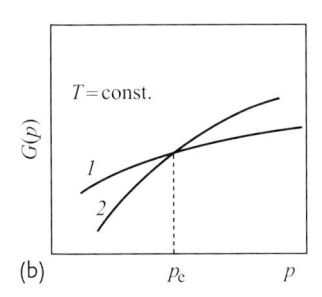

Figure 2.3 Possible dependencies (a) $G = G(T, p = \text{const.})$ and (b) $G = G(p, T = \text{const.})$ for two different phases, specified by curves (1) and (2), respectively. The points of intersection of these curves determine the values of p and T at which an equilibrium coexistence of both phases is possible. Such behavior corresponds to first-order phase transformations according to Ehrenfest's classification.

point of view. The difference of the thermodynamic potentials is a measure of the thermodynamic driving force of the considered process of phase transformation. The particular way and the rate such processes proceed depend on additional thermodynamic and kinetic factors discussed in detail in a variety of books (cf. for example [1]).

Phase transformations of the form as depicted in Figure 2.3a and b are denoted, according to the classification of Ehrenfest [75], as first-order phase transitions. This notation originates from the property that in an equilibrium two-phase state the values of the molar Gibbs free energies, g, of both phases coincide, while the first-order derivatives with respect to p or T differ. It means that the relations

$$g^{(1)}(T, p) = g^{(2)}(T, p), \tag{2.57}$$

$$\left(\frac{\partial g^{(1)}}{\partial T}\right)_p \neq \left(\frac{\partial g^{(2)}}{\partial T}\right)_p, \quad \left(\frac{\partial g^{(1)}}{\partial p}\right)_T \neq \left(\frac{\partial g^{(2)}}{\partial p}\right)_T \tag{2.58}$$

have to be fulfilled. Here and in the following derivations small letters alway refer to molar values of the respective extensive variables. Taking into account Eq. (2.22), Eq. (2.58) is equivalent to

$$s^{(1)} \neq s^{(2)}, \quad v^{(1)} \neq v^{(2)}. \tag{2.59}$$

It follows that, if a system is transferred from one phase to another in a first-order phase transition, the values of the molar entropy and volume are discontinuously changed. This change is connected with a qualitative variation of the structure of the system and the release or adsorption of the latent heat of the transformation (heat of melting, heat of sublimation, heat of evaporation) manifested in the discontinuous change of the molar enthalpy, h, that is, $(h^{(1)} \neq h^{(2)})$.

According to Gibbs's phase rule a one-component two-phase equilibrium system has one degree of freedom. Thus, the equilibrium value of the pressure can be considered as a function of temperature. The type of dependence $p = p(T)$ can be derived from the equilibrium condition $\mu^{(1)}(p, T) = \mu^{(2)}(p, T)$. A derivation of this

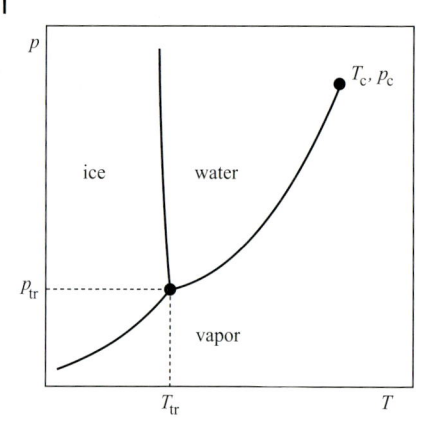

Figure 2.4 Phase diagram for water with triple point ($T_{tr} = 273.16$ K, $p_{tr} = 6.1 \cdot 10^2$ Pa) and critical point ($T_c = 647$ K, $p_c = 22.1 \cdot 10^6$ Pa).

equation with respect to T leads to the Clausius–Clapeyron equation:

$$\frac{dp}{dT} = \frac{s^{(1)} - s^{(2)}}{v^{(1)} - v^{(2)}} . \tag{2.60}$$

The difference of the molar entropies can also be expressed through the molar latent heat of the transformation q as:

$$s^{(1)} - s^{(2)} = \frac{q}{T}, \tag{2.61}$$

which yields

$$\frac{dp}{dT} = \frac{q}{T(v^{(1)} - v^{(2)})} . \tag{2.62}$$

By the determination of all possible $p = p(T)$ curves representing different phase equilibria of the same substance, the well-known phase diagrams are obtained. The point of intersection of three curves (triple point) corresponds to a three-phase system. According to Gibbs's phase rule a three-phase equilibrium is possible only for single points in the (p, T)-plane. This conclusion is illustrated in Figure 2.4 for the classical example of the water–ice–vapor coexistence.

For second-order phase transitions both the molar free enthalpy and their first-order derivatives are equal at the transition point for both phases, while the second-order derivatives differ. In addition to Eq. (2.57), we must write

$$\left(\frac{\partial g^{(1)}}{\partial T}\right)_p = \left(\frac{\partial g^{(2)}}{\partial T}\right)_p , \quad \left(\frac{\partial g^{(1)}}{\partial p}\right)_T = \left(\frac{\partial g^{(2)}}{\partial p}\right)_T , \tag{2.63}$$

$$\left(\frac{\partial^2 g^{(1)}}{\partial p \partial T}\right) \neq \left(\frac{\partial^2 g^{(2)}}{\partial p \partial T}\right) , \quad \left(\frac{\partial^2 g^{(1)}}{\partial p^2}\right)_T \neq \left(\frac{\partial^2 g^{(2)}}{\partial p^2}\right)_T , \tag{2.64}$$

$$\left(\frac{\partial^2 g^{(1)}}{\partial T^2}\right)_p \neq \left(\frac{\partial^2 g^{(2)}}{\partial T^2}\right)_p .$$

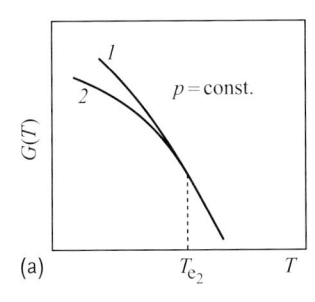

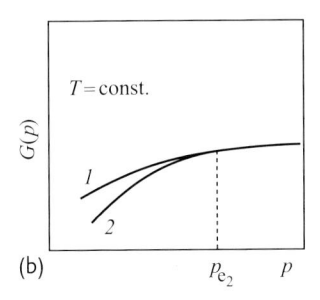

Figure 2.5 Illustration of the (a) $G = G(T, p = \text{const.})$ and (b) $G = G(p, T = \text{const.})$ dependencies for second-order phase transformations according to the classification of Ehrenfest. The different phases are specified again by (1) and (2).

As a consequence of these relations, it follows that in second-order phase transformations no latent heat is released by the system.

The second-order derivatives of the Gibbs free energy with respect to p and T can be expressed through the respective thermodynamic coefficients. As mentioned, thermodynamic coefficients (or susceptibilities, as they are also more generally termed) determine the reaction of a system with respect to the variation of external parameters. An often used set of independent thermodynamic coefficients consists of the heat capacity, C_p (cf. Eqs. (2.12)–(2.14)), the thermal expansion coefficient, α, and the isothermal compressibility, κ. The thermal expansion coefficient, α, and the compressibility, κ, are defined by [62, 63, 76]

$$\alpha = \frac{1}{V}\left(\frac{\partial V}{\partial T}\right)_p, \quad \kappa = -\frac{1}{V}\left(\frac{\partial V}{\partial p}\right)_T. \tag{2.65}$$

Taking into account Eqs. (2.22) and (2.65) the inequalities Eq. (2.64), characterizing second-order phase transitions, may be rewritten as:

$$C_p^{(1)} \neq C_p^{(2)}, \quad \alpha^{(1)} \neq \alpha^{(2)}, \quad \kappa^{(1)} \neq \kappa^{(2)}. \tag{2.66}$$

Equation (2.66) indicate that second-order phase transformations, according to the classification of Ehrenfest, are connected with qualitative changes of the response of the system with respect to a change of the external parameters. An illustration of Eq. (2.64) and thus of second-order phase transitions is given in Figure 2.5a,b. As seen from these two figures, in contrast to first-order phase transitions, the tangents (i.e., the respective first-order derivatives) to the thermodynamic potential curves representing both phases coincide in the transformation point and above it, where also the two potentials are equal. However, the curvatures (i.e., the second-order derivatives) of both the $G = G(T)$ and the $G = G(p)$ curves differ not only below the respective transformation points, but also at T_{e2} and p_{e2}, as required by the above equations.

For second-order phase transformations, relations can be derived in analogy to the Clausius–Clapeyron equation in the form as given by Eqs. (2.60) or (2.62), connecting the equilibrium values of pressure and temperature. Based on Eq. (2.63) or the equivalent relations

$$s^{(1)} = s^{(2)}, \quad v^{(1)} = v^{(2)}, \tag{2.67}$$

by a derivation of these relations with respect to T one obtains

$$\frac{dp}{dT} = \frac{1}{VT} \frac{\left(C_p^{(1)} - C_p^{(2)}\right)}{(\alpha^{(1)} - \alpha^{(2)})}, \tag{2.68}$$

$$\frac{dp}{dT} = \frac{(\alpha^{(1)} - \alpha^{(2)})}{(\kappa^{(1)} - \kappa^{(2)})}. \tag{2.69}$$

Since both these equations are equivalent, after an elimination of (dp/dT) a relation is obtained connecting the changes of the thermodynamic parameters in second-order phase transformations

$$\frac{1}{VT} \frac{\Delta C_p \Delta \kappa}{(\Delta \alpha)^2} = 1. \tag{2.70}$$

Equation (2.70) is called Ehrenfest's equation.

The behavior of different thermodynamic quantities in first- and second-order phase transitions is illustrated in Figure 2.6. As an example for first-order phase transformations the transition melt to crystal is chosen, while as an example for second-order phase transformations an order–disorder transition with a λ-type C_p curve is presented. In first-order phase transformations we have to expect $C_p \to \infty$ for $T \to T_m$ as indicated in the figure by an arrow. T_m is the melting temperature, and T_{e2} the second-order transition temperature. With dashed lines the continuations of the curves into the respective region of stability of the other phase are indicated which, however, can be realized in experiments only for first-order transformations (metastable states).

Formally, Ehrenfest's classification of phase transformations can be extended to define transitions of any arbitrary order depending on the degree of the derivatives, which become discontinuous at the transition point. However, for an experimenter, investigating the structure or the reaction of a system in dependence on the variation of external parameters and identifying the observed qualitative changes with phase transformations, only first-order (change of the structure and thermodynamic properties) and second-order phase transformations (changes of the susceptibilities, that is, the response of the system to parameter changes) of Ehrenfest's classification are of significance. In phase transitions of order equal to or higher than three, changes have to be expected, which correspond neither to qualitative variations of the thermodynamic properties, nor to qualitative changes of the response of the system. So it is no wonder that they have not been verified experimentally up to now.

In the discussion of thermodynamic phases and of the transitions between them, the following comments have also to be added. At the time, when Gibbs [56, 57] introduced into science the notion "thermodynamic phases," the only known and

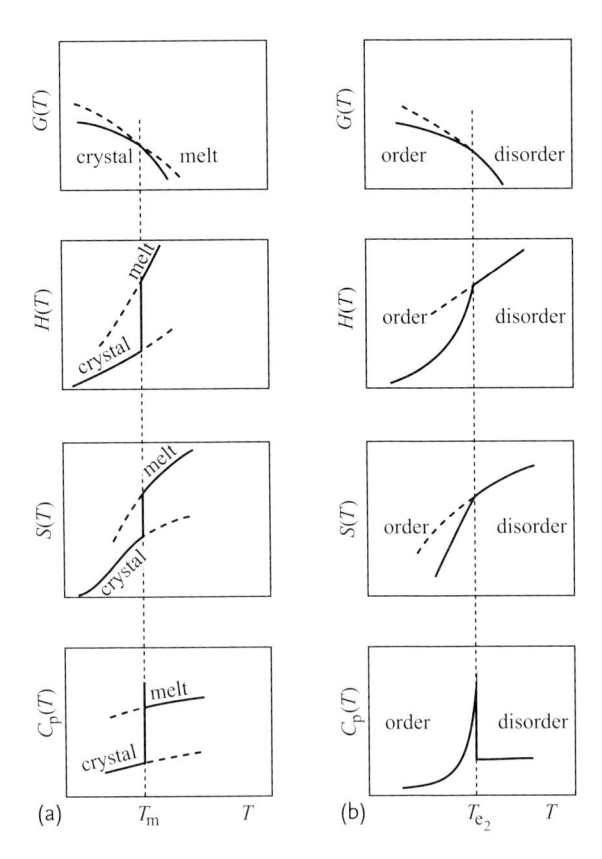

Figure 2.6 Temperature dependence of the thermodynamic functions G, H, S and of the specific heat, C_p, of a system undergoing first (a); second-order (b) phase transformation (see text).

exploited examples of phase transitions were those connected with changes between the three states of aggregation: gaseous, liquid and crystalline. These phase transitions possess a common feature: the change takes place via the formation and growth of new phase clusters in the volume of the initial phase at an excess value of the thermodynamic potential difference, $\Delta g(T)$, which is the driving force of the transformation process. These are the first-order phase transitions in Ehrenfest's classification.

In Gibbs's work and definitions in fact only first-order phase transitions are anticipated. However, Gibbs, as shown in [1], succeeded in fact in finding the thermodynamic measure for the driving force of any phase transformations. He did this for first-order transitions in terms of the above-defined thermodynamic potential difference, $\Delta g(T)$, between the initial phase and newly formed phases, or, generally, between any two possible phases. Deriving the value of the work $W_c(T)$ of formation of a sufficiently large "critical" cluster (the nucleus of the new phase capable of further deterministic growth) Gibbs determined the thermodynamic barrier in the kinetics of this process. Now we know that this procedure is necessary only for the case of first-order phase transitions. However, with this step Gibbs opened the way to calculate the rate of transformation not only in first-order phase transitions, but also for any phase transition, becoming known later. The dependence of $W_c(T)$ on the thermodynamic properties of the system and especially on the specific interfacial energy, σ, explained the specific features in the kinetics of first-order phase transitions. The existence of an interface energy $\sigma > 0$ is determined by the difference of the thermodynamic functions (of the enthalpy difference) at the interphase boundary in first-order phase transitions. Thus the very existence of an interphase boundary, where the thermodynamic functions abruptly change their value, determines the whole thermodynamics and kinetics of first-order phase transitions.

At the beginning of the twentieth century it turned out, however, that phase changes are also possible, which are quite different in both their thermodynamics and kinetics from those known in Gibbs's times. These transitions were to become Ehrenfest's second-order phase transitions. Besides the ferromagnetic behavior and magnetic transformations known from Curie's times (end of nineteenth century) and order–disorder transformations, the most striking examples in this respect were given by the transformations in quantum liquids (fluidity/superfluidity transition of He) and by the conductivity/superconductivity changes in metals, investigated in these days mainly by the Dutch school of physicists (especially by Kamerlingh–Onnes). It was in these surroundings that Ehrenfest proposed his classification.

The first thermodynamic analysis of second- (and third-) order phase transitions were performed in the now somewhat forgotten paper by Justi and von Laue [77] (see also Bazarov [55] for a substantial correction of their initial results, which is accounted for in the present discussion). In Bazarov's book, the temperature course of the thermodynamic properties of a system, undergoing second-order phase transitions, may be found as it is also shown in Figure 2.4. This course is exploited here further on in more details. The comparison between glass formation and thermodynamic phase transitions is based mainly on Bazarov's analysis, and on considerations concerning the glass transition as they have been formulated by Nemilov [78] and more recently by Gutzow and Petroff [79].

Equations (2.63) and (2.64) and Figure 2.6 show the equality of enthalpy and entropy in the two phases, undergoing a second-order phase transition. Thus they indicate also that in this case typically the interfacial energy, σ, should be considered to be equal to $\sigma \cong 0$. Thus the existence of metastable phases, of an interphase boundary and of a thermodynamic barrier in the kinetics of phase transitions determined by a nucleation process of small new-phase clusters in the ambient metastable phase is to be expected only with first-order phase transitions. In this way the zero value of the interphase energy leads to barrier-less kinetics of second-order phase transitions.

In another well-known approach (Landau's phenomenological theory of phase transitions [62, 63, 80, 81]), second and higher-order phase transformations are classified as continuous phase transitions. This is a notion, understandable both because of the equality of thermodynamic functions of the two phases at the equilibrium point and of the *continuous*, barrier-free kinetics of the transition process itself. According to Landau's terminology first-order phase transitions are an example of discontinuous changes. Problems of symmetry, appropriate series expansions and the introduction of an order parameter, φ, are essential elements of this phenomenology. There exist attempts by Gutzow and Petroff [82] and by Landa et al. [83] to use Landau's phenomenology to describe glass formation.

The development of a classification of phase transitions is not an easy task. Thus, although the vapor–liquid phase transition usually has all the typical and classical features of a first-order thermodynamic phase transition in Ehrenfest's terms, it changes to a second-order type transition at temperatures and pressures approaching critical conditions (see [34]). At critical conditions practically any phase change is second-order-like in both its thermodynamics and its kinetics. This statement applies also to processes of phase segregation in two-and multi-component systems, in magnetic transitions, and so on. At critical conditions also typically $\sigma \cong 0$ holds and this property determines the whole process of the continuous, barrierless kinetics of the transformation vapor–liquid there (and the critical opalescence, observed at $T = T_c, p = p_c$).

In many cases it is difficult or even impossible to attribute to a given phase change the distinct features of either first- or second-order phase transition as required in the framework of Ehrenfest's simple but somewhat schematic thermodynamic classification. In addition, with a change of the conditions the very nature of the process of phase transitions can also be altered. Besides the change from first-order to second-order at critical conditions, another classical example of this type is given by the first experiments of Kamerlingh–Onnes: it turned out that in a magnetic field the conductor to superconductor transformation shows typical features not of a second-order, but of a first-order transition. In a magnetic field, the Clausius–Clapeyron equation (2.60) is obeyed; in its absence the Ehrenfest dependencies (Eqs. (2.68)–(2.70)) follow.

Another well-known example of such uncertainties is given by phase transformations in liquid-crystal systems (see [35]), where nucleation seems to take place at very low σ values. Thus it is said that in liquid-crystal systems quasi-first (or quasi-second) order-like transitions are observed. In these and in many other cases, Ehrenfest's classification, although thermodynamically and mathematically strikingly well founded, is in fact far from being sufficient, when directly applied to experiment. Besides these circumstances it also became evident from Onsager's discussion of the Lenz–Ising model (Onsager [84]) that there are also theoretical restrictions to be considered, showing that qualitative variations of the state of a system may exist, which cannot be described in Ehrenfest's general scheme (see, e.g., Gebhardt and Krey [85] and Gunton, San Miguel, and Sahni [86]).

Several alternative empirical or semi-empirical proposals exist to classify phase transitions. Some of these proposals are based on the observed type of change of specific heats at the transition (see [87]), others are based on the dependence of the $\Delta S(T)$-function in the transition region [48, 88], in others, the nature of the $\Delta H(T)$-change is considered as the leading feature. Of interest is also the general thermodynamic approach of Tisza [89], applied to Ehrenfest's classification, as it is summarized also by Callen [43]. In some respect, all these empirically based classifications, operating with different types of $C_p(T)$, $H(T)$ or $G(T)$-changes in real systems, are contained in the classical analysis of Ehrenfest [75] and of Justi and von Laue [77].

Second and higher-order phase transitions require an expansion of the already cited classical definitions of notions like thermodynamic phases, transition points and interphase known from Gibbs's times. Thus in Chapter 14 of the already cited Landau–Lifshitz treatise [62, 63] it is also said: "*While at a first-order transition point in equilibrium are two bodies in different states (in the sense that they are characterized by different thermodynamic functions like V, S, etc.) at a second-order transition point the properties of the two phases are equal.*" The essential point to be remembered is that any thermodynamic phase is an equilibrium state and that any phase change – continuous or discontinuous, first- or higher-order – is a change in and between thermodynamic states, fulfilling the requirements of thermodynamic equilibrium, outlined in the preceding section.

In older and more recent literature there have been attempts to classify glass formation and glass transitions in terms of first [90, 91], second [92–94] and recently also as third- and fourth-order thermodynamic phase transformations. With the above statement and the knowledge summarized in the following paragraphs none of these proposals can be justified. We have to repeat here that glasses are not to be termed thermodynamic phases in the sense of the already adopted definitions and upon vitrification a new and very interesting transition takes place: starting from the thermodynamic state of a metastable system the nonequilibrium state of a glass is formed. Such a transition is out of the scope of both classical thermodynamics and of the generally accepted definitions of phase transformations and of similar ideas referring to changes in between equilibrium states of matter and of symmetry considerations connected with them.

2.3
Crystallization, Glass Transition and Devitrification of Glass-Forming Melts: an Overview of Experimental Results

We now consider processes which take place in the course of cooling of the melt of a substance, which can be transformed, at least, under certain conditions into a glass. The melting temperature of the corresponding crystalline phase we denote by T_m. According to the results outlined in Section 2.2, T_m is at the same time the temperature, at which, at a constant pressure, the liquid and crystalline phases coexist in equilibrium.

Let us assume we are experimenting with a crucible containing a given quantity of a pure glass-forming melt at an initial temperature, T, somewhat above the melting temperature, T_m. Starting from this initial state, the system cools down, that is energy is removed in the form of heat with a nearly constant rate from the melt. The resulting temperature decrease is measured by a thermocouple. Possible $T = T(t)$ curves, which may be obtained in this way, are shown in Figure 2.7.

Part (1–2) of the curve shown in Figure 2.7a describes the process of cooling the melt down to T_m. At this temperature the melt may start to crystallize. If this

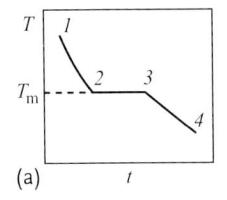

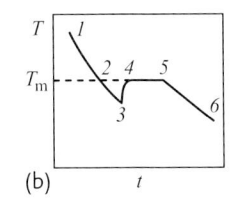

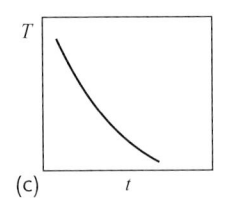

Figure 2.7 Temperature T vs. time t curves of a melt at constant cooling rates for three different cases: (a) the crystallization of the melt proceeds immediately below T_m; (b) a significant crystallization is observed only after a critical value ΔT_{max} of the undercooling is reached; (c) the melt is transformed into a solid without measurable crystallization.

is the case, a time interval is observed, at which cooling caused by the external surrounding is compensated by the release of the latent heat of crystallization (cf. Eq. (2.61)), resulting in a temporary constancy of temperature (part (2–3) of the curve in Figure 2.7a). This horizontal part (2–3) is followed then by the cooling curve (3–4) of the completely crystallized material.

Such a behavior, when the new phase appears immediately without any measurable undercooling, may be observed only in metal melts or when in an oxide melt precautions are undertaken to initiate the crystallization process (i.e., by the introduction of seed crystals of the same substance). In the majority of cases a more or less pronounced degree of undercooling, ΔT_{max}, has to be reached, before an intensive crystallization process starts. As undercooling the temperature difference $\Delta T = T_m - T$ is denoted. Thus, ΔT_{max} has to be understood as:

$$\Delta T_{max} = T_m - T_{min} . \tag{2.71}$$

Here T_{min} indicates the lowest temperature which can be reached without measurable crystallization being initiated.

Figure 2.7b gives an example of a temperature versus time curve for such a case. Again, the part (1–2) corresponds to the cooling curve of the melt, but this time the cooling curve is extended into the metastable region (2–3), where the crystalline state and not the undercooled melt is stable from a thermodynamic point of view. This result is the simplest demonstration of the already mentioned fact that in first-order phase transformations equilibrium has to be more or less exceeded to allow a measurable phase change to take place. In second-order phase transformations the existence of a metastable phase outside the limits of its thermodynamic stability has never been observed. After a critical undercooling ΔT_{max} is reached, spontaneous crystallization takes place. The release of the latent heat accompanying the process of crystallization results in a temperature rise until T_m is again attained. After crystallization is completed, the cooling curve of the crystalline phase is followed (5–6) similarly as in Figure 2.7a.

For a number of substances the maximal undercooling can reach considerable values, for example, 370 K for platinum, 150 K for iron, 20 K for gallium. The values of the undercooling realized experimentally at normal cooling rates (10^{-1} to 10^2 K s^{-1}) are usually found to be of the order $\Delta T_{max}/T_m \approx 0.2$. This value is also predicted by an empirical rule [95, 96], which states that

$$\frac{T_{min}}{T_m} \approx 0.8\text{--}0.9 \,. \tag{2.72}$$

This rule is valid for a large number of substances.

An example of a substance, which can be easily undercooled, is elemental gallium. Gallium has a melting point of about 29 °C. In accordance with the formulated empirical rule, it can be preserved as an undercooled liquid at room temperatures $T_{(room)}$ for practically unlimited periods of time ($T_{(room)}/T_m \approx 0.97$). Other substances, which are often used in laboratory demonstrations as undercooled melts, are sodium thiosulfate ($Na_2S_2O_3 5H_2O$) and salol ($HOC_6H_4COOC_6H_5$). These substances have a melting point of about 50 °C and can be cooled down to room temperatures also without any crystallization being detected [72, 97, 98]. An inspection of available experimental finding shows that for substances with very different compositions and structure (metals, oxides, salts, molecular liquids, polymer melts) the relative critical undercooling, ($\Delta T_{max}/T_m$), which may be achieved by normal cooling techniques, is practically the same. This applies even for melts like $NaPO_3$, lithium disilicate, glycerol, piperine, which are representatives of typical glass-forming substances.

For most of the substances investigated, an undercooling to temperatures below T_{min} leads to immediate crystallization. It is shown further on that the dramatic increase of the viscosity for $T \leq T_{min}$ observed for typical glass-formers has the consequence that crystallization does not occur in the melts of such substances even when they are maintained in this temperature region for prolonged times. Experimental evidence accumulated for more than 150 years with very different classes of substances (cf. the beautiful summary on the history of phase formation processes as it is given in the introduction to Volmer's monograph [99] or, with more details, in Ostwald's classical textbook [100]) allows one to draw the following conclusions concerning the initiation of crystallization processes in undercooled liquids:

1. The highest possible undercooling, T_{min}, can be realized experimentally only if insoluble foreign particles, which may act as centers of crystallization, or certain, specifically acting surface-active substances are removed from the melt. However, foreign particles are not equally active in the induction of crystallization in undercooled melts. The highest activity with respect to the initiation of crystallization show seed particles of the same substance or crystallization cores with structure and cell dimensions close to those of the evolving crystal.
2. In some cases, mechanical effects (vibration, stirring) may initiate crystallization.
3. The achievable undercooling increases by increasing the degree of dispersion of the melt. In the process of quenching of the melt with small droplets significantly larger values of the undercooling may be reached as compared with corresponding data for bulk samples of the same melt, which are to be expected according to Eq. (2.72).

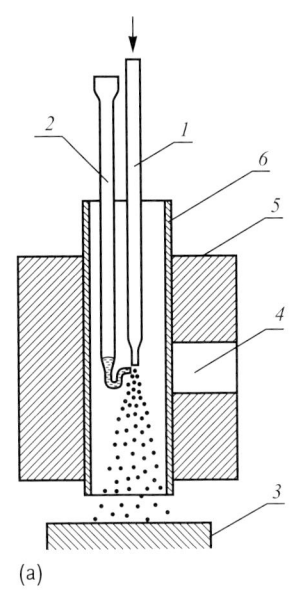

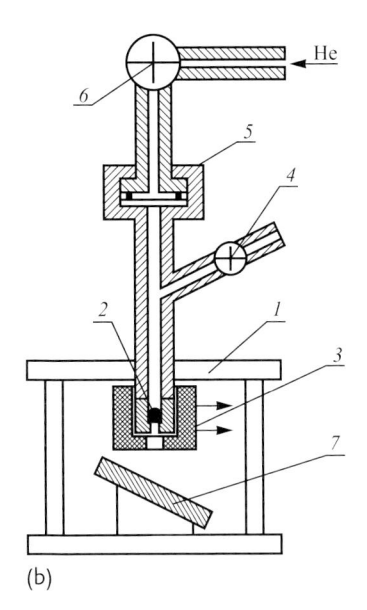

Figure 2.8 (a) Tammann's atomizer for vitrification of molten salts. An inert gas (nitrogen) is passed under pressure through a tube (1) dispersing the molten salt contained in tube (2). The droplets are quenched at the metal plate (3) cooled by liquid nitrogen. The oven tube (5) as well as the tubes (1) and (2) are made of quartz glass, so that the course of the process of dispersion can be followed through the window (4); (b) modification of Tammann's atomizer. A crucible (1) contains a drop (2) of a metallic alloy molten in the oven (3). High- pressure inert gas (He) is supplied with valve (6) rupturing the milarite diaphragm slit (5) and forces the drop at high speed onto a cooled metal plate (7), where it is frozen to a glass. Further metallic alloy samples are introduced into the system by the valve (4) (after Giessen and Wagner [101]).

Thus, undercooling for water droplets of about 10^{-6} m in size may reach the order of $\Delta T \approx 40-50$ K, which is of a particular meteorological importance. For metal melts with a relatively high melting point, $\Delta T \approx 200-300$ K is often reported. A relatively high degree of dispersion of the melt can be achieved, for example, by the process of its pulverization in an air stream. This method, connected with a rapid quenching on cold surfaces, was introduced by Tammann as an effective method for the transformation of substances into the vitreous state (Tammann [24]; see also Figure 2.8a). A modification of this method is shown in Figure 2.8b.

An extension of Tammann's ideas was given by Turnbull [102, 103] (see also Greer [104]) and consists of the emulgation of low temperature melting metals like tin, lead, bismuth, mercury, and so on in silicon oil. Additional emulgators, introduced into the system, prevent the coagulation of the metallic drops. To transfer high temperature melting metals like iron, nickel, platinum or chromium into a dispersed state as an emulgator silicate glass-forming melts have been used. By vibrating and mixing the liquid with another appropriate substance the investigated liquid is dispersed into a large number of small droplets. If impurities are present in the test liquid then some of the drops will contain them while others will remain

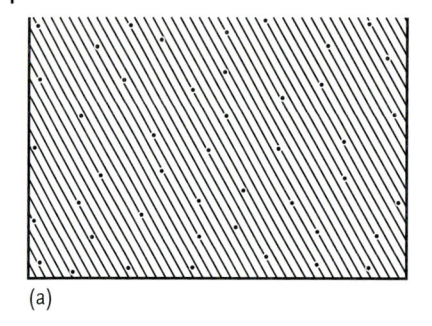

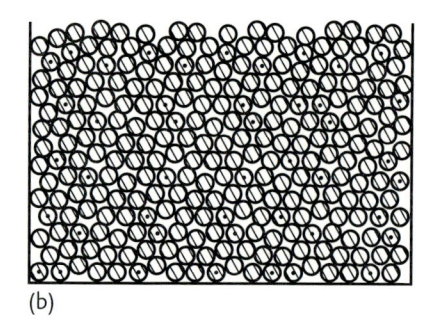

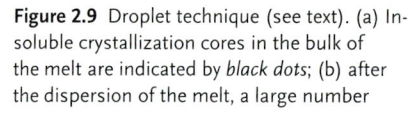

Figure 2.9 Droplet technique (see text). (a) Insoluble crystallization cores in the bulk of the melt are indicated by *black dots*; (b) after the dispersion of the melt, a large number of drops is formed which do not contain insoluble crystallization cores. For such drops, crystallization by homogeneous nucleation may be expected to occur.

unaffected by such foreign crystallization cores and may be undercooled to temperatures at which intensive homogeneous nucleation occurs. Thus the increase in the degree of undercooling in small particles, at which crystallization occurs, is due to the absence of active foreign crystallization cores, at least, in some of the droplets (Figure 2.9a,b).

In Figure 2.10, contemporary methods for super-rapid cooling are schematically illustrated.

The temperature T_g, below which the undercooled melt behaves like a solid, is denoted, according to a proposal by Tammann, as the glass transition (or transformation) temperature. The amorphous solid resulting in a cooling process without perceptible crystallization is a glass. The temperature vs. time curve for the process of glassy solidification is shown in Figure 2.7c. A horizontal part of the curve as in Figure 2.7a,b does not exist. This feature is an indication that the process of vitrification is not connected with the release of any latent heat and thus with a discontinuous change of the structure, the entropy and the enthalpy of the system. The systems remain spatially homogeneous and no crystallization is to be detected. For temperatures sufficiently below T_g, crystallization processes of the already vitrified melt have never been observed. However, crystallization may occur, if the glass is reheated, again, to temperatures above T_g. Such a crystallization process is usually denoted as devitrification.

The existence of the glass transformation temperature T_g, of vitrification and of devitrification processes may be demonstrated more effectively, as in the primitive arrangement shown in Figure 2.7 and the single cooling curves given there, by heating or cooling runs with differential thermal analysis (DTA). The DTA curves of a devitrifying glass at constant heating rates (usually 10 K per minute) show a more or less well-pronounced inflection point, when the temperature T_g is reached. This inflection point can be used both to define (see Chapter 3) and to determine experimentally T_g in a standardized procedure employing DTA measurements.

Somewhat above T_g, an exothermic peak usually occurs, which corresponds to the process of devitrification. For typical glass-forming melts, which can be vitri-

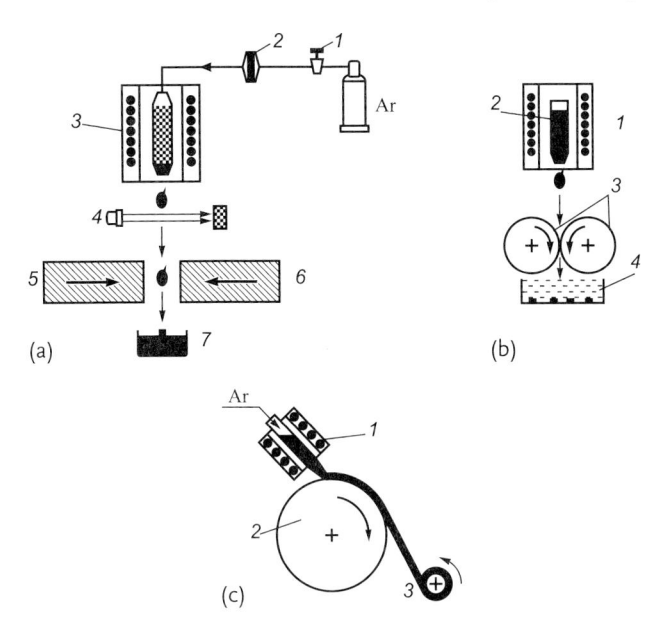

Figure 2.10 Modern methods of super-rapid melt quenching techniques. (a) Anvil and hammer splat-cooling system with cooling rates up to 10^8 K s^{-1} for vitrification of metals and metal-metalloid alloys; (1) pressure valve for initiation of the quenching process, (2) diaphragm (its rupture under argon pressure ensures explosive formation of liquid droplets flowing out of the oven-heated camera (3)), (4) laser-triggered beam initiating the hammer (5) and anvil (6) splat-press quench of the melt droplets. The vitrified material is usually obtained in form of discs of thickness of 10 μm with a diameter of one centimeter. This method is mainly of some historic interest. With this device, the first metallic alloy glasses have been obtained; (b) double roller quenching method employed in the vitrification of oxide, fluoride and halide systems, and with some modifications, also for organic polymer vitrification. Cooling rates reach here typically 10^3–10^5 K s^{-1}: (1) high temperature oven with platinum finger crucible (2) with about a 1 mm orifice, (3) quenching copper rollers rotating with a frequency of 3000–10 000 min^{-1}, and (4) cryogenic bath. The vitrified material is obtained in the form of thin (20–40 μm) flakes. (c) Metal-ribbon spinning quench apparatus utilized in various forms in present-day metallic glass production. Cooling rates of 10^4–10^6 K s^{-1} are reached with it: (1) high-temperature oven with quartz ampule with a rectangular slit orifice, (2) massive copper quenching wheel rotating 1000–3000 min^{-1}, on which the metallic alloy melt is vitrified, and (3) spinner wheel. The vitrified material consists typically of met-glass ribbons with 1–2.5 cm width and 20 μm thickness.

fied at normal cooling rates (i.e., not exceeding 10^{-1} to 10^2 K s^{-1}), devitrification processes begin usually at 30 to 50 K above T_g. For other systems, however, like the already mentioned glass-forming metallic alloys, which are obtained only as the result of extreme cooling rates, the devitrification peak appears just above the T_g-inflection point. In such cases, this crystallization peak itself gives an indirect possibility of determination of T_g (see Figure 2.11).

Figure 2.12 shows the DTA heating curve of a technical silicate glass. In the devitrification process, two different crystalline phases are formed. The respective liquidus temperatures are also clearly seen. Particularly instructive are devitrifica-

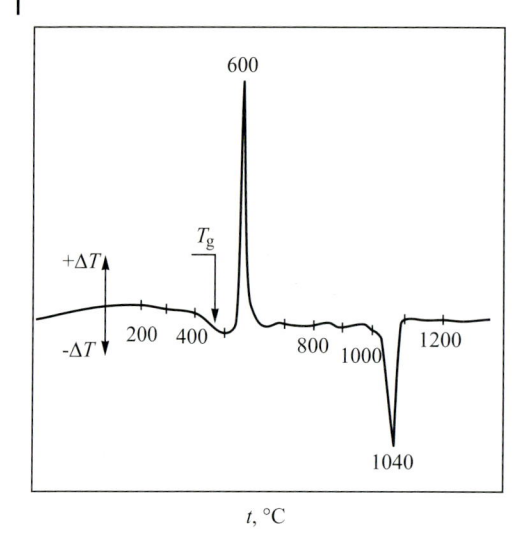

Figure 2.11 Typical DTA curves obtained in the process of heating of a ($Li_2O\ 2SiO_2$)-glass. A crystallization peak at 600 °C is seen followed by an endothermic peak at 1040 °C corresponding to the melting point of crystalline lithium disilicate. T_g is indicated by a turning point in the DTA diagram (see Penkov and Gutzow [105]).

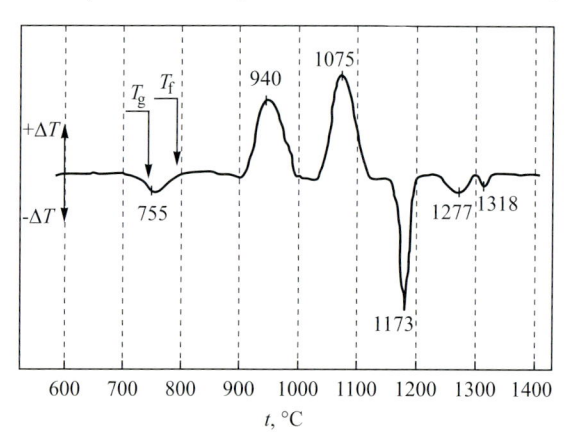

Figure 2.12 Heating curve of a technical glass and the corresponding glass-forming melt (enstatite ceramics precursor glass; after Gutzow et al. [106]). The devitrification process is manifested by two crystallization peaks at 940 and 1075 °C. The three endothermic peaks at higher temperatures correspond to the process of melting of the three different crystalline phases formed in the devitrification process. The transformation temperature and the softening point of the glass are denoted by T_g and T_f, respectively.

tion experiments carried out by differential scanning calorimetry (DSC) (see Figure 2.13). The latent heat of the crystallization or melting process can be directly determined from the areas under the crystallization or melting peaks.

The peculiar course of the DTA experiments and especially the inflection point of DSC curves at T_g is a direct indication that the process of glassy solidification of

2.3 Crystallization, Vitrification and Devitrification | 43

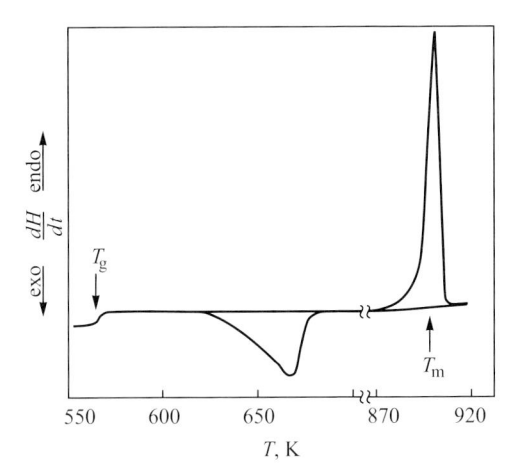

Figure 2.13 Differential scanning calorimetry (DSC) measurements of the devitrification process of a NaPO$_3$-glass heated with a constant heating rate of 10 K min^{-1}. At T_g an inflection point is observed followed by an endothermic devitrification peak. At 898 K the heat of melting of the crystalline α-NaPO$_3$ phase is released. The areas under the crystallization, respectively, melting curves give the latent heat of the transformation (see also Grantcharova, Avramov, and Gutzow [107]).

a melt is connected with an abrupt stepwise or, more precisely, with a sigmoid-like change of the course of the specific heats. From these and similar experimental results it also becomes evident that the solidification of a melt into a glass is not a transformation in the sense of a first-order phase transition with a release of latent heat. It looks at a first glance more similar to a second-order phase transformation (discontinuity in a step-like form in the course of C_p).

A more extended discussion, given in the following section, shows that vitrification is described correctly as a freezing-in process of the undercooled melt. Here we have to mention only that the notation glass transformation (or glass transition) temperature, proposed by Tammann, is to some extent misleading. Correct with respect to the indicated mechanism of vitrification is the proposal, developed by Simon [22], to denote T_g as the freezing-in temperature of the glass (*Glaseinfriertemperatur* $T_e = T_g$). In the English literature the more "neutral" notation temperature of vitrification is also preferred. It corresponds to the word *Glastemperatur* used in the German literature more in the physics of high polymers. However, since in technology of silicate glasses Tammann's notation glass transformation temperature is till now the most common one, it will be preferably applied here, especially, when technical glasses are discussed.

It was also first mentioned by Tammann that the value of the glass transformation temperature varies to some extent in dependence on the method of determination and, what is of even greater physical importance, on the value of the cooling rate $q = (dT/dt)$, reached in the the course of vitrification of the undercooled melt: with increasing absolute values of q higher T_g values are obtained. Moreover, as it was also mentioned by Tammann, we have to speak more precisely of a glass transition temperature range in which the melt is solidified into a glass.

Following again Tammann, we will apply in general the notation T_f (or T_g^+) for the upper and T_g (or T_g^-) for the specification of the lower boundary of the glass transition range. Usually, in technical glass science literature the notation T_g is applied, when a standard method of vitrification is used, giving more or less reproducible values for the temperature of glass transition (for a more detailed discussion of the methods of determination of T_g and the problems involved in this process see the analysis given by Mazurin [108] and the respective discussion in Chapter 3).

For typical one-component glass-forming melts and normal cooling rates the value of the glass transformation temperature, divided by the melting temperature, is usually given by

$$\frac{T_g}{T_m} \approx \frac{2}{3} . \qquad (2.73)$$

This is the so-called Beaman–Kauzmann rule (Kauzmann [109]; for a first derivation of this rule see Gutzow [110]). The Beaman–Kauzmann rule was generalized by Sakka and Mackenzie [111] to the case of multi-component systems. In this case, T_m is to be replaced by the respective liquidus temperature, T_l, of the system. However, for glass-forming metallic alloys the value of the ratio T_g/T_l may be considerably smaller, for example,

$$\frac{T_g}{T_l} \approx \frac{1}{2} . \qquad (2.74)$$

For amorphous thin layers of a number of metals such as tin or gallium even considerably lower values as predicted by Eq. (2.74) are reported.

In Figure 2.14a,b (see also Table 2.1) according to Gutzow and Dobreva [112] distribution histograms for T_g values of a large number of typical glass-forming melts and for vitreous metallic alloys are shown. In accordance with the Beaman–Kauzmann rule for typical glass-formers, indeed, average values about $T_g/T_m \approx 0.65$ are found, while for metallic alloy glasses in addition to the peak at 0.5 a second peak at 0.3–0.4 can be noticed. The substances taken into consideration in Figure 2.14a include representatives of different types of glass-forming melts: oxides (SiO_2, B_2O_3), halides (BeF_2, $ZnCl_2$), simple borate, silicate and phosphate glasses (e.g., $Na_2O\,2B_2O_3$, Na_2SiO_2, $NaPO_3$), glass-forming organic compounds (alcohols, e.g., C_2H_5OH, CH_3OH, glycerol), organic acids and oxiacids as well as a number of more complicated aromatic organic substances. The T_g values of several glass-forming organic polymer melts are also included.

Similar empirical relationships were also proposed by other authors. According to Turnbull and Cohen [113], for example, the following equation, connecting the glass temperature T_g with the boiling temperature T_b of the substance, is valid:

$$T_g \approx \left(\frac{1}{4} - \frac{1}{3}\right) T_b . \qquad (2.75)$$

Table 2.1 Thermodynamic properties of typical glass-forming substances: vitrification temperature T_g, melting temperature T_m, molar melting entropy ΔS_m, and molar frozen-in entropy ΔS_g (both in J K^{-1} mol^{-1}). The ΔS_g values are taken from Simon and Lange [21], Gutzow [164], Nemilov [212], Timura et al. [213], Weyl and Marboe [214], Smith and Rindone [215], Grantcharova et al. [107, 172], Angell and Rao [4], Anderson [216], Tammann [24], Tammann and Jenckel [205], Tammann [204], Greet and Turnbull [217], Simon [59], Bestul and Chang [218], Kelley [219], Chen [220], Chen and Turnbull [221], Dobreva [222].

Substance	T_m (K)	T_g (K)	T_g/T_m	ΔS_m	ΔS_g	$\Delta S_g/\Delta S_m$	References
SiO$_2$	1996	1473	0.73	9.13	3.8	0.42	[212, 223]
GeO$_2$	1386	900	0.65	12.15	3.8	0.31	[212]
BeF$_2$	1076	580	0.54	15.5	4.6	0.30	[212, 213]
H$_2$O 4B$_2$O$_3$	1132	633	0.56	111.4	29.3	0.26	[214, 215]
Na$_2$O 4B$_2$O$_3$	1085	689	0.63	121.1	36.4	0.30	[214, 215]
NaPO$_3$	898	550	0.61	24.7	11.0	0.44	[107]
B$_2$O$_3$	723	521	0.72	31.8	10.6	0.33	[214]
ZnCl$_2$	535	375	0.70	16.6			[215]
H$_2$SO$_4$ H$_2$O	237	157	0.66	102.2	24.72	0.24	[4]
Se	491	303	0.62	10.89	2.9	0.27	[216, 223]
Phenolphtaleine	534	353	0.66	95.5	10.1	0.11	[172]
Betol	368	250	0.68	54.0	19.7	0.36	[24, 205]
Orthoterphenyle	329	245	0.74	51.5	22.6	0.44	[217]
Benzophenone	321	158	0.49	55.3	15.1	0.27	[24, 205]
Glycerol	291	178	0.61	62.85	19.3	0.31	[59, 205]
n-Propanol	146	95	0.65	39.0	15.9	0.41	[218]
Ethanol	156	93	0.6	31.8	10.9	0.34	[4]
Methanol	175			18.1	7.62	0.42	[4]
2 Methylpentane	119	78	0.65	52.8	16.8	0.32	[218]
Butene-1	88	59	0.67	43.6	12.6	0.29	[218]
Poly(propylene)	449	259	0.58	24.3	8.0	0.33	[218]
Poly(ethylen-etherephtalate)	542	342	0.63	47.9	14.3	0.30	[222]
Rubber	301	199	0.66	14.7	5.9	0.40	[218]
Pd$_{0.775}$Cu$_{0.05}$–Si$_{0.165}$		636			3.3		[220]
Pd$_{0.48}$Ni$_{0.32}$–P$_{0.20}$		585			7.5		[220]
Au$_{0.77}$Ge$_{0.136}$–Si$_{0.094}$	624	290	0.46	19.7	6.3	0.32	[220]
Au$_{0.814}$Si$_{0.186}$	636	290	0.45		6.3		[220]

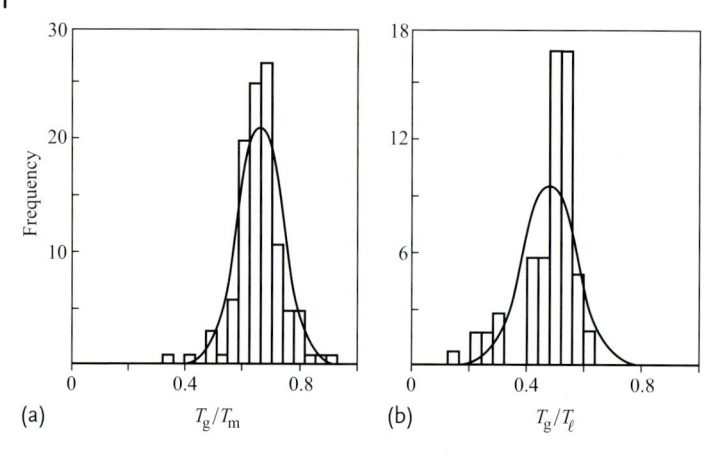

Figure 2.14 (a) Frequency distribution histogram of experimentally observed (T_g/T_m) values for 108 typical glass-formers with different compositions vitrified at normal cooling rates; (b) frequency distribution histogram of 80 experimental (T_g/T_l) values for metallic glass-forming alloys (after Gutzow, Dobreva [112]).

However, if one takes into consideration an additional empirical connection between T_b and T_m of the form

$$T_b \approx \frac{5}{2} T_m, \qquad (2.76)$$

it turns out that Eq. (2.75) and Eqs. (2.73) and (2.74) are to a large extent equivalent.

For organic high polymers the glass temperature T_g depends on the average degree of polymerization, $\bar{x}$. According to Flory [114]

$$\frac{1}{T_g} \approx A + \frac{B}{\bar{x}} \qquad (2.77)$$

holds. A and B are constants, specific for the considered substance.

In the following sections the properties of glass-forming melts in the glass transition range are discussed in more detail. We begin now with the analysis of the temperature dependence of the viscosity of undercooled melts, since primarily the dramatic increase of viscosity determines the transition of the liquid into a solid, the glass.

2.4
The Viscosity of Glass-Forming Melts

2.4.1
Temperature Dependence of the Viscosity

The first systematic studies of the viscosity of glass-forming melts were performed by Tammann. His results are summarized in his well-known monographs *Aggregatzustände* and *Der Glaszustand* [24, 72]. Tammann carried out his investigations

2.4 The Viscosity of Glass-Forming Melts

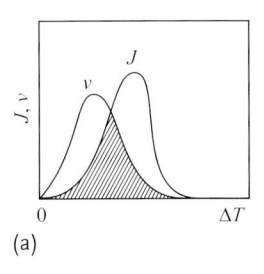

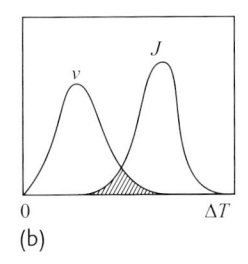

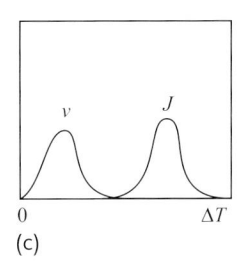

Figure 2.15 Possible temperature dependencies of the nucleation rate, J, and the linear growth velocity, v. According to a proposal by Tammann the curve (a) corresponds to a melt with a low glass-forming ability while (b) and (c) refer to the opposite situation. Tammann's ideas are reformulated today in terms of the so-called TTT (time-temperature-transformation) diagrams (see [1]) and the vitrification criteria developed there.

mainly with low-melting organic model substances like salol, betol, manniol, piperine, natural resins and colophon, any of which can be easily transferred into the vitreous state. Based on the results of his investigations Tammann formulated elementary but very instructive concepts concerning the process of vitreous solidification. He also formulated simple criteria for the conditions under which the more or less rapid quench of a melt will lead to a glass or, vice versa, under which conditions crystallization is preferred.

It was also Tammann, who divided the process of crystallization of an undercooled melt into two consecutive stages: nucleation (characterized by the nucleation rate, J, the rate of formation of centers of the crystalline phase in the bulk of the melt) and their subsequent linear growth characterized by the linear growth velocity, v. It was argued by Tammann that vitrification will occur if, in the range between T_m and T_g, the temperature dependencies of the curves, representing nucleation rate, J, and growth rate, v, respectively, do not show any significant overlapping. In contrast, an overlapping of the nucleation and growth curves will be an indication of the preference of formation of a crystalline phase (see Figure 2.15). According to Tammann both processes, nucleation and growth, are determined by the bulk viscosity of the melt, thus underlining once more the importance of this quantity in the process of glass formation.

Since Tammann's time sophisticated methods for the determination of the viscosity of glass-forming melts and its temperature dependence have been developed and a large number of experimental data have been accumulated. Such data can be found, for example, in the classic reference book by Eitel, Pirani and Scheel [115]. A recent and very excellent summary of properties of glasses and glass-forming melts including viscosity data is given in the series of monographs edited by Mazurin [2, 3] and in the electronic database [116] developed by him with coworkers (cf. also Chapters 6 and 7).

From a physical point of view, viscosity is a measure of the internal friction, which results from the relative motion of different layers of a liquid. If the velocity of the liquid is changed, for example, in the x-direction, then the force F acting

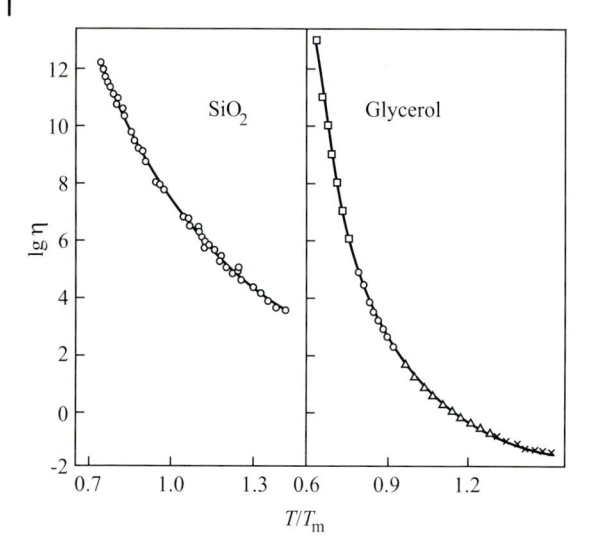

Figure 2.16 Temperature dependence of the viscosity of two typical glass-forming substances (SiO$_2$ and glycerol) according to data from different authors (see Landoldt-Börnstein [117]). The temperature is given in relative units T/T_m, where T_m is the melting temperature.

between the layers is given by

$$F = \eta A \frac{dv}{dx} \, . \tag{2.78}$$

The factor of proportionality η in Eq. (2.78) is denoted as the (shear) viscosity of the liquid, and A is the surface area of the layers. Equation (2.78) was proposed by Newton. Liquids, which can be described by such an equation with a value of η, depending on temperature and the specific properties of the liquid only, are called Newtonian liquids (see also Chapter 4 and [1]). From a thermodynamic point of view, the viscosity η is one of the kinetic coefficients, similarly to the coefficient of diffusion, the coefficient of heat conduction. The kinetic coefficients are, in general, equilibrium properties of the systems considered.

Earlier "Poise" was applied as the unit of the viscosity. This is the natural unit in the so-called CGS (centimeter-gram-second) system. In the international system of units (SI), the corresponding quantity is "Pascal second" (Pa s). Both quantities are related by the conversion factor 1 Poise = 0.1 Pa s. To retain the numerical values, familiar in Poise, in present day literature the unit dPa s is also frequently used.

The first investigations of the viscous properties of glass-forming melts showed that with a decrease of temperature, T, a very steep increase of the viscosity is found. While at the melting temperature, T_m, the viscosity of most liquids seldom exceeds the value $\eta \approx 10^2$–10^3 dPa s, and in the transformation range viscosity is typically of the order $\eta \approx 10^{13}$–10^{14} dPa s. Examples are shown in Figures 2.16 and 2.17 for SiO$_2$ melts, glycerol and for selenium.

For relatively small temperature intervals the viscosity of a glass-forming melt can be described in a good approximation by an equation usually attributed to

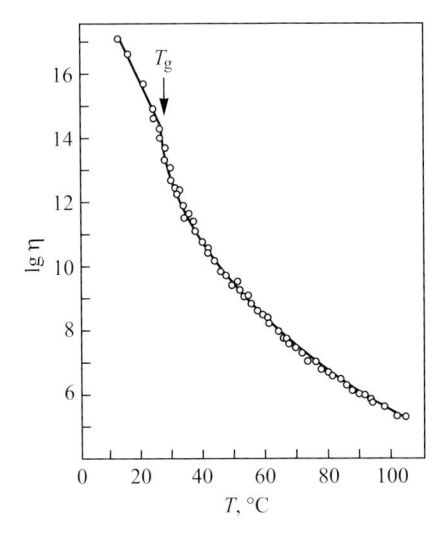

Figure 2.17 Temperature dependence of the viscosity of selenium (Nemilov [118]; see also Rawson [119]). Note the break-point in the viscosity curve near $T = T_g$.

Andrade [120, 121] and Frenkel [25, 122] and denoted as Andrade–Frenkel or Eyring's [123] equation:

$$\eta = \eta_0 \exp\left(\frac{U_0}{k_B T}\right). \tag{2.79}$$

However, as mentioned by Andrade himself, Eq. (2.79) was first proposed as early as 1913 by de Guzman (see Besborodov [124]). Frenkel gave the first elementary molecular-kinetic interpretation of this equation, which was reformulated later by Eyring in terms of the absolute rate theory (see Glasstone et al. [18]).

In Eq. (2.79), U_0 is the energy of activation of the viscous flow process, assumed to be a constant ($U = U_0$), and η_0 is a temperature-independent constant. Experimental results for glass-forming liquids, plotted in so-called Arrhenius coordinates ($\log \eta$ vs. $(1/T)$), do not give, however, a straight line (see Figure 2.18), as expected from Eq. (2.79). Possible generalizations of Eq. (2.79) are connected with the introduction of a temperature-dependent activation energy, $U(T)$, leading to

$$\eta = \eta_0 \exp\left[\frac{U(T)}{k_B T}\right]. \tag{2.80}$$

Experiments show that with decreasing temperature an increase of $U(T)$ is observed, generally, so that the condition

$$\left(\frac{dU}{dT}\right) \leq 0 \tag{2.81}$$

is fulfilled. In Chapters 3 and 4 it is shown that this condition, derived there in a generic phenomenological framework, is of utmost importance in formulating

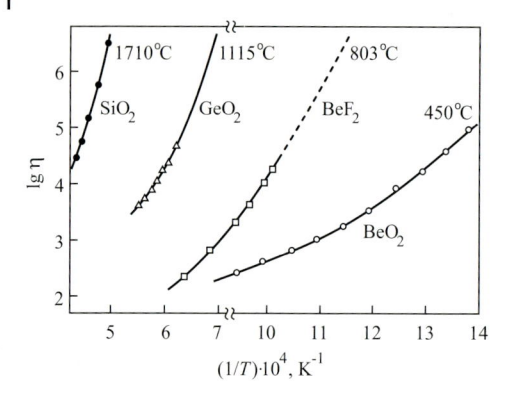

Figure 2.18 Temperature dependence of the viscosity of typical glass-forming melts in log η vs. $1/T$ coordinates (Mackenzie [125]; see also Rawson [119]). The temperature values given to each curve are the melting temperatures of the respective substance.

general physical criteria for the vitrification process. The simplest temperature dependence of the activation energy, satisfying Eq. (2.81), is given by the linear combination

$$U(T) = U'_0 - U_1 T,\qquad(2.82)$$

where U'_0 and U_1 are constants. This dependence was empirically proposed by Kanai and Satoh [126, 127].

According to Eq. (2.79) the slope of the derivative $d(\log \eta)/d(1/T)$ gives directly the activation energy, U_0. However, for temperature-dependent activation energies this is not the case. In contrast, $U(T)$ is determined, now, by the differential equation (2.83):

$$\frac{d\ln\eta}{d(1/T)} = \frac{1}{k_B}\left\{U(T) + \frac{1}{T}\frac{dU(T)}{d(1/T)}\right\}.\qquad(2.83)$$

It is evident that for temperature-dependent activation energies the true value of $U(T)$ cannot be determined directly by an Arrhenius plot but only via Eq. (2.80) as:

$$U(T) = k_B T(\ln\eta - \ln\eta_0).\qquad(2.84)$$

A number of equations of the form of Eq. (2.80) which fulfill the condition in Eq. (2.81) were proposed by different authors. Most of them give a temperature dependence closer to experimental results than the simple linear approximation Eq. (2.82) due to Satoh and Kanai. Some of these dependencies are empirically proposed, others are derived in the framework of more or less thoroughly formulated molecular models. A detailed analysis shows that all these models and empirical relations are in fact based on the general theory of thermodynamic fluctuations, determining the more or less complicated exponential character of known or possible temperature functions of the viscosity of glass-forming liquids.

2.4 The Viscosity of Glass-Forming Melts

Of particular significance in glass-science are viscosity-temperature relations which are distinguished either by the accuracy of description of experimental results or by the possibility of interpreting them in the framework of appropriate more or less general model theories of liquids. The most well-known empirical equation of the first type is the Vogel–Fulcher–Tammann (VFT) equation:

$$\eta = \eta_0 \exp\left(\frac{U_0^*}{k_B(T - T_\infty)}\right). \tag{2.85}$$

Here η, U_0^* and T_∞ are empirical constants specific to the substance considered. The Vogel–Fulcher–Tammann equation was independently proposed by the three men giving their names to this dependence. Vogel [128] developed it based on investigations of the temperature course of the viscosity of technical greases; Fulcher [129] by an analysis of the $\eta(T)$ course of silicate glasses; and Tammann (with Hesse [130]) based on experiments with glass-forming organic substances.

If the choice of the three constants in Eq. (2.85) is carried out appropriately from measurements of values of the viscosity at sufficiently distinct values of temperature and viscosity (e.g., $\log \eta \approx 2\text{--}4$, $\log \eta \approx 6\text{--}8$, $\log \eta \approx 12\text{--}13$), then the VFT equation describes the viscosity in the whole temperature range, characterized by changes of the viscosity by ten orders of magnitude, with an accuracy better than ten percent. For $T \to T_\infty$, according to the VFT equation, the viscosity tends to infinity. This is the reason for the usual notation T_∞ for one of the constants in Eq. (2.85). The VFT equation corresponds to a temperature-dependent activation energy of the form

$$U(T) = U_0^* \frac{T}{(T - T_\infty)}, \tag{2.86}$$

which also tends to infinity for $T \to T_\infty$.

The analysis of a large number of experimental data shows that for most of the typical glass-forming substances (oxide glasses, silicate glasses, organic high polymers) in a good approximation

$$\frac{T_\infty}{T_m} \approx 0.5 \tag{2.87}$$

holds [131]. For glass-forming metallic alloys usually a somewhat lower value

$$\frac{T_\infty}{T_l} \approx 0.33 \tag{2.88}$$

is obtained. If one takes into account Eq. (2.73) then Eq. (2.87) may be written also as:

$$\frac{T_\infty}{T_g} \approx \frac{3}{4}. \tag{2.89}$$

Equivalent to the VFT equation is a dependence, widely used in the rheology of glass-forming organic high polymers named after Williams, Landel and Ferry (WLF equation [132]; see also Ferry [133]). According to the WLF equation the relaxation time, τ, or the characteristic frequency of motion, $\omega = 2\pi/\tau$, of the building units of the melt in the vicinity of T_g can be calculated by (Donth [134]):

$$\log\left(\frac{\omega}{\omega^*}\right) = \frac{C_1(T - T^*)}{(C_2 + (T - T^*))} . \quad (2.90)$$

The parameters ω^* and T^* are arbitrarily chosen reference values, while the constants C_1 and C_2 depend on the choice of the reference state. Equation (2.90) can be transformed into the VFT equation by a simple substitution of variables (see Gutzow [110]).

It is of principal importance that both the VLF and WLF equations can be derived in the framework of the semi-empirical free volume theories of liquids, discussed further on in detail in Chapters 3 and 4. According to one variant of these theories, the free volume $v(T)$ determines the viscosity via

$$\eta = \eta_0 \exp\left(\frac{B_0}{v(T)}\right) . \quad (2.91)$$

Here B_0 and η_0 are constants, again. Equation (2.91) is called the Doolittle equation [135–137]. It can be considered as a generalization of an empirical expression, proposed many years ago by Batchinski [138, 139]:

$$\eta = \frac{\eta_0}{v(T)} . \quad (2.92)$$

Another equation with a temperature-dependent activation energy was proposed by Cornelissen, van Leeuwen and Waterman in 1957 (see Bezborodov [124])

$$\eta = \eta_0 \exp\left(\frac{A}{T^n}\right) , \quad (2.93)$$

corresponding to an activation energy of the form

$$U(T) = \frac{k_B A}{T^{n-1}} . \quad (2.94)$$

Here η_0, A and n are three constants characteristic for a given melt. Evstropev and Skornyakov (see Skornyakov [140]) showed that the temperature dependence of the viscosity can be described in terms of Eq. (2.94) with $n = 2$ for temperatures higher than the liquidus temperature, T_l, with a very satisfactory accuracy. In such a way the three constants in Eqs. (2.93) and (2.94) are reduced to only two. Recently, a derivation of an equation with a temperature dependence of the form of Eq. (2.93) has been developed in the framework of a molecular model of viscous flow in glass-forming melts formulated by Avramov and Milchev [141]. In a series of publications it was successfully employed by Avramov [142–144] to explain the viscous behavior of glass-forming liquids at changing composition, at increased hydrostatic pressure, and so on.

We have to mention here also an equation, proposed by the well-known physico-chemist le Chatelier (1924) and by Waterton (1932) (see, again, Besborodov [124]):

$$\eta = \eta_0 \exp\left[b_0 \exp\left(\frac{U^*}{k_B T}\right)\right]. \qquad (2.95)$$

This equation gives a particularly steep temperature dependence of the viscosity. It was applied by Schischakov [145] to a number of glass-forming systems and is denoted sometimes also as Schischakov's equation. In contrast to Eq. (2.85), Eqs. (2.93) and (2.95) do not predict a divergence of the viscosity for finite values of temperature but only for T tending to zero. A derivation of Eq. (2.95) from a statistical point of view is possible when applying the hole theory of liquids [146].

Sometimes empirical equations are also used for the interpretation of experimental results which represent combinations of the expressions discussed above, for example,

$$\eta = \eta_0 \exp\left(\frac{U_0}{k_B T}\right) \exp\left(\frac{B_0}{T^n}\right), \qquad (2.96)$$

$$\eta = \eta_0 \exp\left(\frac{U_0}{k_B T}\right) \exp\left(\frac{B_0}{(T - T_\infty)}\right), \qquad (2.97)$$

that is, combinations of a Frenkel-type temperature dependence with equations providing a steeper temperature course. Equation (2.96) was proposed by Fulcher [129] while Eq. (2.97) is due to Macedo and Litovitz [147].

Since the number of constants in these latter equations, founded on more or less arbitrary combinations of factors and models, is increased to four, it is not a surprise that, by using them, a description of experimental results can be achieved with a high accuracy. Although the two above-mentioned and similar combined viscosity equations have been formulated in a more or less empirical manner, they nevertheless reflect, from a theoretical point of view, in their combination of two different exponential temperature dependencies the different sides of the mechanisms of viscous flow. These mechanisms may be, for example, connected with the separation of individual structural units from adjacent neighbors as the first step, followed by a translation to an appropriate vacancy (see, e.g., the discussion in Sanditov and Bartenev [146]). In this respect, viscosity-temperature equations of the form of Eq. (2.80) are less general, since they most probably describe (or even exaggerate) only one side of the complicated process of viscous flow.

Of particular significance in the development of ideas in glass science in recent years has been an equation that connects the temperature dependence of the viscosity, $\eta(T)$, of a liquid with its configurational entropy, $\Delta S(T)$. This thermodynamic function, as discussed in more detail later, can be calculated to a first approximation as the entropy difference undercooled liquid-crystal ($\Delta S(T) = S(T)_{liq} - S(T)_{cryst}$). The respective equation was proposed by Adams and Gibbs [148] in the form:

$$\eta = \eta_0 \exp\left(\frac{B^{\#}}{T \Delta S(T)}\right). \qquad (2.98)$$

Here $B^{\#}$ and η_0 are again in a first approximation constants, the first of them according to the original derivation in [148] being, however, in fact a linear function of temperature, T. By introducing appropriate temperature functions for $\Delta S(T)$ in the above equation, approximately exponential temperature dependencies are obtained for $\eta(T)$, corresponding to Eqs. (2.85) and (2.95) obtained in the framework of both free-volume approaches or the Avramov–Milchev model. As far as the mentioned linear dependence of B^* on T is anticipated, Eq. (2.98) can be also written as:

$$\eta = \eta_0 \exp\left(\frac{B_0}{\Delta S(T)}\right), \qquad (2.99)$$

as it was in fact derived in an empirical free volume model by Gutzow et al. [149].

The significance of the Adams–Gibbs viscosity equation stems from the fact that it directly connects a well-defined, structure-dependent thermodynamic function, $\Delta S(T)$, with the most significant kinetic characteristics of the undercooled melt i.e. with its bulk viscosity, $\eta(T)$. It must also be mentioned that in the derivation of the viscosity equation, proposed by Avramov [142–144], in fact the same configurational entropy difference, $\Delta S(T)$, enters. This circumstance most probably explains the general correspondence of results, obtained by applying these two dependencies to various problems in the rheology of undercooled glass-forming liquids (cf. also [150, 151]).

Finally, we would like to mention that the above equations, with a viscosity depending only on temperature and the specific properties of the liquid, are only approximations, strictly valid for medium values of the viscosities and velocities of viscous flow, and most significantly, for liquids in equilibrium states. Generalizations of Eq. (2.80) for nonequilibrium states will be discussed in Chapters 3 (see also [1]), and the significance of non-Newtonian flow in vitrification kinetics and phase transformations in glass-forming melts is discussed in [1]. Here we have to note only that practically all above discussed viscosity versus temperature dependencies, derived (or empirically proposed) for equilibrium melts, cannot satisfactorily predict or describe the change of the temperature course of viscosity in the glass transition region, as it is seen in Figure 2.17 for selenium and, as shown further on, for any other vitrifying liquid at $T \leq T_g$. The real significance of this problem can also be properly formulated and understood only in the framework of the more general, nonequilibrium thermodynamics approach, applied to the description of fluctuations and the kinetic properties of nonequilibrium systems as it is discussed in Chapters 3 and 4.

2.4.2
Significance of Viscosity in the Glass Transition

The temperature dependence of the viscosity of glass-forming melts is of exceptional technological significance, in particular, for processes of purification and homogenization of silicate melts in the glass furnace, in forming, styling and in the additional thermal treatment of the already formed glass products. It determines the temperature range, in which one method of machining or the other may be applied to form a glass, or to use it. In silicate glass technology, traditionally a distinction between *short* and *long* glasses is made in dependence on the steepness of increase of the viscosity near to the transformation range. Usually in technological processes long glasses with a relatively moderate increase of viscosity are preferred, since they are less vulnerable with respect to small temperature changes. Unfortunately, some of the most important silicate technical glasses behave as short glasses, leading to a number of technological difficulties in their industrial fabrication. As extremely short glasses, metallic glass-forming alloys have to be considered, which require the application of specific techniques of vitrification and glass-processing (the already mentioned splat cooling and ultra-rapid spinning methods).

In present-day glass science literature, following a proposal by Angell [5, 52, 53], as fragile liquids glass-forming melts are denoted, more or less dramatically changing their viscosity (as demonstrated in $\log \eta$ vs. T_g/T coordinates, see the final part of this chapter) in approaching T_g. Taking into account the results, outlined in the previous section, this is an indication that for such liquids the activation energy of viscous flow $U(T)$ is significantly temperature-dependent, increasing at $T \to T_g$. As strong liquids, in the framework of the same classification, such glass-forming melts (like SiO_2) are denoted for which $U(T) = U_0 =$ constant holds, giving in the same coordinates a nearly straight line. Fragile liquids tend thus to a viscosity versus temperature behavior, corresponding more to that of the short glasses, while strong liquids are typically long glasses in their viscosity properties. The deeper thermodynamic reasons, determining the differences in the temperature course of the viscosity of glass-forming melts, are also discussed at the end of this chapter.

In glass technology the transformation temperature, T_g, is usually identified with a value of the viscosity of the order of 10^{13} dPa s. The upper value of the vitrification range T_f (Tammann's softening temperature) corresponds to approximately 10^{11} dPa s. In a series of publications of the present authors, summarized in Chapter 3, the upper and lower temperatures of the glass transition interval are denoted by T_g^+ and T_g^-, respectively. For silicate glasses typically $(T_g^+ - T_g^-) \cong 50$ K holds, and for organic and polymeric glasses commonly $(T_g^+ - T_g^-) \cong 20$ K is observed.

In glass processing one of the most important forms of treatment is the annealing of a glass in order to remove strains produced in the course of formation and manufacturing of the glass. This process has to be carried out in a distinct temperature range. T_g corresponds to the upper limit of this interval. It is denoted, therefore, in the technological literature also as the upper annealing point, T_a. Above T_g glass products (silicate glass vessels, sheet window glass, etc.) loose their form, above T_f glasses begin to "flow" and crystallization processes may occur. The lowest temperature at which the annealing process can be still realized in glass-processing, the lower annealing point T_a, corresponds to a viscosity of 10^{14} dPa s.

In a first approximation, the temperatures T_f and T_a roughly correspond to the temperatures T_g^+ and T_g^-. However, while to T_f and T_a a distinct viscosity is as-

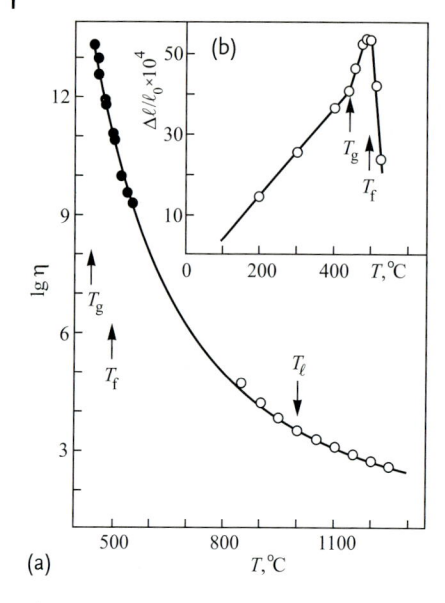

Figure 2.19 Temperature dependence of the viscosity η (a) and dilatometric curve of a lithium disilicate enamel melt (b) (see Penkov and Gutzow [105]). The turning-point in the dilatometric curve determines T_g, while at T_f the sample breaks down. A comparison with the viscosity vs. temperature curve shows that T_g corresponds to $\log \eta \approx 13$ and T_f to $\log \eta \approx 11$ (η in dPa s).

signed, T_g^+ and T_g^- have a more schematic significance specifying the beginning and the end of the glass transition interval from a metastable equilibrium liquid to the frozen-in nonequilibrium solid which is the glass.

At the upper annealing temperature, T_g, strains, existing in the glassy material, relax over a period of about 10–15 min, while at T_a optically measurable strains disappear only after 10–15 h. The temperature T_a for technical silicate glasses is found to be usually 10–15 K below T_g. Below T_g and T_a the highly viscous frozen-in melt can be considered, at least, in technological respects, as a solid. Usually, it is assumed that a solid is characterized by a viscosity of the order 10^{15} dPa s or higher.

The transformation temperature, T_g, and the softening temperature, T_f, corresponding to the lower and upper boundaries of the transformation range, are usually determined by dilatometric measurements (see Figure 2.19). At T_g, a break-point is observed in the dilatometric curves, while at T_f the material begins to flow. At the Littleton softening point, the estimate $\eta \approx 10^6 – 10^7$ dPa s holds. Further temperatures, important from a technological point of view for silicate glass-forming melts, are the flow point, corresponding to a viscosity of $7 \cdot 10^4$ dPa s (the Lillie flow point) or 10^4 dPa s [152, 153], the viscosity range for glass blowing ($\log \eta \approx 3-5$), the seal point ($\log \eta \approx 6$), where an adhesion of the melt on metals becomes possible. The Littleton softening point corresponds to the lower limit of temperature at which crystallization of glass-forming melts may, as a rule, still be observed.

2.4.3
Molecular Properties Connected with the Viscosity

From the viscosity of a melt, which can be measured in a relatively simple way, the temperature dependence of other kinetic coefficients of glass-forming liquids connected with the viscosity can be determined. Though some of the relations, which are discussed in the following, are in part only qualitative estimates, they can be helpful, nevertheless, to obtain a first insight into the temperature dependencies of a number of kinetic coefficients important both for the processes of vitrification and crystallization. Two of these quantities are the self-diffusion coefficient of the building units of the melt, D_0, and the impingement rate, Z, that is, the number of collisions of molecules of the melt with a unit area of a hypothetical surface, embedded in it.

According to Stokes's law the force acting on a sphere with a diameter, d_0, moving with a velocity, v, in a continuum of viscosity, η, can be expressed by

$$F = 3\pi\eta d_0 v . \tag{2.100}$$

If one applies this equation to the motion of a molecule in the melt then, following Einstein's approach for the description of Brownian motion, the following relation is obtained (see Einstein [65], Hodgdon and Stillinger [154]):

$$D_0 = \frac{k_B T}{(3\pi\eta d_0)} . \tag{2.101}$$

In terms of Eyring's absolute rate theory a more correct derivation can be given leading to [18]

$$D_0 = \frac{k_B T}{(\eta d_0)} . \tag{2.102}$$

Equation (2.102) is of the same form as Eq. (2.101), however, with a somewhat different numerical factor (of about one order of magnitude). Applying in addition Einstein's relation

$$|\Delta r|^2 = 6 D_0 t , \tag{2.103}$$

connecting the diffusion coefficient of a Brownian particle with the square of the mean displacement Δr from the initial position occupied at $t = 0$, one may also write:

$$D_0 \approx \frac{d^2}{\tau_R} . \tag{2.104}$$

Here τ_R is Frenkel's average stay time for a particle of the melt at a given position and d is the average displacement connected with a jump to a new position. Usually, it can be assumed that d is of the order of the size of the mean building units of the melt ($d \approx d_0$), which gives

$$D_0 \approx \frac{d_0^2}{\tau_R} . \tag{2.105}$$

Equations (2.102)–(2.105) yield

$$\tau_R \approx \frac{d_0^3 \eta}{(k_B T)} . \tag{2.106}$$

Compared with the exponential dependence of the viscosity on temperature, the terms in the above equations linear in T can be considered practically as constants. Thus, η also determines according to Eqs. (2.105) and (2.106) the temperature dependence of the relaxation time, τ_R, and the self-diffusion coefficient, D_0, of the melt.

From the phenomenological equations, describing the rheological properties of highly viscous liquids, developed by Maxwell, Kelvin and Voigt (see [1]), the macroscopic relaxation time of a melt, τ_R, is connected with the viscosity by

$$\tau_R = \frac{\eta}{G^*} . \tag{2.107}$$

The factor of proportionality G^* in Eq. (2.107) has the dimensions and the physical meaning of a modulus of elasticity (cf. also [155, 156]). A comparison with Eq. (2.106) leads to the conclusion that highly viscous melts can be considered as elastic bodies, characterized by a modulus of elasticity of the order

$$G^* \approx \frac{k_B T}{d_0^3} \approx \frac{k_B T}{v_m} , \tag{2.108}$$

where v_m is the average volume per building unit of the melt.

An elementary estimation of the impingement rate, Z, and its connection with the viscosity, η, can be given in the following way. If $\langle v \rangle$ is the average of the absolute value of the velocity of translation of the molecules of a liquid and c the average number of molecules per unit volume, then Z can be expressed in analogy to the collision frequency in gases as:

$$Z = \frac{1}{4} c \langle v \rangle . \tag{2.109}$$

For condensed systems

$$c \approx \frac{1}{v_m} \approx \frac{1}{d_0^3} \tag{2.110}$$

holds, while the average velocity of molecular translation in a liquid may be written in accordance with Frenkel's model as:

$$\langle v \rangle \approx \frac{d_0}{\tau_R} . \tag{2.111}$$

Equations (2.109)–(2.111) yield

$$Z \approx \frac{1}{d_0^2 \tau_R} \tag{2.112}$$

or

$$Z \approx \frac{k_B T}{d_0^5 \eta}. \tag{2.113}$$

It is obvious that the equations outlined in this section have to be considered only as qualitative estimates. For complex structured molecules, which are far from having a spherical shape, additional steric factors have to be introduced, in particular, for the calculation of the effective value of the impingement rate. The effective number of collisions, $Z^{(\text{eff})}$, can be defined as the ratio, ς, of the total number of molecular collisions, Z, which results in an incorporation of the colliding particle into the aggregate of an evolving crystalline phase. Thus, we may write $Z^{(\text{eff})} = \varsigma Z$. According to its definition, the parameter ς has values in the range $0 \leq \varsigma \leq 1$.

In considering the possible restrictions, imposed on the above written simple rheological relations, it has to be pointed out that they apply strictly speaking only to fluids in thermodynamic equilibrium. It can be shown that in approaching the glass transition region not only the viscosity-temperature dependence changes its character, but also the Stokes–Einstein relation Eq. (2.102) looses its validity. In the vicinity of T_g, a particular process of de-coupling [157–159] of bulk viscosity from the coefficient of self-diffusion takes place. This property is especially significantly demonstrated in the different ("de-coupled") temperature dependencies of bulk viscosity, η, and of the self-diffusion coefficient, $D_0(T)$, in the vicinity of T_g. In this sense, also the determination of impingement rates, Z, and of the $Z(T)$ function by using Eq. (2.113) becomes dubious in the glass transition interval itself and can mislead the experimenter in analyzing, for example, crystallization mechanisms in the glass transition range. However, despite the mentioned limitations, experience shows that the expressions derived to describe the temperature dependence of D_0, Z and τ_R hold with a reasonable accuracy, satisfactory enough for a large number of applications, when temperatures sufficiently above T_g are involved. This statement applies especially for most experiments in crystal growth and nucleation taking place at temperatures sufficiently above the glass transition region.

2.5
Thermodynamic Properties of Glass-Forming Melts and Glasses: Overview on Experimental Results

2.5.1
Heat Capacity

In the glass transition range, with the decrease of temperature not only a dramatic increase of the viscosity of the melt and its peculiar change there (as demonstrated in Figure 2.17) is observed, but also all properties of the undercooled melt change to those of the corresponding glass. Of particular significance in this respect is the change of thermodynamic properties of the vitrifying system, and this section is devoted to the analysis of these changes. In considering this problem, we use both classical and present day-results, first employing one of the most convenient model systems for a study of glass transitions by analyzing the properties of liquid, crystalline and glassy glycerol. This simple glass-forming substance has been a topic of investigation in glass science for more than 80 years.

Of outstanding importance for the understanding of the physical nature of the vitreous state, in addition to viscosity data, are measurements of the change of the caloric properties of glass-forming melts in the glass transition range. It was, again, Tammann who started the investigations of the temperature dependence of the thermodynamic properties of glass-forming melts. He also performed the first measurements of the change of the mechanical properties (thermal expansion coefficient, hardness, etc.) of vitrifying systems applying as usual for this remarkable investigator relatively simple but instructive methods.

As mentioned in Section 2.3, the process of vitrification is not connected with a plateau in the $T = T(t)$ curves (compare Figure 2.7c). This behavior is an indication that the transformation of the melt into a glass is not connected with heat release, that is, with a discontinuous change of the entropy or enthalpy of the system. However, DTA and DSM curves exhibit a turning-point at T_g in the process of glass heating. This property of the DTA curves is an indication for an abrupt change of the specific heat, C_p, the second-order derivative of the thermodynamic potential, G, at T_g. This type of behavior is illustrated in Figure 2.20 for the classical case of glycerol and used also here in several following figures to illustrate the change of thermodynamic properties of glass-forming melts in the glass transition range. These and the subsequent investigations of the temperature dependence of the specific heats of glass-forming substances, carried out in different laboratories in the last 80 years, have shown the same typical sigmoid-shaped decrease of the specific heats of the undercooled melt upon vitrification, as they were first observed for glycerol and silica melts.

The first thorough measurements of the caloric properties of glass-forming substances were initiated by Nernst in connection with his desire to verify the third law of thermodynamics proposed by him in 1906 (see Nernst [60], Simon [22, 23], Eitel [162]). Nernst expected that the third law of thermodynamics and its consequences can be applied to all forms of condensed matter – liquids, crystals and glasses. In the classical studies, carried out by Witzel [163], and Simon and Lange [21], the specific heats of silica glass, of the corresponding crystalline phases (quartz, cristobalite and tridymite) and the glass-forming melt were measured. The choice of SiO_2 as a model for this investigation was not very suitable, since the measurements had to be carried out over a range of temperatures of about 2000 K, that is, from the melting point of cristobalite to temperatures of liquid hydrogen. To comprise this interval, four different calorimetric methods had to be used.

At the same time measurements with the much more convenient substance glycerol were undertaken by Gibson and Giauque [160], and Simon and Lange [21] (see Figure 2.20). The melting point of glycerol is about $T_m \approx 298$ K, while T_g for this substance is $T_g \approx 193$ K and only one crystalline phase of this substance exists

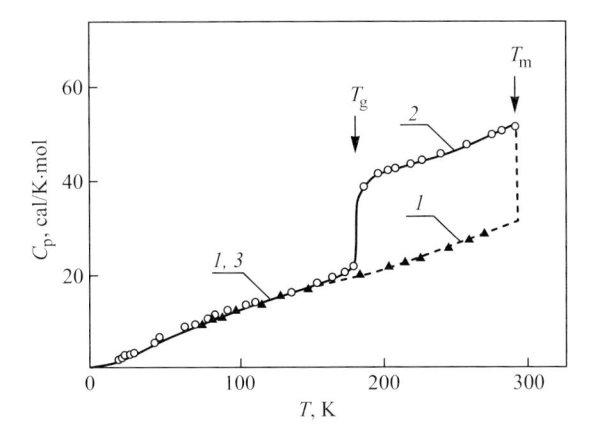

Figure 2.20 Typical $C_p = C_p(T)$ curves for the fluid ($C_p^{(f)}$), crystalline ($C_p^{(c)}$) and vitreous forms of a substance, here presented as an example for glycerol. The values of the specific heats for the crystalline state (1) are specified by *black triangles*, and the curve for the undercooled melt (2) and the glass (3) by *white circles*. Data are taken from the measurements of Gibson and Giauque [160], Simon and Lange [21], as they are summarized by Gutzow and Grantcharova [161].

(silica, in contrast, is known to have six crystalline modifications). In both cases it is thanks to F. Simon's efforts that the $C_p(T)$-measurements were extended down to the lowest temperatures achievable at that time, to those of liquid helium.

Soon after these pioneering investigations similar C_p measurements on several low-melting glass-forming substances were carried out by Tammann and other investigators (organic compounds, ionic melts, metallic alloys, silica, see Figure 2.21). At present, the results of C_p measurements on about 150 substances are known. Qualitatively the same s-shaped curves were always observed for the temperature dependence of this quantity upon vitrification.

Curves of the type given in Figures 2.20 and 2.21 can be obtained reproducibly at moderate cooling rates. With an increase in the cooling rate the typical inflection points in the $C_p(T)$ curves and, consequently, T_g are found to move to higher temperatures. In heating run curves typically an overshot "nose" is observed depending on prehistory and heating rate. An example in this respect is given by the measurements of Zhurkov and Levin (Figure 2.22, see Kobeko [166]). Quantitatively the dependence connecting cooling rate q and glass transformation temperature T_g is given by the Bartenev–Ritland equation

$$\frac{1}{T_g} = A - B \log q, \quad q = -\frac{dT}{dt}, \tag{2.114}$$

where A and B are constants. In Figure 2.23, another kinetic effect in vitrification is demonstrated. The same glass sample, vitrified with different cooling rates, is heated up with a constant heating velocity. The T_g values obtained differ as shown by experiments performed by Moynihan et al. [167] with another simple inorganic glass-forming system.

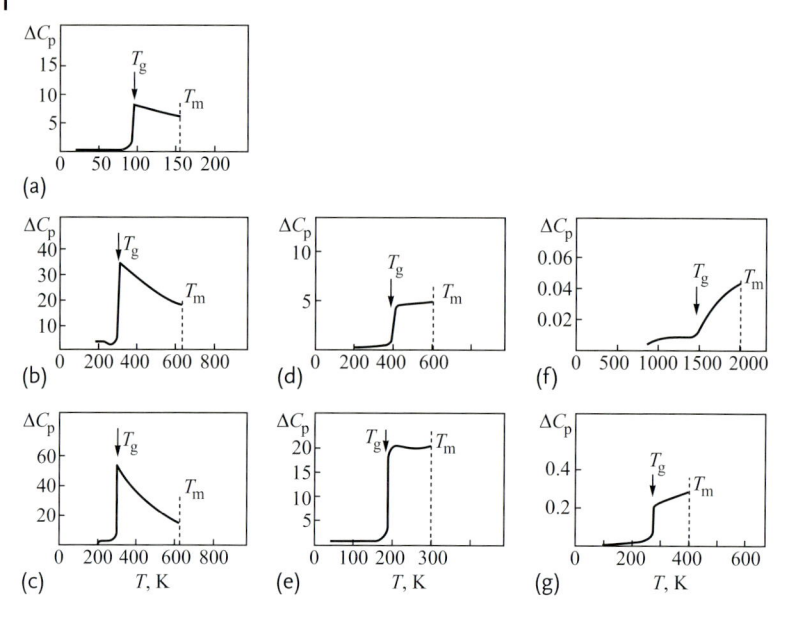

Figure 2.21 Temperature dependence of the difference of the specific heats, ΔC_p, melt crystal of a number of glass-forming melts in the $(T_m - T_g)$ region measured by different authors (see Gutzow [164], and Gutzow and Grantcharova [161]). The substances are: (a) methanol (in cal/mol K); (b) met-glass (Au/Si) (in cal/K g atom); (c) met-glass (Au/Ge/Si) (in cal/K g atom); (d) $ZnCl_2$ (in cal/mol K); (e) glycerol (in cal/mol K); (f) SiO_2 (in cal/g K); (g) glucose (in cal/g K).

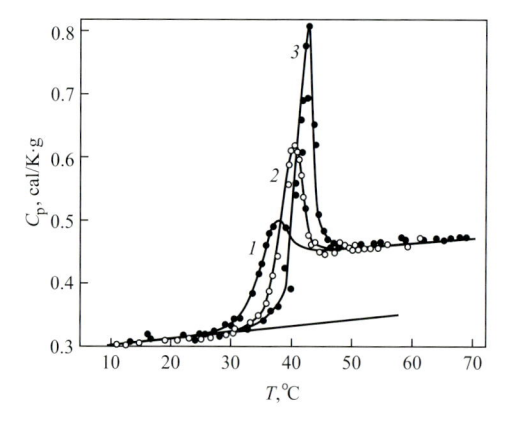

Figure 2.22 $C_p(T)$ curves obtained in the process of heating of a glass-forming organic polymer melt (polyvinyl acetate) for different heating rates according to the measurements of Zhurkov and Levin [165] (see also Kobeko [166]). Curve (1) corresponds to a heating rate of $0.1\,\mathrm{K\,min^{-1}}$, curve (2) to $0.4\,\mathrm{K\,min^{-1}}$, and curve (3) to $1.5\,\mathrm{K\,min^{-1}}$.

The Bartenev–Ritland equation and the experimental results underlying it are of great importance in two different respects. First, it becomes evident that the process of vitrification cannot be considered as an equilibrium phase transformation,

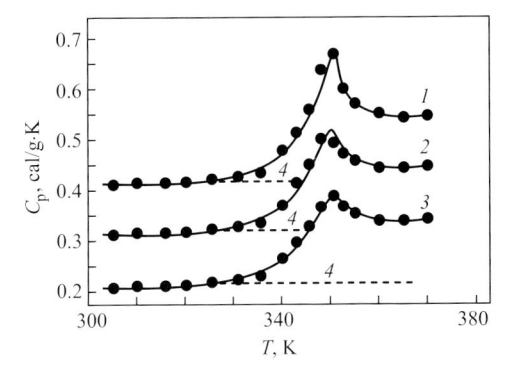

Figure 2.23 Influence of the thermal history on the $C_p(T)$ curves measured in the heating process of a vitrified substance (0.4 Ca(NO$_3$)$_2$ 0.6 KNO$_3$). The melt was quenched with different cooling rates: (1) 0.62 K min^{-1}; (2) 2.5 K min^{-1}; (3) 10 K min^{-1}. In all considered cases, the melt is heated with the same heating rate 10 K min^{-1} (Moynihan et al. [167]; see also Mazurin [70]).

for example, in the sense of the Ehrenfest classification. For equilibrium phase transformations, the transition temperature is a well-defined quantity, which does not depend on kinetic factors. Equation (2.114) shows, moreover that the dependence of the glass transformation temperature upon cooling rate is of a logarithmic form. This result has the consequence that significant variations of the cooling rates are required to reach measurable variations of the properties of the glass, vitrified from a given melt.

The dependence of the glass transition temperature, T_g, on cooling rates (or more generally on prehistory) has also another consequence: the identification of T_g with a viscosity of 10^{13} Poise is, in general, not correct. However, since the dependence of T_g on cooling rate is of logarithmic nature, for most real silicate glass-forming melts T_g is found – as reported by Mazurin [108] – in a range of viscosities from 10^{12} to 10^{14} Poise.

Moreover, experiments show that the height and even the form of the overshot peak in the $C_p(T)$ curves developed in the vicinity of T_g, although showing some resemblance with a second-order type phase transition, is – as it was mentioned above and is also seen in Figures 2.22 and 2.23 – kinetically dependent. The λ-type $C_p(T)$ curves in second-order thermodynamic phase transitions, as they are represented in Figure 2.24 for three typical cases, do not depend on the kinetics of the experiment. Moreover, the position of the respective transition temperatures do not change upon heating or cooling rates as far as sufficiently large samples are studied. They are thermodynamic characteristics of the corresponding materials.

Equation (2.114) was first proposed in a straightforward semi-empirical manner by Bartenev in [169]. Its theoretical derivation and a discussion of the origin and the character of the "nose" peaks upon vitrification is given in Chapter 3 in the framework of the kinetic models of vitrification, developed by a number of authors, and first by Volkenstein and Ptizyn [170] (the "kinetic theory of glass transition" as it was denoted in the Soviet literature at this period [70]; see also Volkenstein [171]).

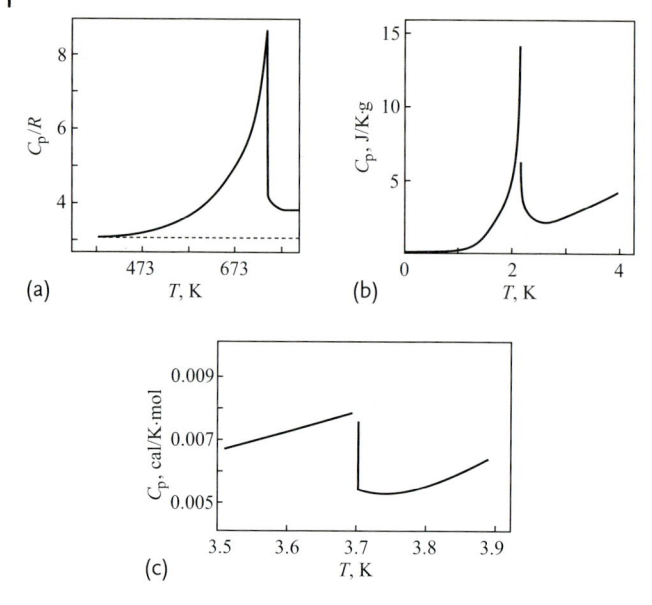

Figure 2.24 The C_p course in second-order λ-type phase transformations. (a) Order–disorder transition in a 50 mol% Au–Cu alloy [14]; (b) fluid–superfluid transition in He^4 [168]; (c) normal conductivity–superconductivity transition in metallic tin [43]. Note the typical λ-like course of the specific heat in cases (a) and (b).

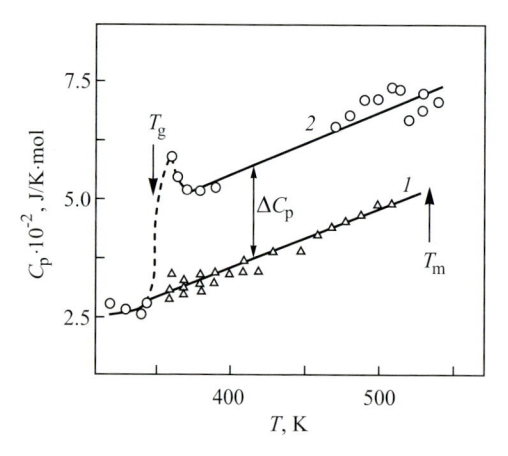

Figure 2.25 Temperature dependence of the specific heats of crystalline (1) and amorphous (2) phenolphtalein as determined by differential scanning calorimetric measurements (after Grantcharova et al. [172]).

The peak as shown in Figure 2.23 is also observed in most $C_p(T)$ determinations made by differential scanning calorimetry (DSC) techniques in heating runs as shown in Figure 2.25. For comparison, the typical course of the $C_p(T)$ curves in second-order phase transitions is shown in Figure 2.24.

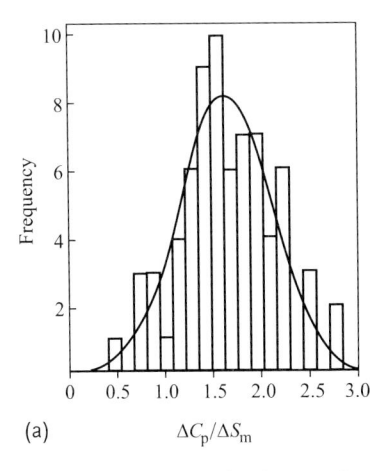

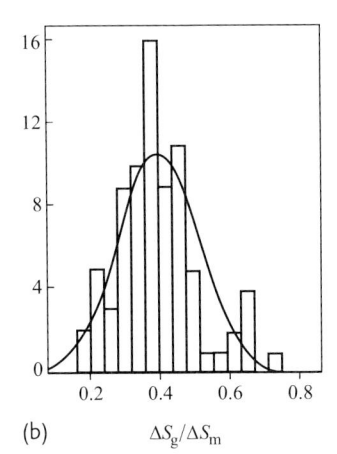

Figure 2.26 Frequency distribution of (a) experimentally determined $\Delta C_p(T_g)/\Delta S_m$ values and (b) $\Delta S_g/\Delta S_m$ values for 80 typical glass-forming systems, for which the respective thermodynamic measurements have been performed up to 1990. Experimental data are taken from Kauzmann [109], Gutzow [98], Privalko [16], and Wunderlich [173]; see Gutzow and Dobreva [112]. According to these results the most probable value of the ratio $\Delta S_g/\Delta S_m$ equals 0.38, while for the ratio $\Delta C_p/\Delta S_m$ the value 1.5 is found (cf. Eq. (2.115)).

It was shown by Wunderlich in 1960 [173] based on the analysis of the then existing experimental results of $C_p(T)$ measurements of about 45 substances (mainly organic and inorganic high polymers) that the ratio $\Delta C_p(T_g)/\Delta S_m$ is, generally, of the order

$$\frac{\Delta C_p(T_g)}{\Delta S_m} \approx 1.5, \tag{2.115}$$

indicating a distinct universality in the caloric behavior of different substances. In Figure 2.26 the validity of Eq. (2.115) is illustrated taking into consideration practically all known data up to 1992 on $C_p(T)$ measurements of glass-forming substances [112].

2.5.2
Temperature Dependence of the Thermodynamic Functions: Simon's Approximation

According to Eqs. (2.28) and (2.29), the temperature dependence of the thermodynamic functions of glass-forming melts can be expressed directly through C_p. In addition, the value of the enthalpy, H, has to be known, at least, for one temperature. Instead of the thermodynamic functions and the potentials H, G or S of the corresponding states we will discuss, however, the differences between these quantities for the liquid or vitreous states and the corresponding crystalline phase.

These differences are denoted in the following by

$$\Delta H = H^{(f)} - H^{(c)}, \quad \Delta H_g = H^{(g)} - H^{(c)}, \qquad (2.116)$$

$$\Delta G = G^{(f)} - G^{(c)}, \quad \Delta G_g = G^{(g)} - G^{(c)}, \qquad (2.117)$$

$$\Delta S = S^{(f)} - S^{(c)}, \quad \Delta S_g = S^{(g)} - S^{(c)}. \qquad (2.118)$$

The subscripts f, g, and c refer to the liquid (f), vitreous (g) and crystalline (c) states of the considered substance, respectively. Denoting further on by ΔH_m and ΔS_m these differences at the melting point (enthalpy and entropy of melting) Eqs. (2.23)–(2.25) may be rewritten (identifying the arbitrary reference value of temperature, T_0, with the melting temperature, T_m) in the form

$$\Delta S(T) = \Delta S_m - \int_T^{T_m} \frac{\Delta C_p}{T} dT, \qquad (2.119)$$

$$\Delta H(T) = \Delta H_m - \int_T^{T_m} \Delta C_p \, dT, \qquad (2.120)$$

and with $G = H - TS$ (Eq. (2.18)) we have:

$$\Delta G(T) = \Delta H(T) - T \Delta S(T). \qquad (2.121)$$

As a result, we obtain (see also Eq. (2.25))

$$\Delta G(T) = \Delta S_m(T_m - T) - \int_T^{T_m} dT \int_T^{T_m} \frac{\Delta C_p}{T} dT. \qquad (2.122)$$

In Eq. (2.122), ΔG_m was set equal to zero as it has to be the case for two thermodynamic phases in equilibrium (compare Section 2.2.3). As a further consequence, we obtain from Eq. (2.121) the relation $\Delta H_m = T_m \Delta S_m$.

In analyzing vitrification we have to determine the course of the differences given above, when one of the states of the system under consideration is not a thermodynamic equilibrium state. It is not a thermodynamic phase but a glass (the respective parameters we specified by (g)), which we expect to be in nonequilibrium. In order to do the necessary calculations in a simple and nevertheless self-consistent way we introduce here into the analysis a remarkable approximation, first proposed by Simon [22, 59] more than 80 years ago. Simon assumed that the glass-forming melt can be treated as an equilibrium system in the whole high temperature region (say beginning from $T = T_m$ or higher temperatures) down to the very temperature of the glass transition, T_g. This temperature Simon considered (and denoted, as we mentioned) as a distinct freezing-in temperature, T_e, where the melt was abruptly frozen-in to a glass. Thus in the nonthermodynamic, frozen-in vitreous state, at any temperature $T \leq T_g$ constant entropy and enthalpy differences $\Delta S(T_g)$ and

$\Delta H(T_g)$ are found. In this way, Simon ignored the glass transition region and the sigmoidal change of the $C_p(T)$-function there, replacing it by a kinetically determined temperature $T_e = T_g$ and by a step-like decrease in $C_p(T)$. Figure 2.20 shows that, in considering the whole temperature interval from absolute zero to T_m, the $C_p(T)$ course of a glass-forming substance can be in fact approximated in this way as done by Simon.

Simon's procedure was adopted in practically all subsequent treatments in the years to come, and especially also in Prigogine's analysis of the glass transition [14], which will be discussed in Chapter 3. The notion of frozen-in entropy and its statistical treatment was proposed by Einstein in 1913 and was mentioned as a possibility even earlier by Boltzmann [174]. The calculation of the thermodynamic properties of vitrifying systems, accounting for the real temperature course of the specific heats in the glass transition region, turned out to be a more complicated task, which can be performed in the framework of the mentioned generic nonequilibrium thermodynamic approach. The respective results are summarized in Chapters 3 and 4.

Here we exploit first Simon's approximation. In doing so, we have to take into consideration the experimental fact that the relation $\Delta C_p \approx 0$ holds for $T < T_g$ (see Figures 2.20 and 2.21). Thus Eqs. (2.119)–(2.122) may be approximated for $T < T_g$ by

$$\Delta S(T) = \Delta S_m - \int_{T_g}^{T_m} \frac{\Delta C_p}{T} dT \quad \text{for} \quad T \leq T_g, \tag{2.123}$$

$$\Delta H(T) = \Delta H_m - \int_{T_g}^{T_m} \Delta C_p \, dT \quad \text{for} \quad T \leq T_g, \tag{2.124}$$

$$\Delta G(T) = \Delta S_m(T_m - T) - \int_{T}^{T_m} dT \int_{T_g}^{T_m} \frac{\Delta C_p}{T} dT \quad \text{for} \quad T \leq T_g. \tag{2.125}$$

Note that in order to compute the frozen-in values of the thermodynamic functions of the glass via Eqs. (2.123)–(2.125), the temperature course of $\Delta C_p(T)$ (or $T_m \Delta S_m$) has to be known from T_m down to as low temperatures as possible. In addition, the enthalpy of melting has to be determined from an independent additional experiment.

Equation (2.123) is equivalent to

$$\Delta S(T) = \Delta S(T_g) \equiv \Delta S_g = \text{const.} \quad \text{for} \quad T \leq T_g, \tag{2.126}$$

that is, the entropy of a glass has a nearly constant value for $T < T_g$ and is not equal to zero for T tending to zero (as required by the third law of thermodynamics in Planck's classical formulation).

Following Simon, the behavior of S as a function of T is shown in Figure 2.27 for glycerol, calculated from the $C_p(T)$ course of this substance (Figure 2.20) in the

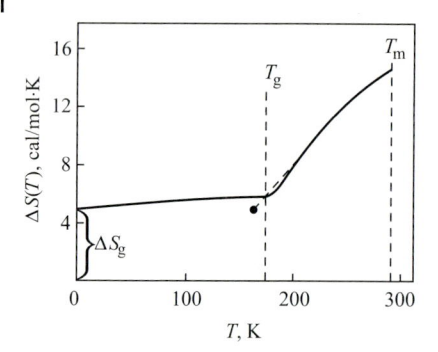

Figure 2.27 Temperature dependence of the entropy difference, $\Delta S(T)$, of vitrifying glycerol (after Simon [59]). In contradiction to Nernst's third law of classical thermodynamics the zero-point entropy difference is not equal to zero but has a finite positive value $\Delta S(0) \approx \Delta S(T_g)$. The *dashed line* indicates the entropy course of the glass as determined by Oblad and Newton [175] by very prolonged annealing. The dot corresponds to a temperature of 174 K.

framework of his approximation. Similarly, we obtain, following Simon's proposal, that for $T < T_g$ the relations

$$\Delta H(T) = \Delta H(T_g) = \Delta H_g = \text{const.} \quad \text{for} \quad T \leq T_g, \tag{2.127}$$

$$\Delta G(T) = \Delta H(T_g) - T\Delta S(T_g) \quad \text{for} \quad T \leq T_g \tag{2.128}$$

are fulfilled (see also Figure 2.28). The constancy of both ΔH and ΔS in the vitrified melt was first recognized by Simon [22, 59].

Figures 2.28 and 2.29 give the essence of the results, following in the framework of Simon's approximate treatment of vitrification and are a convincing demonstration of the nature of vitreous systems as frozen-in, nonequilibrium states. These figures, as well as the above three equations Eqs. (2.126)–(2.128) are in contradiction to the third law of thermodynamics, in particular, with its consequences expressed through Eq. (2.28). In contradiction to Eq. (2.28) and the third law we obtain for the glass

$$\lim_{T \to 0} \frac{\partial}{\partial T}(\Delta G) = -\Delta S(T_g) < 0, \tag{2.129}$$

$$\lim_{T \to 0} \frac{\partial}{\partial T}(\Delta G) \neq \lim_{T \to 0} \frac{\partial}{\partial T}(\Delta H), \quad \lim_{T \to 0} \frac{\partial}{\partial T}(\Delta H) = 0. \tag{2.130}$$

With these considerations and results, the construction shown in Figure 2.29b is drawn, which is to be compared with Nernst's $\Delta G(T)$- and $\Delta H(T)$-curves (Figure 2.29a, Nernst [60]).

All experimental results, known at present, show that, in general, considerable values of the thermodynamic functions H, G and S are frozen-in in the vitreous state. If the glass is solidified at "normal" cooling rates, $q \cong 10^1$–10^2 K s^{-1}, then

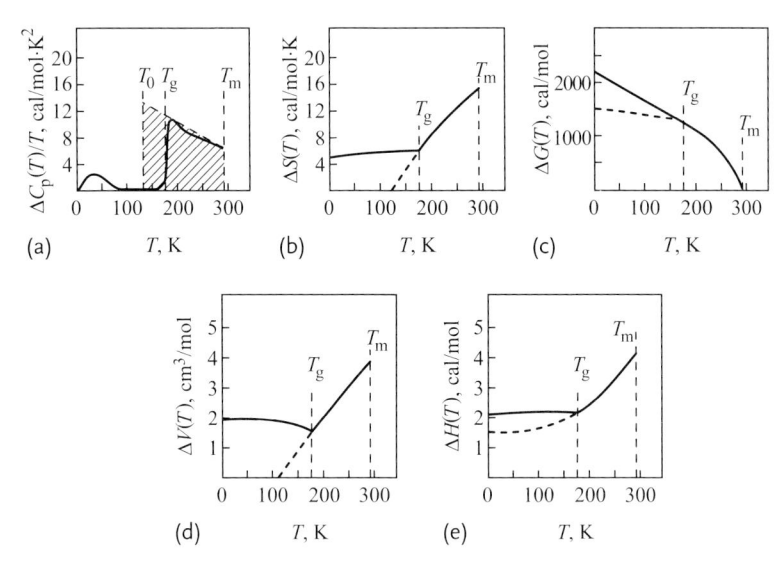

Figure 2.28 Temperature dependence of the thermodynamic functions of glycerol calculated from measurements of the specific heats (compare to Figure 2.20) and of the densities of the liquid, vitreous and crystalline states of glycerol: (a) reduced specific heat difference (the shaded area from T_m to T_0 gives the entropy of melting of glycerol according to Eq. (2.119); (b) entropy; (c) Gibbs free energy; (d) molar free volume (the curve is given based on density measurements); (e) enthalpy. By graphical extrapolations (*dashed lines*), the thermodynamic functions of the undercooled melt into the region below T_g are shown in the range where the system is already frozen-in to a glass. The temperature dependencies for the undercooled melt in thermodynamic equilibrium (above T_g) and the glass (below T_g) are specified by *solid lines*.

on the average the simple relations

$$\frac{\Delta S_g}{\Delta S_m} \approx 1/3, \qquad (2.131)$$

$$\frac{\Delta H_g}{\Delta H_m} \approx 1/2, \qquad (2.132)$$

are approximately fulfilled ([112, 176]; see also Figures 2.26b and 2.30).

Taking into account these experimental data, the following internal inconsistency in Simon's approximation arises. Assuming that the transition to a glass proceeds suddenly at a certain temperature T_g, Eq. (2.125) yields $\Delta G(T) = T_m \Delta S(T_g) - T \Delta S(T_g)$. This result is equivalent to Eq. (2.128) only if $\Delta H(T_g) = T_m \Delta S(T_g)$ holds. However, as a rule, this identity is not fulfilled (cf. Eqs. (2.73), (2.131), and (2.132)). We have to conclude that Simon's approach leads to partly serious conceptional problems in application to the interpretation of experimental data on glass formation. This discrepancy will be overcome in the generic approach given in Chapter 3.

A proof of the above generalizations of experimental results on the thermodynamic properties of different glasses is given by the data obtained with various

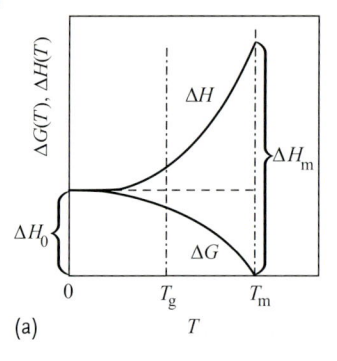

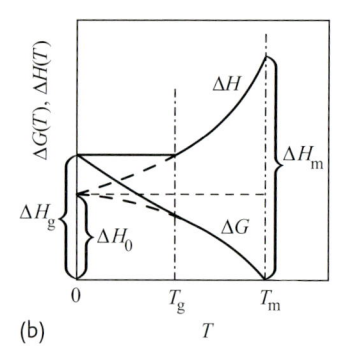

Figure 2.29 (a) Temperature dependence of the enthalpy (ΔH) and the Gibbs free energy (ΔG) difference melt crystal for an equilibrium system obeying the third law of thermodynamics (Nernst's (ΔH, ΔG) diagram); (b) temperature dependence of the enthalpy and the Gibbs free energy for a vitrifying system (*full curves*) according to Simon's approximation. The temperature course, expected from the third law of thermodynamics, is indicated by *dashed curves*. ΔH_0 is the zero-point enthalpy difference for the two different equilibrium states of the same substance (melt and crystal), ΔH_g the zero-point enthalpy difference between the glass and the undercooled melt. Note also that for equilibrium systems $(d(\Delta H)/dT) = (d(\Delta G)/dT) = 0$ is obtained for $T \to 0$, while for a glass the relations $(d(\Delta H)/dT = 0)$ and $(d(\Delta G)/dT) < 0$ hold.

vitrifying systems, as they are summarized in Tables 2.1–2.3. Typical glass-forming melts like SiO_2, GeO_2, Be_2O_3, $NaPO_3$, alkaliborates and ethanol are included in these data as well as glass-forming molecular substances like *n*-propanol, benzophenone, phenolphthalein or metallic alloy systems, which form glasses at higher cooling rates. The following conclusions can be drawn from these experimental results:

1. Even for glass-forming substances with a very different structure the ratio T_g/T_m is given in a good approximation by Kauzmann's rule (compare Eq. (2.73) and Figure 2.26). Exceptions to this rule are found for metallic glass-forming alloys, for which considerably lower (T_g/T_m)-ratios are observed (cf. Figure 2.26b).
2. With respect to the $(\Delta C_p/\Delta S_m)$, $(\Delta S_g/\Delta S_m)$ and $(\Delta H_g/\Delta H_m)$-ratios the relations, expressed through Eqs. (2.115), (2.131) and (2.132) are widely verified. Moreover, also the deviations from the respective average values can be noticed (see also Figures 2.26, 2.29b, and 2.30, where all known experimental data are summarized).

It can be seen that for more complex substances ($NaPO_3$, phenolphthalein) the deviations from the average values are particularly significant. This tendency is a characteristic feature of polymer systems as can also be seen from the experimental data compiled by Privalko [16]. Possible explanations of the relative stability of the ratios discussed above as well as of the observed deviations for some classes of substances are discussed in [1] on the basis of molecular models of glass-forming melts.

Table 2.2 Thermodynamic properties of representative glass-forming substances: Change of the molar specific heat ΔC_p at the vitrification temperature T_g and the ratio $\Delta C_p(T_g)/\Delta S_m$. The $\Delta C_p(T_g)$ values are taken from Wunderlich [173], Grantcharova et al. [107], Angell and Rao [4], Grantcharova et al. [172], Dobreva [222], and Chen and Turnbull [221].

Substance	$\Delta C_p(T_g)$ (J K^{-1} mol^{-1})	$\Delta C_p(T_g)/\Delta S_m$	References for $\Delta C_p(T_g)$
B_2O_3	41.9	1.25	[173]
$H_2SO_4\ 3H_2O$	175.6	1.72	[173]
$NaPO_3$	50	2.02	[107]
$ZnCl_2$	20.9	1.26	[4]
Se	14.7	1.33	[173]
$CaNO_3\ 4H_2O$	230.4	2.40	[4]
Phenolphtalein	190	2.0	[172, 224]
Glycerol	40.2	1.03	[173]
Ethanol	31.8	1.21	[173]
2 Methylpentane	66.6	1.26	[173]
Butene-1	66.2	1.50	[173]
2–3 Dimethylpentane	66.6	1.71	[173]
Poly(stirene)	24.3	1.5	[173]
Poly(ethylenetherephtalate)	76	1.6	[222]
$Au_{0.77}Ge_{0.136}Si_{0.094}$	23.5	1.19	[221]

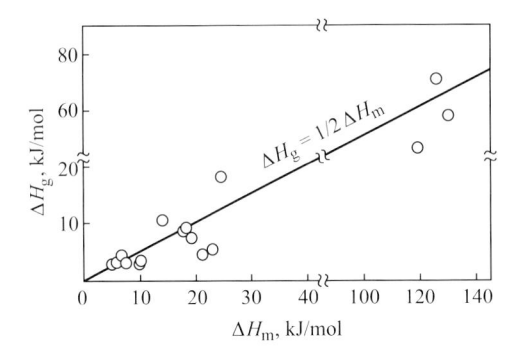

Figure 2.30 Dependence of the molar frozen-in enthalpy, ΔH_g, of glass-forming melts on the molar heat of melting, ΔH_m. The straight line is drawn through the experimental data from Table 2.3 in correspondence with Eq. (2.132).

A natural measure of the instability of the vitreous state with respect to the crystalline phase is the difference in the Gibbs free energy, ΔG, between both states of the same substance. With Eqs. (2.128), (2.131) and (2.132) we obtain

$$\frac{\Delta G(T)}{T_m \Delta S_m} \cong \left(\frac{1}{2} - \frac{T}{3T_m}\right) \quad \text{for} \quad T < T_g. \tag{2.133}$$

Table 2.3 Thermodynamic properties of glass-forming substances: molar enthalpy of melting, ΔH_m, and molar value of ΔH_g for selected glass-forming melts. Data are collected from the following references: Weyl and Marboe [214], Smith and Rindone [215], Eitel [225], Brizke and Kapustinski [226], Anderson [216], Tammann [24], Angell and Rao [4], Chen and Turnbull [221], Mandelkern [227], and Dobreva [222].

Substance	ΔH_m kJ mol^{-1}	ΔH_g kJ mol^{-1}	$\Delta H_g/\Delta H_m$	References
B_2O_3	24.7	18.3	0.74	[214, 215]
SiO_2	14.3	10.5	0.72	[225]
Se	6.7	4.4	0.65	[216, 226]
Ethanol	4.98	2.68	0.54	[24]
n-Propanol	5.72	2.895	0.51	[24]
$Na_2O\ 4B_2O_3$	130.4	58.3	0.45	[214, 215]
$LiO_2\ 2B_2O_3$	120.4	46.4	0.39	[214, 215]
$H_2O\ 4B_2O_3$	126.5	72.6	0.57	[214, 215]
Benzophenone	17.9	8.5	0.48	[24]
Glycerol	18.3	8.7	0.48	[24]
Poly(ethyleneterephtalate)	25.9	12.9	0.50	[222, 227]
Betol	19.2	7.4	0.38	[24]
2 Methylpentane	6.9	3.9	0.46	[4]
$ZnCl_2$	10.0	3.1	0.31	[4]
$H_2SO_4\ 3H_2O$	23.3	5.1	0.22	[4]
$Ca(NO_3)_2\ 4H_2O$	21.6	4.3	0.20	[4]
$Au_{0.77}Ge_{0.136}Si_{0.094}$	10.6	6.3	0.59	[221]

The largest value of ΔG is reached thus for T tending to zero, it is equal to

$$\Delta G(T \to 0) = \frac{1}{2} T_m \Delta S_m . \tag{2.134}$$

With Boltzmann's relation Eq. (2.32) further on

$$\frac{\Delta S_m}{\Delta S_g} = \ln\left(\frac{\Omega_m}{\Omega_g}\right) \cong \frac{1}{3} \tag{2.135}$$

is obtained, resulting with Eq. (2.131) in

$$\Omega_g = \sqrt[3]{\Omega_m} . \tag{2.136}$$

Thus, the glass transition occurs when the number of possible structural configurations Ω, of the melt reaches a value, Ω_g, given by Eq. (2.136). As far as the above relations apply to all glasses, they can be defined as systems with nearly equal relative frozen-in configurational disorder. In a logarithmic scale, the configurational disorder in glasses corresponds thus on the average to one third of the configurational disorder of the melt at the melting temperature, T_m. This state of increased molecular disorder is retained down to temperatures approaching absolute zero. This result is again a violation of the classical formulation of the third law of thermodynamics, according to which, as discussed in Section 2.2, any solid in thermodynamic equilibrium has to approach the absolute zero of temperature with $S(T) = 0$ and with a configurational disorder (in our notations) equal to $\Omega_0 = 1$. Summarizing the above-given thermodynamic results we can conclude:

1. At $T \approx T_g$, in the vitrified melt considerable parts of the values of the enthalpy and entropy of melting (of ΔH_m and of ΔS_m, respectively) are frozen-in. These frozen-in values of these thermodynamic functions remain nearly constant in the further cooling of the substance down to absolute zero. According to Boltzmann's relation (Eqs. (2.32) and (2.135)) this kind of behavior is connected with a relatively high degree of configurational disorder retained in the glass after the vitrification of the melt.
2. The third law of thermodynamics, as it was formulated by Nernst and Planck, is applicable, as proven by many experiments, to any solid, to any form of condensed matter in equilibrium, but it does not hold in its classical formulation for glasses. According to one of the consequences of the third law the temperature derivatives of H and G are nullified at $T \to 0$ for any substance in equilibrium, reaching temperatures $T \to 0$ [55, 62, 63]. This result is also experimentally proven for both crystals and liquids (for liquid helium, giving the only example of a condensed fluid, capable to exist in the vicinity of $T = 0$ in equilibrium and fulfilling there all requirements of the third principle (see [168])). However, for glasses this is again not the case (compare Eqs. (2.129) and (2.130)). Since the temperature derivatives of the thermodynamic potentials and of all thermodynamic functions are nullified at $T \to 0$, it turns out that there no proper temperature measurements can be performed by using the equilibrium forms of any substance. The question arises then whether any thermodynamic property, say the EMF between a nonequilibrium system, a glass (disobeying the third law) and the corresponding crystal (in equilibrium) could be used, at least in principle, to measure temperature differences even for $T \to 0$. Thus the fact that glasses are excluded from the third principle could be of exceptional interest in many possible applications, for example, it could be exploited in temperature measurements, Peltier devices, electrochemical measurements, and so on, as this was discussed in detail several years ago by one of the present authors [177] and outlined here in more detail in Chapter 8.

The thermodynamic results, outlined in this chapter were so surprising that, when they were first reported in the mid-1920s, many distinguished physicists and chemists doubted initially the validity of the experimental findings (see, e.g., Nernst's comments from 1918 on the zero-point entropy of SiO_2 [60] obtained in his own laboratory). However, all the experiments carried out in the years to come confirmed these results with new and new examples of glasses with various structures and compositions. It was Simon [22, 51, 59] who showed the only way out of this dilemma: glasses do not follow the third principle of thermodynamics, because they are nonthermodynamic, nonequilibrium systems. This statement was confirmed by further thermodynamic evidence.

However, a thorough analysis anticipated by classical authors like Nernst, Planck, Simon and Schottky [58, 60, 61, 178] and performed by the present authors [179–182] shows that in frozen-in nonequilibrium states glasses are, although at $T \to 0$ here the inequality $S(T) > 0$ holds, nevertheless the specific heats $C_p(T)$ always tend to zero for $T \to 0$. As a consequence, the absolute zero of temperature cannot be reached when employing systems in nonequilibrium in the respective thermodynamic processes. A detailed discussion of this topic is given in Chapter 9.

2.5.3
Further Methods of Determination of Caloric Properties of Glass-Forming Melts and Glasses

The determination of the thermodynamic functions of glasses based on specific heat measurements is accompanied by tedious experiments over, in general, large temperature ranges. Thus, the application of simpler methods is also desirable both in order to have at ones disposal alternative approaches and as a possibility to exclude systematic errors connected with the employment of only one specific method.

A possible method of direct determination of the enthalpy difference ΔH_g, frozen-in in the glass, is to measure the reaction heat involving a substance, reacting once as a glass and another time as a crystal. Care has to be taken here that the same reaction products are formed in both cases. One well-known example in this respect is to measure the reaction heat of quartz and of the corresponding silica glass in hydrofluoric acid according to the scheme

$$SiO_2 + 4HF \to SiF_4 + 2H_2O + Q. \tag{2.137}$$

ΔH_g can be determined then by

$$\Delta H_g = Q_{glass} - Q_{crystal}, \tag{2.138}$$

where Q_{glass} and $Q_{crystal}$ are the respective reaction heats. Usually the dissolution of silicate glasses in hydrofluoric acid is performed in platinum calorimeters [162]. Similar experiments can be also carried out with water soluble glasses in aqueous solutions or with vitrified metallic alloys, which are dissolved in the calorimeter in acids and other appropriate solvents. Schematically the idea behind these methods is illustrated in Figure 2.31.

An interesting version of the methods of determination of ΔH_g by dissolution experiments was developed by Jenckel and Gorke ([183]; see also Haase [184]). According to the proposal of these authors, not the heats of dissolution or swelling of a crystalline polymer (which is usually difficult to prepare since a complete crystallization is required) are measured, but the heats of dissolution of the glassy samples of the polymer, Δl_g, and of the respective undercooled polymer melt, Δl_e, are compared (see Figure 2.31) at temperatures slightly below T_g. This method can

2.5 Thermodynamic Properties of Glass-Forming Systems

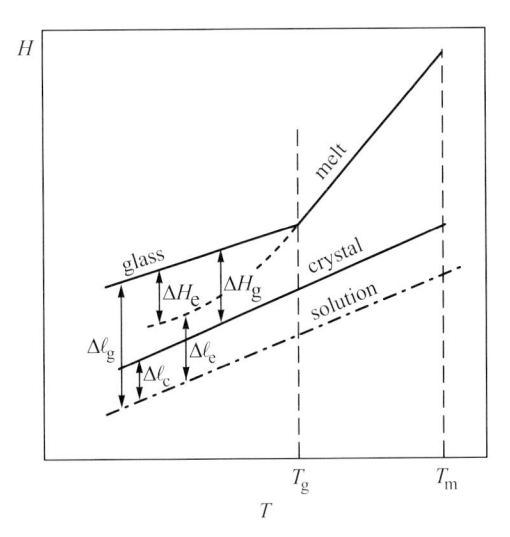

Figure 2.31 Schematic illustration of the methods of determination of ΔH_g by dissolution or reaction experiments with the same substance as a melt, a glass and a crystal. Δl_g and Δl_c are the heats of reaction or dissolution of the glass and the crystal, respectively. Δl_e is the heat of dissolution of the undercooled melt, ΔH_e the enthalpy difference between glass and undercooled melt in the experiment of Jenckel and Gorke [183] (see text).

be used, however, only when practically no energetic difference exists between melt and solution (the enthalpy of dissolution of the polymer melt has to be nearly equal to zero, as it is observed, for example, in polymer solutions in appropriate organic solvents). Only in this case, the relation $\Delta l_g = \Delta H_g$ holds.

A similar method of ΔH_g determinations consists in the measurement of the combustion heats of the crystalline and glassy forms of organic glass-forming substances (or of vitreous carbon materials and of graphite) in a bomb calorimeter. The disadvantage of this method is connected with the necessity of determining ΔH_g as the difference of two comparatively large quantities, which yields relatively large uncertainties in the values of ΔH_g. Comparison of ΔH_g, obtained by $C_p(T)$-measurements, with the above discussed alternative methods, shows a satisfactory coincidence, verifying both approaches.

Two further methods, which can be applied directly for a determination of ΔH_g and ΔS_g, are measurements of the temperature dependence of the solubility and the vapor pressure of glassy and crystalline samples formed out of the same substance. It is to be expected that both the solubility and the vapor pressure of the vitreous material are higher than the corresponding values of the crystalline phase. From the respective differences, measured at several temperatures, the temperature course of the Gibbs's free energy difference glass/crystal can be directly calculated. In a similar way, measurements of the electromotoric force (EMF) of a galvanic cell, that is, an electric element, in which a suitable conducting glass is used as the cathode while the anode consists of the crystalline phase of the same substance, can be exploited for a direct determination of $\Delta G(T)$ at $T < T_g$. An interesting example in this respect gives the galvanic cell consisting of vitreous carbon and graphite. However, these methods are associated also with additional problems, discussed in more details in Gutzow [177], Grantcharova and Gutzow [185].

2.5.4
Change of Mechanical, Optical and Electrical Properties in the Glass Transition Range

Already the first systematic investigations of glass-forming systems showed that similarly to the change of heat capacity, $C_p(T)$, all other thermodynamic coefficients of the vitrifying system, such as the thermal expansion coefficient, $\alpha(T)$, or the compressibility, κ, also exhibit a jump in the mentioned typical s-shaped form upon glass transition. An example in this respect is given in Figure 2.32, where the thermal expansion coefficient α (see Eq. (2.65)) for glycerol is shown as a function of temperature according to the dilatometric measurements of Schulz [186]. The $\alpha(T)$ curves are determined by the temperature dependence of the molar volume, v, or the molar density. These $v(T)$ curves exhibit a typical breaking-point at T_g (see, e.g., Figure 2.32).

Intensive investigations, in particular, of organic polymeric glasses showed that the glass transition is initiated usually at a nearly constant value of the relative free volume of the melt, corresponding approximately to 11–12% of the relative free volume of the melt at the melting temperature. Hereby the relative free volume is determined by $(v_f - v_c)/v_c$ where v_f and v_c are the molar volumes of the fluid and of crystalline phases, respectively. These experimental results encouraged Simha and Boyer [187] to formulate the idea that the process of glassy solidification has to be considered as a freezing-in process taking place at a constant value of the free volume (iso-free volume theory of vitrification). For a whole class of glass-forming melts at normal cooling rates, for example, for SiO_2, GeO_2, BeF_2, Ar_2S_3, also approximately constant but higher values of the relative free volume

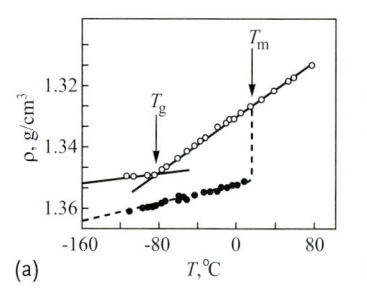

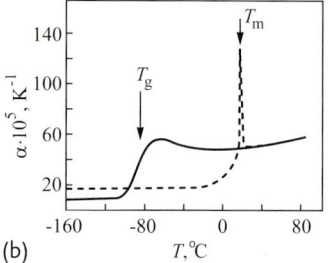

Figure 2.32 Density (a) and coefficient of thermal expansion (b) of vitrifying glycerol as a function of temperature according to the measurements by Schulz [186]. *White circles* in (a) are experimental data for liquid and vitreous glycerol, and *black dots* refer to the crystalline phase. In (b) the full curve represents the temperature dependence of the coefficient of thermal expansion α for the melt, respectively, the glass, while the *dashed curve* refers to the crystal. Note the sigmoidal change of α at T_g and the discontinuity of α_{cryst} at the melting temperature T_m.

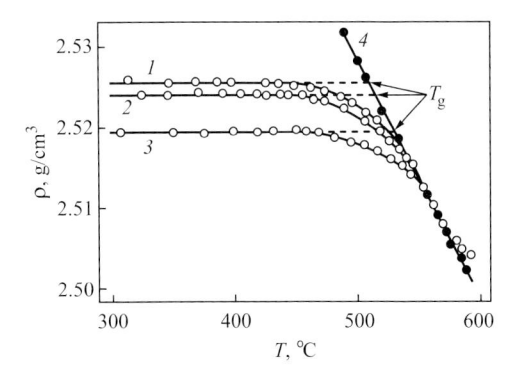

Figure 2.33 Temperature dependence of the density of a borosilicate glass measured at three different cooling rates: (1) 1 K min^{-1}; (2) 2 K min^{-1}; (3) 10 K min^{-1}; (4) annealing curve. It can be seen that T_g increases with increasing cooling rates [194].

(of about 18–20%) are frozen-in in the glass. For vitreous metallic alloys the corresponding value is equal to 8–10%. A summary of results in this respect is given in Table 2.4.

A possible interpretation of these and similar results is indicated by model experiments performed by Scott [188]. Stimulated by a suggestion of Bernal [189, 190], Scott modeled the structure of liquids by a random packing of steel spheres of equal size in containers of various volume (see [1]). The volume occupied by the ensemble of equally sized spheres in the most dense hexagonal crystalline packing is calculated to be about 74% of the total volume available, while for the so-called dense, respectively, loose random packing 68 and 63% occupied volume were observed in Scott's experiments. These values, when referred to the dense hexagonal packing, correspond to relative free volumes between 28 and 17%, that is, to the free volume found in typical inorganic glass-forming substances, when the glasses are referred to the corresponding most stable crystalline modification [191].

The model experiments of Scott are not only instructive from a structural point of view. They show in addition the close connections between mobility and free volume in a system (cf. also the Doolittle or Batchinski equations (2.91) and (2.92)) and give even indications to the possible existence of two- and one-dimensional glasses [192, 193]. Iso-free volumes frozen-in in the glass could explain the result as given by Eq. (2.136) connecting vitrification with a constant value of the relative configurational disorder. As found for the $C_p(T)$ curves, the value of the transformation temperature T_g, determined by volume measurements, also depends on the cooling rate in the way as predicted by the Bartenev–Ritland equation (2.114) (compare Figures 2.22 and 2.33). A similar result of historical significance is shown also in Figure 2.34.

A temperature course, equivalent to the change in $C_p(T)$, is found at $T \cong T_g$ also for the electric properties of glass-forming melts, for example, for the dielectric constant ϵ. Figure 2.35 shows ϵ as a function of temperature for vitrifying glycerol. It is seen from Figure 2.35 that the significant variation in ϵ is found practically at the same temperature as for the coefficient of thermal expansion α and the heat

Table 2.4 Relative occupied volume β of various glasses, calculated from the densities ρ or the packing densities ψ of the glass (g) and the respective crystalline phase (c). The value of β is calculated either by $\beta = \rho_g/\rho_c$ from the density ratio of the glass versus crystal or by $\beta = \psi_g/\psi_c$ from the respective packing densities. The density data are taken from: Simha and Boyer [187], Dietzel and Poegel [228], Sternberg et al. [229], Cargill [230], Chen and Turnbull [231], Zanotto and Müller [232]. For further details see also Gutzow [164].

Substance	Structure of the crystal	ρ_g (g cm^{-3})	ρ_c (g cm^{-3})	ψ_g	ψ_c	β
	Organic linear polymers [187]					
	Various					0.89
	Inorganic glass-forming polymers [187]					
As_2O_3		3.70	4.15			0.89
GeS_2	Hexagonal	2.69	3.0			0.89
Se	Hexagonal	4.28	4.82			0.89
$PbSiO_3$		5.98	6.49			0.92
Li_2SiO_3	Orthorhombic	2.34	2.52			0.93
	Network glass-formers [228, 229]					
SiO_2	Hexagonal Quartz-like	2.20	2.65			0.83
GeO_2	Hexagonal Quartz-like	3.63	4.28			0.85
BeF_2	Hexagonal Quartz-like	1.98	2.37			0.84
$AlPO_4$ [228]	Hexagonal (berlinite)		2.64			0.84
$GaPO_4$	Hexagonal Quartz-like	2.37	2.72			0.87
P_2O_5 [232]		2.37	2.72			0.87
B_2O_3 [232]		1.84	2.46			0.74
	Vitreous metallic alloys [230, 231]					
	(Co–P; Fe–P–C; Pd–Si; Ni–Pd–P; Cu–Pd–Si)			0.68	0.74	0.92
AuGeSi [231] [230]	Hexagonal Close packing			0.64	0.74	0.89
	Bernal–Scott mechanical models of equally sized spheres					
Loose-random packing	Hexagonal Close packing			0.61	0.74	0.82
Dense-random packing	Hexagonal Close packing			0.64	0.74	0.86

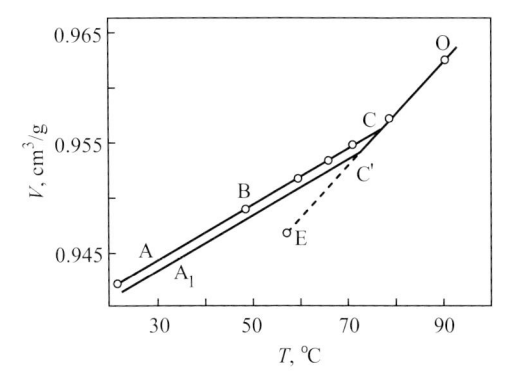

Figure 2.34 Specific volume vs. temperature in the process of vitrification of polystyrene at different cooling rates. The curve (A, B, C, O) corresponds to a process of rapid quenching, the curve (A₁–O) to a cooling rate of 0.2 K min⁻¹, while the results on path (E–O) are obtained after a prolonged annealing [195]. In this way, the points on the curve (E–O) visualize the reality of the equilibrium values of the "fictive" undercooled liquid. Considering the semi-quantitative nature, this picture, together with Oblad's and Newton's caloric determination of entropy relaxation of a glucose glass, Figure 2.27, gives the first direct confirmation of Simon's concepts on the nature of the vitreous state.

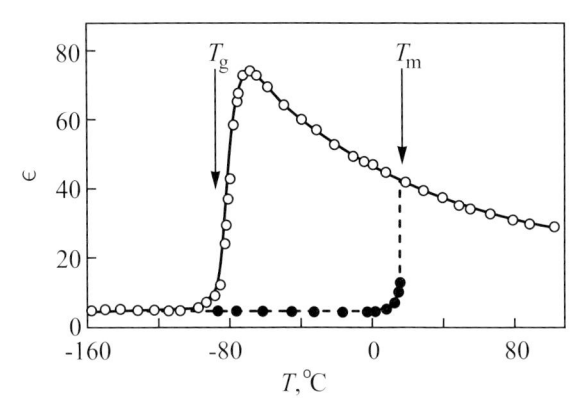

Figure 2.35 Temperature dependence of the dielectric constant ϵ of liquid and vitreous glycerol (*circles*). *Black dots* refer to the crystalline state (after Ubbelohde [95]). Note the analogy in the $\epsilon(T)$-dependence with the $\alpha(T)$- and $C_p(T)$-curves of the same substance.

capacity C_p (Figures 2.20 and 2.33). Since ϵ can be represented as [196]:

$$\epsilon = 4\pi \left(\frac{\partial^2 G}{\partial E^2} \right)_{T,p,\{n_j\}} + 1 , \qquad (2.139)$$

the similarity to the behavior of other thermodynamic coefficients is obvious. The term E in Eq. (2.139) is the absolute value of the electric field vector.

Mechanical, optical and thermodynamic properties, significantly determined by the change of either thermodynamic coefficients or thermodynamic functions, show also changes upon vitrification at temperatures approaching T_g, similar to

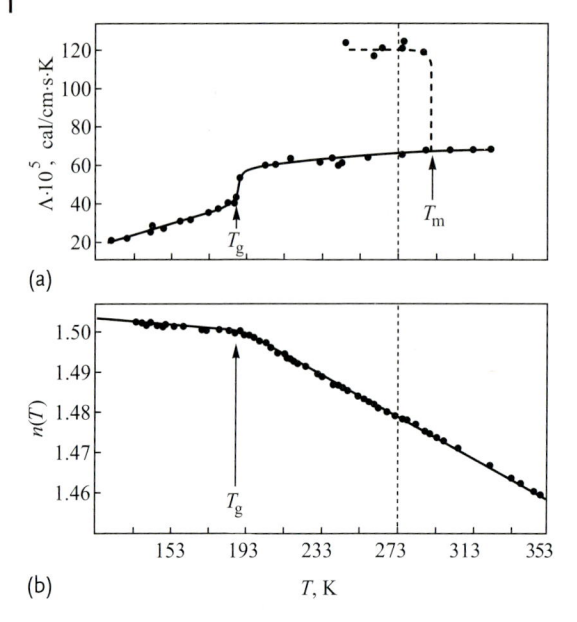

Figure 2.36 Temperature dependence of the properties of glycerol connected either with thermodynamic coefficients or thermodynamic functions: (a) change of the coefficient of thermal conductivity, Λ, for undercooled melts and the glass (bold lines with *black dots*) in dependence on temperature. The behavior of Λ is significantly determined by the specific heat, ΔC_p. Note the jump-like behavior of $\Lambda(T)$ near $T = T_g$ (cf. Figure 2.20). With a dashed line the change of $\Lambda(T)$ upon crystallization is shown; (b) temperature dependence of coefficient of optical refraction, $n(T)$, for liquid (at high temperatures) and glassy glycerol (at low temperatures). At T_g, a breakpoint in the transition from the undercooled melt to the glass can be seen, determined by the respective reciprocal change of the density of glycerol (cf. Figure 2.20; data according to Ubbelohde [95]).

the change of the corresponding dominant factor. Typical in this respect are the temperature changes of both the coefficient of heat conductivity, $\Lambda(T)$, and of the optical coefficient of refraction, $n(T)$, as shown here for glycerol (see Figures 2.36 and 2.37). The heat conductivity coefficient, $\Lambda(T)$, is usually defined via

$$\Lambda = \frac{1}{3}\langle v \rangle \langle l \rangle \rho C_v \qquad (2.140)$$

expressing it via the already discussed mean velocity of molecular translation $\langle v \rangle$, the mean free path, $\langle l \rangle$, in the system, the density, ρ, and the corresponding specific heat at constant volume, C_v. With Eqs. (2.110) and (2.111) and known thermodynamic expressions for the ratio (C_p/C_v), its becomes evident that $\Lambda \propto C_p/\eta$ has to hold [197]. Thus, the significant s-shaped temperature change of $\Lambda(T)$ in the transformation interval is determined by the dramatic decrease of C_p and not by the relatively smooth changes in the $\eta(T)$ curves near T_g (cf. Figure 2.17). This result is in fact found in experiment, as it is illustrated in Figure 2.36a.

According to known dependencies [170, 198], the coefficient of refraction, $n(T)$, is determined by the reciprocal value of the molar volume of the respective sol-

2.5 Thermodynamic Properties of Glass-Forming Systems | 81

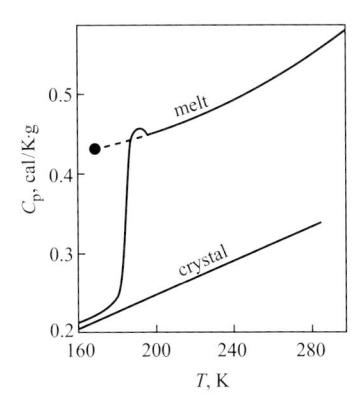

Figure 2.37 Specific heats of undercooled glycerol obtained at normal cooling rates (*full curve*) and after prolonged annealing (72 h at 174 K (*dashed curve*)). With the value of ΔC_p obtained by the annealing procedure the dashed continuation of the entropy vs. temperature curve in Figure 2.27 is calculated.

id. The comparison of Figure 2.32a,b shows that this relationship is indeed fulfilled according to expectation. Moreover, it is seen that all the changes in caloric, mechanical and electric properties in dependence on temperature, including $n(T)$ and $\Lambda(T)$, upon vitrification of glycerol take place in the same glass transition region. Thus, it turns out that the mechanical, caloric and electrical coefficients of vitrifying systems are abruptly changed in the vitrification range.

As already mentioned, the similarity of the behavior of the thermodynamic coefficients to the respective course in second-order phase transformations led Boyer and Spencer [92, 93] to the idea of interpreting vitreous solidification as a second-order phase transformation. However, this suggestion cannot be accepted, since, as observed by Boyer and Spencer themselves, a thermodynamically defined value of T_g does not exist: T_g in all considered cases depends on the cooling rate. The combination of the jumps in the thermodynamic parameters $\Delta\alpha$, $\Delta\kappa$, and ΔC_p at $T = T_g$, written in form of Ehrenfest's equation (2.70), gives a nearly constant value also for vitrification processes (see [19, 20, 199]). However, this constant is not equal to unity as it is the case for second-order equilibrium phase transformations. It is generally greater than one and varies, in dependence on the specific properties of the vitrifying substance, usually between 2 and 5 (see also Chapter 3).

Nevertheless, there exists some deep connection between second-order phase transitions and the glass transition. As mentioned in the discussion of Ehrenfest's classification in Section 2.2.3, second-order phase transitions can be considered as a qualitative change in the response (or susceptibilities) of the respective equilibrium systems with respect to the variation of the external parameters, for example, pressure or temperature. This change of the response (the susceptibilities) is connected with the evolution of some new properties of the system at the transition temperature. Similarly, in glass transitions, the response of the system with respect to variations of the state parameters is also changed. However, in vitrification, the system is transferred not into a different equilibrium state but into a nonequilibrium state. This is the basic feature distinguishing second-order equilibrium phase transitions from the glass transition.

2.6
Thermodynamic Nature of the Glassy State

Summarizing the results of the previous sections, we may formulate the following main conclusions characterizing the glass transition and the nature of the vitreous state:

1. In the transformation range between the undercooled liquid and the glass no heat of transition is released or absorbed. This temperature range is characterized by a steep s-shaped change of all thermodynamic coefficients corresponding to second-order derivatives of the thermodynamic potential, G, to a bending point in the temperature course of the thermodynamic functions S, H and V and by a steep increase of the viscosity followed by a knickpoint of its temperature course.
2. The discontinuities in the values of the thermodynamic coefficients at T_g do not obey quantitatively Ehrenfest's equation (2.70).
3. Vitrification does not take place at a fixed value of external thermodynamic parameters (p, T) but depends on kinetic factors like the cooling rate, q. The glass transition temperature is a function of q (cf. Eq. (2.114)).
4. The third law of thermodynamics in its classical formulation does not hold for the vitreous state. It retains its validity formulated as the Principle of Unattainability of absolute zero temperatures.

The properties 1–4 allow us to conclude that vitrification cannot be interpreted in terms of a thermodynamic phase transformation according to Ehrenfest's classification. Taking into account the considerations of Section 2.1, it is, therefore, not reasonable to consider glasses as a state of aggregation or a phase in the classical sense discussed above. The solution of the problems concerning the nature of the vitreous state and its deviation from the third law of thermodynamics can be explained in a first approximation by Simon's assumption [22]: glasses are kinetically frozen-in, thermodynamically nonequilibrium states. Due to the sharp increase in the viscosity, relaxation processes in the melt towards equilibrium become so slow in the transformation range that they cannot follow the variation of the external parameters. Thus, a certain molecular configuration is frozen-in in the glass, which results in the relatively high values of ΔS_g and ΔH_g, remaining approximately constant for $T < T_g$ (Eqs. (2.126), (2.127) and (2.128)). Simon's proposal excludes glasses from the framework of classical thermodynamics, since thermodynamics is applicable in its classical form only to equilibrium systems. Glasses do not obey, according to this interpretation, the third law because they are nonthermodynamic systems in the classical sense. Thus thermodynamics is not violated by

2.6 Thermodynamic Nature of the Glassy State

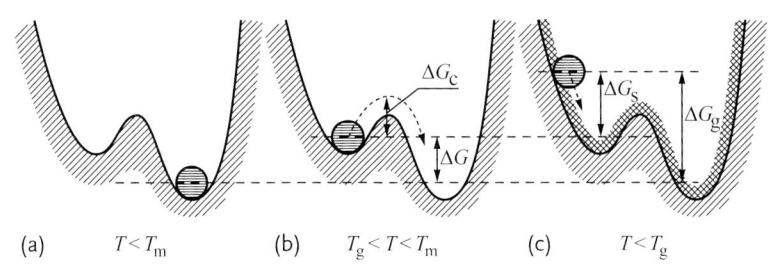

Figure 2.38 Mechanical analogy for an interpretation of the differences between (a) the stable (at $T < T_m$) crystalline state, (b) the metastable melt and (c) the glass below T_g (drawn after Simon and Jones [200]). In this mechanical analogy, the crystalline state corresponds to an absolute minimum of the potential well, the undercooled melt to a higher local minimum. In order to be transferred from the metastable to the stable crystalline state the system has to overcome a potential barrier ΔG_c, the thermodynamic barrier of nucleation. ΔG is the thermodynamic driving force of crystallization. The glass is represented in this analogy by a ball glued to the wall of the potential well above the minimum. Crystallization is commonly preceded by stabilization processes, the thermodynamic force of stabilization, ΔG_s, is also indicated in the figure.

the existence and the particular thermodynamic behavior of glasses, but glasses, as any other frozen-in state, cannot be treated fully quantitatively in the framework of classical thermodynamics. This peculiar nonthermodynamic nature of glasses has to be accounted for in the definition of the vitreous state. However, as shown as well, as far as only the first and second laws of thermodynamics are employed, glasses can be treated in classical thermodynamics terms, at least, approximately, following Simon's proposal. Simon's ideas and the resulting approach prevailed in glass literature for more than 80 years.

Simon's statement represents a generalization of the definition of the vitreous state given first by Tammann. Tammann stated that glasses are solidified (in German: *erstarrte*) undercooled melts (see also Tammann [24]). This definition is correct as a first step but not sufficient since nothing is said about the thermodynamic state of glasses. In discussing Tammann's definition it must also be pointed out that glasses at present are obtained not only by the undercooling of melts but also by a variety of other methods, for example, vapor deposition. Also from such a point of view Tammann's original definition has to be generalized. This generalization is discussed in Chapter 3.

Simon gave an illustration of his ideas on the nature of the vitreous state in a form according to which Figure 2.38 is drawn. This figure demonstrates the difference between the crystalline state (thermodynamically stable at $T < T_m$), the corresponding metastable undercooled melt and the glass. At $T < T_m$ the metastable liquid melt is, as any other metastable system, stable with respect to small fluctuations but unstable with respect to sufficiently large deviations from the initial state. Such sufficiently large changes initiate the transition to the thermodynamically favored crystalline phase. In glasses, the thermodynamic driving force towards the respective equilibrium state is far from being equal to zero. Moreover, it can be shown that the thermodynamic driving force for crystallization in a glass is as a rule even higher than for the respective undercooled melt. The particular nature of glasses as frozen-in nonequilibrium states is represented in Figure 2.38 by a ball glued to the wall of the potential well.

Simon's point of view is verified by a number of additional facts, not mentioned so far. The first of these facts was an experiment proposed by Simon and carried out by Oblad and Newton ([175]; see Figure 2.37). It is based on the following argumentation. If the scenario proposed by Simon is valid, then a decrease in the cooling rate should lead to a decrease of the values of both enthalpy and entropy frozen-in during the transformation to a glass. It was shown by Oblad and Newton with glycerol that slow cooling processes extended over several months led in fact to a decrease of the residual entropy (see Figure 2.27) when it is calculated via Eq. (2.123) from the respective $C_p(T)$-measurements. Due to the steep increase in viscosity and the resulting exponential increase in the time, required to carry out the measurements, such an extension of the experimental investigations of the undercooled melt below the value of T_g, corresponding to normal cooling rates ($q \approx 10^{-2}-10^2 \, \text{K s}^{-1}$), is possible for very narrow temperature intervals, only. Thus, the extension to even lower temperatures can be carried out only based on carefully performed interpolations of the curves, consistent with the principles of thermodynamics, or on statistical mechanical investigations of model systems. For moderate deviations from T_g, the values of $\Delta S(T)$ corresponding to the metastable melt, can be approximated by a linear extrapolation of the curve determined in the range $T_g < T < T_m$ (see Figure 2.29). However, if one extends such a linear interpolation to absolute zero, negative values of the entropy of the fictive undercooled melt below T_0 are obtained (Figure 2.39). Such a possibility is in conflict with classical thermodynamics, in particular, with the third law, according to which the entropy of any real or virtual equilibrium system tends to zero for temperatures approaching zero (cf. Eqs. (2.26)–(2.28)). It also has to be excluded from the point of view of the statistical interpretation of entropy (see Boltzmann's equation (2.32)) since it would correspond to a number of microscopic realizations Ω of such a state less than one. Such virtual extrapolation was carried out by Kauzmann [109]; it is known as Kauzmann's paradox and is repeatedly discussed in the subsequent literature. Kauzmann himself, as it is evident from his paper, was fully aware of the inadmissibility of such a linear extrapolation and proposed it only as a paradoxical possibility [109].

A similar resolution of this problem was given by Mazurin [70]. His argumentation was that if certain assumptions are made leading to paradoxical conclusions, then the assumptions have to be considered as wrong. Mazurin also drew the attention to experimental work performed by Gonchukova [203] showing that the supposed approximation leading to the mentioned paradoxical situation is not correct, experimental data show another kind of behavior not leading to a paradox.

Another, at a first glance, paradoxical result, giving rise to prolonged discussions in the literature (see, e.g., Blumberg [39]), was connected with erroneous $\Delta S(T)$-calculations by Tammann [72], based on $C_p(T)$-measurements of glycerol. Due to an error in the calculations, Tammann was led to the conclusion that there may

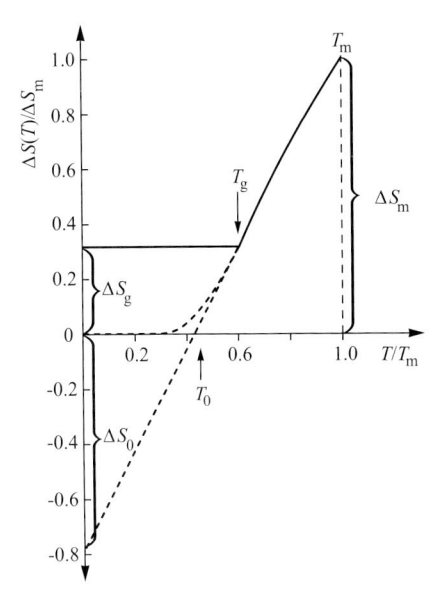

Figure 2.39 $\Delta S(T)$ curves for glycerol and Kauzmann's paradox. The *solid line* represents the undercooled metastable liquid (for $T_g < T < T_m$) and the vitreous state (for $T < T_g$). The *dashed straight line* is an inadmissible linear extrapolation of the $\Delta S(T)$-dependence for the melt down to absolute zero. Only such an unrealistic linear extrapolation yields negative values of the entropy for $T \to 0$. Thermodynamically forbidden consequences and paradoxes do not occur for more realistic thermodynamically self-consistent extrapolations of the properties of the fictive undercooled melt as shown, for example, by the alternative *dashed curve*. The temperature is given in relative units T/T_m. T_0 is the temperature for which ΔS becomes practically equal to zero (cf. also Chapter 3 and [201, 202]).

exist a point of intersection of the $S(T)$ curves of the glass and the undercooled melt. In this way, it seemed to be possible that the thermodynamic potential of the frozen-in glass may be lower for some substances (Tammann mentioned glycerol) as compared with the respective value for the undercooled melt. This led Tammann [204] further to the conclusion that under certain conditions the glass is more stable from a thermodynamic point of view than the undercooled melt. The error in Tammann's calculations was first mentioned by Simon. Tammann agreed with Simon's criticism [205] and gave a correction of his previous incorrect calculations and resulting statements. Despite these facts, Tammann's first erroneous conclusions entered some text-books on glass science and appear in literature from time to time even today.

A next point to be mentioned, giving additional support to Simon's ideas, is the behavior of glasses in devitrification processes. If a glass is heated slowly after reaching sufficiently low values of the viscosity first a relaxation into the actual metastable equilibrium can be observed, before eventually the transition to the stable crystalline state occurs. In glass science, the isothermal relaxation of a glass into the corresponding metastable undercooled liquid is called stabilization. The stabilization of the nonequilibrium glass always precedes the crystallization of the undercooled melt. This is a typical feature of the process of evolution from frozen-in disordered to equilibrium states.

Further arguments supporting Simon's concept is the structural isomorphy of the vitreous state as compared with the liquid melt from which it originated. This isomorphy is supported by X-ray measurements, IR-spectra and other methods of structural investigations. Additional arguments in favor of Simon's interpretation are the possibility to derive the Bartenev–Ritland equation (Eq. (2.114)) and the so-called Prigogine–Defay ratio (cf. Chapter 3), a relation similar to Ehrenfest's equation (Eq. (2.70)), however, derived based on the interpretation of vitrification as a kinetically determined freezing-in process.

As outlined in the previous analysis, glass-forming melts give us the unique possibility to analyze the thermodynamic properties of undercooled liquids. Can one state that these properties are similar or even equal for differently structured liquids? And are the results, described in the present chapter, really representative, typical features of the glass transition, in general? In answering these questions, one can first recall the Beaman–Kauzmann rule, Eqs. (2.73)–(2.75), in its various forms and the relations describing the behavior of the thermodynamic functions at $T = T_g$ (Wunderlich's rule, Eq. (2.115) for the ratio $\Delta C_p(T_g)/\Delta S_m$; the dependencies found by Gutzow and Dobreva for $\Delta S_g/\Delta S_m$, $\Delta H_g/\Delta H_m$ and $\Delta G(T \leq T_g)$, Eqs. (2.131)–(2.133) and Figures 2.14 and 2.26). These results show that, at least,

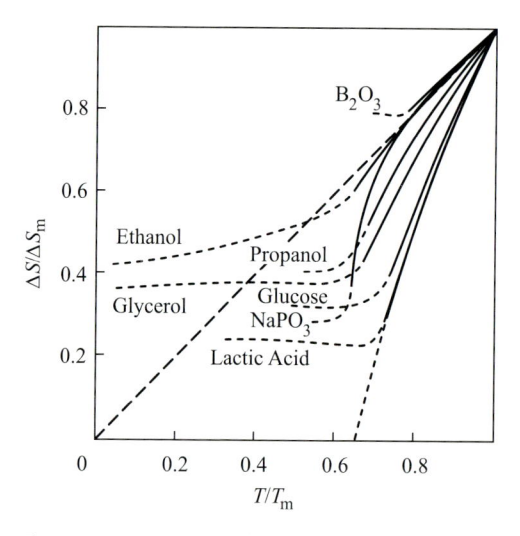

Figure 2.40 Temperature dependence of the configurational entropy $\Delta S(T) = S_{liquid}(T) - S_{crystal}(T)$ and $\Delta S(T) = S_{glass}(T) - S_{crystal}(T)$ (for $T \leq T_g$) for several glass-forming substances in Kauzmann's coordinates ($\Delta S(T)/\Delta S_m$) vs. (T/T_m) [109]. Note the fan-like divergence in the temperature dependence, the similarity in the break-points near $T = T_g$ and the negative values of the extrapolated ΔS values below T_0. Additionally, data for NaPO$_3$ are introduced into Kauzmann's diagram (according to the measurements of Grantcharova et al. [172]) in order to illustrate the "fragile" behavior of a typical inorganic glass-former.

in a first approximation all vitrifying systems are characterized near $T \cong T_g$ by distinct preferred values of certain combinations of thermodynamic properties.

The origin of such similarity in the behavior is most probably determined by the critical values, found by Simha and Boyer [187] and Gutzow [191, 206, 207], of the free volume at vitrification and more generally by the definite value of the configurational disorder, Ω, frozen-in in a glass (cf. Eqs. (2.135) and (2.136) derived here). Moreover, in agreement with the consequences of the theorem of corresponding states [54], the values of the thermodynamic parameters of undercooled melts at the melting point, T_m, can serve, at least, in a first approximation as the normalizing reference factor. In this sense, all above-mentioned empirical rules, connecting the thermodynamic properties at $T = T_g$ and $T = T_m$, are an indication that all undercooled glass-forming liquids display a more or less similar temperature dependence. This result was, in fact, first recognized in the already cited remarkable paper by Kauzmann [109]. In constructing the reduced temperature course of all fifteen glass-forming substances, investigated at his times, he found that they are converging to $\Delta S(T) \cong 0$ at $T_0 \cong 1/2 T_m$ (see Figure 2.39), however, with remarkable exceptions (see Figure 2.40). Using Angell's terminology, we notice that, with respect to the temperature dependence of the entropy, some of the liquids are

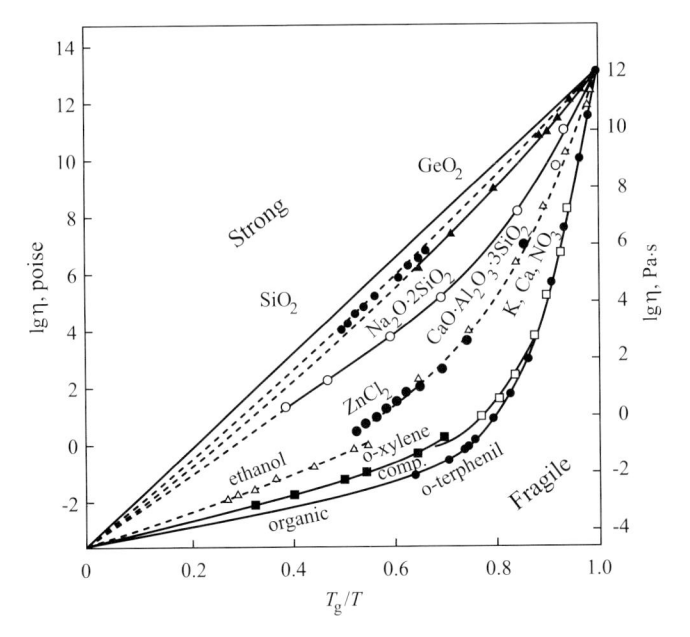

Figure 2.41 Viscosity of several glass-forming melts in Angell's coordinates $\log \eta(T)$ vs. (T_g/T) [52, 53]. Note the "strong"-type rheological behavior of SiO_2, GeO_2 and BeF_2 and the "fragile"-type change of the viscosity in systems with temperature dependent structures and activation energies of viscous flow, $U(T)$, depending on reduced temperature, T/T_g. It is, however, interesting to note [208] (G.P. Johari, personal communication) that similar coordinates have been employed much earlier and partly in a much more detailed description by Oldekop [209] and Laughlin and Uhlmann [210].

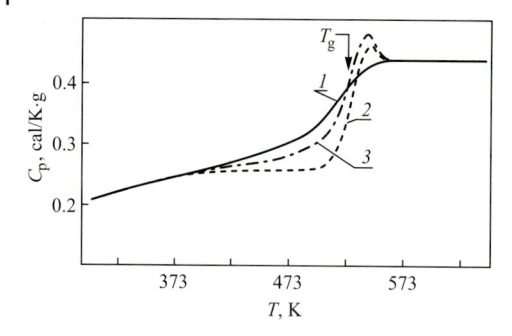

Figure 2.42 The first experiment, showing the real course of the specific heats in the glass transition region: (1) cooling curve, (2) slow heating, and (3) rapid heating (schematically after the measurements of Parks and Thomas [211] with B_2O_3). Note the smooth s-shaped change of the specific heat in the cooling run, the two overshoot "noses" in the heating experiments and the change in the position in the inflection point determining the glass transition temperature, T_g.

more "fragile" than others, which can be called "thermodynamically strong" glass-forming systems. The results, shown in Figure 2.41, and the so-called Kauzmann paradox (Figure 2.39) have strongly influenced the whole course of development of theoretical ideas in glass science.

In the present chapter, it was mentioned that quantities like specific volume differences and the configurational entropy enter the exponent in the dependencies determining the temperature course of the viscosity. Thus, existing structural differences and different temperature dependencies of the configurational entropy enter in the relations governing viscosity in an exponential form and they are amplified in the temperature courses of the viscosity. This property is demonstrated in Angell's diagram Figure 2.41, showing clearly the differences in the temperature course of the viscosities of "strong" and "fragile" liquids. Angell employed T_g as the reference temperature and not the melting temperature, T_m, as required by thermodynamic similarity concepts. Remember, however that both temperatures are connected by equations similar to Eq. (2.73), at least, for the intervals of cooling rates usually employed up to now in transforming a liquid to a glass.

In the next chapter, efforts are undertaken to analyze in more detail the nature of the vitreous state and its structural peculiarities by thermodynamic methods going beyond Simon's approximative treatment. In such a way, the real course of the thermodynamic properties of glass-forming systems may be established as it is in fact observed for many years (cf. Figure 2.42).

2.7
Concluding Remarks

In the present chapter, we have tried to summarize the essentials of both the process of the glass transition and the properties of matter in the unusual state formed

2.7 Concluding Remarks

by this process, the glass, as they appear from experimental evidence. After many years of research by many authors it turns out that the simplest way to analyze and even to explain in a physically reasonable way both glass transition and the nature and the properties of glasses are phenomenological approaches based on the principles of the thermodynamics of irreversible processes. As shown here, already classical thermodynamics applied by utilizing the approximation of Simon, gives a valuable tool to explain the essential features of glasses and the glass transition in an unambiguous way. It is based, however, not only on an approximation, but also on the unrealistic assumption that the glass transition takes place not steadily in the glass transition interval, as this is seen in Figure 2.42, but abruptly at a temperature $T = T_g$. Moreover, in Simon's model relaxation is excluded from the considerations since a metastable equilibrium state of the system goes over at some temperature into a fully frozen-in nonequilibrium system, the glass. Heating and cooling processes have to proceed then in different directions via the same sequence of states. As a consequence, any hysteresis effect cannot be interpreted utilizing such simplifying assumptions. The mentioned assumptions are only some of the limitations of Simon's simplified model.

In the following chapter it is shown that thermodynamics of irreversible processes, introduced in a generic way, gives not only the most simple way of the description of the kinetics of glass transition as it really takes place in a more or less broad temperature range (T_g^+, T_g^-), but it also indicates the possibility to derive the most essential features of the thermodynamics of glass transition and allows one correctly to describe the properties of glasses obtained at different conditions. Thus, it turns out that the thermodynamics of irreversible processes gives, without the need of introducing particular molecular models, a simple point of view of describing vitreous states and their formation. In this way, choosing the thermodynamics of irreversible processes as the tool of the more detailed description of glasses and the glass transition, we tried to follow the advice of the great scientist, Gibbs, who noted in his remarkable letter of 1881 to the American National Academy of Arts and Sciences: "One of the principle objects of theoretical research in any department of knowledge is to find the point of view from which the subject appears in its greatest simplicity."

3
Generic Theory of Vitrification of Glass-Forming Melts

Jürn W. P. Schmelzer and Ivan S. Gutzow

3.1
Introduction

The analysis of the properties of glass-forming systems in vitrification is a topic of principal scientific interest, it gives a deep insight into the very nature of vitrification and the glassy state. On the other hand, the adequate knowledge of the change of the properties of glass-forming systems in vitrification is a basic prerequisite of a correct interpretation of a variety of processes taking place at the glass-transition and, in particular, of crystallization processes which may occur in the course of cooling or heating of glass-forming melts.

Following the classical concepts developed by Simon [1, 59], vitrification of a glass-forming melt in cooling is commonly interpreted as a freezing-in process. Hereby it is assumed that above a given temperature, T_g, the glass-forming melt is able to reach at any moment of time the actual stable or metastable equilibrium state, that is, processes of relaxation of the system to these states are expected to proceed rapidly when compared with changes of the external control parameter (or external state parameter, here temperature). Below the glass transition temperature, the system is assumed to be in a completely frozen-in, nonequilibrium state. Here slow relaxation processes may occur, in general, as well. However, due to very small values of the kinetic coefficients the system cannot reach the respective equilibrium states in the time scales of the experimental investigations. As a result, the system is characterized by excess frozen-in values of entropy, enthalpy, Gibbs' free energy, and so on. as compared with the respective metastable or stable equilibrium states (cf. [1] and also Chapters 2, 5 and 9 in the present monograph).

However, a detailed experimental and theoretical investigation of vitrification shows [1, 11, 12, 233] that in cooling a liquid, the transition from the undercooled metastable equilibrium melt to a frozen-in nonequilibrium state, the glass, does actually proceed in a certain temperature range. Indeed, as already mentioned by Gustav Tammann in his monograph *Der Glaszustand* [24]

> Es ist ... "auch nicht richtig, von einem Transformationspunkt der Gläser zu sprechen; richtiger wäre es, statt Transformationspunkt Transformationsintervall zu sagen ... " (in free translation: It is not correct to talk about a transformation point of the glasses; it would be more correct, to speak about an transformation interval instead of a transformation point...).

The upper and lower boundaries of this temperature range we denote here as T_g^+ and T_g^-, respectively. Inside this temperature range ($T_g^- < T < T_g^+$), the characteristic times,

$$\tau_T = \left(\frac{T}{|\dot{T}|}\right), \tag{3.1}$$

of change of the externally governed parameters (here temperature) are comparable in magnitude with the characteristic times, τ, of relaxation to the respective temporary stable or metastable equilibrium states. Indeed, the definition of the characteristic time of change of temperature implies that the change of temperature in a time interval dt is given by the relation:

$$\frac{dT}{dt} = -\left(\frac{T}{\tau_T}\right). \tag{3.2}$$

Similarly (as will be discussed in detail below), relaxation processes of glass-forming melts to the respective equilibrium states are governed by equations of the type

$$\frac{d\xi}{dt} = -\frac{(\xi - \xi_e)}{\tau}, \tag{3.3}$$

where ξ is the parameter introduced already in Chapter 2, the additional characteristic parameter of the glass-forming melt, the structural order parameter, and ξ_e is its value in the stable or metastable equilibrium state of the system.

Let us consider a glass-forming melt cooled ($q < 0$) or heated ($q > 0$) with a rate $q = (dT/dt)$. For $\tau_T \gg \tau$, the glass-forming system remains over the course of the cooling process in a state of thermodynamic equilibrium, in the opposite case, $\tau_T \ll \tau$, the glass-forming melt is practically frozen-in. Consequently, in the temperature range, defined by $\tau_T \cong \tau$, the system goes over from a metastable equilibrium state to a frozen-in nonequilibrium state, the glass or, vice versa, from a glass to a metastable melt. The vitrification interval (and the range of the reverse devitrification process) can be defined – according to these considerations – generally via the relation

$$\tau_T \cong \tau, \tag{3.4}$$

or, employing the definition of the characteristic time of change of the control parameter, by

$$\left(\frac{1}{T}\left|\frac{dT}{dt}\right|\right)\tau \cong \text{constant} \quad \text{at} \quad T \cong T_g, \tag{3.5}$$

where the constant has a value of the order of unity.

The validity of relations equivalent to Eqs. (3.4) and (3.5) has been established experimentally; they are known as the Frenkel–Kobeko rule (cf. [1]). The theoretical analysis performed here shows that the validity of Eq. (3.5) is a consequence of the very deep nature of vitrification processes (cf. also [1, 11, 233]), they have to be fulfilled consequently for any freezing-in process independent on the rate of change of the external control parameters and the nature of the vitrifying system. In particular, taking pressure instead of temperature as the varying in time external control parameter, similar to Eq. (3.1) a characteristic time of change of pressure can be introduced as $\tau_p = (p/|\dot{p}|)$ and the vitrification range is given similarly to Eq. (3.4) by $\tau_p \cong \tau$.

By the described nature of vitrification, in the vitrification interval processes of change of the external control parameters are always accompanied by irreversible processes of relaxation of the system to the respective equilibrium states. These irreversible processes do occur in the system in cooling as well as in heating, resulting – according to the basic laws of thermodynamics – in the production of entropy. The continuity in the transition from the metastable equilibrium state to the frozen-in nonequilibrium state, the glass, and the resulting entropy production in this temperature range ($T_g^- \leq T \leq T_g^+$) is, consequently, a general feature of any glass transition.

Variations of the rate of change of the control parameters may shift (according to Eq. (3.5)) the position of the vitrification interval, but do not affect the existence of the effect itself. Again, the shift of the glass transition temperature depending on cooling rate has been established first experimentally; it is well-known as the Bartenev–Ritland equation [1]. It can be shown that the Bartenev–Ritland equation is also a direct consequence of Eq. (3.5) (cf. [1, 11, 233] and the further discussion in the present chapter) provided relaxation can be treated as a thermally activated process.

The interpretation of vitrification as proceeding in a finite temperature interval and the resulting necessarily deviation of the affinity from zero values as well as the account of entropy production terms have, as will be shown in the present analysis, significant effects and are of principal importance for the correct understanding of a variety of thermodynamic and kinetic characteristics of glass-forming melts, which are beyond the problems discussed here (cf. [234]). The effect of entropy production in vitrification has not been accounted for in a comprehensive form up to now in the analysis of the thermodynamic properties of glass-forming systems in heating and cooling or in any other possible mechanisms of vitrification, for example, due to pressure changes. It is one of the aims of the present investigation to fill this gap (cf. [19, 20, 235]). Hereby we employ the method developed by De Donder [13, 14] characterizing the system under consideration by a set of structural order parameters introduced into the description in addition to the conventional thermodynamic state parameters.

The present chapter is structured as follows: In Section 3.2, we give an extended, as compared to Chapter 2, summary – following De Donder [13] and Prigogine [14] – of the basic thermodynamic equations utilized in the present analysis. In order to employ these general dependencies, basic general thermodynamic and kinetic properties of glass-forming systems have to be known. In the present analysis, we use for these purposes a simple but qualitatively accurate lattice-hole model of glass-forming systems. The essential features of this model and the basic equations employed for the description of relaxation of glass-forming melts are summarized in Section 3.3. In Section 3.4, the general thermodynamic equations derived are applied to the determination of the expressions for the change of entropy, enthalpy and a variety of other thermodynamic functions in cyclic heating-cooling processes, performed (as the simplest realization) with constant heating and cooling rates. We assume hereby that the cooling process starts at a sufficiently high temperature above T_g at a well-defined stable or metastable equilibrium state. The process is then terminated at a sufficiently low temperature, when the system is completely frozen-in for the experimental time scales considered. Then the system is reheated at the same constant rate until the initial metastable or stable state is reestablished (i.e., it is assumed that the characteristic relaxation times at this state are considerably smaller as compared with the characteristic times of variation of the external parameter). The consequences are discussed in detail, and a first comparison with experimental data is given. Further experimental analysis is suggested allowing one to test specific features of the theory outlined. In addition, a definition of the glass transition temperature is given based on the temperature dependence of the structural order parameter; the Bartenev–Ritland equation is also re-derived and generalized and the limits of validity of this equation are discussed.

The results outlined are employed then for the development of a new theoretical approach to the derivation of the Prigogine–Defay ratio in glass transition in Section 3.5. As a result of this analysis, we show that theoretical estimates of the Prigogine–Defay ratio lead to values not equal to one but yield in vitrification, in general, values larger than one. In contrast to previous approaches in the theoretical analysis of this problem it is shown that this result is obtained already if only one structural order parameter is employed in the description of vitrification. However, in order to be able to arrive at this result, one has to treat vitrification not in terms of Simon's approximation but, as will be elaborated here, as proceeding in a certain transition interval. Experimental data, confirming the theoretical results, and some further possible applications concerning the behavior of thermodynamic coefficients in vitrification are summarized there.

In Section 3.6, the concepts of fictive temperature and fictive (or internal) pressure are analyzed and a general model-independent definition of these structural order parameters is given. The relevance of the method of description of vitrification employed and of the results obtained with respect to the understanding of the behavior of the kinetic parameters of glass-forming melts like viscosity is briefly analyzed in Section 3.7. It is supplemented by a similar analysis of the intensity of fluctuations in cooling and heating of glass-forming systems given in Section 3.8. A summary of the results and their implications and the discussion of future possible developments (Section 3.9) completes the chapter.

3.2
Basic Ideas and Equations of the Thermodynamics of Irreversible Processes and Application to Vitrification and Devitrification Processes

3.2.1
Basic Assumptions

The thermodynamics of irreversible processes, in general, and, in particular, in treating problems of vitrification and devitrification makes the following assumptions, which are additional to the basic concepts of classical thermodynamics:

- Time, t, is introduced as an argument into the equations of thermodynamics. This revolutionary step was first made by De Donder [13, 44].
- Thermodynamics of irreversible processes accepts all the basic derivations and dependencies obtained in the framework of the classical formulations of thermodynamics. However, in considering their applicability it always has to be taken into account that most of the well-known dependencies of classical thermodynamics have to be corrected and enlarged by the introduction of, at least, one additional internal parameter, ξ, which reflects in some generalized way the structure of the system. The difference $(\xi - \xi_e)$ between the actual value of the structural order parameter and its equilibrium value is a measure of the deviation of the system from equilibrium with respect to its structure. In a first approximation, the correction terms thus introduced change linearly with the increase of the deviation from equilibrium. However, as indicated already by Prigogine and Defay [14, 15]: there has always to be expected that from a given more or less well-defined value of deviation from equilibrium $(\xi - \xi_e)$ the corrected dependencies used may loose their linear form. We may be obliged to use more complicated formulations of the nonlinear variants of irreversible thermodynamics. At what value of the deviation from equilibrium this change with its sometimes unforeseeable consequences has to take place can to be decided only by experiment: it determines the limits of applicability of the linear formulation of irreversible thermodynamics. Existing experience collected up to now shows that in treating the problems of glass transition this linear (or quasi-linear) formulation guaranties a sufficient accuracy in the interpretation of experiment.
- The introduction of, at least, one additional internal structural order parameter, ξ, guarantees a strict thermodynamic treatment of any system in nonequilibrium. As already mentioned in Chapter 2, De Donder introduced such an additional parameter ξ initially as a chemical reaction coordinate to open the possibility to describe systems undergoing chemical reactions. Here we treat ξ in a broader sense as a general initially not specified parameter describing in some nonspecified but distinct way the configurationally important features of the structure of the systems investigated. For convenience, we specify the value of ξ to change in the limits from $\xi = 0$ (full order) to $\xi = 1$ (full disorder). The meaning of the structural order parameter as employed in the present analysis will be discussed in detail below.

- Thermodynamics of irreversible processes accepts the three basic laws of classical thermodynamics giving a specific formulation to any of them. The applicability of the third principle of thermodynamics to nonequilibrium systems, in general, and glasses, in particular, is discussed in detail here in Chapter 9.
- Thermodynamics of irreversible processes accepts as a fourth significant principle the already mentioned "phenomenological law," according to which the rates of change (e.g., $(d\xi/dt)$) are proportional to the acting thermodynamic driving forces (X).
- Thermodynamics of irreversible processes accepts the standard values of all basic physical constants and their constancy in any process under investigation.
- Affinity is introduced here in treating the processes in nonequilibrium as a necessary tool of thermodynamics of irreversible processes. It is shown further on in this chapter how affinity can and has to be properly defined guaranteeing its convenient application.

Taking into account the above considerations, which are outlined in more details in existing well-known literature sources (e.g., [14, 43, 47]) we begin here a second stage of treatment of vitrification and devitrification and of the properties of glasses in the framework of a great science, somehow disregarded – mostly due to missing knowledge of it – even by well-known authors in the field of glass science.

3.2.2
General Thermodynamic Dependencies

In treating vitrification processes, we have to consider systems in thermodynamic nonequilibrium states. In such states, the change of the entropy, S, in any process can be expressed, in general, as the sum of two terms [14],

$$dS = d_e S + d_i S, \quad d_e S = \frac{dQ}{T}, \quad d_i S \geq 0. \tag{3.6}$$

Here $d_i S$ describes changes of the entropy connected with irreversible processes taking place in the system under consideration, while the term $d_e S$ describes changes of the entropy connected with an exchange of energy in the form of heat, dQ, with surrounding systems (denoted further as heat bath). With T here originally the temperature of the heat bath is denoted (assumed to be in an internal thermal equilibrium) exchanging energy in form of heat with the system, where the processes under consideration take place. In our analysis, we assume that the heat bath and the system analyzed are always sustained at states near internal thermal equilibrium and that, with respect to temperature, a state near equilibrium between heat bath and system is established at any time. In such cases, the temperature of the system under consideration and the temperature of the heat bath are equal. In other words, we do not consider here changes of the entropy due to the transfer of energy in the form of heat between systems sustained at different temperatures.

Following De Donder [13, 14], we can describe processes of relaxation of glass-forming melts to the respective metastable equilibrium state by the introduction of additional appropriately chosen structural order parameters, ξ_i. Here we assume first that the system can be described with a sufficient degree of accuracy by only one such additional order parameter, ξ. Such an assumption is always suitable when basically qualitative consequences of theoretical concepts are considered. If desired the derivations given below can be easily extended to more complex situations by taking into account, as it is done sometimes in the literature [19, 20, 199, 236], the existence of several independent structural order parameters. However, in the vicinity of the glass transition temperature, where the effects discussed are of major significance, the free volume of the glass-forming melt decreases dramatically with decreasing temperature. As a result, possible alternative structural parameters may become dependent on free volume and cannot be considered then any longer as independent (cf. [19, 20, 199, 236, 237]).

The restriction in the first part of the present chapter to only one structural order parameter is also motivated by the following considerations. In Section 3.5 it will be shown (see also [234, 238, 239]) that the known experimental data on the so-called Prigogine–Defay ratio in vitrification can be interpreted in terms of the theoretical approach developed here by employing only one structural order parameter. This result is considered by the authors as one direct experimental proof of the validity of our approach and confirms, moreover, our belief that the restriction to one structural order parameter may give not only a qualitatively but in a number of cases even a quantitatively correct description of thermodynamic properties of glass-forming melts in vitrification (a detailed discussion of the problem how many structural order parameters may be required in order to appropriately describe a glass-forming system is performed in [193, 234, 240] (see also [19, 20, 78, 167, 236, 237, 241–243] and Chapter 4); by similar considerations as outlined in [241] – not to confuse the issue with the theory of critical phenomena and second-order phase transitions – we employ generally instead of the term "order parameter" the notation "structural order parameter"). Anyway, the developed here theoretical method can be extended easily to the case that several independent structural order parameters have to be utilized for the description of the system under consideration. The respective more general relations are given in the final part of the present chapter (cf. also [1]).

Assuming that a single structural order parameter is sufficient for the description of the degree of deviation of the glass-forming melt from the respective stable or metastable equilibrium state, the combined first and second laws of thermodynamics can be written then in the form [13, 14]

$$dU = TdS - pdV - Ad\xi \ . \tag{3.7}$$

Here U is the internal energy and V the volume of the system, p the external pressure, and A the affinity of the process of structural relaxation considered. Since the number of particles of the different components remains constant (i.e., we consider closed systems), terms describing their possible changes do not occur in Eq. (3.7) in the applications analyzed. Since, for any irreversible processes taking place at constant entropy and volume, the internal energy can only decrease [54, 62, 63], the relation $Ad\xi \geq 0$ has to be fulfilled. A combination of Eqs. (3.6) and (3.7) yields

$$dU = Td_eS + Td_iS - pdV - Ad\xi .\tag{3.8}$$

Alternatively, for any (reversible or irreversible) process of a thermodynamic system, we have according to the first law of thermodynamics [54, 62, 63]

$$dU = dQ - pdV = Td_eS - pdV .\tag{3.9}$$

Consequently, the relations

$$Td_iS = Ad\xi \quad \text{or} \quad d_iS = \left(\frac{A}{T}\right) d\xi \tag{3.10}$$

have to be fulfilled, in general. It follows from the latter equation that equilibrium states are characterized by $A = 0$. Only for such states, the production term of the entropy, d_iS, connected with possible irreversible processes in the system, is equal to zero. Moreover, since the changes of entropy and structural order parameter take place in the same time interval, dt, we can rewrite Eq. (3.10) also as

$$T\frac{d_iS}{dt} = A\frac{d\xi}{dt} \quad \text{or} \quad \frac{d_iS}{dt} = \left(\frac{A}{T}\right)\frac{d\xi}{dt} .\tag{3.11}$$

In this way, we re-derived the basic relationship connecting variations of entropy and structural order parameter [14]. However, in order to employ this relation, we have to express also the affinity, A, via the basic thermodynamic functions.

Employing pressure and temperature as the thermodynamic state parameters and assuming, in addition to thermal equilibrium, that the conditions of mechanical equilibrium are fulfilled (equality of pressure of the heat bath and of the system under consideration), the characteristic thermodynamic potential is the Gibbs free energy, G, defined generally via

$$G = U - TS + pV .\tag{3.12}$$

Equation (3.7) yields

$$dG = -SdT + Vdp - Ad\xi ,\tag{3.13}$$

and the required second necessary relation for the affinity can be obtained in the form

$$A = -\left(\frac{\partial G}{\partial \xi}\right)_{p,T} .\tag{3.14}$$

At constant pressure and temperature, the thermodynamic equilibrium state is characterized by a minimum of Gibbs' free energy, G. Let us denote further by ξ_e

the value of the order parameter, ξ, in this equilibrium state. Since in thermodynamic equilibrium, the state of the considered closed system is determined by two independent variables, the equilibrium value of the structural order parameter can be represented as an unambiguous function of pressure and temperature, that is,

$$\xi_e = \xi_e(p, T). \qquad (3.15)$$

A truncated Taylor expansion of the partial derivative of Gibbs' free energy with respect to the structural order parameter, ξ, in the vicinity of the respective equilibrium state, $\xi = \xi_e$, results in

$$\left(\frac{\partial G}{\partial \xi}\right)_{p,T} \cong G_e^{(2)}(p, T, \xi_e)(\xi - \xi_e), \quad G_e^{(2)}(p, T, \xi_e) = \left(\frac{\partial^2 G}{\partial \xi^2}\right)\bigg|_{p,T,\xi=\xi_e}. \qquad (3.16)$$

The first term in the expansion, $(\partial G/\partial \xi)_{p,T,\xi=\xi_e}$, vanishes since Gibbs' free energy has a minimum at equilibrium ($\xi = \xi_e$). For the affinity, A, we obtain from Eqs. (3.14) and (3.16)

$$A = -G_e^{(2)}(\xi - \xi_e), \quad G_e^{(2)} = G_e^{(2)}(p, T, \xi_e) = \left(\frac{\partial^2 G}{\partial \xi^2}\right)\bigg|_{p,T,\xi=\xi_e} > 0. \qquad (3.17)$$

The parameter $G_e^{(2)}$ has to be positive, since the equilibrium state is characterized by a minimum of Gibbs' free energy [54, 62, 63]. The derivative of the affinity and the second-order derivative of the Gibbs' potential, G, obey at equilibrium, consequently, the conditions

$$\left(\frac{\partial A}{\partial \xi}\right)_{p,T,\xi=\xi_e} = -\left(\frac{\partial^2 G}{\partial \xi^2}\right)_{p,T,\xi=\xi_e} < 0. \qquad (3.18)$$

Assuming an analytic dependence of affinity and thermodynamic potential on the structural order parameter, these inequalities are valid also beyond the equilibrium state. Taking into account the principle of le Chatelier–Braun [54] latter inequality can be expected to hold generally for all nonequilibrium states considered. At least, in the mean-field model, employed later on for quantitative estimates, this inequality holds true independent on the possible values of the order parameter in the temperature range under consideration.

As shown by Prigogine and Defay [14], the change of the affinity as a function of pressure, temperature and structural order parameter can be written generally in the form:

$$dA = \frac{(A + h_{p,T})}{T} dT - v_{p,T} dp + a_{p,T} d\xi. \qquad (3.19)$$

Here the notations

$$a_{p,T} = \left(\frac{\partial A}{\partial \xi}\right)_{p,T} < 0, \quad v_{p,T} = \left(\frac{\partial V}{\partial \xi}\right)_{p,T}, \quad h_{p,T} = \left(\frac{\partial H}{\partial \xi}\right)_{p,T} \qquad (3.20)$$

are used. The inequality $a_{p,T} < 0$ follows directly from Eq. (3.18) and its discussion. Taking into account the identity $G = H - TS$, we arrive with Eqs. (3.1) and (3.14) at

$$h_{p,T} = \left(\frac{\partial G}{\partial \xi}\right)_{p,T} + T\left(\frac{\partial S}{\partial \xi}\right)_{p,T} = -A + T\left(\frac{\partial S}{\partial \xi}\right)_{p,T}. \quad (3.21)$$

By definition of the order parameter, the entropy has to increase (at constant p and T) with increasing ξ, that is, $(\partial S/\partial \xi)_{p,T} > 0$ holds generally. Moreover, for the considered so far cooling processes, we have $A < 0$ (cf. [11, 17, 233, 240]). Consequently, in vitrification the inequality $h_{p,T} > 0$ has to be generally fulfilled. In this way, for vitrification in cooling, we have generally:

$$A + h_{p,T} = T\left(\frac{\partial S}{\partial \xi}\right)_{p,T} > 0, \quad A < 0, \quad h_{p,T} > 0 \quad \text{for cooling runs}.$$
$$(3.22)$$

For devitrification processes (i.e., in heating runs), the inequality $A > 0$ holds in the glass transition interval and the sign of the parameter $h_{p,T}$ depends on the magnitude of both thermodynamic parameters entering Eq. (3.21). In this case, we can write

$$A + h_{p,T} = T\left(\frac{\partial S}{\partial \xi}\right)_{p,T} > 0, \quad A > 0 \quad \text{for heating runs}. \quad (3.23)$$

Above-derived dependencies are of great significance for the theoretical determination of the Prigogine–Defay ratio as performed in the further analysis and, in particular, for the understanding of the differences in the values of this ratio in cooling and heating runs.

3.2.3
Application to Vitrification and Devitrification Processes

Thermodynamics of irreversible processes [43, 47] states that the fluxes (here the change of ξ) are functions of the thermodynamic driving forces (here the affinity, A). In this way, we may write:

$$\frac{d\xi}{dt} = -L f(X), \quad X = \frac{A}{RT}, \quad (3.24)$$

going over from the affinity, A, to the dimensionless thermodynamic driving force, X. Here R is the universal gas constant and L a kinetic coefficient, depending on pressure and temperature but not on the order parameter, ξ. The function $f(X)$ is not known, in general. It can be specified by introducing, in the framework of statistical models, certain assumptions concerning the system under consideration or by employing experimental results [1, 11, 43]. We will return to different methods of specification of this function below. However, first we study several general consequences connected with the definition of this function.

3.2 Basic Ideas and Equations of the Thermodynamics of Irreversible Processes

In thermodynamic equilibrium, the relations

$$A = 0, \quad \xi = \xi_e, \quad \frac{d\xi}{dt} = 0 \qquad (3.25)$$

hold and, consequently, we have to demand $f(X = 0) = 0$. Moreover, employing the methods of the linear thermodynamics of irreversible processes (applicable for sufficiently small deviations from equilibrium), the driving force is expressed as the derivative of the appropriate thermodynamic function (here, the Gibbs free energy, G) with respect to the structural order parameter. The evolution equations are transformed then into the form:

$$\frac{d\xi}{dt} = L\frac{\partial}{\partial \xi}\left(\frac{G}{RT}\right), \quad f(X) = -\frac{\partial}{\partial \xi}\left(\frac{G}{RT}\right). \qquad (3.26)$$

Equations (3.14), (3.24) and (3.26) yield a second condition, the function $f(X)$ has to obey, that is, $f'(X = 0) = (\partial f(X)/\partial X)|_{X \to 0} = 1$ (in an alternative interpretation, possible nonequal to one factors $f'(X = 0)$ in the linear term of the expansion are assumed to be included in the kinetic coefficient, L, that is, $L' f'(X = 0) X = L X$).

By a Taylor expansion of the function $f(X)$ in the vicinity of the thermodynamic equilibrium state ($X = 0$) we obtain (with $f(0) = 0$ and $(\partial f(X)/\partial X)|_{X \to 0} = 1$) the relation:

$$f(X) = X + \frac{1}{2}f''(0)X^2 + \ldots, \quad f''(0) = \left.\frac{d^2 f(X)}{dX^2}\right|_{X=0}. \qquad (3.27)$$

Retaining in the Taylor expansion of the function $f(X)$ terms of up to the second order in X, only, Eq. (3.24) can be written in the form:

$$\frac{d\xi}{dt} \cong -\frac{L G_e^{(2)}}{RT}\left(1 - \frac{f''(0) G_e^{(2)}}{2RT}(\xi - \xi_e)\right)(\xi - \xi_e). \qquad (3.28)$$

Here, in addition, Eq. (3.17) has been employed.

Introducing a characteristic time-scale of the relaxation processes, τ, via

$$\frac{1}{\tau} = \frac{L G_e^{(2)}}{RT}\left(1 - \frac{f''(0) G_e^{(2)}}{2RT}(\xi - \xi_e)\right), \qquad (3.29)$$

Eq. (3.28) yields

$$\frac{d\xi}{dt} = -\frac{1}{\tau}(\xi - \xi_e). \qquad (3.30)$$

For very small deviations from equilibrium ($\xi \cong \xi_e$), quadratic terms in the difference $(\xi - \xi_e)$ vanish and we obtain

$$\frac{d\xi}{dt} = -\frac{1}{\tau_e}(\xi - \xi_e), \qquad (3.31)$$

with a characteristic relaxation time,

$$\tau_e = \left(\frac{RT}{LG_e^{(2)}}\right), \qquad (3.32)$$

depending on pressure and temperature but not on the order parameter, ξ. Retaining terms of second order in the difference $(\xi - \xi_e)$, we can write, consequently,

$$\begin{aligned}\tau(p, T, \xi) &= \tau_e \left(1 - \frac{f''(0) G_e^{(2)}}{2RT}(\xi - \xi_e)\right)^{-1} \\ &\cong \tau_e \left(1 + \frac{f''(0) G_e^{(2)}}{2RT}(\xi - \xi_e)\right). \end{aligned} \qquad (3.33)$$

Not only did we specify in this way the relaxation law but we also introduced a dependence of the characteristic relaxation time on the value of the order parameter, ξ. Since the parameters $f''(0)$ and $G_e^{(2)}$ depend on the properties of the system in the respective equilibrium state, exclusively, it follows from Eq. (3.33) that the relaxation times are different, in general, depending on whether the inequalities $(\xi - \xi_e) > 0$ or $(\xi - \xi_e) < 0$ hold, that is, from which side the system tends at isothermal conditions to the respective equilibrium state.

The kinetic equation, formulated with Eqs. (3.30) and (3.33), can be denoted as a quasi-linear relaxation law. It is linear with respect to the difference, $(\xi - \xi_e)$, but with a characteristic relaxation time, τ, depending also on this difference. Note that this quasi-linear character of the relaxation law can be obtained also in the general case not involving any assumptions, that is, if we do not expand the function $f(X)$ into a Taylor series. Equations (3.24) and (3.30) lead, employing further the expression for τ_e derived, to the conclusion that generally the quasi-linear relaxation law holds where the relaxation time can be expressed – in the most general form – as:

$$\frac{d\xi}{dt} = -\frac{1}{\tau}(\xi - \xi_e), \quad \frac{1}{\tau} = \frac{Lf(X)}{(\xi - \xi_e)} = \frac{1}{\tau_e}\left\{\frac{RT f(X)}{G_e^{(2)}(\xi - \xi_e)}\right\}. \qquad (3.34)$$

So, there exists a uniquely defined interrelation between the kinetic coefficient, L, the function $f(X)$ in Eq. (3.24) and the relaxation time, τ. Employing the Taylor expansion of the function $f(X)$ (cf. Eq. (3.27)), Eq. (3.29) is reestablished from this general relation as a special case, again.

Equations (3.11), (3.17) and (3.30) yield the following expression for the entropy production in the system due to irreversible relaxation

$$\frac{d_i S}{dt} = \frac{G_e^{(2)}}{T\tau}(\xi - \xi_e)^2. \qquad (3.35)$$

As should be the case, the entropy production term is always a positive quantity independent of the sign of the deviation of the order parameter from its equilibrium value, $(\xi - \xi_e)$, since the relaxation time, τ, is a positive quantity. This condition has to be fulfilled in any model considerations concerning the function $f(X)$ in order to lead to physically meaningful results.

In establishing the basic equations of the theory, we employed – as evident from the derivations – exclusively basic laws of classical thermodynamics and thermodynamics of irreversible processes. The theoretical method is consequently model-independent and generally applicable. Only in one case is an approximation employed (a truncated Taylor expansion is performed) leading to the expression for the affinity (Eq. (3.17)).

In order to derive the principal consequences of the theory outlined, in the next section thermodynamic and kinetic properties of glass-forming melts are specified in more detail. In particular, a lattice-hole model of glass-forming melts is introduced allowing us to describe the most essential qualitative features of any vitrification processes. For a wide class of glass-forming melts, it is even able to describe the behavior quantitatively. In the course of specification of this model, the definition of the structural order parameter we used in our computations is specified. As will be shown, with such choice all above-given relations are fulfilled in the vicinity of ξ_e and even beyond. However, for particular applications other selections of the structural order parameters may be useful or desirable and can be employed since the above-mentioned relations are exclusively consequences from very basic and general thermodynamic laws valid for any appropriate definition of the order parameter. This freedom ensures that the equations cited above hold for any particular realizations of glass transitions, since these equations are of general thermodynamic nature.

3.3
Properties of Glass-Forming Melts: Basic Model Assumptions

3.3.1
Kinetics of Relaxation

A large amount of experimental data – based mainly on measurements of the temperature dependence of the viscosity – shows that the characteristic relaxation times of glass-forming melts in stable or metastable equilibrium states can be generally expressed in the form [1]

$$\tau_e = \tau_0 \exp\left(\frac{U_a(p, T)}{RT}\right), \quad \tau_0 = \frac{h}{k_B T}, \tag{3.36}$$

where the activation energy, U_a, of the relaxation processes obeys the inequality

$$\frac{d U_a(p, T)}{d T} \leq 0. \tag{3.37}$$

As a typical example, we will employ here the Vogel–Fulcher–Tammann (VFT) equation [1, 70, 108], where the activation energy can be written as:

$$U_a(p, T) = U_a^*\left(\frac{T}{T - T_0}\right). \tag{3.38}$$

Here U_a^* and T_0 are empirical constants specific for the substance considered. The parameter T_0 we identify with Kauzmann's temperature (cf. [1]) and set it equal in the following model computations to $T_0 = T_m/2$, where T_m is the melting or liquidus temperature of the liquid. At the glass transition temperature (reached at conventional cooling rates and corresponding under such cooling conditions to a viscosity $\eta \cong 10^{13}$ Poise), T_g, the activation energy, U_a, is of the order $(U_a(T_g)/RT_g) \cong 30$. With $(T_g/T_m) \cong 2/3$, we obtain as an estimate $(U_a^*/RT_g) = 7.5$. The preexponential term, τ_0, in Eq. (3.36) is determined by the frequency of molecular oscillations. It is practically not affected by deviations of the system from the respective equilibrium state. For an estimate of its value, we employ here the relation $\tau_0 = (h/k_B T)$, where h is Planck's constant.

As will be discussed in detail below (see figures below in Sections 3.7 and 3.8, and the accompanying discussion, as well as Chapter 4 in the present book and our monograph [1]), not only the thermodynamic but also the kinetic parameters of glass-forming systems depend, in general, on the value of the structural order parameter. In order to account for the effect of deviations of the system from equilibrium on its relaxation behavior, the activation energy, U_a, has to be considered, consequently, as a function not only of pressure and temperature, but also of the structural order parameter, ξ. By means of a Taylor expansion of the thus introduced function $U_a(T, p, \xi)$ in the vicinity of the respective equilibrium state, $\xi = \xi_e$, we obtain in a linear approximation with respect to the difference $(\xi - \xi_e)$

$$U_a(p, T, \xi) = U_a(p, T) + \left(\frac{\partial U_a}{\partial \xi}\right)\bigg|_{p,T,\xi=\xi_e} (\xi - \xi_e). \tag{3.39}$$

Instead of Eq. (3.36), we may write then the following dependence for the relaxation time of a glass-forming melt in a thermodynamically nonequilibrium state

$$\tau = \tau_0 \exp\left(\frac{U_a(p, T, \xi)}{RT}\right) = \tau_e \exp\left(\frac{1}{RT}\left(\frac{\partial U_a}{\partial \xi}\right)\bigg|_{p,T,\xi=\xi_e}(\xi - \xi_e)\right). \tag{3.40}$$

A Taylor expansion of the exponential term in Eq. (3.40) results in

$$\tau \cong \tau_e \left(1 + \frac{1}{RT}\left(\frac{\partial U_a}{\partial \xi}\right)\bigg|_{p,T,\xi=\xi_e}(\xi - \xi_e)\right). \tag{3.41}$$

A comparison of this approximation with the result derived here earlier in an alternative way and given by Eq. (3.33) shows that they are identical when the condition

$$\frac{f''(0) G_e^{(2)}}{2} = \frac{1}{RT}\left(\frac{\partial U_a}{\partial \xi}\right)\bigg|_{p,T,\xi=\xi_e} \tag{3.42}$$

holds. In this way, we have found a third requirement the function $f(X)$ has to fulfill in order to describe adequately relaxation processes in glass-forming melts.

In addition to the discussed coincidence in the expressions for the relaxation times, derived here in two different ways, Eqs. (3.39) and (3.40) can be given additional support based on well-known ideas of the theory of absolute reaction rates [18] and Prigogine's analysis of the probability of fluctuations in thermodynamic nonequilibrium systems [15]. An analysis of the respective concepts in application to the problems considered is given in [11, 233].

3.3.2
Thermodynamic Properties: Generalized Equation of State

The equations outlined in the previous section hold generally. In order to apply them to glass-forming melts, the essential properties of glass-forming melts have to be introduced based on some model considerations, that is, we have to connect the conventional thermodynamic parameters of the glass-forming system with the structural order parameter. The respective relationships we denote as generalized equations of state.

For such specification of the thermodynamic properties of glass-forming melts, we employ here relations derived from a simple lattice-hole model of liquids discussed in detail in [1]. The structural order parameter is connected in the framework of this model with the free volume of the liquid and defined via the number of unoccupied sites (or holes), N_0, per mole of the liquid where each of them have a volume, $v_0(p, T)$, identical to the volume of a structural unit of the liquid at the same values of pressure and temperature. According to this model, the molar volume of the liquid is determined as [1]:

$$V(p, T, \xi) \cong N_A v_0(p, T)(1 + \xi), \quad \xi = \frac{N_0}{N_A + N_0} \cong \frac{N_0}{N_A}. \tag{3.43}$$

Here N_A is Avogadro's number.

The thermodynamic functions of the system are described in the framework of this model by the sum of contributions resulting from the thermal motion of the molecules of the liquid and, in addition, from the configurational contributions described by the structural order parameter, ξ. The configurational contribution to the volume is given, consequently, by

$$V_{\text{conf}} \cong N_A v_0(p, T) \xi. \tag{3.44}$$

The configurational contribution to the enthalpy, H_{conf}, of one mole of the liquid is described in the framework of this lattice-hole model via the molar heat of evaporation, $\Delta H_{\text{ev}}(T_m)$, of the liquid at the melting temperature as:

$$H_{\text{conf}} = \chi_1 \Delta H_{\text{ev}}(T_m) \xi. \tag{3.45}$$

The parameter χ_1 will be determined later. Experimental data show that the molar heat of evaporation can be expressed for a wide class of liquids in accordance with Trouton's rule (cf. Figure 3.1) as:

$$\Delta H_{\text{ev}}(T_m) \cong \chi_2 R T_m \quad \text{with} \quad \chi_2 = 20. \tag{3.46}$$

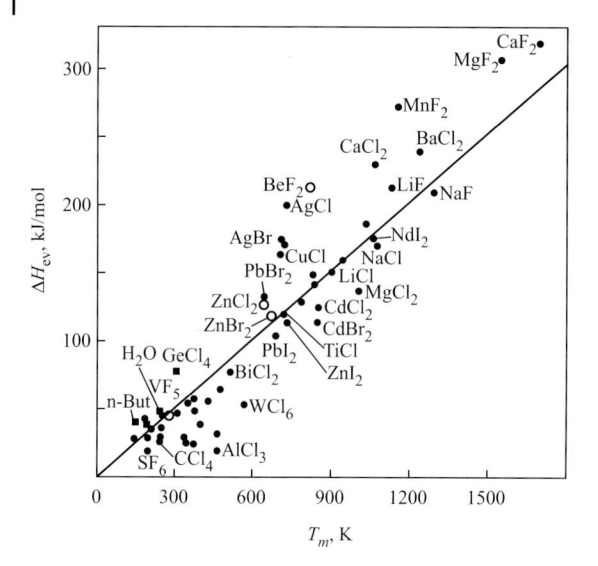

Figure 3.1 Molar heat of evaporation, ΔH_{ev}, in dependence on melting temperature, T_m, for different halide and oxide substances with very different compositions (*white circles* refer to typical glass-formers, and *black circles* to substances not vitrified as yet). *Black squares* represent data for simple nonhalide glass formers [149]. The straight line is given as a confirmation of Eq. (3.46) employed in the analysis.

The configurational part of the entropy per mole is described in this model via an expression similar to the conventional mixing term (here referred to N_A particles of the liquid and N_0 holes):

$$S_{\text{conf}} = -R\left(\ln(1-\xi) + \frac{\xi}{1-\xi}\ln\xi\right). \tag{3.47}$$

With $H = U + pV$, we obtain for the configurational contribution to the internal energy the relation

$$U_{\text{conf}} = \left[\chi_1 \Delta H_{ev}(T_m) - p v_0(p, T)\right]\xi. \tag{3.48}$$

Finally, employing the definition of Gibbs' free energy, Eq. (3.12), we arrive at

$$G_{\text{conf}} = \chi_1 \Delta H_{ev}(T_m)\xi + RT\left(\ln(1-\xi) + \frac{\xi}{1-\xi}\ln\xi\right). \tag{3.49}$$

The equilibrium value of the structural order parameter, $\xi = \xi_e$, is determined via the relation $(\partial G/\partial \xi)_{p,T} = (\partial G_{\text{conf}}/\partial \xi)_{p,T} = 0$. With Eqs. (3.46) and (3.49), we obtain the following result:

$$\frac{(1-\xi_e)^2}{\ln \xi_e} = -\frac{1}{\chi}\left(\frac{T}{T_m}\right) \quad \text{where} \quad \chi = \chi_1 \chi_2. \tag{3.50}$$

Knowing the value of χ_2 (cf. Eq. (3.46)), we determine the value of the parameter χ_1 demanding that at $T = T_m$ the value of ξ_e should be approximately equal to 0.05 (corresponding to experimentally observed density differences between liquid and crystal at the melting temperature, T_m). In the computations performed here we set $\chi_2 = 20$ and $\chi_1 = 0.166$ resulting in $\chi = 3.32$.

As should be the case, in the vicinity of the state of configurational equilibrium, we obtain from Eq. (3.49) after performing a truncated Taylor expansion, the following result:

$$G_{\text{conf}}(p, T, \xi) \cong G_{\text{conf}}(p, T, \xi_e) + \frac{1}{2}\left(\frac{\partial^2 G_{\text{conf}}}{\partial \xi^2}\right)\bigg|_{p,T,\xi=\xi_e} (\xi - \xi_e)^2 . \quad (3.51)$$

The value of $G_e^{(2)} = (\partial^2 G_{\text{conf}}/\partial \xi^2)|_{\xi=\xi_e}$ at equilibrium and, consequently, the affinity of the process of structural relaxation (cf. Eq. (3.17)) can now be easily calculated based on Eqs. (3.49) and (3.50). For small values of ξ, we get as an estimate

$$G_e^{(2)} \cong \left(\frac{RT}{\xi_e}\right). \quad (3.52)$$

Moreover, knowing the dependence of the structural order parameter on temperature (cf. Eq. (3.50)), one can establish the deviations of all thermodynamic functions discussed above from the respective equilibrium values. For this reason, the determination of the function $\xi = \xi(T)$ is of basic importance for the understanding of the behavior of the thermodynamic properties of vitrifying melts. The method of determination of this dependence is discussed in the next section.

3.4
Kinetics of Nonisothermal Relaxation as a Model of the Glass Transition: Change of the Thermodynamic Functions in Cyclic Cooling-Heating Processes

3.4.1
Description of the Cyclic Processes under Consideration

In the further analysis, we consider cyclic processes of the following type: at constant pressure, we cool the system with a constant rate,

$$q = \frac{dT}{dt} = \text{const.} \quad (3.53)$$

Since, according to Eq. (3.53), dT and dt are linearly dependent, we can always go over from a differentiation or integration with respect to time to the respective procedures with respect to temperature $(dT = qdt)$ and vice versa $(dt = dT/q)$ with different signs of the parameter q ($q < 0$ for cooling and $q > 0$ for heating processes). Equation (3.30) then takes the form [11, 233]:

$$\frac{d\xi}{dT} = -\frac{1}{q\tau}(\xi - \xi_e) . \quad (3.54)$$

Of course, Eq. (3.54) retains its validity also in the more general cases when constancy of cooling and heating rates is not assumed. Equation (3.54) was used in 1956 (without detailed derivation, and assuming $\tau = \tau_e$ with $U_a(p, T) = U_0 =$ const.; cf. Eq. (3.36)), in application to vitrification by Volkenstein and Ptizyn [170] and even earlier (in 1934) with the same restrictions by Bragg and Williams [244] in the analysis of formation of metastable alloys. For this reason, we denote Eq. (3.54) as the Bragg–Williams equation (cf. [11, 233]).

For the cyclic process we start at some well-defined stable or metastable equilibrium state. We denote the temperature of this equilibrium state of the liquid by T_l. In the cooling process, the internal order parameter, ξ, cannot follow, in general, the change of temperature of the system. It cannot be represented, consequently, as a function only of pressure and temperature, but depends also on the rate of change, q, of temperature, that is,

$$\xi = \xi(p, T; q). \tag{3.55}$$

From a mathematical point of view, Eq. (3.55) is a direct consequence of the differential equation, Eq. (3.54). Since the derivative $(d\xi/dT)$ depends on the heating and cooling rates, q, the solution has to depend on these parameters as well.

Reaching temperatures sufficiently below the temperature of vitrification, T_g, the relaxation time becomes extremely large (cf. Eq. (3.36)), and the value of ξ becomes frozen-in, that is, it practically does not change any more in the time scales of the experiment. At and below this temperature, which we denoted as T_g^-, the glass behaves as a solid body. Reversing now the process (again the pressure being fixed) and heating-up the system with the same heating rate (with respect to its absolute value), we arrive after some time at the initial state, again. We assume in this way that the initial state is sufficiently near the melting temperature so that the characteristic times of relaxation become small as compared with the characteristic times of heating of the sample. In the course of the heating process, the structural order parameter is an unambiguous function of the heating rate, again, similar to the case of cooling. Note, however, that the function will be different for heating and cooling. Mathematically, this difference is expressed in Eq. (3.55) by the differences in the sign of the rate of change of temperature, being negative for cooling ($q < 0$) and positive for heating ($q > 0$) processes.

3.4.2
Temperature Dependence of the Structural Order Parameter in Cyclic Cooling and Heating Processes

As mentioned in the previous section, the knowledge of the temperature dependence of the structural order parameter is of essential significance in order to determine the properties of glass-forming melts in vitrification and in the process of reheating of the glass. For this reason, we start the analysis of the behavior of thermodynamic functions in vitrification with the discussion of this thermodynamic parameter.

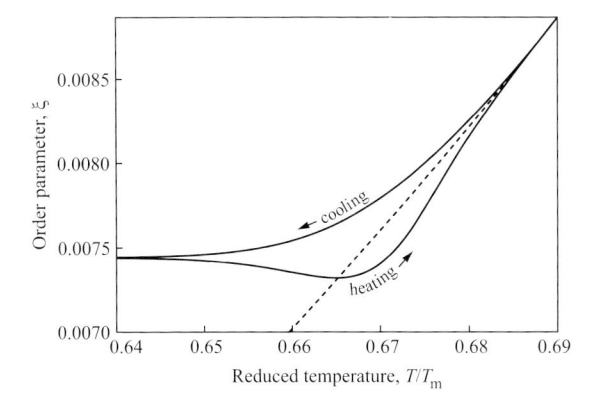

Figure 3.2 Dependence of the structural order parameter, $\xi = \xi(p, T; q)$ (or $\xi = \xi(\theta)$) with $\theta = T/T_m$), in a cyclic cooling-heating process (*solid line*). Arrows indicate the direction of change of temperature. By a dashed curve, the equilibrium curve $\xi_e = \xi_e(\theta)$ is shown. In undercooling a liquid, the order parameter cannot follow generally the change of the external parameter and deviates from the equilibrium value, ξ_e, resulting in $\xi > \xi_e$. Such kind of behavior is found in the vicinity of the glass transition temperature when the time scales of change of the external parameter become comparable with the characteristic relaxation times of the system. In heating, ξ decreases first until, after intersection with the equilibrium curve, it rapidly approaches it from below due to the exponential decrease of the relaxation times with increasing temperature. For the computations, we used here $T_m = 750$ K, $T_0 = T_m/2$, $T_g \cong (2/3)T_m$, $\chi_1 = 0.166$, $\chi_2 = 20$, $\chi = 3.32$ and the relations $U_a(T_g)/RT_g = 30$ and $U_a^*/RT_g = 7.5$. The values of the other parameters employed are given in the text (cf. also [240]).

An example for the change of the structural order parameter, ξ, in dependence on temperature, obtained via the numerical integration of Eq. (3.54), is shown in Figure 3.2. We start the process at $T_l = T_m$ and cool the system down to $T_s = T_0 = T_m/2$. For convenience, we introduce the reduced temperature, $\theta = (T/T_m)$. Equation (3.54) yields then

$$\frac{d\xi}{d\theta} = -\frac{1}{q_\theta \tau}(\xi - \xi_e), \quad q_\theta = \frac{d\theta}{dt}, \quad \theta = \frac{T}{T_m}. \quad (3.56)$$

Consequently, in terms of the reduced temperature, θ, the considered temperature range is $0.5 \leq \theta \leq 1$.

The relaxation time, τ (cf. Eq. (3.40)), we express in performing the computations in the form

$$\tau = \left(\frac{h}{k_B T}\right) \exp\left(\frac{U_a^*}{RT_m} \frac{1}{(\theta - \theta_0)}\right) \exp\left(\frac{\chi_1 \Delta H_{ev}}{RT_m} \frac{(\xi - \xi_e)}{\theta}\right). \quad (3.57)$$

In this way, it is assumed for the computations that the configurational part of the activation energy behaves similarly to the configurational part of the internal energy of the glass-forming melt. Independent of this particular assumption, the relaxation time, τ, increases exponentially with decreasing temperature. It follows that the characteristic times of relaxation of the system to the respective equilibrium states become, in the course of cooling, first comparable (in the vitrification interval) and then considerably larger (below it) than the characteristic times of change of the external parameters, here the temperature. As a result, the order parameter, ξ, retains higher values as compared with its equilibrium value, ξ_e, at the vitrification interval and below it.

The values of the characteristic temperatures T_g^+ (upper boundary of the vitrification interval), T_g (glass transition temperature) and T_g^- (lower boundary of the vitrification interval) depend on the cooling rate. The absolute value of the cooling rate is determined in the present computations via the condition (cf. Eq. (3.5))

$$\left|\frac{d\theta}{dt}\right|\tau \cong 10^{-2} \quad \text{at} \quad T = T_g. \tag{3.58}$$

This condition implies in agreement with the Kauzmann–Beaman rule [1, 11] that the temperature range of vitrification is located around the temperature $T_g \cong (2/3)T_m$ or $\theta_g \cong 2/3$ as it is in fact observed in Figure 3.2.

Similar computations in a wide range of cooling and heating rates have been performed, and are reported in [245, 246]. As it turns out the shape of the curves remains qualitatively the same but the amplitude of the hysteresis loops of the $\xi = \xi(T)$ curves increases with increasing heating rates. Vice versa, by lowering the absolute value of the rate of change of temperature the hysteresis loops are diminished and the glass transition approaches the model as proposed by Simon. These features are reflected in a similar way in the dependence of the intensity of entropy production on cooling rates, as will be discussed below.

After the completion of a cyclic cooling-heating process, the initial state of the system is reestablished, again. Consequently, we have $\oint d\xi = 0$. Here $\oint$ denotes the integration over the whole cycle. It follows as a first consequence (cf. Eq. (3.56)) that the relation

$$\oint \frac{1}{q_\theta \tau}(\xi - \xi_e) d\theta = 0 \tag{3.59}$$

has to be fulfilled.

3.4.3
Definition of the Glass Transition Temperature via the Structural Order Parameter: the Bartenev–Ritland Equation

Knowing the temperature dependence of the structural order parameter ξ, the derivatives of $\xi(T)$ with respect to temperature may be computed. The results are shown in Figure 3.3. In the framework of the lattice-hole model, the definition of the glass transition temperature, Eq. (3.5), can be given a more precise meaning (establishing the value of the constant) identifying it with the inflection point of the $(d\xi(T)/dT)$ curve [11, 17, 179]. As a consequence, the glass transition temperature T_g may be defined in terms of the structural order parameter via

$$\left.\frac{d^3\xi(T)}{dT^3}\right|_{T=T_g} \cong 0. \tag{3.60}$$

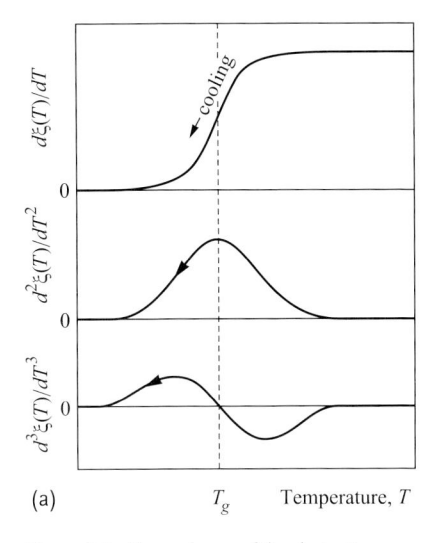

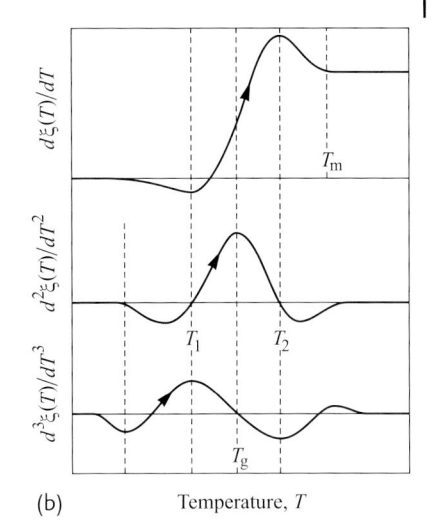

3.4 Nonisothermal Relaxation and Glass Transition

Figure 3.3 Dependence of the derivatives of the structural order parameter, ξ, with respect to temperature, T, for the cooling (a) and heating (b) runs. At $T = T_g$, the third order derivative of $\xi(T)$ is by definition nullified which corresponds to an inflection point in the configurational contributions to the specific heat [11, 17, 179].

As will become evident in further analysis, this definition identifies T_g with the inflection point of the temperature dependence of the configurational contributions to the specific heat (cf. Eq. (3.77)). Vice versa, taking Eq. (3.60) and the resulting relation for the specific heats as the primary definition of the glass transition temperature (as done by us in [11, 17, 179]), the Frenkel–Kobeko relation follows as a consequence. The definition of the glass transition temperature via Eq. (3.60) (or the equivalent to them in such approach consequences, Eqs. (3.4) and (3.5)) have the additional advantage that they are equally applicable for both heating and cooling. Moreover, they connect the definition of T_g with a central temperature in the transition interval and not a value near its boundary (for a recent overview on different methods of specification and experimental determination of T_g, see [108]).

As discussed above, Eq. (3.60) (and the resulting relation Eq. (3.77) for the configurational specific heat discussed later) allow one to derive, employing the particular generalized equation of state discussed here, the Frenkel–Kobeko relation. Despite that, we started here with the definition of vitrification via Eqs. (3.4) and (3.5) in order to underline the basic kinetic character of vitrification and devitrification processes. Moreover, these relations are valid independent of any particular model of the glass-forming melts and of the number of structural order parameters which may eventually be required in order to describe correctly glass-forming melts, that is, in addition to its basic character it is a general relation independent of any specific models employed for the description of the system under consideration and the kind of control parameter which is changed (see the discussion to Eqs. (3.4) and (3.5)). Note as well that, in order to obtain a kind of behavior as given in Figure 3.3 (resulting in the definition of T_g via Eq. (3.60) or the equivalent one for the specific heat) or in order that Eqs. (3.4) and (3.5) can be fulfilled for a wide spectrum of cooling and heating rates, a steep exponential dependence of the relaxation time on temperature (or appropriate alternative control parameters) is required. In this way, dependencies of the time of relaxation of the type as given by Eq. (3.36) and their generalizations have to be expected also from a fundamental theoretical point of view to be an essential prerequisite for glass-formation (cf. [11, 17, 179]).

Employing Eq. (3.5) with the specification as discussed above for the determination of T_g and expressing the relaxation time via

$$\tau \cong \tau_e = \tau_0 \exp\left(\frac{U_a}{RT}\right), \quad \tau_0 = \frac{h}{k_B T}, \quad (3.61)$$

with a constant value of the activation energy U_a, then an expression of the form

$$\frac{1}{T_g} \cong C_1 - C_2 \log|q| \quad (3.62)$$

is obtained, where C_1 and C_2 are parameters depending weakly on temperature only. This is the so-called Bartenev–Ritland equation [169, 194]. It is thus a direct consequence of the basic definition of the glass transition range, Eq. (3.5), provided the relaxation behavior can be expressed in the simple form as given with Eq. (3.61) involving the additional approximation of constant activation energy of relaxation processes.

Utilizing the more correct VFT equation for the description of the relaxation behavior

$$\tau \cong \tau_e = \tau_0 \exp\left(\frac{U_a^*}{R(T - T_0)}\right) \quad (3.63)$$

with a constant value of the activation energy U_a^*, then an expression of the form

$$\frac{1}{(T_g - T_0)} \cong C_1 - C_2 \log|q| \quad (3.64)$$

is found instead. Taking into account even more correct relations of the type given by Eq. (3.57), these dependencies will be further modified (cf. also [245, 246]).

The existence of such dependencies of glass transition temperature and cooling and heating rates implies that – taking into account that as a rule viscosity and relaxation time behave at least similarly – the definition of the glass transition as proceeding at a viscosity of 10^{13} Poise is valid only for a very restricted range of rates of change of temperature. By changing the rate of change of temperature, the glass transition temperature is also changed. As noted by Mazurin [108], for inorganic nonmetallic and, in particular, silicate glass-forming systems at the accessible rates of change of temperature, the glass transition proceeds at viscosities in the range 10^{12}–10^{14} Poise. This range of viscosities can be considerably enlarged, in principle, increasing significantly the rates of change of temperature so that even glass transitions originating from a thermodynamically stable liquid (i.e., referring to viscosities at temperatures higher than the melting temperature) are possible (for details see [245, 246]).

3.4.4
Structural Order Parameter and Entropy Production

As evident from Figure 3.2, the inequality $(\xi - \xi_e) \geq 0$ holds in the whole course of the cooling process. Inside the vitrification interval, both the differences $(\xi - \xi_e)$ and the relaxation times increase with decreasing temperature, however, with very different rates: the relaxation times increase exponentially (cf. Eq. (3.57)) and the difference $(\xi - \xi_e)$ with a rate near a linear one (cf. Figure 3.2). As a result from these considerations and Eq. (3.35) for the entropy production, we obtain a single maximum of the entropy production in cooling, shown in Figure 3.4 by a solid line.

After completion of the cooling process, the initial state can be reestablished only if, in the course of the subsequent heating, the order parameter ξ becomes less than ξ_e at a certain value of temperature and then approaches ξ_e from below. It follows that, in heating runs, a point of intersection of the curves $\xi = \xi(p, T; q)$ and $\xi_e = \xi_e(p, T)$ has to exist. At this point, the curve $\xi = \xi(p, T; q)$ has – according to Eq. (3.56) – a minimum. At this minimum, the system is in equilibrium and the entropy production drops to zero (cf. Eq. (3.35)) being different from zero below and above this particular temperature. As a direct consequence, the entropy production, $(d_i S/d\theta)$, has two maxima in heating processes in the glass transition interval. Provided heating would be performed with a different rate, the location of the minimum would be shifted, however, the general shape of the curve will remain the same. This general result is illustrated in Figure 3.4.

In this figure, the entropy production in cooling (showing one single maximum) is given by a solid line, the two maxima of the entropy production in heating are given via a dashed curve. As evident from the previous discussion, this result is valid independent of the value of the cooling and heating rates and of the particular model assumptions utilized here for the computation of Figure 3.4: for any model

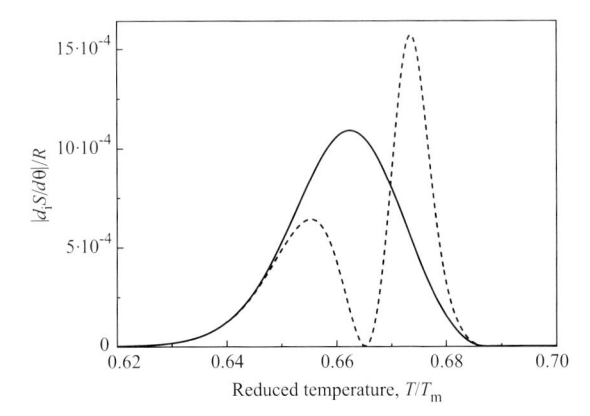

Figure 3.4 Entropy production (Eq. (3.65)) in vitrification and devitrification in a cyclic cooling-heating run experiment performed with the same absolute value of the rate of change of temperature. The entropy production has one maximum for cooling (*solid line*) and two maxima for the heating processes (*dashed curve*) (cf. also [240, 245, 246]).

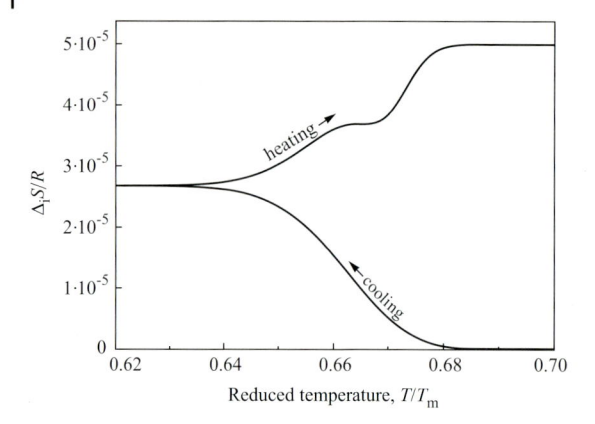

[Figure: plot of $\Delta_i S/R$ vs reduced temperature T/T_m showing heating and cooling curves between 0.62 and 0.70]

Figure 3.5 Total entropy, $\Delta_i S$, produced in the system in the cyclic cooling-heating process. In the cooling run, the entropy production becomes effective in the vitrification range, it drops then to zero due to the large relaxation times. In the subsequent heating, again, entropy is produced in the same temperature range as in the cooling process provided, as it is done, the absolute value of the rate of change of temperature is the same in both cooling and heating. Note that the entropy produced in cooling is larger than the entropy generated in heating. This result is due to the differences in the values of $(\xi - \xi_e)$ in the glass transition interval for cooling and heating, respectively (cf. also [240, 245, 246]).

of glass-forming melts employing one structural order parameter, where ξ_e is a monotonous function of temperature, the entropy production term has to have one maximum in cooling and two in heating.

Figure 3.5 shows the total entropy produced due to irreversible relaxation processes proceeding in the system in the glass transition interval. For the computations leading to Figures 3.4 and 3.5, we write Eq. (3.35) in the form (cf. also [240])

$$\frac{d_i S}{d\theta} = \frac{G_e^{(2)}}{T_m \theta q_\theta \tau} (\xi - \xi_e)^2 \,. \tag{3.65}$$

For the determination of $G_e^{(2)}$, we employed the relation (cf. Eq. (3.49))

$$G_e^{(2)} = \left.\frac{\partial^2 G}{\partial \xi^2}\right|_{\xi=\xi_e} = \left.\frac{\partial^2 G_{\text{conf}}}{\partial \xi^2}\right|_{\xi=\xi_e}$$
$$= R T_m \left[\frac{\theta}{\xi_e(1-\xi_e)^2} \left(1 + 2\frac{\xi_e \ln \xi_e}{(1-\xi_e)}\right) \right], \tag{3.66}$$

where ξ_e is determined via Eq. (3.50). Note that the product $(T_m \theta)$ cancels when Eqs. (3.65) and (3.66) are combined.

For an analytical analysis of the possible magnitude of the effect of entropy production on the thermodynamic functions, we can use the approximative result $G_e^{(2)} \cong RT/\xi_e$, derived earlier (cf. the discussion following Eq. (3.51) and Eq. (3.52)). Equation (3.65) can be written then in the form

$$\frac{d_i S}{d\theta} \cong \frac{R}{q_\theta \tau} \frac{(\xi - \xi_e)^2}{\xi_e} \,. \tag{3.67}$$

In the vicinity of the glass temperature, independent of the substance considered and the rate of change of temperature (or, more generally, the external control parameter), the relation $q_\theta \tau \cong$ constant holds (cf. Eq. (3.5), again). Consequently, the magnitude of the entropy production is determined, in general, basically by the ratio $(\xi - \xi_e)^2/\xi_e$ in the glass transition interval. Due to this peculiarity, entropy production terms gain in importance with increasing rates of change of temperature and, vice versa, become more and more negligible with decreasing heating and cooling rates. Such behavior was predicted in [240] and proven by numerical computations in [245, 246]. A general estimate of the possible magnitude of the effect of entropy production, based on Eq. (3.67), is performed for the model employed here in [180]. It is shown there that the effect is in this case small.

3.4.5
Temperature Dependence of Thermodynamic Potentials at Vitrification

3.4.5.1 Configurational Contributions to Thermodynamic Functions

As discussed in detail in the previous sections, in the cyclic processes considered, the order parameter is a well-defined function of pressure, temperature and the respective cooling and heating rates, that is, $\xi = \xi(p, T; q)$ (cf. Eq. (3.55)). The knowledge of this dependence allows us to determine the values of the thermodynamic functions. In particular, employing the model considerations outlined above, the configurational contributions to volume, enthalpy, entropy, internal energy and Gibbs' free energy can be determined via Eqs. (3.43)–(3.49). For the specification of volume and internal energy in terms of the model employed, the dependence of the volume per particle of the liquid in the absence of holes, $v = v_0(p, T)$ must also be known. The kind of temperature course of the volume (and the internal energy) of the system under consideration in a cyclic cooling-heating process depends on the relative significance of two terms. Indeed, employing Eq. (3.43), we obtain (at constant pressure) the following expression for the changes of the volume depending on temperature

$$\frac{dV(p, T, \xi)}{dT} = \left(\frac{\partial V}{\partial T}\right)_{p,\xi} + \left(\frac{\partial V}{\partial \xi}\right)_{p,T} \frac{d\xi}{dT}$$

$$\cong N_A \left(\frac{\partial v_0(p, T)}{\partial T} + v_0(p, T)\frac{d\xi}{dT}\right)_p. \quad (3.68)$$

The temperature dependencies of the configurational contributions to the entropy (Eq. (3.47)) and Gibbs free energy (Eq. (3.49)) are shown in Figures 3.6 and 3.7 both for Simon's model (i.e., $\xi = \xi_e$ for $T \geq T_g$ or $\theta \geq (2/3)$ and $\xi = \xi(T_g)$ for $(1/2) \leq \theta < (2/3)$) and the model of a continuous transition to the glass as discussed here ($\xi = \xi(p, T; q)$). The configurational contribution to the enthalpy is not drawn separately since its dependence on temperature repeats the respective curves drawn for the structural order parameter, ξ, shown in Figure 3.2. A comparison of Figures 3.5 and 3.6 shows that – at least for the model parameters considered – the entropy production term is small as compared with the character-

116 | *3 Generic Theory of Vitrification of Glass-Forming Melts*

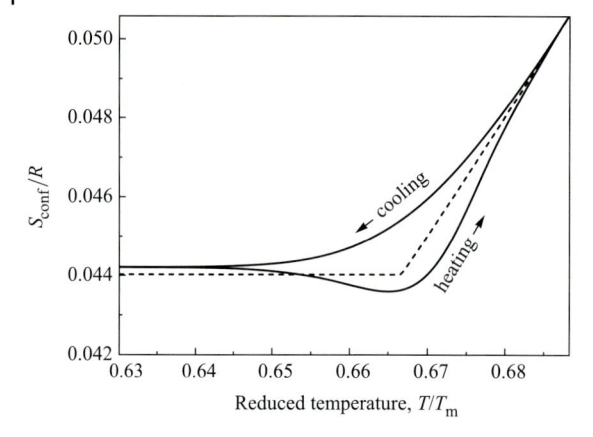

Figure 3.6 Configurational contribution to the entropy, S_{conf} (Eq. (3.47)), determined in accordance with Simon's model of vitrification (*dashed curve*) and the model of a continuous transition as employed here (*solid curve*). Note that in the latter case the entropy behaves differently in cooling and heating runs due to differences in the temperature dependencies of the structural order parameter, $\xi(p, T; q)$, for cooling and heating (cf. Figure 3.2).

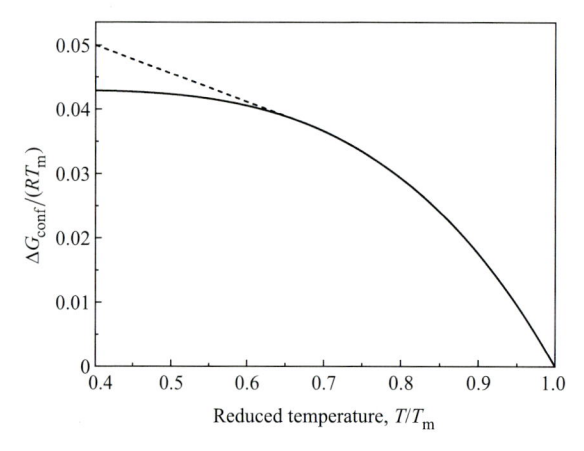

Figure 3.7 Configurational Gibbs' free energy difference, $\Delta G_{conf} = G_{conf}(T) - G_{conf}(T_m)$ (Eq. (3.49)), determined in accordance with Simon's model of vitrification and the model of a continuous transition as employed here. The *solid line* corresponds to the state of the metastable liquid. For the parameters considered and in the scale used for the presentation, the results of Simon's model and of the generic model of glass transition, employed for the analysis in the present contribution, coincide, they are given by a dashed curve (cf. also [240, 245, 246]).

istic values of the structural contributions to the entropy shown in Figure 3.5. This result is reflected also in Figure 3.7.

Finally, the specific heat along the path, given by $\xi = \xi(p, T; q)$, is a function of pressure, temperature and the order parameter $C_p = C_p(p, T, \xi)$. We have, by definition,

$$C_p(p, T, \xi) = \left.\frac{dQ}{dT}\right|_{p=\text{const}} . \tag{3.69}$$

From the first law of thermodynamics ($dU = dQ - pdV$, Eq. (3.9)) and the definition of the enthalpy, $H = U + pV$, we obtain

$$dH = dQ + Vdp . \tag{3.70}$$

This relation yields, with the condition $p = \text{const.}$

$$dH = C_p(p, T, \xi)dT . \tag{3.71}$$

In a more extended form, Eq. (3.71) can be written as the sum of two terms (cf. [1, 14, 234]):

$$C_p(p, T, \xi) = \frac{dH}{dT} = \left(\frac{\partial H}{\partial T}\right)_{p,\xi} + \left(\frac{\partial H}{\partial \xi}\right)_{p,T} \frac{d\xi}{dT} . \tag{3.72}$$

The first term on the right hand side of above equation reflects the contribution to the specific heat due to thermal motion of the molecules (e.g., the phonon part), while the second term refers to configurational contributions.

Employing the model of glass-forming melts discussed above, the configurational contribution to the specific heat, $C_{p,\text{conf}}$, is given with Eq. (3.45) by

$$C_{p,\text{conf}} = \left(\frac{\partial H_{\text{conf}}}{\partial \xi}\right)_{p,T} \frac{d\xi}{dT} = \chi_1 \Delta H_{\text{ev}}(T_m) \frac{d\xi}{dT} . \tag{3.73}$$

In the model of vitrification, developed by Simon, the glass-forming system remains in a metastable state until a certain temperature, T_g, is reached in the cooling process (i.e., $\xi = \xi_e$ for $T \geq T_g$ or $\theta \geq (2/3)$). At this temperature, the system becomes frozen-in and the structural order parameter does not change any more in the further cooling process (i.e., $(d\xi/dT) = 0$ for $T \leq T_g$ or $\theta \leq (2/3)$). The configurational contribution, $C_{p,\text{conf}}$, to the specific heat resulting from processes of structural reorganization, has, consequently, the form

$$C_{p,\text{conf}}^{(\text{Simon})} = \begin{cases} \chi_1 \Delta H_{\text{ev}}(T_m) \left(\frac{d\xi_e}{dT}\right) & \text{for } T_g \leq T \leq T_m \\ 0 & \text{for } 0 \leq T < T_g \end{cases} \tag{3.74}$$

with (cf. Eq. (3.50))

$$\frac{d\xi_e}{dT} = \chi \left(\frac{T_m}{T^2}\right) \frac{\xi_e(1-\xi_e)^3}{1 - \xi_e + 2\xi_e \ln \xi_e} \tag{3.75}$$

or

$$\frac{d\xi_e}{d\theta} = \frac{1}{\chi} \frac{\xi_e(\ln \xi_e)^2}{(1 - \xi_e + 2\xi_e \ln \xi_e)(1 - \xi_e)} . \tag{3.76}$$

3 Generic Theory of Vitrification of Glass-Forming Melts

In contrast, considering vitrification as a continuous transition from a metastable to a frozen-in state, we have generally

$$C_{p,\text{conf}} = \chi_1 \Delta H_{\text{ev}}(T_m) \frac{d\xi}{dT} \cong \chi R T_m \frac{d\xi}{dT}, \quad T_0 \leq T \leq T_m. \quad (3.77)$$

The configurational specific heat is proportional (at least, for the model employed) to the derivative of the structural order parameter with respect to temperature. Due to this correlation, as mentioned earlier in this chapter, the definition of T_g is equivalent to the point of inflection of the configurational specific heat.

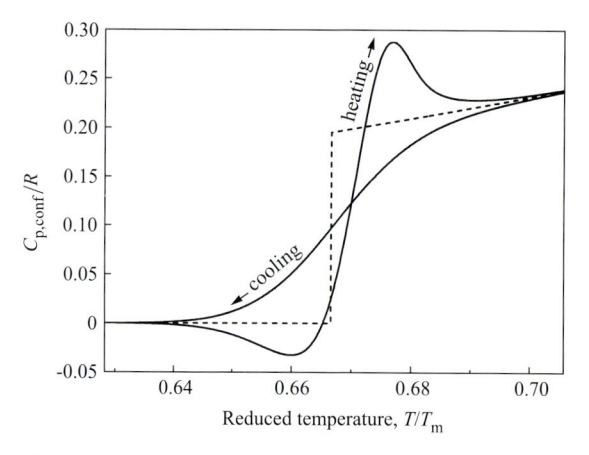

Figure 3.8 Configurational contribution, $C_{p,\text{conf}}(p, T, \xi)$, to the specific heat, $C_p = C_p(p, T, \xi)$, as obtained via Simon's model approach (*dashed curve*) and the model of a continuous transition as employed here (*solid curve*). Note that, in describing vitrification more appropriately in terms of a continuous transition, the specific heats turn out to be different for cooling and heating runs (cf. experimental data shown in Figure 3.9).

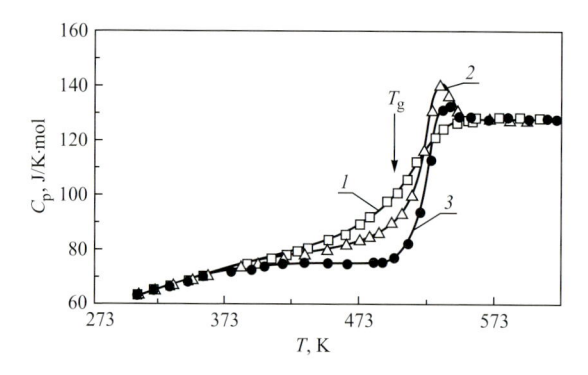

Figure 3.9 Experimental heat capacity curves of B_2O_3-melts upon vitrification after Thomas and Parks [211] (see also [247]): (1) cooling run curve, (2) heating curve after slow cooling run, (3) heating curve after fast cooling run. Note that in this figure the full specific heat is shown (cf. Eq. (3.72)) and not only the configurational contribution presented as in Figure 3.8. In a more schematic form, this figure (Figure 2.42) is shown in Chapter 2.

Moreover, the configurational contribution to the specific heat becomes also different in dependence on whether we consider cooling or heating processes. Results of the theoretical computations of the configurational contributions to the specific heats are shown in Figure 3.8. In Figure 3.9, experimental results are presented, giving a confirmation of the theoretical considerations outlined here. It is evident that experimental data for the heat capacity of glass-forming systems in cooling and heating behave qualitatively in the same way as shown in Figure 3.8 representing

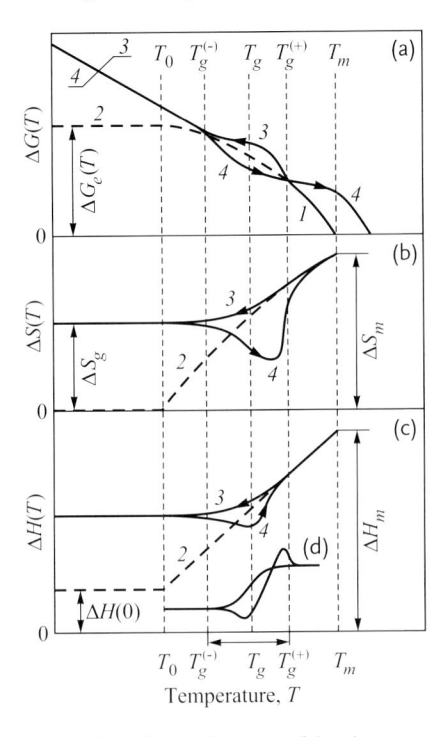

Figure 3.10 Thermodynamics of the glass transition in terms of the generic theory of vitrification [11, 17, 179]. (a) Temperature dependence of the configurational part of Gibbs' thermodynamic potential difference, $\Delta G(T)$, in terms of the differences liquid/crystal or glass/crystal: (1) equilibrium course ΔG_e in cooling in the range from $T = T_m$ to temperatures tending to zero; (2) $\Delta G(T)$ course upon vitrification in a cooling run; (3) $\Delta G(T)$ course in heating from zero temperatures to temperatures near T_g; (4) $\Delta G_g(T, \xi)$ of the vitrified melt representing the tangent to $\Delta G_e(T)$ of the undercooled melt. T_m, T_g and T_0 are melting, glass transition and Kauzmann temperatures, respectively. The glass transition interval is determined by the steep change of the specific heat, ΔC_p, and is bounded to the range $T_g^- \leq T \leq T_g^{(+)}$. With $\Delta G(0)$, the extrapolated zero-point potential difference of the metastable liquid is specified; (b) temperature dependence of the entropy difference, $\Delta S(T)$, liquid/crystal (2) and of glass/crystal in cooling (3) and heating (4). ΔS_m is the entropy of melting and ΔS_g is the frozen-in entropy of the glass; (c) change of the enthalpy difference, ΔH, liquid/crystal (3, 4). $\Delta H(0)$ is the zero-point enthalpy corresponding to the metastable undercooled liquid; (d) liquid/crystal or glass/crystal specific heat differences. In cooling, ΔC_p decreases monotonically with decreasing temperature, and in heating a minimum and a maximum is observed (cf. Figure 3.9 and [12]).

the results of theoretical computations. This coincidence is considered by us as one of the straightforward verifications of the theoretical approach employed here.

The thermodynamics of the glass transition in terms of the generic theory of vitrification is illustrated in more detail in Figure 3.10. Here the configurational contributions to several thermodynamic functions are shown. All results are given schematically, computed with typical values of the respective constants (for details see [11, 17, 179]).

The course of the functions $\Delta S(T, \xi)$ and $\Delta G(T, \xi)$ determined theoretically is verified by experimental data (cf. [185, 248]). In Figure 3.11, the temperature course of $\Delta S(T)$ is given for a typical glass-forming system (As_2Se_3) as it is calculated in two ways. For one case this procedure is performed employing $\Delta C_p(T)$ measurements by Moynihan et al. [250]. For another case this is done using measurements of the vapor pressure of As_2Se_3-glass by Kinoshita upon cooling and heating runs

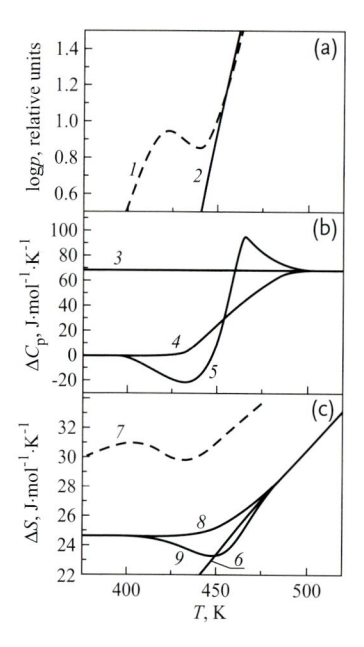

Figure 3.11 Coincidence of theory and experiment in the determination of $\Delta G(T, \xi)$ and $\Delta S(T, \xi)$ as shown in Figure 3.10 and employing vapor pressure, $p(T, \xi)$, and specific heat, $\Delta C_p(T, \xi)$ data reported by Kinoshita [249] and Schnaus et al. [250] for As_2Se_3 for the liquid, glassy and crystalline forms of this substance (according to [185, 248]). (a) Temperature dependence of the vapor pressure of As_2Se_3 at heating (curve 1) and very slow cooling (curve 2) of the melt; (b) temperature dependence of the specific heat difference, ΔC_p, of As_2Se_3 for liquid/crystal at equilibrium (curve 3), for cooling (4) and heating (5) of the undercooled liquid/crystal or the glass; (c) computed values of the entropy difference, $\Delta S(T)$: undercooled liquid/crystal (curve 6 obtained from curve 3); glass-liquid/crystal (curve 7 obtained from the experimental $p = p(T)$ curve (1)); curve 8 obtained from curve 4 and curve 9 from curve 5 (from calorimetric data). Note the similar $\Delta S(T)$ course of curve 7 obtained from tensimetric measurements and of curve 9 following from calorimetric measurements on samples with similar thermal history (cf. also [251]).

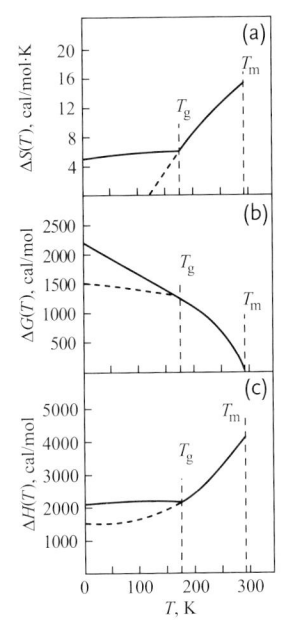

Figure 3.12 (a–c) Temperature dependence of the thermodynamic functions of glycerol computed by Simon employing his model of the glass transition [1, 23]. For the computations performed by him, Simon employed $C_p(T)$ data measured by several authors and his own measurements.

(see [249]). As far as

$$\ln\left(\frac{p_{\text{glass}}}{p_{\text{cryst}}}\right) = \frac{\Delta G(T, \xi)}{RT} \qquad (3.78)$$

holds, from such measurements the value of $\Delta G(T, \xi)$ can be determined. It is seen that in both cases a similar $\Delta S(T)$ course is obtained. Tensimetric measurements directly confirm thus the peculiar "humps" seen in the temperature course of $\Delta G(T)$ (cf. Figure 3.11a) as they follow from the integration of the $\Delta S(T)$ and $\Delta H(T)$ curves.

The course of $\Delta G(T)$, $\Delta H(T)$ and $\Delta S(T)$, constructed applying the model developed by Simon, is shown in Figure 3.12 (see [1]). Its comparison with Figure 3.10 shows the new features introduced by the generic model of relaxation and glass transition in a way corresponding to experiment.

3.4.5.2 Some Comments on the Value of the Configurational Entropy at Low Temperatures and on the Kauzmann Paradox

In the framework of the model analyzed, the configurational entropy is, according to Eq. (3.47), uniquely determined by the order parameter, ξ. Since the relation

$$\frac{\partial S_{\text{conf}}}{\partial \xi} = -R\frac{\ln \xi}{(1-\xi)^2} > 0 \qquad (3.79)$$

holds generally, we get

$$\begin{aligned} S_{\text{conf}} &\geq S_{\text{conf}}^{(e)} \quad \text{for} \quad \xi \geq \xi_e, \\ S_{\text{conf}} &\leq S_{\text{conf}}^{(e)} \quad \text{for} \quad \xi \leq \xi_e. \end{aligned} \qquad (3.80)$$

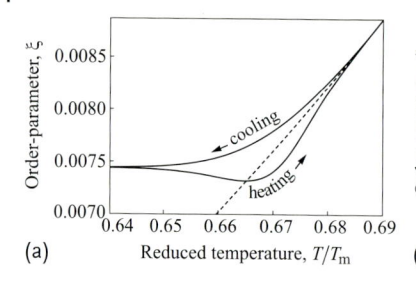

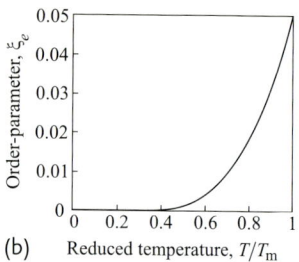

Figure 3.13 (a) Typical behavior of the structural order parameter, ξ, in dependence on temperature in the vicinity of the glass transition range; (b) temperature dependence of the equilibrium value of the structural order parameter as obtained in the framework of the lattice model employed here.

Here $S_{\text{conf}}^{(e)}$ denotes the configurational contribution to the entropy referring to $\xi = \xi_e$. For $T \to 0$, the relations $\xi_e \to 0$ and $S_{\text{conf}}^{(e)} \to 0$ are fulfilled for the (hypothetical) extension of the metastable liquid to zero temperatures, while for the frozen-in liquid, the glass, a configurational contribution $S_{\text{conf}}(T \to 0) > 0$ is obtained (cf. Figure 3.13a). The configurational contribution $S_{\text{conf}}(0)$ is identical to the respective value frozen-in in the vitrification interval, it depends on the rate of cooling, that is, on the value of the structural order parameter, ξ, frozen-in in the vitrification interval. Based on the model employed we arrive – in contrast to [252–254] – at the "conventional" classical conclusion [1, 237] that glasses and not only glasses but also disordered crystals and similar systems [1] do have a nonzero configurational entropy at absolute zero. This point of view has been reconfirmed also in a recent studies of these problems by Goldstein [255] and Johari [256, 257]. A comprehensive analysis of experimental data confirming the results of our analysis (the traditional or "conventional" point of view) is given in [182] and here in Chapters 5, 8 and 9 (for some recent investigations leading to similar results see also [243, 258]).

Note that only in terms of Simon's model are the following relations

$$\Delta H(T_g) = 0, \quad \Delta S(T_g) = 0, \quad \Delta V(T_g) = 0, \tag{3.81}$$

$$\Delta G(T_g) = 0, \quad \Delta U(T_g) = 0$$

(Δ denotes the difference between glass and liquid) fulfilled. Treating vitrification and devitrification in terms of the generic approach, the respective differences are, in general, not equal to zero since, in general, $\xi \neq \xi_e$ holds at T_g or, more generally, in the vitrification interval and below it (cf. [252, 254]).

Note as well the following aspect: a linear extrapolation of the temperature dependence of the structural order parameter – as indicated by a dashed curve in Figure 3.13 – to temperatures below the vitrification range would lead to an intersection with the abscissa at some finite temperature T^*. In contrast, the correct (for the considered model) dependence $\xi_e(T)$, given by Eq. (3.50), does not cross the abscissa but approaches it from above at $T \to 0$. Quite interestingly, the temperature change of the structural order parameter can be divided into two regions, a

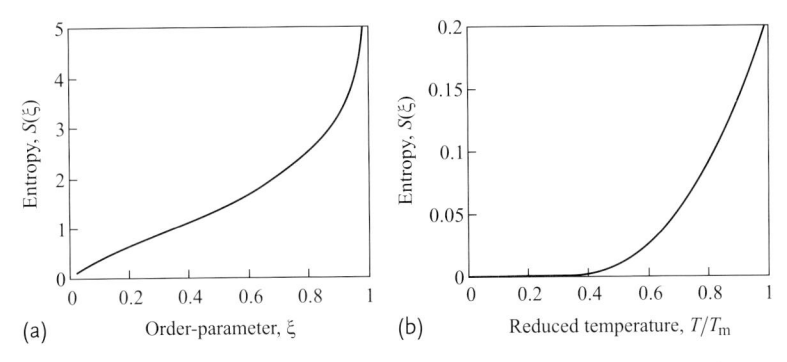

Figure 3.14 Configurational contributions to the entropy of the metastable glass-forming melt resulting from a combination of (a) Eq. (3.47) and (b) Eq. (3.50).

high-temperature one, where the structural order parameter changes significantly with the change of temperature and a low-temperature region, where the respective changes are of minor relevance. The behavior of the structural order parameter is reflected also in a similar kind of behavior of the configurational entropy (cf. Figure 3.14). In this way, employing not the simplified linear extrapolations but the correct (for the considered model) result, Kauzmann type paradoxes [109] do not occur here.

Moreover, we find here a deep analogy in the consequences to the statistical model considerations performed by Gibbs and Di Marzio [259, 260]. Based on a statistical model analysis, they showed that the entropy of the liquid approaches the entropy curve of the crystal at a temperature T_2. At lower temperatures both quantities coincide more or less. It was assumed first that there is some sharp change at T_2, this requirement has been later formulated less strictly [261]. As evident from Figures 3.13 and 3.14, this particular temperature T_2 is equal in our model to $T_2 \cong 0.4 T_m$. Both models show that there are very specific states at low temperatures where changes of temperature (and pressure) lead only to such small changes of the state of the liquid that they can be widely ignored (cf. also Mazurin [70]).

3.4.6
Cyclic Heating-Cooling Processes: General Results

So far, we have determined the thermodynamic functions based on the knowledge of the temperature dependence of the structural order parameter, $\xi = \xi(p, T; q)$, employing results of statistical-mechanical model considerations. Alternatively, one can determine the thermodynamic functions based on measurements of the specific heats at a given rate of cooling or heating. For example, the enthalpy can be determined then via Eq. (3.71). Similarly, we obtain for the entropy (cf. Eq. (3.6))

$$C_p(p, T, \xi) = \frac{dQ}{dT} = T\frac{d_e S}{dT} = T\frac{dS}{dT} - T\frac{d_i S}{dT} \qquad (3.82)$$

$$\Rightarrow \quad T\frac{dS}{dT} = C_p(p, T, \xi) + T\frac{d_i S}{dT}.$$

Furthermore, employing the heating and cooling law, $dt = dT/q$, along with Eq. (3.35) we arrive at:

$$dS = \frac{C_p(p, T, \xi)}{T} dT + \frac{G_e^{(2)}}{Tq\tau}(\xi - \xi_e)^2 dT . \tag{3.83}$$

Remember that the structural order parameter, ξ, can be expressed in all equations given above, Eqs. (3.69)–(3.83), uniquely via pressure, temperature and cooling, respectively, and heating rates. The respective dependencies differ, in general, for cooling and heating (i.e., for negative and positive values of the heating rate, q). Consequently, the thermodynamic functions are different in cooling and heating as well. Having at one's disposal the dependence $S = S(p, T, \xi)$ or (with Eq. (3.55)) $S = S(p, T, q)$, one can immediately compute also other thermodynamic parameters like the change of the volume of the system with respect to temperature via

$$\left(\frac{\partial V}{\partial T}\right)_{p,\xi} = -\left(\frac{\partial S}{\partial p}\right)_{T,\xi} . \tag{3.84}$$

Completing the analysis, we finally derive some integral characteristics of the cyclic processes analyzed which do not depend on particular properties of the systems under consideration but are consequences from the basic laws of thermodynamics (cf. also [240]). Since we start at and return to equilibrium states, the thermodynamic state functions like entropy, enthalpy, and Gibbs' free energy recover after the completion of the cyclic process the same values they had at the starting point. In this way, we may write, for example,

$$\oint dS = \oint d_e S + \oint d_i S = 0 . \tag{3.85}$$

Since

$$\oint d_i S \geq 0 \tag{3.86}$$

holds, we get

$$\oint d_e S = \oint \frac{dQ}{T} = \oint \frac{C_p(p, T, \xi)}{T} dT \leq 0 . \tag{3.87}$$

Note that the specific heat depends not only on (the constant) pressure and (the changing in time) temperature but also on the structural order parameter, ξ. Only for reversible cyclic processes, the equality sign in Eq. (3.87) holds, but this is not the case in vitrification, where cooling and heating is accompanied, in general, by nonequilibrium relaxation and entropy production processes.

Similarly, Eq. (3.70) yields (at assumed constancy of pressure)

$$\oint dH = \oint dQ = \oint C_p(p, T, \xi) dT = 0 . \tag{3.88}$$

Furthermore, from Eq. (3.13) we obtain, with Eq. (3.10) and assuming, again, constancy of the pressure, the relation

$$-\oint dG = \oint S\,dT + \oint A\,d\xi = \oint S\,dT + \oint T\,d_i S = 0. \tag{3.89}$$

Since $\oint T d_i S \geq 0$ holds, we arrive at

$$\oint S\,dT \leq 0. \tag{3.90}$$

Note that all relations derived in this section are valid independent on any particular assumptions about cooling or heating rates employed in the realization of the cyclic processes discussed.

3.5
The Prigogine–Defay Ratio

3.5.1
Introduction

One of the basic characteristics of the glass transition is the so-called Prigogine–Defay ratio, Π. It gives a general correlation between the jumps of compressibility, $\Delta\kappa$, thermal expansion coefficient, $\Delta\alpha$, and isobaric heat capacity, ΔC_p, near the glass transition temperature, $T = T_g$, and is defined as:

$$\frac{1}{VT}\left\{\frac{\Delta C_p \Delta \kappa}{(\Delta\alpha)^2}\right\}\bigg|_{T=T_g} = \Pi. \tag{3.91}$$

Here V is the volume of the system under consideration. A similar relation is of significance in the description of second-order equilibrium phase transitions, where it is denoted as Ehrenfest's ratio (cf. Eq. (2.65)).

In the existing theoretical analysis, the Prigogine–Defay ratio is treated commonly in two ways (cf. [1] for an overview), considering vitrification either as a particular form of a second-order equilibrium phase transition or based on the concept of vitrification as developed by Simon. According to Simon [59], in the process of undercooling a liquid below the melting or liquidus temperature, T_m, the system may remain in a state of metastable thermodynamic equilibrium until, at a certain temperature T_g the glass transition temperature, the system becomes suddenly kinetically frozen-in. At and below the glass transition temperature, the process of structural reorganization cannot follow any more the change of temperature and the undercooled liquid system is transformed into a glass. Consequently, according to the concept of vitrification developed by Simon, in vitrification a state of (metastable) thermodynamic equilibrium is suddenly transformed into a frozen-in thermodynamically nonequilibrium state of the liquid under consideration. In the present analysis, we will consider mainly temperature as the external control parameter, however, having in mind that similar processes with similar consequences may occur also in response of the systems under consideration to changes of a variety of other external control parameters like, for example, pressure.

As a consequence of vitrification, the mode of reaction of the system is changed qualitatively and any thermodynamic coefficients exhibit a step-like (or more precisely, an s-shaped) dependence on temperature or other appropriate control parameters in the glass transition region. The situation is here similar to some extent (but, as we will see, not identical) to second-order equilibrium phase transitions in the classical Ehrenfest sense [1, 54, 62, 63]. In both cases (second-order equilibrium phase transitions on one side and glass transition on the other side), the response is changed qualitatively while the state of the system (structure and thermodynamic functions) remains the same at the transition point. However, in one case, the system goes over from one equilibrium state into the other (second-order equilibrium phase transitions), while in the glass transition the system is transferred from a metastable equilibrium into a nonequilibrium state, which is the glass.

This qualitative difference is reflected in a particular way in the so-called Prigogine–Defay (in vitrification) and Ehrenfest (in second-order equilibrium phase transition) ratios. Both formally identical relations give a general correlation between the jumps of compressibility, thermal expansion coefficient and isobaric heat capacity in vitrification, respectively, in second-order equilibrium phase transitions. The respective ratio is commonly denoted as Π. For the Ehrenfest ratio, describing second-order equilibrium phase transitions, the value $\Pi = 1$ is theoretically predicted and experimentally verified [54, 62, 63]. For the Prigogine–Defay ratio, in contrast, commonly values $\Pi > 1$ are found experimentally [1]. This discrepancy in the values of Π is a verification of the point of view that vitrification cannot be treated as a second-order equilibrium phase transition.

Employing the structural order parameter concept by De Donder [13] to describe in an unified form both thermodynamic equilibrium and nonequilibrium states of glass-forming systems, attempts have been developed to derive the value of the Prigogine–Defay ratio treating vitrification as a freezing-in process in accordance with Simon's model. A first attempt in this direction was performed by Prigogine and Defay [14] and repeated in several modifications by Davies and Jones [19, 20], Moynihan, Gupta [199, 236], Kovacs [262], Nemilov [118, 237] and others (see also [1]). In such treatment, employing one structural order parameter for the description of the system under consideration, again, a value $\Pi = 1$ is theoretically obtained in contrast to above-mentioned experimental findings. This result led some authors to the conclusion that more than one order parameter must be introduced into the description to obtain a theoretical result for the Prigogine–Defay ratio in agreement with experimental data (cf. the discussion in [1]). Here we will show that there exists an alternative straightforward solution of this problem treating vitrification in a more correct way as done by utilizing Simon's simplified model approach (cf. also [234, 239, 245, 246]).

Indeed, a detailed experimental and theoretical analysis shows – as discussed in detail in the previous sections of the present chapter – that vitrification does not proceed at some sharp temperature, T_g, the glass-transition temperature but in a certain temperature interval, $(T_g^- \leq T \leq T_g^+)$ [11, 233, 236, 240]. In this temperature range, the characteristic times of change of the external control parameters like temperature are comparable in magnitude with the characteristic relaxation times of the system to the respective metastable equilibrium states. For this reason, the glass-forming system is not in (metastable) equilibrium but in a nonequilibrium state, however, not kinetically frozen-in fully so far. For this reason, processes of structural relaxation may occur with finite values of the affinity of the respective process of structural change. As will be shown in the present analysis, such an accurate description of the process of vitrification is sufficient for an adequate interpretation of experimental results on the value of the Prigogine–Defay ratio even when only one order parameter is employed describing the structural or configurational disorder of the vitrifying melt. In this way, in addition to the theoretical determination of the Prigogine–Defay ratio, the results are expected to stimulate, hopefully, the resolution of some controversies in the continuing discussion on the nature of the glassy state and the glass transition and the appropriate methods of their theoretical description.

3.5.2 Derivation

3.5.2.1 General Results

For closed systems in thermodynamic equilibrium, only two of the three thermodynamic state parameters (pressure, p, volume, V, and temperature, T) are independent variables and the structural order parameter is an unambiguous function of these parameters independent of the prehistory of the system. For such equilibrium states, the thermodynamic coefficients (like thermal expansion coefficient, α, isothermal compressibility, κ, and isobaric molar heat capacity, C_p)

$$\alpha = \frac{1}{V}\left(\frac{\partial V}{\partial T}\right)_p, \quad C_p = \left(\frac{\partial H}{\partial T}\right)_p, \quad \kappa = -\frac{1}{V}\left(\frac{\partial V}{\partial p}\right)_T, \quad (3.92)$$

can be expressed, consequently, as functions of pressure, p, and temperature, T, exclusively. The above definitions retain their validity also in the more general case, when the system is in a thermodynamically nonequilibrium state characterized by one (or several) additional structural order parameter(s), ξ, being not uniquely defined by pressure and temperature alone but depending also on the prehistory. In application to vitrification, we have come across the following three cases: (i) the system is in a stable or metastable thermodynamic equilibrium state. Then, the order parameter is uniquely defined as a function of pressure and temperature, that is, $\xi = \xi_e(T, p)$, independent of the prehistory. (ii) The system is in a completely frozen-in, nonequilibrium state. Then we have to demand $\xi = $ constant. (iii) The system is cooled down with a given rate, $q = (dT/dt) < 0$. Then, we transfer the system into a glass along one of the possible trajectories of cooling giving rise to different dependencies, $\xi = \xi(T, p; q)$. Here $q = q(t)$ specifies the chosen cooling law (i.e., the solution of the equation $q = (dT/dt)$ with appropriately chosen initial conditions, for example, $T(t = 0) = T_\mathrm{m}$, where T_m is the melting temperature; note that q is not necessarily constant but may depend on time (or, equivalently, temperature)).

For example, if we assume that the system is cooled down with a constant cooling rate, $q = (dT/dt) = \mathrm{const.} < 0$, then Eq. (3.30) can be written in the form [11, 17, 170, 233, 236, 240, 262]:

$$\frac{d\xi}{dT} = -\frac{1}{q\tau}(\xi - \xi_\mathrm{e}) \,. \tag{3.93}$$

This equation (which holds also at much less restrictive conditions concerning cooling and heating rates, for example, as mentioned before or – in a modified version replacing T by p – if not temperature but pressure is changed) allows us to determine in a unified way the dependence of the structural order parameter, ξ, on temperature for equilibrium, nonequilibrium and frozen-in nonequilibrium states. Since the differential equation depends on the cooling rate, q, the solution has to depend, in general, also on the cooling rate resulting in the already mentioned dependence, $\xi = \xi(p, T; q)$.

A typical example of such dependence for cooling is illustrated in Figure 3.15 (cf. [11, 17, 233, 240]). For states near the melting temperature, the characteristic times of relaxation of the system to the respective equilibrium states are small as compared with the characteristic time of variation of the external control parameter, here the temperature. For this reason, the system follows here the equilibrium curve, $\xi = \xi_\mathrm{e}(p, T)$. With decreasing temperature, the relaxation times increase exponentially. As a result, the function $\xi = \xi(p, T; q)$ starts to deviate at some temperature $T = T_\mathrm{g}^+$ from the equilibrium curve. However, relaxation processes continue to proceed until, at $T = T_\mathrm{g}^-$, they become completely frozen-in with respect to time scales of the experiment. As a consequence, for temperatures below T_g^-, the structural order parameter retains a constant value. Both the characteristic temperatures, T_g^+, T_g, and T_g^- as well as the course of the curve, $\xi = \xi(p, T)$, depend quantitatively on the value of the cooling rate, however, qualitatively, we observe always a scenario similar to the one discussed above [245, 246]. For comparison, in Figure 3.15 the course of the function $\xi = \xi(p, T)$ according to the model of vitrification as proposed by Simon is also shown. In such an approximation, the system remains in a state of thermodynamic equilibrium (ξ is equal to its equilibrium value, ξ_e, that is, $\xi(T \geq T_\mathrm{g}) = \xi_\mathrm{e}(p, T)$) for temperatures down to the glass transition temperature, T_g. At $T = T_\mathrm{g}$, the system becomes suddenly completely frozen-in and the structural order parameter retains in further cooling constant values equal to $\xi(T \leq T_\mathrm{g}) = \xi_\mathrm{e}(p, T_\mathrm{g})$.

The generic treatment of vitrification as taking place in some temperature interval gives a more detailed and correct description of the real process. It contains Simon's model as a first approximation as the limiting case when both T_g^- and T_g^+ tend to the glass transition temperature, T_g. Moreover, the detailed treatment followed here leads to a number of additional important consequences for an adequate understanding of the glass transition. With respect to the analysis of the behavior of the thermodynamic coefficients and their ratio in vitrification, it is the

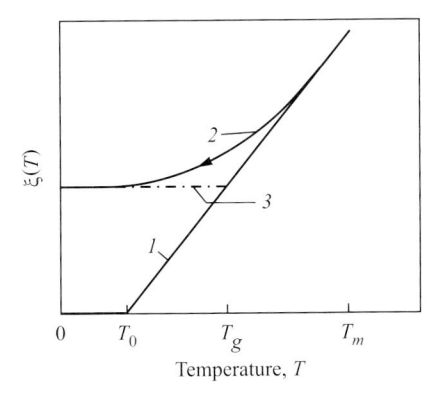

Figure 3.15 Qualitative illustration of possible dependencies of the order parameter, ξ, on temperature in cooling of glass-forming melts. Curve (1) corresponds here schematically to the dependence of the equilibrium value of the structural order parameter on temperature, that is, $\xi_e = \xi_e(p, T)$. In cooling a liquid, the order parameter cannot follow, in general, the change of the external parameter and deviates from the equilibrium value, ξ_e (curve 2), resulting in $\xi > \xi_e$. Such kind of behavior is found in the vicinity of the glass transition temperature when the time scales of change of the external parameter become comparable with the characteristic relaxation times of the system. According to Simon's classical model of vitrification, the order parameter is identical to ξ_e down to $T = T_g$ and at this temperature, the structure becomes suddenly frozen-in (curve 3).

conclusion that the affinity, A, in the glass transition interval (and, in particular, near and at T_g) is not equal to zero as supposed in Simon's simplified treatment. Indeed, Figure 3.15 shows that the structural order parameter, ξ, in the vitrification range is not equal to its equilibrium value, ξ_e. Taking into account Eq. (3.17), we arrive at the conclusion that $A(T_g) \neq 0$ holds (and for the so far considered cooling processes, more precisely, $A(T_g) < 0$ is fulfilled, cf. Eq. (3.22)).

For any of above-defined trajectories, $\xi = \xi(p, T; q)$, of vitrification (including Simon's simplified model), the thermodynamic coefficients can be expressed as a sum of two terms. For example, the change of the volume in dependence on pressure at constant temperature can be written in the form:

$$\left(\frac{\partial V}{\partial p}\right)_T = \left(\frac{\partial V}{\partial p}\right)_{T,\xi} + \left(\frac{\partial V}{\partial \xi}\right)_{p,T}\left(\frac{\partial \xi}{\partial p}\right)_T. \tag{3.94}$$

With the definition of the compressibility κ (Eq. (3.92)), this equation can be transformed to (cf. [1, 14])

$$\kappa = \kappa_\xi - \frac{v_{p,T}^2}{V a_{p,T}}. \tag{3.95}$$

In this equation, the compressibility of the glass below the glass transition interval is given by κ_ξ (describing changes of the volume in dependence on pressure at frozen-in, that is, fixed values of the structural order parameter) and the compressibility of the metastable melt in thermodynamic equilibrium by κ (including changes of the structure due to variations of pressure). The step-like change in the compressibility in glass transition can be expressed therefore as:

$$\Delta \kappa = \kappa - \kappa_\xi = -\frac{1}{V}\left(\frac{v_{p,T}^2}{a_{p,T}}\right) > 0. \tag{3.96}$$

The inequality sign in Eq. (3.96) is a consequence of the inequality $a_{p,T} < 0$ (cf. Eq. (3.20)). It follows that the compressibility of the glass is lower as compared with the metastable liquid.

Similarly to Eq. (3.94), we can express the isobaric heat capacity of the undercooled metastable melt as

$$C_p = \left(\frac{\partial H}{\partial T}\right)_p = C_{p,\xi} + \left(\frac{\partial H}{\partial \xi}\right)_{p,T}\left(\frac{d\xi}{dT}\right)_p \tag{3.97}$$

or

$$C_p = C_{p,\xi} - \frac{(A + h_{p,T})h_{p,T}}{T a_{p,T}}. \tag{3.98}$$

For the thermal expansion coefficient (defined by Eq. (3.92)), the analogous expression

$$\alpha = \frac{1}{V}\left(\frac{\partial V}{\partial T}\right)_p = \alpha_\xi - \frac{v_{p,T}(A + h_{p,T})}{V T a_{p,T}} \tag{3.99}$$

can be derived easily (cf. [1, 14]). The changes of the specific heat, $\Delta C_p = C_p - C_{p,\xi}$, and the thermal expansion coefficient, $\Delta \alpha = \alpha - \alpha_\xi$, upon vitrification can be expressed, consequently, similarly to Eq. (3.96) as:

$$\Delta C_p = -\frac{(A + h_{p,T})h_{p,T}}{T a_{p,T}} > 0, \quad \Delta \alpha = -\frac{v_{p,T}(A + h_{p,T})}{V T a_{p,T}} > 0. \tag{3.100}$$

In both cooling and heating runs, we have $\Delta C_p > 0$ and $\Delta \alpha > 0$ (cf. Eqs. (3.20), (3.22) and (3.23)) in agreement with experimental data (cf. [1]). By a combination of Eqs. (3.21), (3.96) and (3.100), we obtain the following expression for the Prigogine–Defay ratio:

$$\Pi(T_g) = \frac{1}{VT}\left\{\frac{\Delta C_p \Delta \kappa}{(\Delta \alpha)^2}\right\}\bigg|_{T=T_g} = \frac{h_{p,T}}{A + h_{p,T}}\bigg|_{T=T_g} = \frac{\left(\frac{\partial H}{\partial \xi}\right)_{p,T}}{T\left(\frac{\partial S}{\partial \xi}\right)_{p,T}}\bigg|_{T=T_g}. \tag{3.101}$$

Note that the derivation is widely independent from the value of temperature analyzed. However, in order to apply this result to vitrification, the differences in the thermodynamic coefficients in Eq. (3.101) have to be determined according to the approach employed for a temperature $T = T_g$ (a comprehensive derivation of this

3.5 The Prigogine–Defay Ratio

expression for the Prigogine–Defay ratio is given by us in [245, 246]). The details on how this result must be applied to the interpretation of experimental data on the Prigogine–Defay ratio will be discussed below.

The result of our analysis, expressed by Eq. (3.101), is similar in its form to Ehrenfest's equation describing the jumps of the thermodynamic coefficients in second-order equilibrium phase transitions (cf. [1, 54]). However, in latter case the parameter Π is always equal to one. In contrast, Eq. (3.101) leads to the result $\Pi(T_g) = 1$ only if Simon's simplified model of vitrification is employed (implying $A = 0$ at $T = T_g$ [1, 14]). Treating vitrification in terms of the generic kinetic model of relaxation and vitrification [11, 17, 233, 240] (employing the concepts discussed above which result in $A < 0$ for cooling processes in the glass transition interval), we obtain – for vitrification processes by cooling of glass-forming melts – the general theoretical estimate $\Pi(T_g) > 1$ (due to Eq. (3.22)).

So far, we have analyzed changes of the thermodynamic parameters in vitrification in the course of cooling of glass-forming melts. Variations of thermodynamic coefficients can be measured similarly for heating runs. In contrast to cooling, here one significant difference has to be accounted for. This difference is illustrated in Figure 3.16. In this figure, the dependence of the structural order parameter is shown not only for cooling (curve 2) but also for a process of heating (curve 3). The curve $\xi = \xi(p, T; q)$ has one minimum at the intersection with the equilibrium curve, $\xi = \xi_e(p, T)$ (cf. Eq. (3.93) and [240]) and approaches then the equilibrium curve from below. As a consequence, in heating runs the relations ($\xi \leq \xi_e$) and $A > 0$ (cf. Eq. (3.17)) hold in the vicinity of the glass temperature, T_g. Consequently, for heating runs we have to expect, according to Eq. (3.101), that the inequality $\Pi(T_g) < 1$ has to be fulfilled. The question then arises which one of the respective values (either for cooling or for heating) has to be employed in order to compare the theory with experimental data.

In order to compare the results of the theory with experiment, we have to remember what meaning is usually assigned to the changes in the thermodynamic coefficients in vitrification employed then in the computation of the Prigogine–Defay ratio. According to the theoretical approach utilized, the change of the thermodynamic coefficients in vitrification is determined basically by the change of the configurational contributions to the thermodynamic coefficients. As an example, in Figure 3.8 results of computation of the configurational specific heats are given for a model system [240] analyzed here briefly already earlier. The results, obtained by treating the process as proceeding in some given temperature interval (solid curves), are compared with the respective course if Simon's simplified model is used for the theoretical explanation of vitrification (dashed curve). The values of the changes in the thermodynamic coefficients employed in the determination of the Prigogine–Defay ratio based on experimental data ($\Delta C_p^{(ex)}$, $\Delta \kappa^{(ex)}$, $\Delta \alpha^{(ex)}$) are determined commonly replacing the real course by a simplified model curve in terms of Simon's model of vitrification (cf. dashed curve in Figure 3.8). This dashed curve is different from the real course and and the real course differs for heating and cooling runs. Consequently, we obtain different values of the changes of the thermodynamic coefficients and the Prigogine–Defay ratio in cooling and

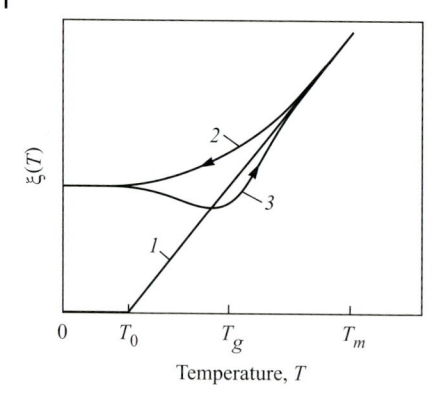

Figure 3.16 Qualitative illustration of possible dependencies of the order parameter, ξ, on temperature both in cooling (curve 2) and heating runs (curve 3). In addition, by curve 1 the equilibrium value of the structural order parameter in dependence on temperature is shown schematically, again. At the points of intersection of curves 1 and 3, curve 3 has a minimum and approaches with increasing temperature curve 1 from below. Consequently, for heating processes the inequality $\xi < \xi_e$ holds in the vicinity of the glass transition temperature.

heating, when the respective values of the thermodynamic coefficients at $T = T_g$ are substituted into Eq. (3.101).

In order to interpret the experimental data for the Prigogine–Defay ratio in the commonly discussed form, we have to supplement the analysis by an additional consideration based on the results presented in Figure 3.8. Figure 3.8 shows (for the considered example of the configurational specific heat) that, in cooling processes, the real course for the configurational contributions to the thermodynamic coefficients is always of s-shaped form [240]. These theoretical results are confirmed by a huge variety of experimental observations. Consequently, the changes of the configurational contributions to the thermodynamic coefficients at $T = T_g$, as derived in the model of a continuous transition to the vitreous state for cooling runs, are by a factor of the order $\omega \cong 1/2$ smaller than the values as determined experimentally employing above-described procedure based on Simon's model. Making the replacement in Eq. (3.101) of $\Delta C_p(T_g) = \omega \Delta C_p^{(ex)}$, $\Delta \kappa(T_g) = \omega \Delta \kappa^{(ex)}$, and $\Delta \alpha(T_g) = \omega \Delta \alpha^{(ex)}$, the value of Π is not changed. Consequently, the experimentally determined values of the Prigogine–Defay ratio, Π,

$$\Pi = \frac{1}{V(T_g) T_g} \left\{ \frac{\Delta C_p^{(ex)} \Delta \kappa^{(ex)}}{[\Delta \alpha^{(ex)}]^2} \right\} \geq 1 \qquad (3.102)$$

are given also by Eq. (3.101) if in the determination of $\Delta C_p(T_g)$, $\Delta \kappa(T_g)$, and $\Delta \alpha(T_g)$ the respective theoretical values for the cooling processes are used.

Summarizing this part, we conclude the following: the values of the Prigogine–Defay ratio in vitrification obey generally the inequalities $\Pi(T_g) > 1$ for cooling and $\Pi(T_g) < 1$ for heating processes. In order to compare the theoretical results with experimental findings, however, the estimates obtained for cooling processes have to be utilized in order to reflect appropriately the procedure of determination of Π as employed in the analysis of experimental data. Consequently, for the Prigogine–Defay ratio determined experimentally in the described form, we always get $\Pi > 1$ (cf. Eq. (3.102)).

A detailed comparison of the theoretical results with available experimental data is given in Section 3.5.3. However, before going over to such a comparison, in Section 3.5.2.2 we discuss briefly the basic consequences one may derive with respect to the theoretical estimates of the ratio Π as expressed via Eq. (3.101) if the lattice-hole model is employed for the description as discussed in Section 3.3.2.

3.5.2.2 Quantitative Estimates

In order to have at our disposal theoretical estimates of the value of the partial derivatives of enthalpy and entropy with respect to the order parameter, here we will employ a classical lattice-hole model for the description of glass-forming systems discussed earlier in this chapter. According to this model approach, the configurational or mixing contribution to the entropy (referred to one mole), S_{conf}, as compared with the state with ($N_0 = \xi = 0$) is given in this model by the conventional mixing term, Eq. (3.47). In a linear approximation with respect to the order parameter, ξ, we obtain

$$S_{conf}(\xi) - S_{conf}(\xi = 0) \cong \frac{5}{3} R \xi \ . \tag{3.103}$$

According to the same model considerations, the configurational contribution to the enthalpy, H_{conf}, reads (cf. Eqs. (3.45) and (3.46))

$$H_{conf} \cong \chi_1 \Delta H_{ev}(T_m) \xi \ , \tag{3.104}$$

and the partial derivative $(\partial H_{conf}/\partial \xi)_{p,T}$ is proportional to the molar heat of evaporation of the liquid under consideration, $\Delta H_{ev}(T_m)$, at the melting temperature, T_m. Employing Eq. (3.103) for the determination of the partial derivative, $(\partial S_{conf}/\partial \xi)_{p,T}$, we get as an estimate for the Prigogine–Defay ratio the result

$$\Pi \cong \left(\frac{3\chi_1}{5}\right) \frac{\Delta H_{ev}(T_m)}{R T_g} \ . \tag{3.105}$$

Utilizing further Trouton's rule (cf. Figure 3.1)

$$\Delta H_{ev}(T_m) \cong \chi_2 R T_m \quad \text{with} \quad \chi_2 \cong 20 \ , \tag{3.106}$$

we arrive, finally, at

$$\Pi \cong \left(\frac{3\chi}{5}\right) \frac{T_m}{T_g} \quad \text{with} \quad \chi = \chi_1 \chi_2 \ . \tag{3.107}$$

From a more general point of view, we can derive another estimate for the value of the Prigogine–Defay ratio in vitrification. The analysis of a variety of alternative model approaches to the description of glass-forming melts shows that generally the relations

$$\left(\frac{\partial H}{\partial \xi}\right)_{p,T} \propto \Delta H_{ev}(T_m) \, , \quad \left(\frac{\partial S}{\partial \xi}\right)_{p,T} \propto \Delta S_m(T_m) \qquad (3.108)$$

are fulfilled in a good approximation. For this reason, we arrive with Eq. (3.108) at a widely model-independent expression for the Prigogine–Defay ratio of the form:

$$\Pi \propto \frac{1}{T_g}\left(\frac{\Delta H_{ev}(T_m)}{\Delta S_m(T_m)}\right). \qquad (3.109)$$

Connecting the melting entropy with the complexity of the liquid under consideration and the evaporation enthalpy with the bonding strength, we may conclude that the Prigogine–Defay ratio is determined by the interplay of bonding strength and structural complexity of the glass-forming melt. With Eq. (3.104) and $T_g \cong (2/3)T_m$, we get then

$$\Pi \propto \frac{30R}{\Delta S_m}. \qquad (3.110)$$

The two methods given above that estimate the ratio Π can serve as a guide in order to understand possible values of the magnitude of the Prigogine–Defay ratio, $\Pi(T_g)$, determined via Eq. (3.101). However, these model considerations have one common significant deficiency, they do not account in the given form for the differences in the value of the Prigogine–Defay ratio on whether we are considering cooling or heating runs and at which rate the temperature is changed. This gap is connected – in the framework of the simplified model discussed – with the application of the linearized approximation for the excess entropy, Eq. (3.103), instead of the original expression, Eq. (3.47). Equations (3.47), (3.101), and (3.104) yield, without any further approximations,

$$\Pi(T_g) = \left.\frac{\left(\frac{\partial H}{\partial \xi}\right)_{p,T}}{T\left(\frac{\partial S}{\partial \xi}\right)_{p,T}}\right|_{T=T_g} = \left.\frac{\chi_1 \Delta H_{ev}}{-RT\frac{\ln \xi}{(1-\xi)^2}}\right|_{T=T_g} . \qquad (3.111)$$

A Taylor expansion of the denominator in Eq. (3.111) in the vicinity of $\xi = \xi_e$ results in

$$\frac{\ln \xi}{(1-\xi)^2} \cong \frac{\ln \xi_e}{(1-\xi_e)^2} + \frac{G_e^{(2)}}{RT}(\xi - \xi_e) ,$$

$$\frac{G_e^{(2)}}{RT} = \left[\frac{1}{\xi(1-\xi)^2} + \frac{2\ln \xi}{(1-\xi)^3}\right]_{\xi=\xi_e} > 0 . \qquad (3.112)$$

Taking into account Eqs. (3.104), (3.106) and (3.50), Eqs. (3.111) and (3.112) yield

$$\Pi(T_g) = \left.\frac{1}{1 - \left(\frac{G_e^{(2)}}{\chi RT_m}\right)(\xi - \xi_e)}\right|_{T=T_g} \cong \left.\frac{1}{1 - \frac{\theta}{\chi}\left(\frac{\xi-\xi_e}{\xi_e}\right)}\right|_{T=T_g} . \qquad (3.113)$$

For $\xi = \xi_e$, the affinity is equal to zero and the Prigogine–Defay ratio has to be equal to one. For processes of cooling, the inequalities ($\xi > \xi_e$) and $\Pi(T_g) > 1$ hold, while for heating we have ($\xi < \xi_e$) and $\Pi(T_g) < 1$. The actual value of $\Pi(T_g)$ depends hereby on the value of the order parameter, $\xi = \xi(p, T; q)$, near T_g and is determined consequently by the heating and cooling routes used to prepare the samples for which the values of the thermodynamic coefficients are measured. In this way, the deviation of the order parameter from its equilibrium value near T_g determines uniquely the deviation of the Prigogine–Defay ratio, $\Pi(T_g)$ (determined via Eq. (3.101)), from the value $\Pi = 1$.

As discussed here in detail earlier, the experimentally determined values of the Prigogine–Defay ratio, Π (cf. Eq. (3.102)), are obtained from the theoretical result, Eq. (3.101), for the case considering cooling processes. The model discussed above approaches the determination of this ratio and contains implicitly an estimate of the typical values of the difference ($\xi(T_g) - \xi_e(T_g)$) in cooling processes. Provided that this difference is larger than the typical values, then also large values of Π are to be expected. Vice versa, knowing the value of Π, the difference ($\xi(T_g) - \xi_e(T_g)$) can be uniquely determined based on Eq. (3.113). Results of computations of the Prigogine–Defay ratio in dependence on the cooling rate are given in [245, 246]. It is shown there that the Prigogine–Defay ratio increases with increasing cooling rate.

3.5.2.3 An Alternative Approach: Jumps of the Thermodynamic Coefficients in Vitrification

Equations (3.114)–(3.116)

$$\Delta \kappa = \kappa - \kappa_\xi = -\frac{1}{V}\left(\frac{\partial V}{\partial \xi}\right)_{p,T}\left(\frac{\partial \xi}{\partial p}\right)_T, \tag{3.114}$$

$$\Delta C_p = C_p - C_{p,\xi} = \left(\frac{\partial H}{\partial \xi}\right)_{p,T}\left(\frac{\partial \xi}{\partial T}\right)_p, \tag{3.115}$$

$$\Delta \alpha = \alpha - \alpha_\xi = \frac{1}{V}\left(\frac{\partial V}{\partial \xi}\right)_{p,T}\left(\frac{\partial \xi}{\partial T}\right)_p \tag{3.116}$$

supply us with the configurational contributions to the thermodynamic coefficients

$$\alpha = \frac{1}{V}\left(\frac{\partial V}{\partial T}\right)_p, \quad C_p = \left(\frac{\partial H}{\partial T}\right)_p, \quad \kappa = -\frac{1}{V}\left(\frac{\partial V}{\partial p}\right)_T, \tag{3.117}$$

both for the real course of vitrification and devitrification but also for the changes of the state of the metastable equilibrium melt. In the latter case, the affinity A is equal to zero and above-employed equations (Eqs. (3.19)–(3.23)) and their consequences resulting in the estimate of the Prigogine–Defay ratio given by Eqs. (3.101) and (3.102) cannot be used any more. However, the configurational contributions to the equilibrium thermodynamic coefficients can be determined also directly in an alternative way knowing the configurational contributions to the enthalpy, the volume and the partial derivatives $(\partial \xi_e/\partial T)_p$ and $(\partial \xi_e/\partial p)_T$ for the respective metastable equilibrium states. Similarly to Eq. (3.101), we get

$$\Pi(T_g) = \frac{1}{VT}\left\{\frac{\Delta C_p \Delta \kappa}{(\Delta \alpha)^2}\right\}\bigg|_{T=T_g} = -\frac{\left(\frac{\partial H}{\partial \xi}\right)_{p,T}\left(\frac{\partial \xi_e}{\partial p}\right)_T}{T\left(\frac{\partial V}{\partial \xi}\right)_{p,T}\left(\frac{\partial \xi_e}{\partial T}\right)_p}\bigg|_{T=T_g} \quad (3.118)$$

this time directly for the experimentally measured values of the Prigogine–Defay ratio.

In order to compute these quantities, we employ here again the mean-field model leading to Eqs. (3.43)–(3.52). As should be the case, in the vicinity of the state of configurational equilibrium, we obtain from Eq. (3.49) after performing a truncated Taylor expansion the following result:

$$G_{\text{conf}}(p, T, \xi) \cong G_{\text{conf}}(p, T, \xi_e) + \frac{1}{2}\left(\frac{\partial^2 G_{\text{conf}}}{\partial \xi^2}\right)\bigg|_{p,T,\xi=\xi_e}(\xi - \xi_e)^2. \quad (3.119)$$

The value of

$$\frac{\partial^2 G}{\partial \xi^2}\bigg|_{\xi=\xi_e} = RT\frac{\partial}{\partial \xi}\left[\frac{\ln \xi}{(1-\xi)^2}\right]_{\xi=\xi_e} > 0 \quad (3.120)$$

at equilibrium can now be easily calculated based on Eqs. (3.49) and (3.121),

$$\frac{\partial G}{\partial \xi} = \chi_1 \Delta H_{\text{ev}}(T_m) + RT\frac{\ln \xi}{(1-\xi)^2} = 0. \quad (3.121)$$

For physically reasonable small values of ξ, we get as an estimate

$$G_e^{(2)} \cong \frac{RT}{\xi_e}. \quad (3.122)$$

With the above relations, we get with Eqs. (3.44) and (3.45)

$$\left(\frac{\partial V}{\partial \xi}\right)_{p,T} = N_A v_0(p, T), \quad \left(\frac{\partial H}{\partial \xi}\right)_{p,T} = \chi RT_m, \quad (3.123)$$

and, with Eq. (3.124)

$$\frac{(1-\xi_e)^2}{\ln \xi_e} = -\frac{1}{\chi}\left(\frac{T}{T_m}\right) \quad \text{where} \quad \chi = \chi_1\chi_2, \quad (3.124)$$

and assuming $\chi = $ constant,

$$\left(\frac{\partial \xi_e}{\partial T}\right)_p = -\frac{1}{\chi T_m}\frac{1}{\frac{\partial}{\partial \xi}\left[\frac{(1-\xi)^2}{\ln \xi}\right]_{\xi=\xi_e}}, \quad (3.125)$$

$$\left(\frac{\partial \xi_e}{\partial p}\right)_T = \frac{T}{\chi T_m^2}\frac{1}{\frac{\partial}{\partial \xi}\left[\frac{(1-\xi)^2}{\ln \xi}\right]_{\xi=\xi_e}}\frac{dT_m}{dp}. \quad (3.126)$$

According to Eq. (3.120), here the inequality

$$\frac{\partial}{\partial \xi}\left[\frac{(1-\xi)^2}{\ln \xi}\right]_{\xi=\xi_e} < 0 \qquad (3.127)$$

has to be fulfilled.

A substitution of Eqs. (3.123) and (3.125) into Eq. (3.118) yields

$$\Pi = \frac{\chi R}{N_A v_0(p_g, T_g)} \frac{dT_m}{dp} . \qquad (3.128)$$

Employing the Clausius–Clapeyron equation [263, 264] for the description of the change of the melting temperature with pressure

$$\frac{dT_m}{dp} = T_m \frac{\Delta V(T_m)}{\Delta H(T_m)} , \qquad (3.129)$$

where $\Delta V(T_m)$ and $\Delta H(T_m)$ are the change of molar volume and of molar enthalpy in the liquid–solid phase transformation at the melting temperature, T_m, we arrive finally at:

$$\Pi = \frac{\chi R T_m}{N_A v_0(p_g, T_g)} \frac{\Delta V(T_m)}{\Delta H(T_m)} . \qquad (3.130)$$

A more detailed discussion of this alternative approach to the determination of the Prigogine–Defay ratio is given in [265].

3.5.3
Comparison with Experimental Data

3.5.3.1 The Prigogine–Defay Ratio

Experimental data on the Prigogine–Defay ratio for various glass-formers with quite different structure and composition are reported by Davies and Jones [19, 20], Moynihan et al. [199], Moynihan and Gupta [236], Nemilov [78, 237] and Donth [134]. They are summarized in Table 3.1 and Figure 3.17. Particular often cited examples are B_2O_3 with a value of the Prigogine–Defay ratio equal to $\Pi = 4.7$, 40 mol% $Ca(NO_3)_2$ 60 mol% KNO_3 with $\Pi = 4.5$, glycose ($\Pi = 3.7$), rubber ($\Pi = 8.3$), polyvinylacetate ($\Pi = 2.2$) and polystyrene ($\Pi = 16$). As it is evident, these values are (with one exception, see Table 3.1) larger than one confirming qualitatively our theoretical findings. Even quantitatively, the estimates, involving simple lattice-hole concepts and leading to the theoretical estimate of the Prigogine–Defay ratio as given by Eq. (3.107), can be considered as quite acceptable. Note that Π-values in the range $\Pi \cong 2$–4 are most frequently observed (cf. Figure 3.17).

Of particular interest are also data showing the change of the ratio Π in dependence on the composition of Na_2O–SiO_2 and K_2O–SiO_2 melts studied by Nemilov [78, 237] (see Figure 3.18). Employing well-known models of two-component

Table 3.1 Values of the Prigogine–Defay ratio for different glass-forming substances. Under the name technical glasses, here window glasses, soda lime silicate glasses, crown glasses, flint glasses, and so on, are summarized.

Substance	T_g, K	Π	Reference
Technical glasses	727–921	2.0–2.7	[237]
Se	300	2.4	[134, 237]
Glucose	300	3.7	[237]
Rubber	320	8.3	[237]
Polystyrene	350	16	[237]
B_2O_3	550	4.7	[134, 236]
$0.4 Ca \cdot NO_3 \cdot 0.6 KNO_3$	340	4.5	[134, 236]
Polyvinylacetate	304 (303)	2.2 (1.7)	[134, 236]
Pure SiO_2-glass		$10^3 - 2 \cdot 10^5$	[237]
Glycerol	183	9.4	[134]
n-Propanol	95	1.9	[134]
PVC	350	1.7	[134]
Polyisobutene	198	0.9	[134]
Metallic glasses (Cr, Ti, Cu, Ni, Be)	610	2.4	[134]

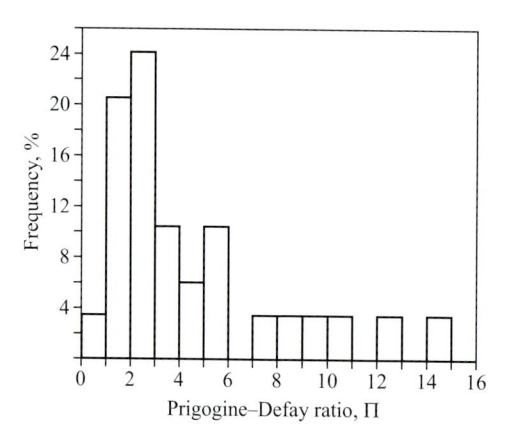

Figure 3.17 Frequency distribution of the values of the Prigogine–Defay ratio as obtained from the results summarized in Table 3.1 and Figure 3.18. Note that most frequently Π-values are found in the range $\Pi = (2-4)$ in agreement with theoretical estimates (cf. also [234]).

solutions, the melting entropy, ΔS_m, of the solution can be expressed via the melting entropy of a system consisting of component one, ΔS_{m1}, and the molar fraction of the second component, x_2, in the form

$$\Delta S_m \cong \Delta S_{m1} \left(1 + \frac{R}{\Delta S_{m1}} x_2 \right). \tag{3.131}$$

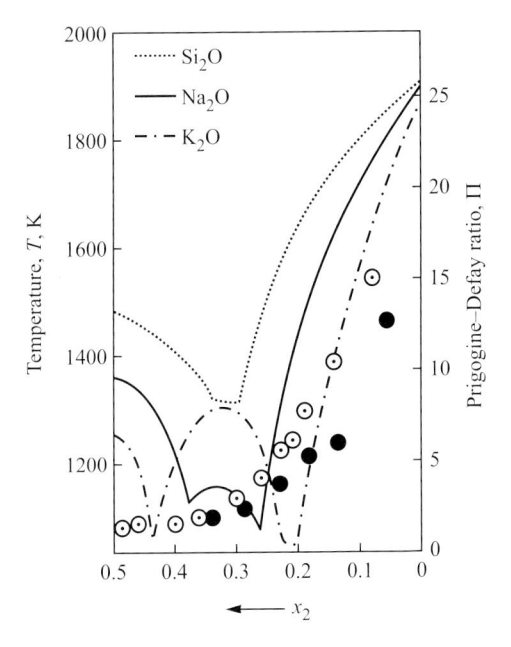

Figure 3.18 Results of measurements of the Prigogine–Defay ratio on two alkali-silicate glasses with different compositions [237] (circles refer to SiO$_2$–Na$_2$O; full points to SiO$_2$–K$_2$O). The *solid, dotted and dashed-dotted lines* represent the phase diagrams, x_2 is the content of the alkali component. Note that in agreement with Eq. (3.132) a decrease of Π is found with increasing entropy of melting. The value of Π for pure SiO$_2$ is not given since it is based on controversial data sets (cf. also [234]).

Employing Eq. (3.131), Eq. (3.110) yields

$$\Pi \propto \frac{30R}{\Delta S_m} \cong \frac{30R}{\Delta S_{m1}\left(1 + \frac{R}{\Delta S_{m1}}x_2\right)}. \qquad (3.132)$$

This equation allows us a qualitative interpretation of the data shown in Figure 3.18. For SiO$_2$, characterized by an extremely low molar entropy of melting (even lower than R) and unusually high bonding strength, relatively high values of the Prigogine–Defay ratio should be expected based on Eq. (3.109). Neither this exceptional case nor the prediction of the commonly found values of Π in the range $\Pi \cong (2\text{--}4)$ or the dependence of the Prigogine–Defay ratio on composition can be interpreted in existing attempts to correlate Π with the number of structural order parameters required for the description of glass-forming systems. In this way, treating vitrification as a freezing-in process proceeding in a certain temperature interval allows us to give an adequate theoretical interpretation of the values of the Prigogine–Defay ratio for vitrifying systems employing only one order parameter connected with the free volume of the system.

3.5.3.2 Change of Young's Modulus in Vitrification

The method of derivation of the Prigogine–Defay ratio, employed here, can be used also for the establishment of a variety of additional correlations between the jumps or s-shaped changes of the thermodynamic coefficients in the glass transition. In particular, we conclude that the inequalities Eqs. (3.96) and (3.100) have to be fulfilled. These theoretical predictions are in agreement with experimental data (cf. [1]). Starting with Eqs. (3.100), employing the definitions, Eq. (3.20), and the results of the lattice-hole model of glass-forming melts discussed in Section 3.5.2.2, we obtain further:

$$\frac{\Delta C_p}{\Delta \alpha} = V \left. \frac{h_{p,T}}{v_{p,T}} \right|_{T=T_g} = V \left. \frac{\left(\frac{\partial H}{\partial \xi}\right)_{p,T}}{\left(\frac{\partial V}{\partial \xi}\right)_{p,T}} \right|_{T=T_g} \cong \chi R T_m . \qquad (3.133)$$

Equations (3.101) and (3.133) yield then immediately

$$\frac{\Delta \kappa}{\Delta \alpha} \cong \frac{2V}{3\chi R} \Pi . \qquad (3.134)$$

It follows that a linear correlation has to exist between $\Delta \kappa$ and $\Delta \alpha$. Note that the changes of the thermodynamic coefficients discussed here refer to the respective values at $T = T_g$. These values can be replaced by the differences of these coefficients as determined experimentally employing the Simon model of vitrification (see the discussion of Eq. (3.101)).

Completing the discussion on this topic, we would like to draw the attention to the behavior of one thermodynamic coefficient, the knowledge of which is of particular importance for the understanding of crystallization processes in glass-forming melts near the glass transition interval. In the analysis of crystallization processes in glass-forming liquids, it is so far widely assumed that elastic stresses are of negligible significance both for crystal nucleation and growth. This assumption is based on the argument that elastic stresses relax too fast as compared with the characteristic times of critical crystal nucleus formation or crystal growth. However, this argumentation is based on one implicit assumption, that is, it is assumed that the Stokes–Einstein equation – connecting diffusion coefficient and Newtonian viscosity – is fulfilled not only for sufficiently high temperatures but also for temperatures near the glass transition interval and below. A detailed analysis [266] shows, however, that the latter assumption is not correct and that elastic stresses may be of significance both for crystal nucleation and growth (see [159, 267–269] for an overview).

In order to quantitatively describe the effect of stresses on nucleation and growth, the temperature dependence of Young's modulus, E, or of the compressibility, κ, has to be known. They are correlated by the equation [76]

$$\frac{1}{\kappa} = \frac{E}{3(1 - 2\gamma)} , \qquad (3.135)$$

where γ is the Poisson ratio. As far as temperatures sufficiently above the temperature of vitrification are concerned, following, for example, the classical work of Frenkel [25], Young's modulus can be approximated by $E = RT/v_m$, where v_m is the molar volume of the liquid. It is evident that in this range of temperatures (where the Frenkel approximation holds, at least, approximately) Young's modulus, E, can change only slightly depending on temperature. In general, one can also expect a moderate change of Young's modulus with temperature for glasses. Here the expected values of E are near the values of the respective crystal. In this way, the knowledge of the temperature dependence of Young's modulus or compressibility in the vicinity of the glass transition temperature, T_g, is of particular importance.

Direct experimental measurements of Young's modulus of glass-forming systems in the glass transition range are rare, nevertheless, some examples can be found in the literature. Measurements of the temperature dependence of Young's modulus, E, for B_2O_3-glass and melts upon vitrification have been performed by Tarassov [270]. They demonstrated clearly a considerable step-like increase of E with decreasing temperature in the vicinity of the glass transition temperature. The measurements of Tarassov were performed based on the analysis of the temperature dependence of the velocity of ultrasound.

A similar experimental example, employing a direct method, consists in the determination of the temperature dependence of Young's modulus of soda lime silica glass performed by McGraw [271] (see Figure 3.19). Note that, in agreement with Eq. (3.96), the compressibility of the glass is lower as compared with the respective undercooled liquid and, consequently, taking into account Eq. (3.135) Young's modulus is higher for the glass.

Other examples demonstrating the same behavior of Young's modulus – that is, a step-like or s-shaped increase with decreasing temperature in the vitrification interval – can be found in the books by Kobeko [166], Bartenev [272] and in the PhD thesis by Reinsch [273]. These experimental results confirm (at least, qualitatively) the validity of Eqs. (3.96) and (3.101). Conversely, these relations can be employed in order to determine the function $E = E(T)$ near the temperature of vitrification if the temperature dependence of Young's modulus is not directly available for the system under consideration.

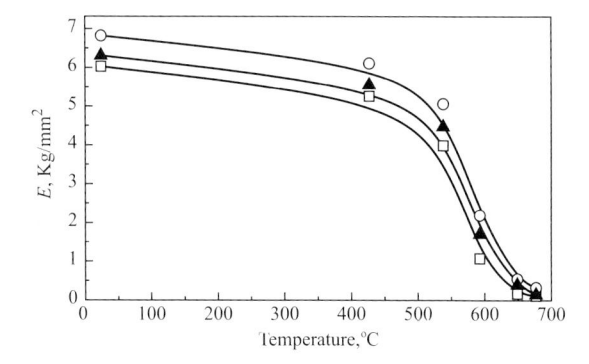

Figure 3.19 Temperature dependence of the modulus of elasticity, E, of three soda lime silicate glasses in the glass transition region according to the measurements of McGraw [271].

3.5.4
Discussion

In the present analysis, we employed a very general and well-established thermodynamic method for the theoretical determination of the Prigogine–Defay ratio in vitrification developed in its basic form by De Donder [13], Prigogine and Defay [14]. Treating vitrification not in terms of Simon's model but as a process proceeding in some finite temperature interval it can be shown without further assumptions and/or approximations that (i) the Prigogine–Defay ratio has to have in vitrification values not equal to one ($\Pi \neq 1$), (ii) the actual ratio in dependence on pressure and temperature is larger one for cooling ($\Pi(T_g) > 1$) and less than one for heating ($\Pi(T_g) < 1$). As discussed in detail, in order to compare theoretical estimates of the changes of the thermodynamic coefficients near T_g with experimental data used commonly in the computation of the Prigogine–Defay ratio, (iii) theoretical estimates for the changes of the thermodynamic coefficients in cooling processes have to be used resulting in $\Pi > 1$. We show further that, employing in addition a mean-field model of glass-forming systems able to describe large classes of glass-forming melts qualitatively and partly even quantitatively [1], (iv) experimentally measured values of the Prigogine–Defay ratio can be reestablished. Moreover, employing only general premises of a wide class of models of glass-forming systems, (v) a definite physical meaning may be assigned to the Prigogine–Defay ratio as determined by the interplay of bonding strength and structural complexity of the liquids under consideration. Based on these results, (vi) the behavior of the thermodynamic coefficients in vitrification is discussed in detail being of considerable significance for the understanding of a variety of processes in glass transition. Hereby, as shown, (vii) the introduction of only one structural order parameter is sufficient in order to arrive at an agreement between theory and experimental data.

In the application of the structural order parameter concept to the description of the glass transition, the problem was posed several times, how many independent order parameters are required for an adequate description of this process. Here different opinions have been expressed. From the experience with more or less realistic lattice-hole models of vitrifying melts (see, for example [223, 259, 260, 274, 275]) the conclusion was drawn that, in general, three structural parameters are needed to define, with a sufficient degree of quantitative accuracy, the thermodynamic state of a simple or polymer glass-forming melt. One parameter is needed for the description of the topological disorder (i.e., the free volume of the melt), a second parameter describes the degree of complexity of the building units of the system (i.e., the degree of polymerization or aggregation in the melt), while a third parameter reflects the probability of the building units to exist in different conformations. Latter parameter is connected with the degree of flexibility, for example, of the polymer chains or other molecular building units of the system. In most statistical models, these parameters are considered as independent variables. However, it has to be accounted for that in the general case flexibility or the configuration of the units constituting the melt depend on the free volume of the liquid. In such cases, connections may exist between the three above-mentioned parameters and they cannot be considered then any more as independent quantities.

The experience with such statistical models shows that it is difficult to give a final answer to the question of how many structural order parameters, ξ_i, have to be introduced for a proper description of the state of the considered system (see, e.g., Moynihan [199] and Nemilov [78]). Even the opinion has been expressed that it is, in principle, impossible to have more than one independent order parameter for a description of a glass, in particular, in the vicinity of T_g. This statement is based on the argument that different equilibration processes in glasses are interconnected in this temperature range. On the other hand, a general purely thermodynamic analysis of the pressure dependence of viscosity leads to the conclusion ([276] and Chapter 4) that the introduction of only one order parameter (connected with the free volume) is not sufficient in order to describe adequately the experimental data when extended temperature (or pressure) intervals are considered.

The necessity of introduction of more than one structural order parameter seemed to be evident from another point of view. The introduction of more than one order parameter was considered as an essential requirement in order to obtain realistic values for the Prigogine–Defay ratio. Deviations from the value $\Pi = 1$ (resulting from Eq. (3.101) when $A = 0$ is assumed) have been explained earlier by the introduction of more than one order parameters into the description of freezing-in phenomena as performed by Meixner [277], Davies and Jones ([19, 20]; see also Moynihan [199], Moynihan, Gupta [236], Nemilov [78], Rehage and Borchardt [278]). Here we have shown that realistic values for this ratio can be derived theoretically employing only a structural order parameter. Consequently, the additional argument in favor of the necessity to employ several order parameters is no longer valid.

No theoretical explanation for the mentioned deviation of the Prigogine–Defay ratio from unity can also be given, if vitrification is considered as a particular example of second-order equilibrium phase transformations. Consequently, experimental data concerning the Prigogine–Defay ratio, giving the result $\Pi > 1$ and its theoretical foundation by the treatment of vitrification as a freezing-in process in terms of the generic model of vitrification and relaxation as performed here, give a convincing additional proof that the glass transition is not a second-order equilibrium phase transition but a transition from a thermodynamically metastable to a frozen-in thermodynamic nonequilibrium state as proposed long ago first by Simon.

3.6
Fictive (Internal) Pressure and Fictive Temperature as Structural Order Parameters

3.6.1
Brief Overview

For the first time, the concept of a structural order parameter was introduced into glass science by Tool [69, 279, 280] in terms of the fictive temperature concept.

Davies and Jones [19, 20] noted that instead of fictive temperature one can employ equivalently the concept of fictive pressure as the determining structural order parameter. Gupta [242] developed a theoretical approach treating both fictive temperature and fictive pressure as independent parameters. There are a variety of further attempts to employ these concepts in the treatment of glass-forming systems, some of them are discussed below.

All the consequences discussed above in the present analysis can be derived without relying on the concept of an internal (fictive) pressure or fictive temperature as a structural order parameter. Anyway, it is of interest to see in which way such concepts may be introduced in the framework of De Donder's approach avoiding as far as possible any simplifying assumptions, and what results from this procedure. This task will be performed in the present section.

In the already cited paper, Gupta [242] introduced two structural order parameters $\xi_i, i = 1, 2$ (in our notations) assumed to be chosen in such a way that for stable or metastable equilibrium

$$\left.\frac{\partial G}{\partial \xi_i}\right|_{T,p} = 0, \quad i = 1, 2 \tag{3.136}$$

holds and that the second-order derivative matrix is positive definite. Similarly to Eq. (3.15), in equilibrium these parameters are considered to be functions of pressure and temperature, that is, $\xi_{ie} = f_i(p, T)$ for $i = 1, 2$. Having at one's disposal the functions $f_i(p, T)$ connecting the equilibrium values of ξ with pressure and temperature, fictive pressure, p_f, and fictive temperature, T_f, are determined generally via

$$\xi_i = f_i(T_f, p_f), \quad i = 1, 2. \tag{3.137}$$

Employing certain approximations (e.g., a linear dependence of f_i on pressure and temperature, linear relaxation laws with kinetic coefficients being independent of the structural order parameters) the relaxation behavior in such systems is studied.

Landa et al. developed an interpretation of the glass transition and accompanying effects employing the concept of negative internal pressure as the structural order parameter [83, 281, 282]. In [83], it was stated by Landa et al. that their approach is based on De Donder's principle. This statement is expressed by mentioned authors in the form "*Non-equilibrium systems can be described by equalities that include the notion of affinity and internal parameter,*" continuing then in the following way: "*In case of the glass-forming systems, their excess volume is affinity to glass transition and relaxation, and negative pressure – internal parameter,*" [83, p. 8]. However, a proof of this statement is missing or, at least, not given in a convincing form.

As will be shown here the concept of internal pressure can be introduced straightforwardly, employing the results outlined above, as follows. According to Eq. (3.44), the excess volume V_{conf} (or the configurational contributions to the volume) is (in a good approximation) linearly connected with the structural order parameter, ξ. Considering $v_0(p, T)$ in an also good approximation as nearly

3.6 Fictive (Internal) Pressure and Fictive Temperature as Structural Order Parameters

constant, Eq. (3.7) can be rewritten as:

$$dU = TdS - pdV - \frac{A}{N_A v_0(p,T)} dV_{conf}. \tag{3.138}$$

By introducing or, more precisely, defining the internal pressure via

$$\Upsilon = \frac{A}{N_A v_0(p,T)}, \tag{3.139}$$

we obtain with $A = -G_e^{(2)}(\xi - \xi_e)$ (Eq. (3.17)) and the estimate $G_e^{(2)} \cong (RT/\xi_e)$ [240] the result

$$\Upsilon = -G_e^{(2)} \frac{(\xi - \xi_e)}{N_A v_0(p,T)} \cong -\frac{(\xi - \xi_e)}{\xi_e} \frac{RT}{N_A v_0(p,T)}. \tag{3.140}$$

The sign of the internal pressure is thus directly determined by the sign of the difference $(\xi - \xi_e)$ or of the ratio $(\xi - \xi_e)/\xi_e$, its magnitude by the ratio $(RT/(N_A v_0(p,T)))$.

So, indeed, a parameter having the dimension of pressure can be introduced, having the essential properties as assigned to them by Landa et al. [83]; for example, it is equal to zero in stable or metastable equilibrium (here $\xi = \xi^{(e)}$ holds) and has negative values in the glass transition range and below at cooling. However, in the subsequent heating it can and even has to assume a positive value until in the further heating stable or metastable equilibrium states are reached again. Moreover, as it turns out, in contrast to the mentioned at the beginning statement of Landa et al. [83], the internal pressure, defined in such a way, is proportional to the affinity while the structural order parameter is connected with the excess volume. Some further points of agreement and disagreement with the statements of Landa et al. formulated in [83, 281, 282] are discussed in detail in [238].

In a series of papers (see, e.g., [283–285]), Nieuwenhuizen attempted to develop a thermodynamic description of the glassy state introducing the concept of an effective temperature similar to the concept of fictive temperatures developed originally by Tool [69, 279, 280]. In a recent analysis of the concept of fictive temperatures, Garden et al. [286] came to the conclusion that the approach by Niewenhuizen is equivalent to De Donder's approach. This could be the case (neglecting for a while the other problems discussed in detail in [238, 239]) if Eq. (2.1) in [285] and not Eq. (6.1) would be considered as generally valid. This is, as it follows from cited papers, however, not the original intention of the mentioned author.

Based on this approach, Niewenhuizen [283–285] also discussed in recent years the possible origin of the deviation of the Prigogine–Defay ratio from unity. He utilizes the effective temperature, T_{ef}, as an additional parameter for the description of vitrifying melts and assumes correctly that, for any chosen law of heating and/or cooling, this additional parameter is an unambiguous function of pressure and temperature. In re-analyzing the data of Rehage and Oels [287] on vitrification in cooling of atactic polysterene (the only example discussed in his papers), Niewenhuizen found, in contrast to the authors reporting $\Pi \cong 1.09$, a value $\Pi \cong 0.77$.

Note that the original estimate of Rehage and Oels is in full agreement with our predictions leading generally to the conclusion that $\Pi \geq 1$ has to be fulfilled for cooling processes. A detailed comparison of both approaches is given in [238, 239].

3.6.2
Model-Independent Definition of Fictive (Internal) Pressure and Fictive Temperature

As outlined above, the concept of fictive or internal pressure can be introduced employing appropriate models of glass-forming systems. However, starting with Eq. (3.7) a model-independent definition of fictive pressure and fictive temperature can be given. For this purpose it is required that the variation of the external parameters is performed in such a way that the structural order parameter can be represented as a differentiable function of the two independent thermodynamic state parameters (we consider closed systems). As such state parameters, we chose total entropy, S, and the total volume, V, of the system. Then we may write

$$\xi = \xi(S, V) \implies d\xi = \left(\frac{\partial \xi}{\partial S}\right)_V dS + \left(\frac{\partial \xi}{\partial V}\right)_S dV . \tag{3.141}$$

In order to avoid confusion, we would like to stress once more the basic assumption underlying Eq. (3.141). In general, the structural order parameter is an independent parameter required for the description of the deviations of the system from equilibrium. However, if some well-defined heating or cooling rules (and/or changes of pressure) are assumed leading to a dependence like $\xi = \xi(S, V)$ (cf. Figures 3.2 and 3.16) then for the given trajectory of vitrification or devitrification we have the right to formulate Eq. (3.141). With Eq. (3.7) we obtain

$$dU = \left\{T - A\left(\frac{\partial \xi}{\partial S}\right)_V\right\} dS - \left\{p + A\left(\frac{\partial \xi}{\partial V}\right)_S\right\} dV . \tag{3.142}$$

As a consequence, the fictive pressure and the fictive temperature can be determined similarly to the definition of the respective parameters in classical equilibrium thermodynamics via

$$T_{\text{fictive}} = \left(\frac{\partial U}{\partial S}\right)_V = T - A\left(\frac{\partial \xi}{\partial S}\right)_V , \tag{3.143}$$

$$p_{\text{fictive}} = -\left(\frac{\partial U}{\partial V}\right)_S = p + A\left(\frac{\partial \xi}{\partial V}\right)_S . \tag{3.144}$$

For stable and metastable thermodynamic equilibrium states, the so-defined parameters are equal to normal pressure and temperature (since for these states $A = 0$ holds) while in general these parameters depend on the path the system is transferred into the respective state.

Equations (3.143) and (3.144) are general thermodynamic relations independent on any model assumptions concerning the system under consideration except the assumption (which will be later removed) that one structural order parameter is sufficient for the description. However, in order to obtain quantitative results – as

3.6 Fictive (Internal) Pressure and Fictive Temperature as Structural Order Parameters

it is generally the case in thermodynamics – equations of state of the system have to be known. Employing here again results of the model of glass-forming melts discussed earlier, that is,

$$A \cong -RT\left(\frac{\xi - \xi_e}{\xi_e}\right) \tag{3.145}$$

$$V \cong N_A v_0(p, T)(1 + \xi), \tag{3.146}$$

$$S_\xi = -R\left(\ln(1 - \xi) + \left(\frac{\xi}{1 - \xi}\right)\ln \xi\right), \tag{3.147}$$

we obtain

$$\left(\frac{\partial V}{\partial \xi}\right)_S = N_A v_0(p, T), \quad \left(\frac{\partial \xi}{\partial V}\right)_S = \frac{1}{N_A v_0(p, T)}, \tag{3.148}$$

$$\left(\frac{\partial S}{\partial \xi}\right)_V = \left(\frac{\partial S_\xi}{\partial \xi}\right)_V = -R\frac{\ln \xi}{(1 - \xi)^2}, \tag{3.149}$$

$$\left(\frac{\partial \xi}{\partial S}\right)_V = -\frac{(1 - \xi)^2}{R \ln \xi}. \tag{3.150}$$

A substitution into Eqs. (3.143) and (3.144) yields finally

$$T_{\text{fictive}} = T\left\{1 - \left(\frac{\xi - \xi_e}{\xi_e}\right)\frac{(1 - \xi)^2}{\ln \xi}\right\} \cong T\left(\frac{\xi}{\xi_e}\right), \tag{3.151}$$

$$p_{\text{fictive}} = p - \frac{RT}{N_A v_0(p, T)}\left(\frac{\xi - \xi_e}{\xi_e}\right). \tag{3.152}$$

Note that fictive pressure and temperature are thus uniquely determined via the knowledge of the value of ξ. Moreover, in such general definition (fulfilling the usually assumed limiting condition that in stable and metastable equilibrium fictive pressure and temperature are equal to external pressure and temperature), the fictive pressure (in contrast to the result following from the definition given via Eqs. (3.139) and (3.140)) may be not less than zero in the glass transformation range and below even in cooling a melt. In the opposite limiting case, for temperatures tending to zero (since for $T \to 0$, we have $\xi_e \to 0$ and $(T/\xi_e) \to \infty$), the relations

$$A \to -\infty, \quad \frac{T_{\text{fictive}}}{T} \to \infty, \quad \frac{p_{\text{fictive}}}{p} \to -\infty \quad \text{for} \quad T \to 0 \tag{3.153}$$

hold.

The definition of fictive temperature and fictive pressure can be generalized easily to the case of an arbitrary number of structural order parameters. Indeed, in this more general situation, when, for example, f structural order parameters have to be introduced into the description, we have to replace Eq. (3.7) by

$$dU = TdS - pdV - \sum_{i=1}^{f} A_i d\xi_i. \tag{3.154}$$

Along a given path of evolution of the system, the different order parameters are functions of the external control parameters chosen here to be entropy and volume. Instead of Eqs. (3.142)–(3.144), we get then

$$dU = \left\{ T - \sum_{i=1}^{f} A_i \left(\frac{\partial \xi_i}{\partial S}\right)_{V,\{\xi_j : j \neq i\}} \right\} dS$$
$$- \left\{ p + \sum_{i=1}^{f} A_i \left(\frac{\partial \xi_i}{\partial V}\right)_{S,\{\xi_j : j \neq i\}} \right\} dV . \quad (3.155)$$

As a consequence, the fictive pressure and the fictive temperature can be determined similarly to the definition of the respective parameters in classical equilibrium thermodynamics via

$$T_{\text{fictive}} = \left(\frac{\partial U}{\partial S}\right)_V = T - \sum_{i=1}^{f} A_i \left(\frac{\partial \xi_i}{\partial S}\right)_{V,\{\xi_j : j \neq i\}}, \quad (3.156)$$

$$p_{\text{fictive}} = -\left(\frac{\partial U}{\partial V}\right)_S = p + \sum_{i=1}^{f} A_i \left(\frac{\partial \xi_i}{\partial V}\right)_{S,\{\xi_j : j \neq i\}} . \quad (3.157)$$

Equations (3.156) and (3.157) represent – as we believe – the most general and appropriate definition of these internal parameters: (1) it is a straightforward generalization of the basic thermodynamic definitions of temperature and pressure as known from classical equilibrium thermodynamics (e.g., [54, 62, 63]). For stable and metastable thermodynamic equilibrium states, the so-defined parameters are again equal to normal pressure and temperature (since for these states $A_i = 0$, $i = 1, 2, \ldots, f$ holds). In general, these parameters depend on the path the system is transferred into the respective state reflecting in this way the nonequilibrium nature of the system. (2) Having at one's disposal the generalized caloric equation of state $U = U(S, V, \xi_1, \xi_2, \ldots, \xi_f)$ (i.e., the dependence of the internal energy on the external control parameters S and V and the chosen set of structural order parameters $\{\xi_i\}$ to be obtained from statistical mechanical model computations or experiment), fictive temperature and fictive pressure can be determined uniquely via Eqs. (3.156) and (3.157) independent of the number of structural order parameters required for the description of the system under consideration. (3) Employing the definition of fictive pressure and fictive temperature as developed by us (Eqs. (3.156) and (3.157)), the fundamental law of thermodynamics for closed systems and arbitrary numbers of structural order parameters can be written as:

$$dU = T_{\text{fictive}} dS - p_{\text{fictive}} dV . \quad (3.158)$$

In such a definition, fictive pressure and fictive pressure have a much wider meaning as to represent structural order parameters, they represent the generalizations of the external control parameters (external pressure, p, and external temperature, T) governing the thermodynamic behavior of closed systems for equilibrium states and reversible processes proceeding in between them.

The property listed above as (2) – general validity of the definition of fictive pressure and temperature, Eqs. (3.156) and (3.157), independent of the number of structural order parameters – is not fulfilled, for example, for the definition of these parameters (Eq. (3.137)) given by Gupta [242]. In particular, the latter relation does not allow one uniquely to determine p_f and T_f for the case that only one structural order parameter is required. The extension of Eq. (3.137) to f independent structural order parameters is as a rule also impossible since f sets of equations of the form as given by Eq. (3.137) will not have, in general, a solution. Note as well that even in the case that only one structural order parameter is required for the description of the system, both T_f and p_f are different, in general, from T and p (cf. Eqs. (3.151) and (3.152)).

Finally, we would like to note that the above-given definitions of fictive temperature and fictive pressure can be generalized – if required – also to the case of open systems in a similar form as done here.

3.7
On the Behavior of the Viscosity and Relaxation Time at Glass Transition

As already accounted for in analysis of the kinetics of relaxation, not only thermodynamic but also the kinetic parameters of glass-forming systems may depend on the structural order parameters. Experimental data, verifying this statement with respect to the viscosity, are presented in Figure 3.20.

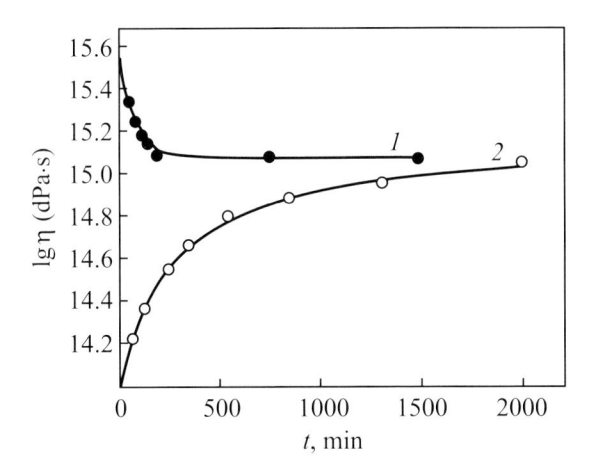

Figure 3.20 Viscosity vs. time curves for two samples of a silicate glass with the same composition but different thermal histories heat-treated at 486 °C (after Lillie [288]): (1) glass sample equilibrated at a lower temperature (478 °C), (2) glass sample transferred from temperatures considerably above 486 °C to room temperatures and brought to 486 °C afterwards. Note that in both cases from different sides the same equilibrium value of the viscosity is approached corresponding to the equilibrium value of the structural order parameter.

As shown in this figure, isothermal relaxation processes, connected with changes of the structural order parameter, may result in variations of the viscosity, and the value of the viscosity at some given pressure and temperature may depend, in general, on the prehistory of the system. For this reason, the theoretical approach developed is of essential significance not only for the comprehensive understanding of thermodynamic but similarly of the kinetic parameters of glass-forming melts in cooling and heating processes. This statement will be elaborated on briefly in the present section and in application to viscosity as one of the basic kinetic parameters of glass-forming melts in more detail in Chapter 4.

Assuming, again, the existence of only one structural order parameter, ξ, we may write generally the following expression for the temperature dependence of the viscosity (at constant pressure):

$$\frac{d\eta}{dT} = \left(\frac{\partial \eta}{\partial T}\right)_\xi + \left(\frac{\partial \eta}{\partial \xi}\right)_T \frac{d\xi}{dT}. \tag{3.159}$$

For systems in stable or metastable thermodynamic equilibrium, ξ is an unambiguous function of pressure and temperature. In general, ξ is a function also of the cooling rate. For this reason, close to the glass transition temperature, equilibrium (here the derivative $(d\xi/dT)$ has to be determined for the curve $\xi = \xi_e(p, T)$) and nonequilibrium (here the derivative has to be determined for the curve $\xi = \xi(p, T; q)$ depending on cooling or heating rates) viscosity may be distinguished.

Explicitly, this effect may be interpreted employing the Macedo–Litovitz equation [147]

$$\eta(p, T, \xi) = \eta_0(p, T) \exp\left(\frac{U_a}{RT}\right) \exp\left(\frac{B}{\xi}\right), \tag{3.160}$$

giving a dependence of the viscosity on free volume. Postulating that this relation holds for both equilibrium and nonequilibrium states, it can be employed to reflect the dependence of both equilibrium and nonequilibrium viscosity on temperature. Since, for cooling processes, the relation $\xi_e \leq \xi$ holds, for cooling we have to expect a behavior as shown in Figure 3.21 (see also [1]). A detailed experimental analysis for a number of glasses was performed by Mazurin et al. [290–292]. Some of the results are shown in Figure 3.22.

By the way, the Macedo–Litovitz equation is in agreement with the expressions employed here for the relaxation time (cf. Eq. (3.40)). Indeed, we can write the second term in the Macedo–Litovitz equation as:

$$\frac{B}{\xi} = \frac{B}{\xi_e + (\xi - \xi_e)} = \frac{B}{\xi_e \left(1 + \frac{\xi - \xi_e}{\xi_e}\right)} \cong \frac{B}{\xi_e}\left(1 + \frac{\xi - \xi_e}{\xi_e}\right). \tag{3.161}$$

Consequently, the Macedo–Litovitz equation can be written approximately in the form

$$\eta(p, T, \xi) = \eta_0(p, T) \exp\left(\frac{U_a}{RT}\right) \exp\left[\frac{B}{\xi_e}\left(1 + \frac{\xi - \xi_e}{\xi_e}\right)\right] \tag{3.162}$$

in, at least, qualitative agreement with Eq. (3.40).

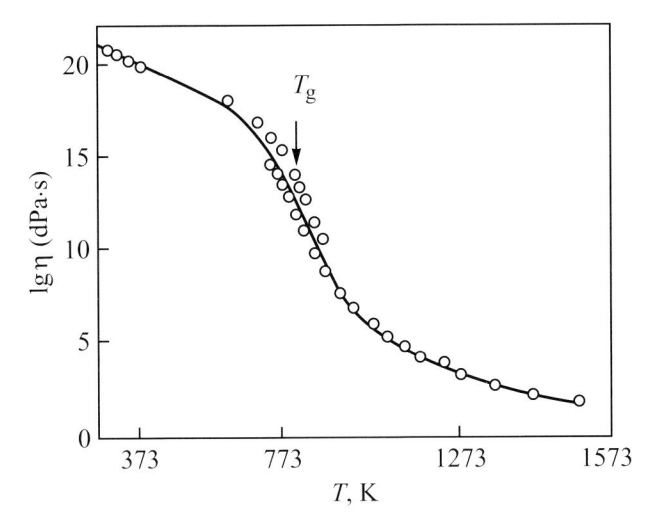

Figure 3.21 Temperature dependence of the viscosity of a classical glass-forming melt: the change of the slope of the $\log \eta(T)$ curve upon vitrification for a soda lime silicate glass in a very broad viscosity interval (after Winter's analysis [289], based on experimental data of several authors and compiled by Gutzow et al. [179, 180]).

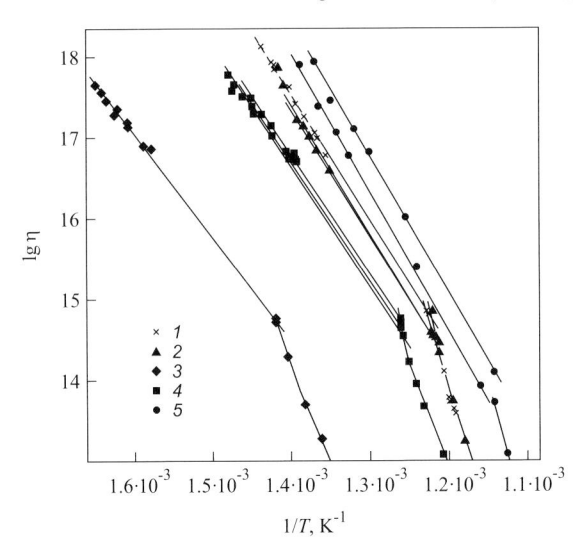

Figure 3.22 Temperature dependence of the viscosity of five technologically relevant silicate glass-forming melts of different compositions as obtained by Mazurin et al. (for details see [291, 292]).

For heating processes, the order parameter behaves differently (see Figure 3.2). For this reason, in heating runs we have to expect some kind of oscillatory approach of the viscosity to the equilibrium viscosity, that is, the curve corresponding to the nonequilibrium viscosity will intersect the equilibrium curve and tend to it with

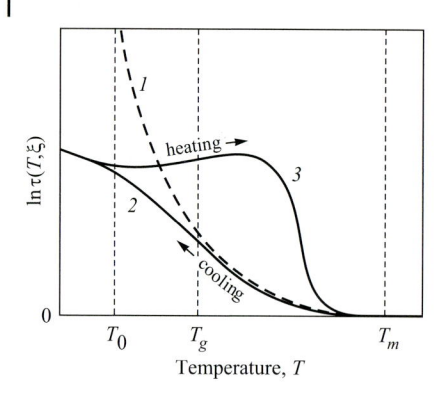

Figure 3.23 Results of theoretical computations of the time of molecular relaxation, $\tau(T, \xi)$, on temperature: (1) temperature course of $\tau_e(T)$ determined employing a temperature dependence of the activation energy as in the Vogel–Fulcher–Tammann equation (i.e., $U(T) = \text{const}/(T - T_0)$); (2) normal cooling (cf. curve 2 in Figure 3.2); (3) heating run (cf. also [181]).

increasing temperature from above. These results can be given a straightforward theoretical interpretation in terms of the generic approach employed here analyzing the dependence of the relaxation time on temperature in cooling and heating. In Figure 3.23, results of such theoretical computations of the time of molecular relaxation, $\tau(T, \xi)$, on temperature are presented (cf. [11, 17, 179]). Note that for cooling the relaxation time increases more slowly as compared with the value corresponding to the metastable equilibrium state while upon heating, a well-expressed overshoot in the $\tau(\xi, T)$ curves occurs.

3.8
On the Intensity of Thermal Fluctuations in Cooling and Heating of Glass-Forming Systems

3.8.1
Introduction

Glasses exhibit highly interesting and peculiar properties not only with respect to the (average) bulk properties but also with respect to the the spectrum of fluctuations of the respective properties frozen-in in a glass. As will be shown in the present section, the approach developed here allows one to understand not only the peculiarities in the bulk properties of glasses as compared to systems in thermodynamic equilibrium but also the origin of the observed in experiments increased intensity of fluctuations in glasses (cf. also [181]). The origin of this effect can be interpreted qualitatively twofold: (i) remaining inside Simon's model of vitrification, not only the (average) bulk properties of the undercooled liquid become frozen-in suddenly at the glass transition temperature, T_g, but also the developed spectrum of density and/or composition fluctuations. Since the spectrum of fluctuations in an equilibrium system depends on temperature, the intensity of fluctuations in a glass – resembling the respective frozen-in at T_g spectrum of fluctuations – has to be different (the intensity has to be higher) as compared with fluctuations in a stable or metastable equilibrium system at the same temperature $T \leq T_g$. (ii) Going beyond Simon's model and treating vitrification in terms of the generic model employed in the present chapter, one has to determine, in principle, also the change of the spectrum and intensity of fluctuations in cooling and heating of glass-forming melts. The intensity of fluctuations will be here as a rule higher since there is no response force trying to restore the initial state the fluctuation has evolved from. Since there is no such response trying to restore the initial state, the degree of heterogeneity of glass-forming systems has expected to be higher as compared with systems in thermodynamic equilibrium, where local density fluctuations will be damped out restoring in this way the homogeneity of the system.

A detailed description of the evolution of the spectrum of fluctuations in a glass-forming system in dependence on its properties and the cooling or heating conditions in terms of statistical physics is a highly complicated task. First attempts in this direction have been developed by Volkenstein [198, 293] applying the statistical physics approach to the determination of the spectrum of fluctuations in a glass and its possible verification by appropriate optical methods of measurement. Quite recently, Berthier *et al.* [294] have also analyzed the possibility of evolution of fluctuations in the framework of statistical and dynamic theories of glass formation (the mode coupling theory) developed by Goetze *et al.* [26, 27] and by other authors.

The analysis of such approaches is out of the scope of the present considerations. Here we would like to give a summary of some research examples exhibiting the peculiarities in the spectrum of fluctuations in glasses in order to show then that the generic approach developed here allows one to understand basic features of these effects.

3.8.2
Glasses as Systems with Frozen-in Thermodynamic Fluctuations: Mueller and Porai-Koshits

It seems that it was Mueller [295] who first considered glasses as solid bodies with a considerable intensity of frozen-in fluctuations. These fluctuations are considered by him as remnants of the previous states of the system, when it was in equilibrium as a metastable undercooled liquid. Such fluctuations, together with the distorted volume of the glass around them, determine, according to Mueller, the highly increased scattering losses of light, when compared with the expected value of losses in a translucent body, in which only fluctuations, equilibrium in number and size, exist, that is, when the glass is considered to be at the given room temperature in equilibrium. This increase in the intensity of fluctuations is, according to Mueller, responsible for the approximately 30 times higher losses in light, transmitted through any glass. The technical significance of such an optical problem is obvious.

A variety of outstanding experimental analysis accompanied by the development of new ideas and results on the nature of glasses and the processes proceeding in them is connected in with the achievements of Porai-Koshits, one of the outstanding representatives of Russian and international glass science. Porai-Koshits's long period of successful work in the field of glass structure, of liquid phase separation and crystallization in silicate melts was determined mostly by the methods of small angle X-ray diffraction developed by him to a remarkable perfection. The beginning of his analysis of glass structure is indicated by the insight he has given to the succession of crystallization processes in the system $Na_2O–PbO–SiO_2$; crystallization was thus used to initiate and follow structural changes, which are the main topic of the present chapter. The results obtained gave the basis for Porai-Koshits' structural approach, uniting elements of Lebedev's "crystallite hypothesis" with the Zachariasen–Warren "continuous network" theory and with Hägg's ideas on the existence of distinct anionic structural units in glass (cf. [1]). Porai-Koshits's crystallization studies led him and his glass science school to a new understanding of silicate glass ceramics. The small angle X-ray diffraction method, employed by Porai-Koshits and his coworkers, enabled them measurements on the kinetics of liquid phase separation in glasses and silicate melts. Porai-Koshits connected these processes with the initiation of crystallization and with the definition of glasses as systems with frozen-in high temperature homophase density fluctuations. He outlined both the possibilities and the limitations of experimental proof in this field, the expected interference with heterophase fluctuations, changing composition, impurities, and so on. These investigations opened a new chapter in understanding the structural nature of glasses and gave an impetus to further developments in many fields of glass science.

In considering the specific achievements of Porai-Koshits, it is usually pointed out [296] that he has developed a new, more consistent approach to the structure of silicate glasses, a model in between Lebedev's crystallite hypothesis and the Zachariasen–Warren continuous network theory. Porai-Koshits was, as it seems, also considerably influenced by Hägg's ideas on the existence in glasses of distinct anionic structural elements. A thorough description of the story of these structural developments may be found in [1]. Lebedev's hypothesis [297, 298] corresponded in fact to the earlier X-ray results of Randall, Ruxby and Cooper [299]. Optical and infrared findings of changes in the optical properties of glasses in line with the $(\alpha-\beta)$-transitions of three of the modifications of silica, which were demonstrated in [300] and elsewhere, exaggerated the possible influence of crystalline remnants on the properties of glasses. In fact, simple calculations made by Appen (see again [1]) had shown that the changes of optical properties in glasses, corresponding to the $(\alpha-\beta)$-transformation of quartz crystals, are caused by crystalline matter of not more than a very small percentage, a_x, of the volume of the investigated glasses. Such an effect, which could be caused by random recrystallization, even by contamination from the precursor raw materials, could and cannot influence the overall thermodynamic properties of the system. In the configurational entropy contribution, this would give effects proportional to $a_x \ln a_x$, which is negligible as also $(1 - a_x)\ln(1 - a_x)$ in the corresponding entropy of mixing contribution [206, 207]

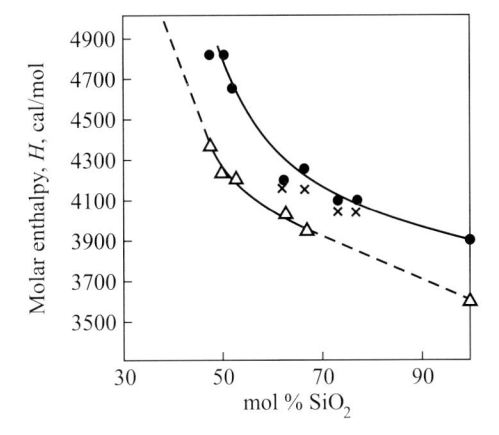

3.8 On the Intensity of Thermal Fluctuations in Cooling and Heating of Glass-Forming Systems | 155

Figure 3.24 Molar enthalpy, H, of crystalline (lower curve, *triangles*) and of glassy (upper line, *black dots and crosses*) samples from the system Na_2O/SiO_2 according to Evstropiev and Skornyakov [301] obtained from dissolution heats. Note the considerable difference ΔH_g between the enthalpy of glass and crystal (approximately 1/3 to 1/2 of the respective melting entropy ΔH_m), precluding the notion that the glassy samples are crystalline-like (e.g., built of 85% volume crystallites) (according to [181]).

given by $\Delta S_x = -R[a_x \ln a_x + (1-a_x) \ln(1-a_x)]$. Thus, "crystallites," even if they could exist, could not explain the enhanced entropy values, $\Delta S(0)$, frozen-in in glasses, which even in the simplest glass-forming systems (SiO_2, GeO_2, BeF_2) were found to be of the order of $\Delta S(0) \cong R$ (see [1, 206, 207]).

The followers of the crystalline hypothesis in its initial formulations assumed on the other hand that may be up to 85% of the volume of the glass is crystalline-like. This effect, if it really exists, should influence significantly frozen-in enthalpy (i.e., dissolution heats) and zero-point entropy of glasses, $\Delta S(0)$. It is strange that so little attention was given at that time to thermodynamic reasoning in considering structural problems. In fact even the earliest thermodynamic results on the enthalpy difference, ΔH, glass/crystal, obtained by the Soviet school (see Evstropiev and Skornyakov's paper from 1949 [301]) showed a considerable, nearly equal difference of the enthalpy, ΔH_g, frozen-in in the glass and the melting enthalpy, ΔH_m, of the crystal ($\Delta H_g/\Delta H_m \cong 0.5$) which excluded a crystal-like or "crystallite" structure of any of the glasses in the system Na_2O/SiO_2 (Figure 3.24). Only X-ray analysis seemed to be of significance for the authors of this period.

Porai-Koshits preferred the direct, structural way of analyzing things. However, even his first results (with Valenkov, also from the year 1949 [302]) showed the gradual change of the bulk structure of heat-treated glasses towards the crystalline structure expected for the respective glass in accordance with the corresponding phase diagram. So depending on the pre-treatment of the glass, different structures were obtained and had to exist for different glasses. If crystallites really would have existed, then, as this was shown by Valenkov and Porai-Koshits (Figure 3.25) [1, 302], they would have to have dimensions not exceeding 10–15 Å. This result was

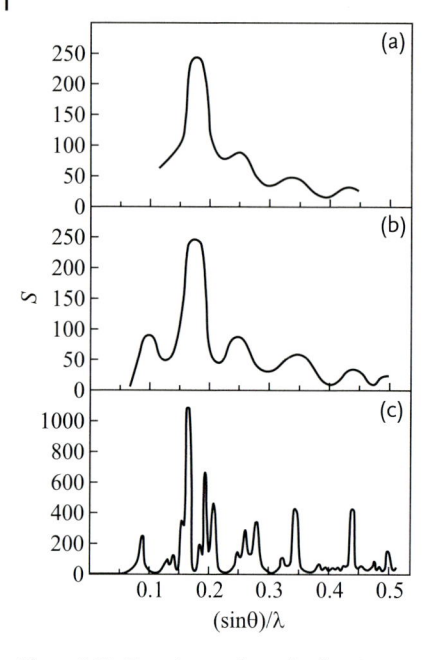

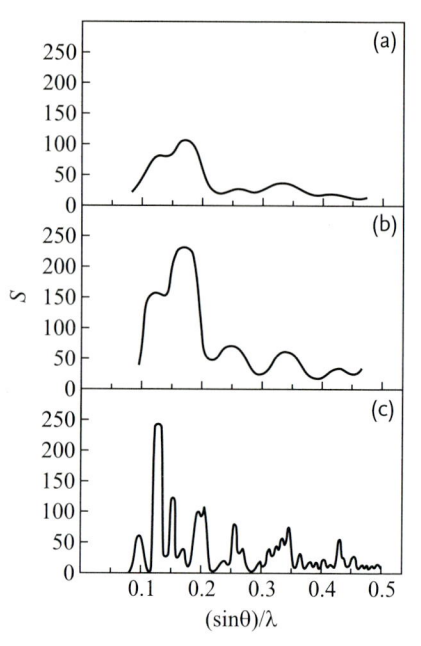

Figure 3.25 Two pictures from the first (Russian) publication of the article of Valenkov and Porai-Koshits in 1949 [302]. The (a) graphs indicate the diffraction pattern of the initial glass of two compositions (1: left row and 2: right row), from the Na_2O/SiO_2 system, respectively, with 50.0 and with 37.5 mol% Na_2O. The (b) graphs indicate the diffractograms after heat treatment (without crystallization) for 2 h at 430 °C and for 10 min at 800 °C for the two compositions (1) and (2). The (c) graphs indicate the fully devitrified samples. Note the gradual change from (a) to (c) indicating the intermediate structural agglomeration in the devitrifying, but still amorphous system.

in fact a confirmation of the generally amorphous, continuous network model of the Zachariasen–Warren theory.

However, there was an essentially new element in the experimental proof given by Valenkov and Porai-Koshits in their first work and in the following investigations by Porai-Koshits and his team of coworkers in the Institute of Silicate Chemistry in St. Petersburg: there exists a continuous disordered SiO_2 network in glasses, as required by Warren, but the degree of disorder of this structure was by far lower than expected. The Na-ions and especially the silicate anions in this disordered lattice were by far more ordered than anticipated originally. The anions had even a distinct "chemical" structure.

In fact, in silicate glasses of different heating prehistory a different stage of pre-ordering of the Na-ions was proven by Porai-Koshits and the structural changes in the anionic part resembled in many respects the ideas of Hägg [303]. Hägg proposed glass structures, which could be more or less exactly described as a mixture of anions with cationic arrangements around them, mimicking structural and "chemical" compositions, possible according to the corresponding phase diagrams. These investigations were extended by Porai-Koshits to borate glasses in the follow-

3.8 On the Intensity of Thermal Fluctuations in Cooling and Heating of Glass-Forming Systems

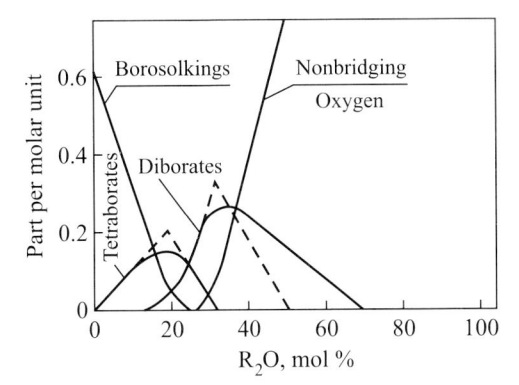

Figure 3.26 Anionic composition of alkali borate glasses according to one of the latest publications of Porai-Koshits [304]. Different methods have been used by several authors to construct a picture corresponding to the fully developed ideas of Porai-Koshits, with oxide glasses being a complicated solution of several structural anionic units.

ing years and here these ideas of continuously changing amorphous anionic structures were confirmed by several methods of investigations as summarized in one of the review papers, finalizing his work (Figure 3.26 and [304]). These structural ideas and results of Porai-Koshits gave the stimulus for a new development, the formulation of a new approach to the chemical thermodynamics of glass-forming systems. This task was performed mainly by Vedishcheva, Shakhmatkin, Shultz (see [305, 306] and the references cited there) and by Mazurin and Priven ([307] and Chapters 6 and 7 of the present book).

It is of particular interest to mention that the processes of crystallization, as analyzed via small angle X-ray diffraction by Porai-Koshits and his coworkers, not only indicated significant structural features, but also led to substantial developments in theory and experiment of induced nucleation and to the formation of new glass-ceramic materials. Here of particular interest was the work of both Porai-Koshits and Filipovich [308, 309] who later on, also in collaboration with Porai-Koshits, investigated thoroughly the thermodynamics and kinetics of liquid phase separation in glasses. The results obtained in this respect were summarized in the monograph [310] which is still the leading textbook in the field, interested in the formation of secondary amorphous phases in the bulk of glasses. In addition, the interplay between liquid phase separation and crystallization was investigated by Porai-Koshits *et al.* [310] and later on, the common features of liquid phase separation and other processes, caused by "heterophase" fluctuations, and the formation of "homophase" fluctuations in glasses [1, 304, 311–313] were analyzed.

There is another important effect to be mentioned in the present overview, the remarkable "flare up" of fluctuations in heating up experiments of glasses reported by Vasilevskaya, Titov, Golubkov and Porai-Koshits [311, 314, 315], that is, an increased probability of density fluctuations in the vicinity of T_g. This is another example how close Porai-Koshits and his colleagues were, in investigating structural problems, also to indicate and experimentally verify another problem of the physics of glass: of the fluctuation nature of optical effects in both undercooled liquids and glasses and of their possible role in the glass transition (see also [181]).

3.8.3
Final Remarks

The given overview on the experimental results by Mueller and Porai-Koshits and coworkers shows that glasses are indeed systems with an increased intensity of density and composition fluctuations caused by the peculiar properties of glasses as summarized in the introduction to this section. Even the observed by Porai-Koshits and coworkers "fluctuation flash" in heating can be given a straightforward explanation. Indeed, having a look at Figures 3.16 and 3.23, in heating processes a range of temperatures is observed, where the average value of the order parameter changes with a relatively high rate in order to approach the respective equilibrium value. The change of the order parameter proceeds hereby locally and results correspondingly in an increasing degree of heterogeneity until the system becomes close to equilibrium. The fluctuation flash becomes, as pointed out in [181], possible due to the rapid decrease of the characteristic relaxation time in this temperature interval as evident from Figure 3.23. This effect cannot be given an interpretation in terms of Simon's simplified model but only in terms of the generic approach as employed in the present section.

Finally, in analyzing the mentioned "fluctuation flash," the general fluctuation theory of second-order phase transitions should be recalled. However, in glasses, the fluctuation flash occurs only at heating indicating again the similarity but also the existing differences with respect to second-order equilibrium phase transitions.

3.9
Results and Discussion

In Chapter 2, we have summarized some of the most significant properties of glass-forming melts, as they are exhibited by experiment, and in an empirical phenomenological approach we have also described the main characteristic features of the liquid to glass transition as they appear in experimental evidence. We have also outlined the simple, but extraordinary instructive approximate way in which known thermodynamicians, like Simon, have tried to treat the glass transition. In the framework of his ideas vitrification was described as a process leading the system out of both thermodynamics and of proper thermodynamic treatment. And really from Simon's approximation, the results of his considerations and his exact measurements of the specific heats for the two already mentioned cases – for crystalline, vitreous and liquid glycerol and for the crystalline and amorphous forms of silica, and for the many additionally investigated cases by many authors to come – it followed that glasses are nonequilibrium, even nonthermodynamic systems from the point of view of classical thermodynamics. To make the necessary calculations, leading to this conclusion, Simon's approximation had to be used in order to apply classical thermodynamics. Nevertheless, as shown in the preceding chapter, Simon's approximation gave the possibility to *calculate* the values of the frozen-in thermodynamic parameters of glasses employing the accumulated experimental evidence. The possible errors, introduced into the theoretical description by following Simon's approximation, are discussed here and in some other respect in Chapter 8 and in [316]. We arrived, based on the generalized nonequilibrium form of thermodynamics, at the conclusion that the error in the determination of the thermodynamic functions of glass-forming melts and glasses, utilizing Simon's approximation, are so small that they can be safely neglected, at least, in most cases (cf. [316]).

The formalism employed in the present chapter, i.e. the quasi-linear treatment in the framework of thermodynamics of irreversible processes seems to us at present to give the most secure way of analysis of glass transition. Other possible ways are connected with the application of different model approaches of statistical physics. Many attempts have been made in this respect beginning with the approaches and results described both in early (see Tammann [72]) and present-day glass literature (e.g., by Binder and Kob [317] or Nieuwenhuizen [283]) to develop new specialized thermodynamic approaches and especially molecular statistical models to describe glasses and the glass transition. A particular approach in this direction, based on the so-called energy landscape statistical model, has been developed by Mauro and Gupta [252]. Although the results of the latter authors give in some respects dubious results, even directly contradicting experimental evidence, they have to be mentioned here as giving the first model-calculated temperature course of the thermodynamic functions of glasses. In some respects, however, as with the so-called nullification of the configurational entropy of glasses near T_g, this model approach led to a direct failure (see, e.g., the critical analysis of this model in [255–257] and in [316]).

Similar statements can be made with respect to other glass transition models and results of newly developed model approaches and even with respect to many of the results following from the mode-coupling theory: although formulated on a safe and very general basis, the theory, even according to the assessment of the authors which have contributed to its development, has provided in its best applications only results, which are *in accordance* with well-known empirically or semi-empirically established dependencies. It has given essentially no new results with respect to experimental investigations nor in possible technical applications. On the contrary, in many other statistical models, even serious efforts, as was mentioned with the example of the energy landscape model, brought about inconsistent results. This statement refers especially to statistical and computational models, in which necessary approximations and assumptions have to be introduced which are, however, sometimes of dubious, difficult to control or even noncontrollable validity.

Note further that in our approach it is the interplay between the relaxation time and the characteristic time of change of the external parameters (and not of an observation time (cf. [252, 254])), which determines the glass transition. This distinction is by no means irrelevant. One of the basic problems in equilibrium statistical physics is how averaging over time and averaging over ensembles can be reconciled. In order to give some foundation of the equivalence of both methods, the ideas of ergodicity have been introduced without a final resolution of the problem (recurrence times, etc.). The times required for the system to pass all or, at least, near to all possible states of the constant energy hypersurface in phase space is much too large in order to give a sound foundation of the equivalence of averaging in time and averaging over the respective ensemble. Considering systems at time-dependent conditions, in addition to the observation time, one has to account for the characteristic times of change of the thermodynamic control parameters. Consequently, observation times and time of change of the control parameters are by no means equivalent quantities. The introduction of the observation time as an essential time scale of vitrification is employed in [252, 254] as the starting point for the derivation of the conclusion of zero values of the configurational entropy. This conclusion is in contrast with the classical "conventional view" we arrived at in the present analysis and which is expressed also in the already mentioned recent analysis by Goldstein [255], Johari [256, 257] and Gutzkow et al. [316]. Consequently, to stress that it is the characteristic time of change of external control parameters (as we interpret it) and not of the observation time (cf. [252, 254]) gives an additional support to the conventional point of view of a nonzero value of the configurational entropy. Note as well the following aspect: Ergodicity is by definition the property of a system to have the ability to access all possible states of the constant energy hypersurface in phase space. So, the notation of a continuously broken ergodicity lacks any sense in the original meaning of this word. In the original meaning, the property of ergodicity can be lost or retained but not in a continuous fashion.

In this respect it may be of use to remember a statement by J.W. Gibbs, one of the founders of statistical physics. In the introduction to his fundamental work *Elementary Principles of Statistical Physics* he wrote [318]: "*Even if we confine our attention to the phenomena distinctly thermodynamic, we do not escape difficulties ... Certainly, one is building on an insecure foundation, who rests his work on hypotheses concerning the constitution of matter ... Difficulties of this kind have deterred the author from attempting to explain the mysteries of nature, and have forced him to be contended with the more modest aim of deducing some of the more obvious propositions relating to the statistical branch of mechanics.*" Gibbs mentioned the problems concerning advancing correct hypotheses with respect to thermodynamic systems in equilibrium and not the much more complicated nonequilibrium systems that glasses are.

The general premises of the phenomenological thermodynamic theory, developed first by De Donder, Prigogine, and Defay and Onsager give an extraordinarily interesting, more secure way to treat systems in nonequilibrium, even frozen-in thermodynamic states, of processes in nonequilibrium systems or processes leading out of equilibrium. First efforts in this respect have been undertaken by Prigogine himself to apply directly the derivations and the way of thinking of thermodynamics of irreversible processes to the glass transition. This task was performed by him, however, in the framework of Simon's assumption that glass transition takes place abruptly at a given temperature, the freezing-in temperature, T_E, of the system, which Simon identified with the glass transition temperature, T_g. Another important development in this respect was made by Davies and Jones on relaxation in the framework of nonequilibrium thermodynamics employing the concept of internal or structural order parameters. As the next step, the treatment of the problems of vapor pressure and the solubility of glasses performed by Gutzow et al. have to be mentioned [172, 185]. All these efforts were, however, also made in the mentioned limits of Simon's approximation.

In the present chapter, we have employed the thermodynamics of irreversible processes as a tool to describe the real course of vitrification, as it takes place in the glass transition region, using a simple kinetic model, first proposed by Bragg and Williams in 1934 [244] in analyzing freezing-in processes in metastable metallic alloy systems. A similar kinetic model was also developed in the same *ad hoc* manner – as performed by Bragg and Williams – also by Volkenstein and Ptizyn in 1956 [170] in order to describe the kinetics of glass transition. In Russian literature of those days this kinetic approach was denoted as the kinetic theory of glass transition (cf. also Mazurin [70]). Here we have shown that this kinetic model can be in fact *derived* in the framework of the thermodynamics of irreversible processes, that it follows directly from one of the basic principles of thermodynamics of irreversible processes: from its so-called phenomenological law. In a first move into this direction it was shown by Gutzow and Schmelzer [1] how the isothermal kinetics of relaxation, that is, De Donder's law follows from the *linear* formulation of the phenomenological law. It was also indicated in [1] how the more general, nonlinear formulations of the relaxation kinetics follow from a more general formulation of the phenomenological law, first proposed by Callen [43]. In a second move, Gutzow et al. [11, 12, 17, 319] could show that the kinetic theory of vitrification follows from the thermodynamics of irreversible processes in a generic derivation, in which glass transition is considered as a nonisothermal process of relaxation. In a third step, applying known results and notions of the thermodynamics of irreversible processes, the thermodynamics of glass transition was constructed in a generic way out of the process of nonisothermal cooling or heating runs of a glass-forming liquids (see the series of papers by Gutzow, Schmelzer et al. [179, 180, 182, 240]) reviewed and extended in the present chapter. In the present chapter, it is also already indicated how in terms of the same kinetic model of glass transition the temperature dependence of the viscosity of glass-forming melts can be derived (see [11, 181]). A more detailed analysis of this circle of problems is given in Chapter 4.

In a series of subsequent publications, Gutzow, Schmelzer et al. applied this generic thermodynamic approach to some of the most essential problems of glass transition and to the description of the very nature of glasses and the glassy state. It was shown that in this way a phenomenological formalism can be derived from a new, broader standpoint giving the temperature course of the thermodynamic functions over the whole glass-transition region, the change of thermodynamic and kinetic coefficients in this region in both cooling and heating runs. Of particular significance was that in this series of investigations it turned out that both the *freezing-in and the production of entropy* can be calculated, and that *in both heating-up and cooling entropy production* has to take place [240]. In a following development the necessary corrections to Simon's approximate approach were determined.

Any isothermal measurements of relaxation processes in glass-forming systems represent – in accordance with the basic laws of thermodynamics – a direct experimental proof that entropy production effects occur and have to be accounted for as an inherent feature of any glass transition. However, following Simon's model, the relevance of such effects in vitrification has been ignored so far widely both in theoretical and experimental investigations. For this reason, direct experimental analysis allowing one to analyze the effect of entropy production and its consequences for the understanding of the properties of glasses are relatively scarce. For the model considered entropy production effects are small, this may be different for other classes of glass-forming melts.

Moreover, the value of an important combination of thermodynamic coefficients (the so-called Prigogine–Defay ratio), having an exclusive cognitive significance in revealing the nature of glass, found its relatively simple explanation in this new theoretical approach [234]. The same approach also elucidated the significance in terms of the thermodynamics of irreversible processes of notions like *fictive temperatures* and *fictive or negative pressures* in glass transition and in the definition of the vitreous state itself. The phenomenologically bounded treatment of glass-transition gives, as it also turned out, the only possible method to investigate this process in the way as it proceeds: in a naturally broad interval where entropy and structure are continuously fixed and changed.

Summarizing, we may conclude the following. The generic approach to the description of the glass transition has, as shown, a number of advantages in addition to the possibility of an account of entropy production terms. Some of them are: (i) the possibility of derivation of the Frenkel–Kobeko and Bartenev–Ritland equations based on Eq. (3.5) and [11, 12, 17, 320], (ii) the qualitative identity of experimental and theoretical $C_p(T)$ curves (cf. Figures 3.8 and 3.9 and [251], (iii) the possibility of a straightforward theoretical interpretation of the Prigogine–Defay ratio [234], (iv) the possibility of interpretation of peculiarities in the temperature dependence of the viscosity and characteristic relaxation times near and below T_g (cf. Figures 3.20 and 3.21), (v) the mere existence of isothermal relaxation processes, and so on. These results give a convincing proof that (i) vitrification has to be explained theoretically as a process proceeding in some finite interval of the external control parameters, (ii) the vitrification interval is characterized by an interplay of relaxation and change of external control parameters which determines the properties of the glasses, (iii) the freezing-in process is accompanied by entropy production, (iv) it allows one to understand an increased intensity of fluctuations in glasses (the spectrum of fluctuations has to be different as compared with fluctuations in a stable or metastable equilibrium system, the intensity of fluctuations will be as a rule higher since there is no restoring force [181]), and (v) the theoretical approach developed here gives an appropriate tool to treat such effects.

Of course, a more detailed direct experimental and theoretical analysis of entropy production and its effect on the properties of the resulting glasses for different classes of glass-forming melts is highly desirable. One of the possible circle of analysis could consist here in the measurement of heat effects in isothermal relaxation. A more straightforward circle of experiments allowing one to verify directly the theoretical predictions outlined here consists in the analysis of thermal effects of relaxation in cyclic heating and cooling processes. Here, as one of the possible tasks, a direct proof of the general thermodynamic consequences outlined in Section 3.4.6 could be performed. Another topic of interest is, to our opinion, the analysis of the magnitude of entropy production effects in dependence on the rate of change of the external control parameters. Some first attempts in these directions are performed already in [245, 246, 265].

According to the theoretical concepts followed in the paper, the glass transition temperature, T_g, is defined by the relation (cf. Eq. (3.5))

$$|q_\theta|\tau(T_g) \cong \text{const.}, \quad \theta = \frac{T}{T_m}, \quad q_\theta = \frac{d\theta}{dt}. \tag{3.163}$$

In verbal form, the vitrification temperature is reached, when the characteristic relaxation times are comparable in magnitude with the characteristic times of change of temperature [1]. It follows as one consequence that – describing the viscosity via the VFT equation – the glass transition temperature has to always be located above the Kauzmann temperature, since the assumption $T_g = T_0$ would lead to the necessity of employing a zero cooling rate, that is, to a process which cannot be realized. On the other hand, Eq. (3.67), supplemented by the definition of the glass temperature Eqs. (3.5) and (3.163), shows that the rate of approach of the structural order parameter, ξ, to its equilibrium value, ξ_e, is, close to the glass transition temperature, independent of the cooling rate. For this reason, with increasing cooling rates higher values of the difference $(\xi - \xi_e)$ are expected to be found, in general, close to T_g: the rate of approach of ξ to ξ_e is constant near T_g but the rate of change of ξ_e increases with increasing cooling rates (similar considerations can be made also with respect to the heating runs). As the result, an increased effect of the entropy production term has to be expected, in general, with increasing rates of change of the control parameters (cf. also [245, 246, 265]). Vice versa, with a decrease of the cooling rate the entropy production term should loose its importance. This result is in qualitative agreement with expectations that close to the Kauzmann temperature, a so-called ideal glass transition has to take place, where entropy production terms are negligible. However, as mentioned already such an ideal glass transition can be realized only by zero cooling rates, or for infinite times (cf. also [150, 321, 322]).

In the present analysis, we considered mainly cyclic processes of vitrification and devitrification, where both cooling and heating proceed with the same constant absolute value of the rate of change of temperature. Most of our conclusions retain their validity if we assume that this condition of constant heating or cooling rate is not fulfilled but the temperature is a strictly monotonic function of time in both heating and cooling. However, the respective results are more complicated in their interpretation and in discussing new effects, one should always start, as we did here, with the simplest realizations. In addition, we considered cyclic processes, where the initial state is always reestablished after the completion of the process of cooling and subsequent heating. Provided one starts the cycle at temperatures inside the vitrification interval, ($T_g^- \leq T_g \leq T_g^+$), then the behavior of the system should depend on the number of the cycle in a series of cyclic heating-cooling processes. It would be of interest to also perform experiments in this respect and to compare them with the theoretical predictions obtained in the framework of the theory presented here.

The theoretical approach developed in the present analysis leads to a variety of additional results in the theoretical interpretation of vitrification. Some of them are discussed in [245, 246]. However, the potential applications of the theoretical concepts outlined here are expected to be of much wider applicability (cf., e.g., [323, 324]) and will be explored in future analysis.

Finally, we would like to mention that the approach followed here is compatible with amorphous polymorphism. Different amorphous states could be described in the theoretical model as attractors the order parameter can evolve to, in dependence on the initial conditions, when the model exhibits more than one metastable equilibrium state.

Simon's model of glass transition, employed so far in most analyses of glass formation, allows one to perform a sufficiently accurate determination of a certain spectrum of properties of glasses. In particular, it allows one to compute the thermodynamic functions of glass-forming melts in cases where entropy production terms are of minor significance. It does not cover, however, adequately a variety of other features (in particular, relaxation, hysteresis effects and the Prigogine–Defay ratio). Our generic phenomenological vitrification, based upon De Donder's structural order parameter approach, is able to give a comprehensive description both of thermodynamic and kinetic properties of glasses and glass-forming melts. Concentrating the analysis in the present chapter on thermodynamic aspects, in the next chapter, we will apply the method to a detailed analysis of the temperature dependence of the viscosity and, related to it, relaxation kinetics of glass-forming melts.

4
Generic Approach to the Viscosity and the Relaxation Behavior of Glass-Forming Melts

Jürn W. P. Schmelzer

4.1
Introduction

In the present chapter, we apply the generic approach to the analysis of some further kinetic properties of glass-forming systems, the pressure dependence of the viscosity and the relaxation behavior both at isothermal conditions. In particular, we show how the pressure dependence of the viscosity can be expressed provided free volume dominates the behavior and how the well-known stretched exponential relaxation behavior can be understood in terms of the approach developed by us and described in Chapter 3.

While the type of temperature dependence of viscosity at isobaric conditions is qualitatively well-established (the viscosity increases with decreasing temperature), current discussion on the dependence of viscosity on external pressure (at isothermal conditions) shows a spectrum of controversial statements (for the details, see [276]). In a variety of analysis it is concluded that the viscosity must increase with increasing pressure and this expectation has been confirmed in a large number of experimental studies. However, opposite results were experimentally found as well, for example, for some glass-forming silicate liquids, such as basalt, dacite, jadeite and albite. In these cases, a decrease of the viscosity of about one order of magnitude has been reported for pressures increasing by 2 to 5 times. For a number of silicate and aluminosilicate melts, an even more complex nonmonotonic behavior of viscosity in dependence on pressure was found at high pressure in the range of GPa.

The mentioned decrease of viscosity with increasing pressure, obtained in certain ranges of pressure and temperature, was denoted as *anomalous* and being in contradiction with the results of "free volume" theories of viscosity [325, 326]. This specification of the negative pressure dependence of viscosity as anomalous indicates that an increase of viscosity with increasing pressure is considered as the rule, but that deviations are also possible. In this way, the question arises on how such kind of behavior can be explained theoretically and whether they are, indeed, exceptions.

Glasses and the Glass Transition, First Edition. Jürn W.P. Schmelzer and Ivan S. Gutzow.
© 2011 WILEY-VCH Verlag GmbH & Co. KGaA. Published 2011 by WILEY-VCH Verlag GmbH & Co. KGaA.

One attempt in this direction of analysis was put forward by Gupta [327]. Later on, Bottinga and Richet [325] also derived an equation for the pressure dependence of viscosity based on a generalization of the Adam–Gibbs equation [148]. Gupta concluded that, in most cases of interest, an increase of viscosity with increasing pressure must be expected, although exceptions are possible (according to his approach) when the relaxation expansion coefficient, $\Delta \alpha$ – the difference between the thermal expansion coefficient of the equilibrium liquid and the glass at given values of pressure and temperature – is negative. However, to compute the pressure dependence of viscosity the resulting equation requires the knowledge of the differences between the entropies of equilibrium liquid and glass, and the relaxation expansion coefficient. The respective data are seldom available and for temperatures above the glass transition temperature not available, in principle [325]. By this reason, it is of interest to express the pressure dependence of the viscosity by more accessible thermodynamic coefficients, which refer exclusively to the liquid in the actual state considered avoiding in this way, in addition, some problems inherent in approaches like those followed in [325] and [327].

4.2
Pressure Dependence of the Viscosity

4.2.1
Application of Free Volume Concepts

We first consider liquids of constant composition in (stable or metastable) thermodynamic equilibrium states. According to the Gibbs phase rule, the number of degrees of freedom of the system is equal to two, and thus one can choose temperature and pressure as the independent variables determining the properties of the system. In such cases, the viscosity, η, can be considered as a function of pressure and temperature, that is,

$$\eta = \eta(p, T). \tag{4.1}$$

In their extended investigation of similarities and differences of liquid–vapor and liquid–solid phase transitions, Skripov and Faizullin [328, 329] analyzed the dependence of viscosity, η, on temperature, T, and pressure, p, for several classes of liquids. They restricted their analysis to the cases where the thermal expansion coefficient of the liquid is positive (this property is fulfilled at atmospheric pressure for most but not all liquids). The analysis of both literature data and their own results led mentioned authors to conclude that the following relations must be fulfilled:

$$\left(\frac{\partial \eta}{\partial T}\right)_p < 0, \tag{4.2}$$

$$\left(\frac{\partial \eta}{\partial p}\right)_T > 0. \tag{4.3}$$

These relations imply that the viscosity must decrease with increasing temperature (for isobaric processes), and must increase with increasing pressure (at isothermal conditions). Moreover, considering the viscosity as a function of pressure and temperature, that is, $\eta = \eta(p, T)$, they wrote down the following identity:

$$\left(\frac{\partial \eta}{\partial T}\right)_p \left(\frac{\partial T}{\partial p}\right)_\eta \left(\frac{\partial p}{\partial \eta}\right)_T = -1. \tag{4.4}$$

Equation (4.4) follows from purely mathematical considerations and does not involve originally any physics except the specification of the number of degrees of freedom.

Taking into account the viscosity dependencies given by Eqs. (4.2) and (4.3), they concluded that the inequality

$$\left(\frac{\partial p}{\partial T}\right)_\eta > 0 \tag{4.5}$$

must be fulfilled. This equation – appearing here first as a purely mathematical relation – has, of course, a quite definite physical meaning. As mentioned earlier, the viscosity of liquids of constant composition in (stable and metastable) thermodynamic equilibrium states can be considered as a function of two state variables, pressure and temperature, that is, $\eta = \eta(p, T)$. However, if one considers viscosity as constant (i.e., $\eta = \eta(p, T) = \text{const.}$), then this relation gives a dependence between pressure and temperature (at constant viscosity). Therefore, Eq. (4.5) implies that – in order for the viscosity to remain constant – an increase of temperature leads to effects which can be compensated by an increase of pressure.

It should be stressed that Eqs. (4.2) and (4.3) are corroborated by a variety of experimental results and lead to the theoretical consequence given by Eq. (4.5). However, one can easily reverse the above-mentioned argumentation. Indeed, taking into account general concepts connecting viscosity with the "free volume" in the liquid [1, 25, 328, 329] we can analyze theoretically how the partial derivative $(\partial p/\partial T)_\eta$ should behave. Taking into account, additionally Eq. (4.2), one can arrive then at the desired theoretical prediction concerning the pressure dependence of the viscosity. Indeed, following the classical work of Frenkel [25], the essence of "free volume" concepts can be expressed as follows: (i) free volume uniquely determines the value of the viscosity, η; (ii) free volume is uniquely determined by the total volume, $V(p, T)$, of the liquid and not by pressure, p, and temperature, T, separately. Frenkel mentioned Batchinskij's equation as one example for such type of dependence and described the necessity of free volume as a prerequisite of flow via an analogy to social behavior. He wrote: "A highly densely packed crowd of people cannot leave a room even if the doors are open," (in the English translation [25], this phrase has been modified to "A highly densely packed crowd of people cannot leave a room until the doors will be opened," giving a slight but significant deviation of the content due to the personal view of this statement by the translator). The respective ideas have been developed later on by different authors advancing particular models of free volume theories for the dependence of the viscosity on the thermodynamic state parameters. In contrast to such earlier attempts, here we employ only the above-given essence of free volume concepts without specifying any particular models for viscous flow.

In order to proceed we realize that the free volume of liquids, as a rule, decreases with pressure. In this way, in order to reestablish the value of the free volume (and the resulting value of viscosity), one has to vary the temperature to such extent as to compensate the changes of free volume due to the effect of pressure. Consequently, from such general theoretical considerations, one arrives directly at the inequality Eq. (4.5), but now independently of the knowledge of experimental data. Taking exclusively the dependence given by Eq. (4.2) from experiment, we then arrive, utilizing Eq. (4.4), at Eq. (4.3). In this way, employing only general concepts connecting viscosity with free volume of the liquid (and assuming the absence of any other structural changes in addition to densification) we can conclude that, in isothermal conditions, the viscosity has to increase, as a rule, with increasing pressure. However, exceptions from this general rule are possible, as will become evident from the following quantitative analysis.

In addition to the qualitative conclusions given above, one can easily formulate a method to quantitatively estimate the pressure dependence of viscosity (at isothermal conditions) provided the temperature dependence of viscosity (at constant pressure) and some other purely thermodynamic characteristics of the liquid are known. Indeed, according to above analysis, we may connect variations in viscosity with variations of "free volume" and suppose that the free volume is uniquely connected with the total volume of the system. Then, in order to secure constancy of viscosity, one has to demand that the total volume of the system is kept constant if both pressure and temperature are varied slightly by dT and dp, respectively. From Eq. (4.6),

$$dV(p,T) = \left(\frac{\partial V}{\partial T}\right)_p dT + \left(\frac{\partial V}{\partial p}\right)_T dp = 0, \tag{4.6}$$

we then get the following result:

$$\left(\frac{\partial T}{\partial p}\right)_\eta \cong \left(\frac{\partial T}{\partial p}\right)_V = -\frac{\left(\frac{\partial V}{\partial p}\right)_T}{\left(\frac{\partial V}{\partial T}\right)_p}. \tag{4.7}$$

With Eq. (4.4), we finally obtain

$$\left(\frac{\partial \eta}{\partial p}\right)_T = \left(\frac{\partial \eta}{\partial T}\right)_p \frac{\left(\frac{\partial V}{\partial p}\right)_T}{\left(\frac{\partial V}{\partial T}\right)_p} = -\frac{\kappa(p,T)}{\alpha(p,T)}\left(\frac{\partial \eta}{\partial T}\right)_p, \tag{4.8}$$

where κ is the isothermal compressibility and α the isobaric thermal expansion coefficient, that is,

$$\kappa = -\frac{1}{V}\left(\frac{\partial V}{\partial p}\right)_T, \quad \alpha = \frac{1}{V}\left(\frac{\partial V}{\partial T}\right)_p. \tag{4.9}$$

Let us, now, analyze the possible consequences from Eq. (4.8). For systems in (stable or metastable) thermodynamic equilibrium, the condition of thermodynamic stability

$$\kappa = -\left(\frac{\partial p}{\partial V}\right)_T > 0 \tag{4.10}$$

(or $\kappa > 0$) must be fulfilled. Taking into account Eq. (4.3), it follows that the signs of the derivatives $(\partial T/\partial p)_\eta \cong (\partial T/\partial p)_V$ and $(\partial \eta/\partial p)_T$ are determined by the sign of the thermal expansion coefficient, which is (at atmospheric pressure), as a rule, a positive quantity. In these cases, Eq. (4.8) predicts an increase of viscosity with increasing pressure. Skripov and Faizullin [328, 329] restricted their analysis to such cases. Consequently, their conclusion is correct as far as this condition is fulfilled.

Summarizing, we first conclude that free volume concepts do not, in general, forbid a decrease of viscosity with increasing pressure. They lead to such results as well, if the thermal expansion coefficient of the liquid is negative in the respective ranges of pressure and temperature. However, at low and moderate pressures, the thermal expansion coefficient is, in general, a positive quantity and, in such cases, the viscosity should increase with increasing pressure provided the behavior is determined by free-volume effects.

As a second conclusion, we can state: If the viscosity of a liquid decreases with increasing pressure and the thermal expansion coefficient is positive, then the pressure dependence for these anomalously behaving systems cannot be explained, in principle, exclusively by free volume concepts. One example in this respect will be analyzed in detail in the subsequent Section 4.2.2.

4.2.2
A First Exception: Water

In Figure 4.1, density and thermal expansion coefficients of ordinary water are shown as functions of temperature for different values of pressure in the range 0–20 MPa. At 0 °C, water has negative thermal expansion coefficients in the range of pressures from 0–20 MPa. According to Eq. (4.8), one expects in this range a decrease of viscosity with increasing pressure. This is indeed the case, however, it turns out that the viscosity of water at 0 °C decreases with increasing pressure also in the range where the isothermal expansion coefficient becomes positive. Consequently, in this range of pressures, Eq. (4.8) cannot describe the experimental data even qualitatively correctly.

Figure 4.2 shows the viscosity of water as a function of pressure for wider ranges of the thermodynamic state parameters (pressure values in the range from 0–100 MPa and temperatures 0 and 25°C). As evident from the figure, with an increase of temperature, the dependence of the viscosity on pressure changes qualitatively and a minimum occurs (illustrated in the figure for the temperature 25 °C). For values of pressure above those corresponding to this minimum, the viscosity behaves in the "normal" way, that is, increases with increasing pressure. With a further increase of temperature, this minimum is shifted to lower values of pressure and disappears at all. Such situation is found, for example, at a temperature 40°C. At these temperatures, the thermal expansion coefficient of water is positive and the viscosity of water increases with increasing pressure. In this

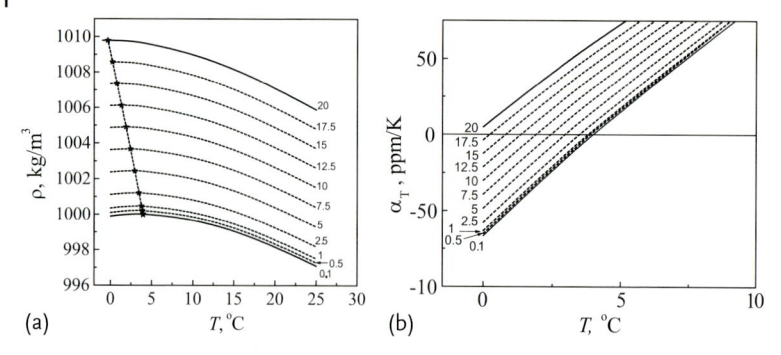

Figure 4.1 (a) Density of water vs. temperature at different pressures (in MPa) as specified in the figure from [330]. The line crossing the curves obtained for different pressures specifies the location of the density maximum. At the left side of this line, the thermal expansion coefficient is negative. To the right hand side of this line, the thermal expansion coefficient is positive; (b) thermal expansion coefficient of water as a function of temperature at different pressures (in MPa) as specified in the figure from [330].

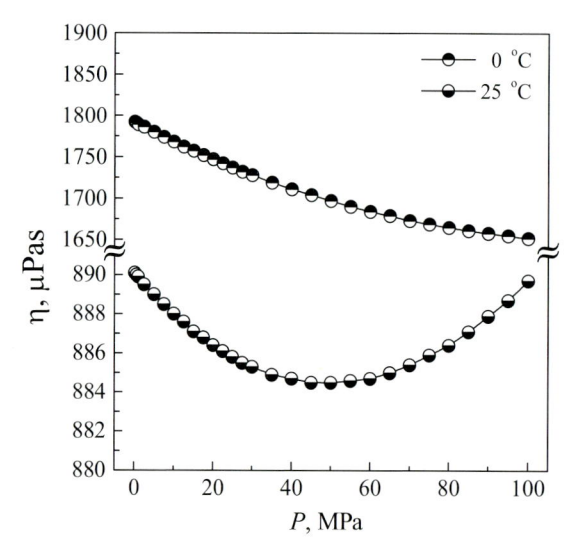

Figure 4.2 Viscosity of water as a function of pressure (in the range from 0–100 MPa) at 0 and 25°C.

temperature range, Eq. (4.8) gives a qualitatively correct description of the pressure dependence of the viscosity also for water. By this reason it is of interest to check to which extent Eq. (4.8) can give a quantitatively correct description.

As it turns out [276], for water the theoretical expression, based exclusively on free volume concepts as discussed above, leads to quantitative deviations with experimental data up to one order of magnitude. However, these deviations decrease with increasing pressure and temperature. Such behavior can be interpreted as a consequence of diminishing importance of structural rearrangements of water with increasing pressure and temperature in favor of free volume effects.

Summarizing the results of the analysis for the pressure dependence of the viscosity of water, we conclude that free volume effects alone are obviously not sufficient to give a correct interpretation of the dependencies observed. Consequently, one can expect the occurrence of similar effects also for other classes of liquids and, in particular, for glass-forming melts. By these reasons, in the subsequent sections we generalize Eq. (4.8) to account for the effects of variations of other structural parameters (in addition to free volume effects) on the pressure dependence of the viscosity.

4.2.3
Structural Changes of Liquids and Their Effect on the Pressure Dependence of the Viscosity

As outlined in a variety of papers (e.g., [326, 331–335]), liquids, in general, and silicate melts, in particular, can exhibit a variety of additional mechanisms of structural adjustments (change in the degree of polymerization, coordination numbers, frequency of distribution of different structural units, formation of five- and six-fold coordinated Si, etc.), when subjected to high isostatic pressures, as compared exclusively with variations of free volume. Such changes occur most easily at sufficiently high temperatures when the liquid is in stable or metastable equilibrium states. For such cases, we have to introduce, at least, one additional state variable describing the change of the mentioned structural properties with pressure.

In order to treat theoretically the dependence of the viscosity on pressure for the case that additional order parameters have to be introduced, Eq. (4.1) must be modified. In the simplest approach, we can assume that the viscosity depends, in addition to pressure and temperature, on one additional order parameter, ξ, describing structural reorganization in addition to free volume concepts. In such cases, we have instead of Eq. (4.1)

$$\eta = \eta(p, T, \xi). \tag{4.11}$$

Considering changes of the state parameters T, p and η of the system, by keeping the value of the order parameter, ξ, constant, we then get

$$\left(\frac{\partial \eta}{\partial T}\right)_{p,\xi} \left(\frac{\partial T}{\partial p}\right)_{\eta,\xi} \left(\frac{\partial p}{\partial \eta}\right)_{T,\xi} = -1. \tag{4.12}$$

Instead of Eqs. (4.6)–(4.8), we then obtain

$$dV(p, T, \xi) = \left(\frac{\partial V}{\partial T}\right)_{p,\xi} dT + \left(\frac{\partial V}{\partial p}\right)_{T,\xi} dp + \left(\frac{\partial V}{\partial \xi}\right)_{p,T} d\xi = 0, \tag{4.13}$$

$$\left(\frac{\partial T}{\partial p}\right)_{\eta,\xi} \cong \left(\frac{\partial T}{\partial p}\right)_{V,\xi} = -\frac{\left(\frac{\partial V}{\partial p}\right)_{T,\xi}}{\left(\frac{\partial V}{\partial T}\right)_{p,\xi}}, \tag{4.14}$$

$$\left(\frac{\partial \eta}{\partial p}\right)_{T,\xi} = \left(\frac{\partial \eta}{\partial T}\right)_{p,\xi} \frac{\left(\frac{\partial V}{\partial p}\right)_{T,\xi}}{\left(\frac{\partial V}{\partial T}\right)_{p,\xi}} = -\frac{\kappa_\xi(p, T, \xi)}{\alpha_\xi(p, T, \xi)} \left(\frac{\partial \eta}{\partial T}\right)_{p,\xi}. \tag{4.15}$$

The derivations can be performed similarly with identical results if ξ is replaced by a set of order parameters $\{\xi_i\}$.

Equation (4.15) is widely identical to Eq. (4.8) with the only difference that all quantities entering Eq. (4.15) now depend on one or a set of additional order parameters. Assuming that Eq. (4.2) remains valid in a modified form as

$$\left(\frac{\partial \eta}{\partial T}\right)_{p,\xi} < 0, \qquad (4.16)$$

the sign of the pressure dependence of the viscosity is determined, again, by the signs of thermal expansion coefficient and compressibility.

The change of viscosity with pressure has now to be written – similarly to Eq. (3.159) – in the form:

$$\left(\frac{d\eta}{dp}\right)_T = \left(\frac{\partial \eta}{\partial p}\right)_{T,\xi} + \left(\frac{\partial \eta}{\partial \xi}\right)_T \left(\frac{d\xi}{dp}\right)_T. \qquad (4.17)$$

With Eq. (4.15) we arrive at

$$\left(\frac{d\eta}{dp}\right)_T = -\frac{\kappa_\xi(p,T,\xi)}{\alpha_\xi(p,T,\xi)}\left(\frac{\partial \eta}{\partial T}\right)_{p,\xi} + \left(\frac{\partial \eta}{\partial \xi}\right)_T \left(\frac{d\xi}{dp}\right)_T. \qquad (4.18)$$

The first term on the right hand side of Eq. (4.18) describes variations of the viscosity with increasing pressure provided additional structural rearrangements of the melt can be neglected. The latter effect is expressed primarily by the second term and determines the pressure dependence of viscosity for water at sufficiently low temperatures as discussed in detail in Section 4.2.2.

The above-given equations can be generalized to the case when several independent structural rearrangements of the liquid occur as a result of a pressure increase. In such cases, we have to introduce a set of additional order parameters $\{\xi\} = \xi_1, \xi_2, \ldots$ and Eq. (4.18) takes the form:

$$\left(\frac{d\eta}{dp}\right)_T = -\frac{\kappa_{\{\xi\}}(p,T,\{\xi\})}{\alpha_{\{\xi\}}(p,T,\{\xi\})}\left(\frac{\partial \eta}{\partial T}\right)_{p,\{\xi\}} + \sum_i \left(\frac{\partial \eta}{\partial \xi_i}\right)_T \left(\frac{d\xi_i}{dp}\right)_T. \qquad (4.19)$$

Employing Eq. (4.17) and a similar relation for the derivative $(d\eta/dT)_p$ and utilizing, in addition, Eq. (4.15), we can rewrite Eq. (4.18) as:

$$\left(\frac{d\eta}{dp}\right)_T = -\frac{\kappa_\xi(T,p,\xi)}{\alpha_\xi(T,p,\xi)} \frac{\left\{1+\left[\left(\frac{\partial \eta}{\partial \xi}\right)_{p,T}\Big/\left(\frac{\partial \eta}{\partial T}\right)_{p,\xi}\right]\left(\frac{d\xi}{dT}\right)_p\right\}}{\left\{1+\left[\left(\frac{\partial \eta}{\partial \xi}\right)_{p,T}\Big/\left(\frac{\partial \eta}{\partial p}\right)_{T,\xi}\right]\left(\frac{d\xi}{dp}\right)_T\right\}} \left(\frac{d\eta}{dT}\right)_p. \qquad (4.20)$$

This equation can be generalized similarly to Eq. (4.19), when several order parameters have to be incorporated into the description. It shows that, in the general case, the pressure dependence of viscosity is not exclusively expressed by compressibility and thermal expansion coefficient (reflecting free volume concepts), but also by the reaction of the system to other types of structural adjustments to pressure variation. In case that such additional processes do not occur, we obtain from Eq. (4.20), as a special case, Eq. (4.8), again.

As may be verified easily (cf. for the details [276]), similar dependencies govern the behavior of the pressure dependence of liquids not only for liquids in stable and metastable equilibrium, but they can be applied also for glasses (liquids in frozen-in nonequilibrium states).

4.2.4
Discussion

We have shown here that free volume concepts allow one to predict both an increase and decrease of viscosity with increasing pressure. The type of dependence is hereby determined by the sign of the thermal expansion coefficient. Since, in most cases of interest (at atmospheric pressures), the thermal expansion coefficient is positive, commonly the "normal" behavior, that is, an increase of viscosity with increasing pressure is observed. However, the situation can become quite different for complex liquids and, in particular, at pressures in the GPa range and sufficiently high temperatures. Here, reliable experimental data on the value of the thermal expansion coefficients are, to our knowledge, so far not available.

As a generalization, equations have been developed allowing one to incorporate the effect of additional structural variations of the liquids under consideration, which are not reflected by free volume concepts, in the form of the pressure dependence of the viscosity. These relations may result in a negative pressure dependence of the viscosity even in cases if the thermal expansion coefficient α_ξ is positive. They contain additional terms, which are effective, however, only in ranges of pressures and temperatures, where such additional structural changes indeed occur. By this reason, the respective equations allow one to predict an increase, a decrease and a nonmonotonic behavior of viscosity with pressure, and cover all types of dependencies found in the experimental analysis. Vice versa, the latter behavior can be considered therefore as a strong indication of the existence of structural transformations in the liquid in the considered range of thermodynamic state parameters.

Summarizing the results of the analysis, we have shown from general assumptions concerning the dependence of free volume of liquids and viscosity that – at isothermal conditions and moderate pressures – the viscosity of simple liquids typically increases with pressure. In contrast, complex liquids, such as molten silicates, either show a negative pressure dependence starting from small pressures or change from a positive to a negative dependence at some sufficiently high threshold pressure. An equation is derived that allows one to determine the dependence of viscosity on pressure, provided (in the simplest case) that the temperature dependence of viscosity, the isothermal compressibility and the isobaric thermal expansion coefficient of the liquid are known. This equation can be generalized to situations, when, in addition to free volume effects, other mechanisms of structural reorganization of the liquid with respect to pressure variations exist. In such situations, an increase, decrease and/or nonmonotonic behavior of viscosity with respect to pressure variations are possible.

4.3
Relaxation Laws and Structural Order Parameter Approach

4.3.1
Basic Equations: Aim of the Analysis

In the present section, we consider isothermal relaxation and try to find out what kind of laws for the description of relaxation can be obtained from the structural order parameter approach developed in Chapter 3. Here we consider, again, the case that one structural order parameter, ξ, is sufficient for the description of relaxation processes. Since pressure and temperature are assumed to be kept constance, $\xi^{(e)} = \xi(p, T)$ is a constant as well. At given values of pressure and temperature, the relaxation behavior can be expressed as (cf. Eqs. (3.30) and (3.34)):

$$\frac{d\xi}{dt} = -\frac{1}{\tau(p, T, \xi)} \left(\xi - \xi^{(e)} \right). \tag{4.21}$$

In order to simplify the notations, we will use in the following derivation the dimensionless reduced quantity

$$\tilde{\xi} = \frac{\xi - \xi^{(e)}}{\xi^{(e)}} \tag{4.22}$$

resulting in the following relation

$$\frac{d\tilde{\xi}}{dt} = -\frac{1}{\tau(p, T, \xi)} \tilde{\xi}, \tag{4.23}$$

instead of Eq. (4.21). In this definition, the equilibrium state to which the system eventually tends, corresponds to $\tilde{\xi}(t \to \infty) = 0$.

In a first approximation, one could suggest to replace the characteristic relaxation time, $\tau(p, T, \xi)$, by the Maxwell relaxation time, $\tau_e(p, T)$, depending only on pressure and temperature but not on the order parameter. Such an approximation would lead to the simplified equation

$$\frac{d\tilde{\xi}}{dt} = -\frac{1}{\tau_e(p, T)} \tilde{\xi}, \tag{4.24}$$

and an exponential relaxation behavior of the order parameter given by

$$\tilde{\xi}(t) = \tilde{\xi}(0) \exp\left(-\frac{t}{\tau_e}\right). \tag{4.25}$$

However, already more than 150 years ago, R. Kohlrausch and F. Kohlrausch were aware (cf. [233]) that the simple relaxation law, as given by Eq. (4.25), is not sufficient for the description of relaxation processes in glass-forming melts, replacing them (in above notations) by dependencies of the form

$$\frac{d\xi}{dt} = -\frac{\psi(\xi)}{\tau_e(p, T)} \xi , \qquad (4.26)$$

or

$$\frac{d\xi}{dt} = -\frac{\varphi(t)}{\tau_e(p, T)} \xi . \qquad (4.27)$$

Employing different assumptions for the dependencies of either $\psi(\xi)$ on ξ or $\varphi(t)$ on t, different relaxation laws can be obtained, for example, Kohlrausch's stretched exponent formula (for $\varphi(t) \propto 1/t^n$ with $n < 1$; for more details see [233]):

$$\frac{\xi}{\xi(0)} = \exp\left\{-\left(\frac{t}{\tau_K}\right)^\beta\right\}, \quad \beta = 1 - n, \quad \tau_K = \tau_0(1-n)^{1/(1-n)} . \qquad (4.28)$$

In order to establish the type of dependence $\varphi = \varphi(t)$ or $\psi = \psi(\xi)$ one can – as one possible approach – develop statistical mechanical models of relaxation processes appropriate for the desired classes of systems under consideration. Alternatively, one can start with Eq. (4.23) and try to find out which kind of relaxation behavior one may expect making only very general assumptions in the analysis of this relation. In this approach, one can either understand the origin why certain classes of relaxation laws have been appropriate in describing relaxation or suggest dependencies eventually capable for a detailed description. The analysis of this second approach is the aim of the following section.

4.3.2
Analysis

In the analysis of Eq. (4.23) we assume (i) that $\tau(p, T, \xi)$ can be expanded into a Taylor series with respect to the structural order parameter, ξ, and (ii) that in the limit $\xi \to 0$ Eq. (4.24) holds, that is, $\lim_{\xi \to 0} \tau(p, T, \xi) = \tau_e(p, T)$. Introducing a dimensionless time scale, $t' = t/\tau_e$, Eq. (4.23) can be written then as:

$$\frac{d\xi}{dt'} = -\frac{1}{1 + a_1\xi + a_2\xi^2 + a_3\xi^3 + \ldots} \xi , \quad t' = \frac{t}{\tau_e(p, T)} . \qquad (4.29)$$

Here $a_i = a_i(p, T)$ are the (dimensionless) expansion coefficients of $\tau(p, T, \xi)$ at the limit $\xi = 0$.

Equation (4.29) can be easily rewritten as:

$$\left(\frac{1}{\xi} + a_1 + a_2\xi + a_3\xi^2 + a_4\xi^3 + \ldots\right) d\xi = -dt' . \qquad (4.30)$$

Integration in the limits $(\xi(0), \xi(t'); 0, t')$ yields

$$\left(\ln \xi + a_1\xi + \frac{a_2}{2}\xi^2 + \frac{a_3}{3}\xi^3 + \ldots\right)\bigg|_{\xi(0)}^{\xi(t')} = -t'. \tag{4.31}$$

In general, knowing the expansion coefficients, $\xi(t')$ can be determined from Eq. (4.31).

However, in the particular cases that one of the terms in Eq. (4.30) respectively Eq. (4.31) dominates, we get some spectrum of particular simple differential equations and solutions

$$\frac{d\xi}{\xi} = -dt', \quad \xi(t') \cong \xi(0)\exp(-t'), \tag{4.32}$$

$$a_k \xi^{k-1} d\xi = -dt', \quad \xi(t') \cong \left\{\xi^k(0) - \frac{k}{a_k}t'\right\}^{1/k}, \quad k=1,2,\ldots. \tag{4.33}$$

Let us now consider the particular case that the relaxation behavior is governed by the term containing the expansion coefficient a_j with $j = k+1$. According to Eq. (4.33), the dependence $\xi(t)$ for this mode is given by

$$\xi(t') \cong \left\{\xi^j(0) - \frac{j}{a_j}t'\right\}^{1/j}. \tag{4.34}$$

In the course of time, the term containing the expansion coefficient a_k may replace this mode since ξ is decreasing with time and the neighboring terms in the expansion may become dominating. The differential equation of this mode, we can write as:

$$\frac{a_k \xi^k}{\xi} d\xi = -dt' \quad \text{or} \quad \frac{d\xi}{\xi} = -\frac{1}{a_k \xi^k} dt'. \tag{4.35}$$

Replacing $\xi(t)$ on the right hand side by its solution for the case $j = k+1$ (cf. Eq. (4.34)), we obtain

$$\frac{d\xi}{\xi} = -\frac{1}{a_k \left(\xi^{k+1}(0) - \left(\frac{k+1}{a_{k+1}}\right)t'\right)^{k/(k+1)}} dt'. \tag{4.36}$$

In other words, we arrive at Kohlrausch (or Jenckel-type; cf. Eq. (4.28) and [233]) relaxation equations where the parameters $n = k/(k+1)$ and $\beta = 1-n$ will have the values

$$n = \frac{1}{2}, \quad \beta = \frac{1}{2} \quad \text{for} \quad k = 1,$$
$$n = \frac{2}{3}, \quad \beta = \frac{1}{3} \quad \text{for} \quad k = 2,$$
$$n = \frac{3}{4}, \quad \beta = \frac{1}{4} \quad \text{for} \quad k = 3\ldots \tag{4.37}$$

Let us suppose now that the term a_{k+1} is very small, then the relaxation behavior governed by the term in the expansion with $j = k+2$ may go over into a behavior governed by the term a_k and the respective differential equation. Instead of Eq. (4.36) we then obtain

$$\frac{d\xi}{\xi} = -\frac{1}{a_k \left(\xi^{k+2}(0) - \left(\frac{k+2}{a_{k+2}}\right) t'\right)^{k/(k+2)}} dt' . \tag{4.38}$$

Instead of Eq. (4.37), we have then

$$\begin{aligned} n &= \frac{1}{3}, \quad \beta = \frac{2}{3} \quad \text{for} \quad k = 1, \\ n &= \frac{1}{2}, \quad \beta = \frac{1}{2} \quad \text{for} \quad k = 2, \\ n &= \frac{3}{5}, \quad \beta = \frac{2}{5} \quad \text{for} \quad k = 3\ldots \end{aligned} \tag{4.39}$$

This analysis can be extended to even higher numbers of eventually missing terms in the expansion of the relaxation time, τ. In general, we would have then (for $a_k \neq 0$, $a_{k+1} = \ldots = a_{j-1} = 0$, $a_j \neq 0$)

$$n = \frac{k}{k+j}, \quad \beta = 1 - n = \frac{k}{k+j} . \tag{4.40}$$

The respective differential equation would read as

$$\frac{d\xi}{\xi} = -\frac{1}{a_k \left(\xi^{k+j}(0) - \left(\frac{k+j}{a_{k+j}}\right) t'\right)^{k/(k+j)}} dt' \tag{4.41}$$

with the general solution

$$\ln\left(\frac{\xi(t')}{\xi(0)}\right) = \frac{a_{k+j}}{j a_k} \left\{ \left[\xi^{k+j}(0) - \left(\frac{k+j}{a_{k+j}}\right) t' \right]^{j/(k+j)} - \xi^j(0) \right\} . \tag{4.42}$$

4.3.3
Discussion

The approach outlined above predicts that, in addition to the exponential decay, near equilibrium, relaxation may be governed in intermediate stages by laws of the form $\xi(t') = (\xi^k(0) - (k/a_k) t')^{1/k}$. In particular, for $k = 2$, we obtain a relaxation behavior $\propto t^{1/2}$. Such kind of relaxation behavior was distinguished already long ago by Kauzmann [109] and recently found to dominate the dielectric α relaxation process in viscous organic liquids [336]. In addition, it allows one to understand the origin of stretched exponential-type relaxation processes, and it gives estimates of the coefficient β in agreement with experimental findings ($0.3 \leq \beta \leq 0.75$) supplying us in this way with the understanding of the origin of the widely employed stretched exponential relaxation law not having received so far a theoretical foundation [70] except the one outlined here.

Finalizing this chapter, we can conclude, consequently, that the structural order parameter approach allows a detailed and appropriate interpretation not only of thermodynamic but also of kinetic properties of glass-forming systems, and, of course, the possibilities of this approach are by far not exhausted. In this connection it is interesting also to remember a statement by Oleg Mazurin in this respect, which, in free translation from the Russian source [70], reads: *"Very often attempts to find a better solution are an enemy of finding a good one. So, let us be engaged in the may be less prestigious but more useful task to improve the relaxation theory of glass transition. Here a huge amount of things has to be and can be done."*

5
Thermodynamics of Amorphous Solids, Glasses, and Disordered Crystals

Ivan S. Gutzow, Boris P. Petroff, Snejana V. Todorova, and Jürn W. P. Schmelzer

5.1
Introduction

In Chapter 2, we have outlined the characteristic properties of typical glass-forming systems, their change upon glass transition, the thermodynamic and kinetic peculiarities of the process of vitrification, the mean values of the essential properties of the systems, which they take over, when the undercooled liquid has formed a glass. In Chapter 3, a generic phenomenological approach is outlined, based on the thermodynamics of irreversible processes, which gives the possibility to describe and predict in a quantitative or, at least, in a semi-quantitative way, these changes both in their magnitude and evaluate them in their thermodynamic and kinetic significance. In Chapter 4 the main kinetic characteristics of any glass-forming liquid, its shear viscosity, η, is considered in details both in its dependence on temperature and pressure. Similar methods can be employed to account for the effect of external electric or magnetic fields applied on vitrification [337]. In Chapter 8 the ways to employ in different fields of applications the particular properties of glasses will analyzed, and in Chapters 6 and 7 it is shown, how the properties of the desired glasses can be forecasted, using existing data: both from literature sources and from own experiments. Lastly, in Chapter 9 we consider the particular changes of the properties of glasses, when they are considered at very low temperature, that is, when approaching the zero of absolute temperature.

In all these cases (with the exception of some of the derivations made in Chapter 8, where we consider the possible technical application also of defect crystals), we are mainly concerned with the properties and processes concerning glasses, in the sense of their appropriate definition, given in Chapter 2. In Chapter 8 it will turn out, however, that amorphous solids (in the more general meaning of the word: as solids with any form of frozen-in disorder) and other defect solids, and in particular defect crystals "pathologically born" with a frozen-in structure, or "glassy crystals" which have formed and preserved their disorder at a glass transition-like change, may be also of technical importance as accumulators of potential energy. Moreover, defect crystal states have been historically of exceptional significance in understanding one of the basic problems of irreversible thermodynamics: the definition of residual entropy in kinetically frozen-in thermodynamic systems. This is the reason why they are still of significance in formulating the thermodynamics of irreversible systems and of glasses in particular.

In the present chapter both the generic thermodynamic phenomenology and the possibilities and methods of measuring and calculating the thermodynamic properties of systems with frozen-in disorder using caloric and tensimetric experimental results are discussed in details. This possibilities of thermodynamic experimentation and calculus are also of general significance in analyzing nonequilibrium states and could even have some technical perspectives. Following this general aim, in the present chapter we give a summary on experimental evidence on the thermodynamic properties of solids with frozen-in configurational disorder: topological, orientational and conformational. Such a categorization of disorder into three main types corresponds both to Ziman's models [193] and to the hierarchy of disordered structures the present authors have introduced and used in previous publications [1, 112]. In so-called structural (real or ordinary or typical) glasses which are molecular and polymeric representatives of organic substances or of oxide, silicate, halide or chalcogenide compounds stemming from different fields of inorganic chemistry (including the so-called metallic glasses, i.e., vitrified metallic alloy systems) all three types of disorder are frozen-in. The terms "real" or "structural" glasses, attributed to both technical and laboratory model glass-forming systems (i.e., to solids with frozen-in amorphous structure), are usually employed to distinguish them from both computer-modeled states of frozen-in disorder or from spin glasses. In the latter mentioned systems (having either an amorphous or a crystalline topological structure) a magnetic, orientational type of disorder is frozen-in [338]. Although usually considered in various aspects from the standpoint of elaborated statistical physics models (see [338, 339]) relatively little experimental evidence is accumulated on the caloric properties of spin glasses. Besides spin glasses there exist also many other crystalline substances, in which different types of topological or orientational disorder is frozen-in. Simple molecular disordered crystals (sometimes somewhat exaggeratedly and, as shown here, improperly called "glassy crystals") like CO, N_2O_2 or ice were investigated beginning from 1920s. In the early 1980s, it was found, as mentioned above, that several so-called plastic crystals have also thermodynamic properties similar to those of glasses and can be with more justification called glassy crystals.

Most results on the thermodynamics of systems with frozen-in disorder, as it is also seen from the present analysis, come from hundreds of investigations, from calorimetric measurements, performed since 1925 with the already (more or less loosely) defined real glasses. These caloric investigations gave in fact the first possibility for a detailed consideration of the thermodynamics of solids with frozen-in disorder and the general aspects of the kinetics of vitrification. In this sense, undercooled glass-forming liquids and the "real" glasses they form gave an instructive and general example of the freezing-in process in metastable systems, both amorphous and crystalline. Many different structures have to be considered in this respect.

A particularly interesting representative of vitrified melts, as it turns out from our analysis, are glass-forming liquid crystals: these systems, which are in fact partly oriented liquids with complicated structures (somewhat inappropriately called crystals) can be frozen-in to highly ordered glasses with peculiar thermodynamic properties. Thus, glassy liquid crystals give an example of frozen-in, partially ordered amorphous solids. All frozen-in systems are nonequilibrium, nonergodic, even nonthermodynamic bodies. Thus, they do not obey the third principle of thermodynamics in its classical formulation as discussed here in Chapter 9.

As seen in the next paragraphs, the first caloric experiments with "real" glass-forming systems gave the first experimental clues how to treat a very general problem. The first result that turned out after a thorough caloric analysis of the initially examined cases was that in real glasses a nonvanishing zero-point entropy, $\Delta S_g^0 > 0$, is frozen-in. Nearly of the same significance was also the evidence obtained with the mentioned molecular crystals in which a comparison of caloric and tensimetric measurements with the statistically calculated entropy (the so-called spectroscopic entropy of the respective gas because the statistical calculations are based on spectroscopic evidence) gave the proof that here also an orientational disorder and values $\Delta S^0 > 0$ exist in solids with seemingly perfect crystalline structure. Similar results were also obtained with the plastic crystals, in which by thorough caloric measurements it is verified that the orientational disorder is frozen-in in the crystal at a glass transition-like process at a temperature, T, considerably below the melting temperature, T_m.

In the present chapter, we are trying to find the most significant common features and the differences, characterizing both real glass-forming systems and the other mentioned forms of disordered solids. This task can be easily performed when glasses and systems with other forms of frozen-in disorder are described in terms of simple *thermodynamic invariants*, for example, in connecting the properties of the system as a liquid at melting temperature, $T = T_m$, and the ones at (or below) the freezing-in temperature (e.g., the temperature of vitrification, $T = T_g$, for real glass formers). In more detail, the term *thermodynamic invariants* we use here to denote dimensionless numbers, connecting (in dimensionless reduced coordinates) characteristic properties of the system at a reference temperature, for which the melting point, T_m, is chosen. Such a procedure resembles in some respect the basic ideas of the corresponding state approach in classical thermodynamics, connecting and equalizing (or trying to equalize) in the framework of an appropriate generalized equation of state (e.g., according to the van der Waals equation) the properties of all substances in dimensionless coordinates. Here, as discussed in more detail in Chapter 8, we use as the reducing coordinates not the (mostly unknown) critical constants of the substances vitrified, but the respective T_m values and thermodynamic property change at this same temperature. Thus we use here as a secondary corresponding and reference state the melting temperature, T_m, which at least for simple, nonassociating liquids is empirically connected with the critical temperature, T_c, of the same substance by dependencies like $T_c \cong (5/3) T_m$.

We have organized the present chapter in the following way: first existing experimental evidence on the process of vitrification in the case of real glasses is illustrated by several examples and critically summarized. In performing this task, we use from one side data on really typical cases already treated in Chapters 2 and 3, but introduce also less typical cases, to show existing deviations from the average, mean and typical behavior. Then, in the following sections, evidence on spin glasses and on the vitrification of liquid crystals is compared with the vitrification of "real" glasses. After that results on the thermodynamic properties of crystals with frozen-in orientational disorder are discussed, summarized and compared with vitrification characteristics of real glasses and liquid crystals. First we use the vitrification invariants as typical empirical connections, as this was first approached in 1948 by Kauzmann [109]. Then, however, we make an attempt to derive the vitrification invariants in the framework of the generic phenomenological approach, summarized in Chapter 3. At the end of the present chapter, an attempt is made to formulate several consequences from the results outlined showing the common features and the specific differences for various systems with frozen-in disorder: of undercooled liquids vitrified to glasses, of defect crystals, of glassy crystals and of liquid crystals.

5.2
Experimental Evidence on Specific Heats and Change of Caloric Properties in Glasses and in Disordered Solids: Simon's Approximations

The first indications on the peculiar state of typical real glasses as kinetically frozen-in, thermodynamically nonstable systems were obtained in the early 1920s as a result of thorough investigations of the temperature dependence of the specific heats, $C_p(T)$, of two typical glass-forming substances, of SiO_2 and of glycerol as crystals, undercooled liquids and glasses. The subsequent measurements were initiated independently in two of the best calorimetric laboratories of those times: in Nernst's institute in Berlin (see [21, 59, 60, 163]) and in Giauque's laboratory in the United States [160, 340]. They were performed under the guidance of the two above-mentioned scientists, both to become Nobel Prize laureates for their achievements, who have to be cited among the most famous thermodynamicians of the twentieth century. These investigations were initially implemented in order to give experimental proof of the applicability of the third principle of thermodynamics (in the form of Nernst's heat theorem [1, 51, 55, 58, 60, 112]) to noncrystalline solids. The results obtained indicated, however, that in contradiction to Nernst's expectation (see [59, 60]) of both these glass-forming systems investigated (SiO_2 [21, 58, 59, 163], glycerol [58, 160], and soon for ethanol glass [51, 219]), a nonvanishing, nonzero value of the zero-point entropy of the glass (cf. also Chapter 9)

$$\Delta S(T)|_{T \to 0} = \Delta S_g(0) = \Delta S_g(T_g) > 0 , \tag{5.1}$$

remained as temperatures approach absolute zero. These results are presently confirmed for more than 120 glass-forming substances experimentally investigated (see [1, 16, 51, 58, 112, 237, 341]) with compositions ranging from organic high polymers [16] through network glasses like SiO_2, GeO_2 and BeF_2 [1, 237], with phosphates, borates, and so on, to the metallic alloy glass-forming systems summarized in [1, 231, 237, 342].

The first recognition of the exceptional significance of the nonvanishing zero-point entropy of glasses for the whole development of thermodynamics was given by the well-known twentieth century thermodynamician Simon [21, 23, 51, 58], who at that time was one of Nernst's closest collaborators (see [1, 60]). Chapter 9 gives a detailed historic background and the general thermodynamic significance of these investigations. The initial results of this development were first summarized by Kauzmann [109] and by Eitel [225]. In our monograph [1] (as well as in [16, 51, 112, 185, 237, 341]) the ΔS_g values of glasses are compiled for a great variety of substances and especially for those with particular structural or more general methodical significance. An extension to newer problems (spin glasses, etc.) may be found in [343–345].

In the last paper mentioned above, as well as in [341, 346, 347], the first data on $C_p(T)$ measurements and on zero-point entropies in vitrified liquid crystals are given. The first $\Delta S(T)$ determinations for this particular class of substances were performed by Sorai *et al.* [347–349] in 1973 and 1983. In [1, 112], we have given a graphical analysis of the results obtained for ΔS_g with typical glass formers summarized here in Chapter 2. A historical survey of the theoretical approaches and ideas connected with the development of the problems of the zero-point entropy of glasses may be also found in [1, 237] and in Chapter 9. An essential point is that Einstein [350] had predicted, from the very first discussions on the statistical foundations of the third principle (as early as 1914), that for glasses and for solid solutions as states of constant (i.e., frozen-in) statistical disorder, as a rule, zero-point entropies ($\Delta S(T \to 0) > 0$) have to be expected. A further development of these ideas and of the possible limits of the values of the entropy for glasses as predicted by statistical physics was given later on by Pauling and Tolman [351] using general thermodynamic considerations and then by Gutzow [191], who employed a combination of structural models to derive these models in a more concrete way.

The first determinations of zero-point entropies, $\Delta S^0 > 0$, for disordered molecular crystals were performed by Clusius [352] and again in Giauque's laboratory (see [353]). It was Giauque [353] and another Nobel laureate, Pauling [354], who gave the first structural models (1930–1935) of orientational disorder in molecular crystals and especially in H_2O (ice) and in the structurally corresponding crystalline deuterium oxide, D_2O. Kaischew [355] demonstrated experimentally in 1938 that in CO-crystals two structural arrangements with very narrow potential energy have to exist. According to his measurements even lowest possible crystal growth rates at CO-crystal synthesis or prolonged annealing (as demonstrated by Eucken [356], see the respective summary in [352]) could not remove or at least decrease the zero-point entropy, ΔS^0, in disordered molecular crystals: the CO-crystals were "born" with their "pathologically" defect structure in the immediate vicinity below their melting point, T_m. The results of Eucken and Kaischew thus turned out to be of great significance in treating the thermodynamics of disordered crystals.

The $C_p(T)$ measurements performed with glass-forming melts in the early days of the mid-1920s as well as further investigations of later times, summarized in [1, 112, 185, 237, 341], showed that upon vitrification the specific heats $C_p(T)$ of the undercooled melts of typical glass-formers drop via a sigmoid dependence in a relatively narrow interval (the glass transformation range, in Tammann's somewhat nonadequate terminology (see [1, 24])) to low values, alike to those of the corresponding crystal. A typical picture in this respect is given in the present book with Figure 2.42 for the case of B_2O_3 melts, vitrifying at different cooling rates (see also Figure 5.1). A similar $C_p(T)$-jump was observed also in other thermodynamic properties, which are to be considered as thermodynamic coefficients of the vitrifying system, that is, as second-order derivatives of the corresponding thermodynamic potential (at a pressure $p = $ const.: of the Gibbs free energy difference, ΔG). Typical examples in this respect, discussed already in Chapter 2, are besides the already mentioned specific heats, $C_p(T)$,

$$\Delta C_p(T) = -T \left(\frac{d^2 \Delta G(T)}{dT^2} \right)_p , \quad (5.2)$$

the thermal expansion coefficient,

$$\alpha = \frac{1}{V} \left(\frac{\partial V}{\partial T} \right)_p = \frac{1}{V} \frac{\partial}{\partial T} \left[\left(\frac{\partial G}{\partial p} \right)_T \right]_p = \frac{1}{V} \frac{\partial^2 \Delta G}{\partial T \partial p} , \quad (5.3)$$

the coefficient of compressibility,

$$\kappa = -\frac{1}{V} \left(\frac{\partial V}{\partial p} \right)_T = -\frac{1}{V} \frac{\partial}{\partial p} \left[\left(\frac{\partial G}{\partial p} \right)_T \right]_T = -\frac{1}{V} \frac{\partial^2 \Delta G}{\partial^2 p} , \quad (5.4)$$

the modulus of elasticity, $E(T)$,

$$E(T) = -V \left(\frac{\partial p}{\partial V} \right)_T , \quad (5.5)$$

which is in fact the reciprocal of the coefficient of compressibility (see [55]), the dielectric constant, which can be represented as

$$\varepsilon = \left(\frac{\partial^2 \Delta G}{\partial^2 E} \right)_{T,p} , \quad (5.6)$$

where E is the electric field strength (see [1, 62, 63]).

The change of $C_p(T)$, $\alpha(T)$ and $\varepsilon(T)$ for one of the most thoroughly investigated model glass-forming systems – glycerol – upon vitrification is shown in Figure 5.1, where we combined the results of two previously given figures from Chapter 2. The change of the modulus of elasticity, $E(T)$, of silicate glasses is evident from Figure 3.19. It is interesting to note that as a reciprocal of the corresponding thermodynamic coefficient (κ) the temperature course of $E(T)$ shows in fact also a dependence reciprocal to the temperature course of $\alpha(T)$ and $C_p(T)$ given in Figure 5.1.

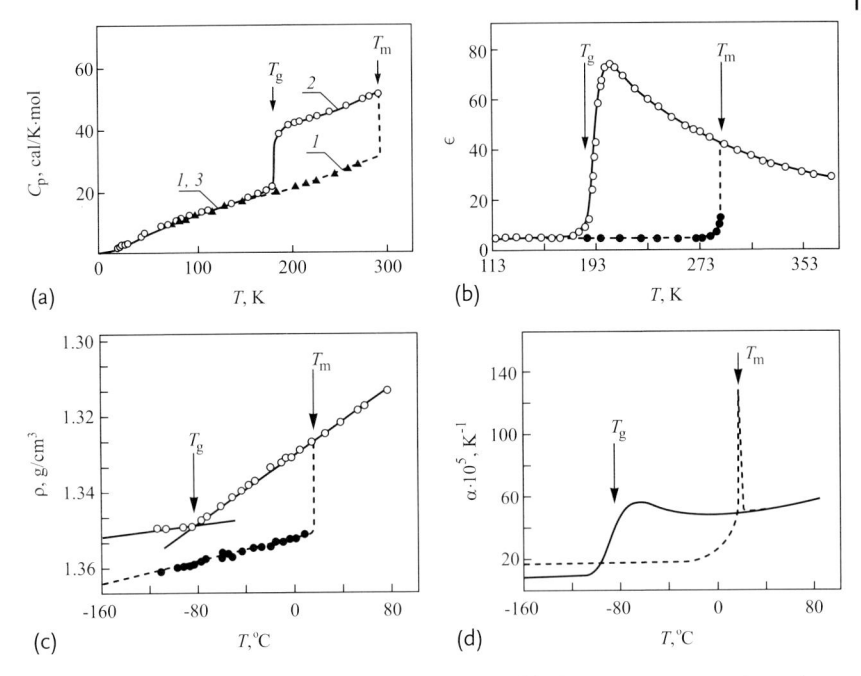

Figure 5.1 Temperature dependence of thermodynamic properties of glycerol: (a) specific heats of liquid (○), crystalline (▲), and vitreous (●) glycerol; (b) dielectric constant of liquid (○), crystalline (▲), and vitreous (●) glycerol; (c) density of glycerol. The *white* circles of the upper curve refer to liquid and vitreous glycerol, respectively, while the *black* circles of the lower curve are experimental data for crystalline glycerol; (d) thermal expansion coefficient, α, for glycerol. The *solid line* corresponds to the liquid and the vitreous state while the *dashed line* refers to the crystal (literature is cited in the text, the data are taken from Simon [23], and Schulz [357]).

It also turned out (cf. Chapters 2 and 3) that the thermodynamic functions, which have to be considered as the first-order derivatives of the thermodynamic potential (e.g., molar volume, $\Delta V = (\partial \Delta G / \partial p)_T$, molar enthalpy, $\Delta H = -1/T(d\Delta G/dT)$, molar entropy, $\Delta S = -(\partial \Delta G / \partial T)_p$), with respect to temperature, T, and other external parameters of state (pressure, p, electric, E, or magnetic, M, field strength, etc.) display a break point in their temperature course when vitrification takes place in cooling run experimentation. A full picture of the vitrification process in its thermodynamic aspect, using Simon's approximation, was discussed in Chapter 2 and illustrated with Figure 2.28.

The temperature course of the change of volume (and density, ρ, as they are given in Figure 5.1) can be directly measured: enthalpies and entropies have to be calculated from $C_p(T)$ measurements and melting enthalpy data, ΔH_m (see, however, also the subsequent analysis for alternative ways of direct caloric determination of $\Delta H_g (T = 0)$). Out of existing experimental evidence summarized in previous publications [1, 16, 109, 112, 185, 212, 237, 341] for the more than one hundred glass-forming substances, we would like to discuss here in more detail the results, obtained for the typical glass-former of glycerol, and only to mention a number of

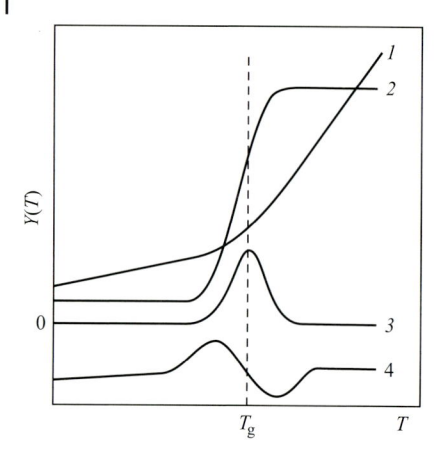

Figure 5.2 Tammann's definition of the caloric glass transition temperature, T_g, and of the glass transition region: (1) Change of thermodynamic functions with temperature; (2) change of the first derivatives of the thermodynamic functions (i.e., the thermodynamic coefficients, e.g., $C_p(T)$); (3) change of the second-order derivatives of the thermodynamic functions. According to Tammann [24], T_g is defined by the maximum in the $(dC_p(T)/dT)$ curve. A similar but more distinct definition of T_g can be provided connecting it with the nullification of the second-order derivative of the specific heats (i.e., at $(d^2 \Delta C_p(T)/dT^2) = 0$) and the corresponding inflection point is given with (4), as proposed by Gutzow et al. [11] (cf. also Chapter 3).

other cases. We would also like to focus some of our attention on various exotic laboratory models of glass-forming substances, some of them having also a distinct technical significance.

Glycerol is probably the thermodynamically most thoroughly investigated glass-forming substance. With this model, measurements have been performed on the temperature dependence of specific heats, $C_p(T)$ [21, 59, 160], of the dielectric constant, ε [95], and of the density ρ [357, 358], of the crystalline, liquid and vitreous states of the substance as they are summarized in Figure 5.1. It is seen that in the glass transition region in the vicinity of the glass transition temperature, the glass transition temperature, T_g, is manifested at cooling as the inflection point of a sigmoidal curve of the $C_p(T)$ and $\varepsilon(T)$-dependencies and also of the temperature functions of the coefficient of thermal conductivity, $\lambda(T)$ (see also Figure 2.36). The inflection point of the s-shaped $C_p(T)$ curve has been proposed by Tammann [24] as the caloric definition of the glass transition temperature, T_g (see Figure 5.2). In this figure, the uppermost three curves 1, 2, 3 are drawn according to a picture in Tammann's book. Curve 1 describes the change of an appropriately chosen thermodynamic function Y (e.g., $H(T)$), with the sigmoid shape in curve 2 as its first temperature derivative (i.e., $C_p(T)$). Following that, curve 3 shows the change of the second-order derivative of Y (i.e., $[\partial^2 H(T)/\partial T^2]_p$) upon vitrification. In a more recent series of publications Gutzow et al. [11, 12], following Tammann's proposal, defined T_g mathematically more precisely and more conveniently as the nullification of the second-order temperature derivative of $C_p(T)$ (i.e., from a third-order derivative of $\Delta H(T)$) given as curve 4 in Figure 5.2. In this way, at $T = T_g$ the sigmoidally shaped $C_p(T)$ curve has an inflection point, which is defined mathematically, when $(dC_p(T)/dT) > 0$ at $(d^2C_p(T)/dT^2) = 0$. The latter condition makes it on one side easy to define exactly T_g from experimental $C_p(T)$ curves, and on the other side it gives the possibility for deriving the basic kinetic condition of vitrification, introduced empirically in both Chapters 2 and 3.

In the temperature dependence of the density, $\rho(T)$, of the same substance as presented in Figure 5.1c at $T \to T_g$, a typical break point is observed. This break point gives, as shown in Figure 5.1c, another classically employed possibility to define the glass transition temperature, T_g: by the prolongation of the two parts of the $\rho(T)$ curve above and below T_g, as indicated in this figure. The differentiation of the corresponding $\rho(T)$ curve gives the temperature dependence of the thermal expansion coefficient, $\alpha(T)$, for vitrifying glycerol (Figure 5.1d). Thus, similar to the $C_p(T)$ and $\varepsilon(T)$ dependencies a jump also appears in the $\alpha(T)$ curve for vitreous glycerol, practically at a temperature close to the "caloric" glass transition temperature, T_g, defined by the previously discussed $C_p(T)$ course.

A similar sigmoidal temperature course in the $\alpha(T)$- or $C_p(T)$-curves is also observed in heating run experiments, when the already vitrified glass is reheated and then again transformed into the undercooled liquid (see the two heating $C_p(T)$ curves for B_2O_3 glass in Figure 2.42, given according to [211] and data first reported in [359]). Here, however, a typical overshot "nose" maximum in the $C_p(T)$ heating curves is found (see Figures 2.42, 3.9, and 5.3). This "nose" part is increased with

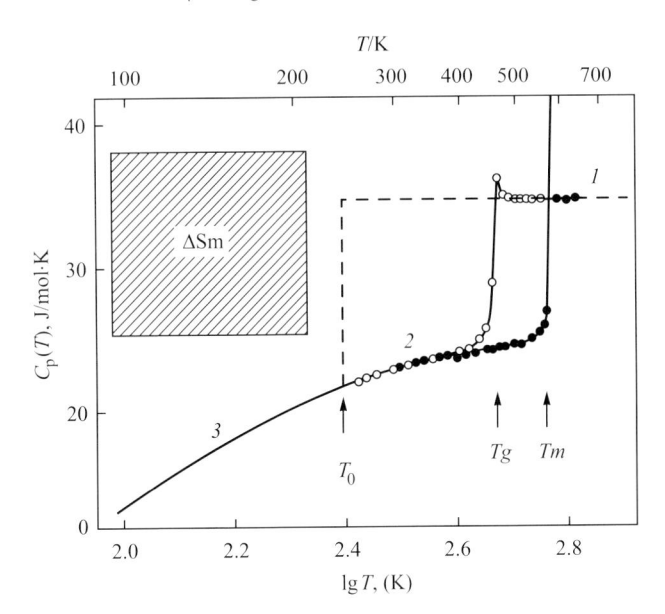

Figure 5.3 Temperature course of the specific heats, $C_p(T)$, of the glass to melt transition in As_2S_3. Heating curves 2, and 1 with a typical "nose" have to be compared with curve 3 of the crystal. The value $\Delta S(T_m)$ of this substance, given as a square, in $C_p(T)$ vs. log T-coordinates. Below T_0 in the undercooled fictive melt $\Delta C_p(T) = 0$ (dotted line) and $\Delta S(T) = 0$ is to be expected (according to data by Blachnik and Hoppe [360]).

increasing heating rate (see Figures 2.22, 2.42, and 3.9 and further experimental evidence in this respect, collected in [1] and Chapter 2).

Glycerol gives also an interesting example in the way the course of the temperature dependence of the properties of the system and the T_g value, which are experimentally determined, are changed upon glass transition when determined either by the measurements of a thermodynamic function or a thermodynamic coefficient. Thus, the coefficient of optical refraction, $n(T)$ (which according to a well-known dependence is determined by the molar volume, V, of the vitrifying melt), displays a break-point, analogous to the one observed in the $\rho(T)$ curve (compare Figures 5.1 and 2.36). On the other hand the coefficient, λ, of heat transfer, which is determined significantly by the change of specific heats, $C_p(T)$, of the vitrifying system (see [1]) displays at $T \approx T_g$, the typical jump observed for thermodynamic coefficients (see again Figures 5.1 and 2.36).

The viscosity, η, of vitrifying undercooled liquids dramatically changes in the glass transition range. In the vicinity of the calorically defined T_g value, a break-point of the $\eta(T)$-dependence is observed (see Figures 3.21 and 5.4) illustrating this property for both an organic glass (see Figure 5.4 and [361]) and for a typical silicate glass-forming melt [289]. It is to be noted that the course of the log $\eta(T)$ vs. T curve near the break-point is reciprocal to the course of the $\Delta S(T)$ curve. In [10] and in Section 5.9, this peculiar behavior of the log $\eta(T)$ curve finds its explanation in terms of an interesting theoretical viscosity model proposed by Adams and Gibbs [148] and in the change of the entropy temperature functions upon vitrification.

In the transformation range, the viscosity of the undercooled melts usually changes from 10^{10} to 10^{14} dPa·s and at temperatures approaching T_g this kinetic parameter has (at conventional cooling rates) values of about 10^{13} dPa·s. The break

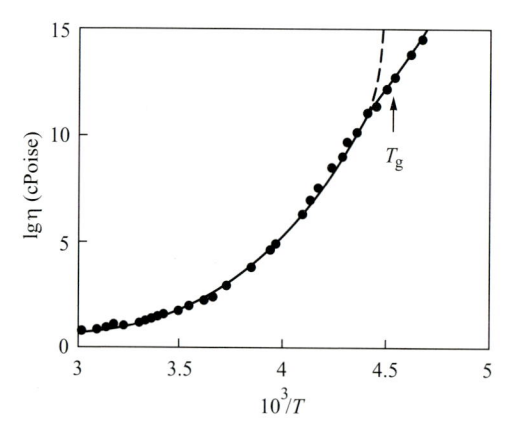

Figure 5.4 Viscosity $\eta(T)$ vs. temperature curve for a typical organic glass-forming model system (salol) in the temperature range of vitrification. Note the break point of the $\eta(T)$ curve in the vicinity of T_g, indicated (according to caloric measurements) by an arrow (see [361]). The *dashed line* gives the expected increase in $\eta(T)$ of the fictive undercooled melt.

point in the $\eta(T)$ curve for $T \to T_g$ is a very significant finding, as it demonstrates the change of a kinetic property upon vitrification. As far as viscosity, η, is directly connected with the time of molecular relaxation, $\tau(T)$ (see [1]), of the melt or with the coefficient of self-diffusion, $D(T)$, of the building units of the system, a similar behavior is also to be expected (and observed) for these kinetic characteristics of the vitrifying system. According to classical free volume models of viscosity of glass-forming liquids, mentioned in Chapter 2 and discussed in detail in Chapters 3 and 4, the temperature course and the break-point in the $\log \eta(T)$ curve is caused by the change of relative free volume, $[V_{\text{liquid}}(T) - V_{\text{crystal}}(T)]/V_{\text{crystal}}(T)$, of the melt, where the indices $l(iquid)$ and $c(rystal)$ here and below denote the molar volumes of the liquid and the crystal. The index $g(lass)$ in the following discussions indicates properties referring to the glass.

Knowing the $C_p(T)$-dependence for liquid and crystalline glycerol and its enthalpy of melting, ΔH_m (from direct determinations at the temperature of melting, T_m; thus the entropy of melting, $\Delta S_m = \Delta H_m/T_m$, is also known), the temperature course of the difference of the thermodynamic functions of liquid and crystalline glycerol can be established by using a known classical thermodynamic formalism (see [1]). The result is given on Figure 2.28 using Simon's thermodynamic data and his respective calculations. Here we repeat the essence of these calculations, as they are already given in Chapter 2.

In order to construct the entropy difference, $\Delta S(T) = S_l(T) - S_c(T)$, we have to write according to classical thermodynamics (see, e.g., [1, 55]) in the range of its expected applicability, that is, for equilibrium

$$\Delta S(T) = \Delta S_m - \int_T^{T_m} \left(\frac{\Delta C_p}{T}\right) dT, \qquad (5.7)$$

where $\Delta S_m = S_f(T_m) - S_c(T_m)$ is the entropy of melting (the enthalpy of melting being $\Delta H_m = \Delta S_m T_m$). Assuming as indicated in Figures 5.1, 5.3, and 5.5 that at $T < T_g$ for various glass-forming systems (including typical glass-formers like glycerol (Figure 5.1) or less known examples like chalcogenides, soluble glasses or metallic alloy glasses, see Figure 5.6), we can write $\Delta C_p(T) = C_{p,g}(T) - C_{p,c}(T) = 0$, it follows that the entropy value

$$\Delta S_g = \Delta S(T)|_{T<T_g} \approx \Delta S_m - \int_{T_g}^{T_m} \left(\frac{\Delta C_p}{T}\right) dT \cong \text{const.} \qquad (5.8)$$

is frozen-in below T_g. The corresponding calculations (remembering that $(dT/T) = d(\ln T)$) are easily performed graphically with a logarithmically scaled argument, for example, in $\Delta C_p(T)$ vs. $\log T$ coordinates shown in Figure 5.3 for As_2S_3 [360] and in Figure 5.5 for lithium acetate, analyzed in [362]. For this latter mentioned substance an unusually low value (only about $0.2\Delta S_m$) of the entropy of melting is frozen-in. In this way, Figures 5.3 and 5.6a give in $C_p(T)$ vs. $\ln T$ representation a graphical illustration of Eq. (5.8). Note that according to both Figures 5.3 and 5.6,

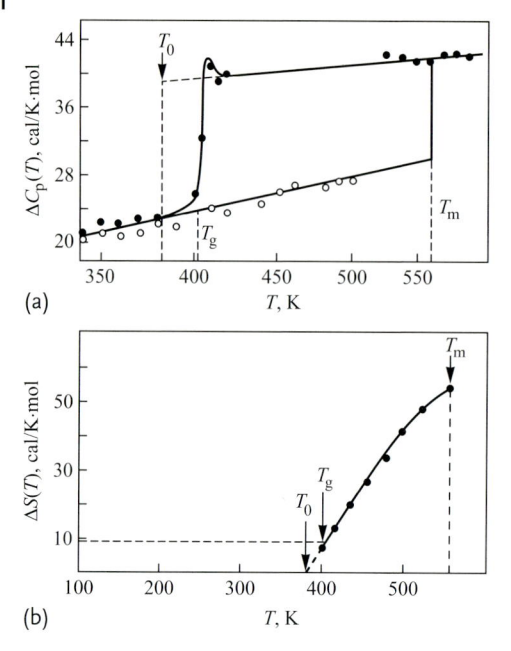

Figure 5.5 Heat capacity, $C_p(T)$ (a), and the entropy (b) of lithium acetate as a liquid (black points above T_g) and as a glass (below T_g) according to [362]. Open circles refer to crystalline samples (after Wong and Angell [362]). This is the $C_p(T) - \Delta S(T)$ diagram for a nontypical glass former (with a relatively low frozen-in ΔS_g value, that is, $\Delta S_g \approx (1/5)\Delta S_m$). Note that here as in Figure 5.3, a logarithmically scaled abscissa is used for temperature, T, and that T_0 indicates again the temperature at which for the fictive undercooled melt, $\Delta C_p(T) = 0$ and thus $\Delta S(T) = 0$ holds. In (b) the temperature dependence of the entropy difference, $\Delta S(T)$, of the undercooled melt and the glass with respect to the crystal is given.

$\Delta S(T) \approx 0$ should be expected for the fictive undercooled melt at temperatures $T \approx T_0$ where $T_0 \cong 1/2 T_m$. Thus, the Kauzmann temperature, T_0, introduced in Chapter 2, has for the considered cases and for all typical glasses considerably higher values than the absolute zero of temperature ($T = 0$).

This extrapolation of the $\Delta S(T)$ curve and the mentioned value of T_0 are again of exceptional theoretical significance: the third principle of thermodynamics (or Nernst's theorem) requires for the metastable (equilibrium) liquid $S(T) = 0$ at $T = 0$, however, it cannot and does not specify the temperature, beyond which the fictive metastable liquid should reach $S(T) \to 0$ (see [363]). All glass-forming liquids indicate for T_0 relatively high temperatures, as can be also expected from simple mean-field models of undercooled glass-forming liquids [364]. The result of this extrapolation is of considerable significance for the whole theory of liquids. However, the real course of the $\Delta S(T)$ curves upon vitrification and the result for ΔS_g calculated by Eq. (5.8) contradict the third principle of classical thermodynamics. According to it, we should always expect $\Delta S(T = 0) = 0$.

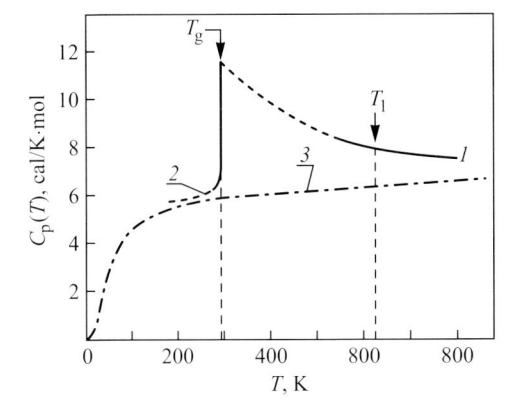

Figure 5.6 Temperature dependence of $C_p(T)$ for a typical metallic alloy glass-forming system (after Chen and Turnbull [231]): (1) undercooled melt, (2) glass, (3) crystal. T_l is the liquidus temperature of the respective metallic melts. *Solid lines* indicate the temperature regions where measurements were possible. *Dashed lines* indicate extrapolation into the crystallization region of the undercooled melt. The *dashed-dotted line* shows results calculated according to Debye's law for the crystalline alloy.

Particular attention deserves here both Simon's way of thinking and of resolving the contradiction between thermodynamic expectation and caloric experiment and the approximations reintroduced by Simon to make nevertheless his calculations possible in the framework of classical thermodynamics. As far as both thermodynamic phenomenology and experiments are correct, argued Simon [51, 58, 59], the only possible assumption is that glasses do not obey thermodynamic laws because they are outside thermodynamic considerations. They are nonequilibrium systems with frozen-in structure and entropy and classical thermodynamics does not (and cannot) treat such states. This is in fact the *first* of Simon's assumptions. His *second* assumption is that the glass transition takes place at a distinct temperature, $T = T_g$. In this way, *classical thermodynamics* can be applied from temperatures, $T > T_m$, down to $T = T_g$ in order to calculate the frozen-in enthalpy, ΔH_g, and entropy, ΔS_g values of the glass. This brought with it Simon's *third* necessary assumption: vitrification is a process in which entropy is only frozen-in, no entropy production was considered by him.

We know now that, as discussed in Chapter 3, in the glass transition interval, $(T_g^{(+)} - T_g^{(-)})$, for example, symmetrically situated around T_g, we have to expect and can determine also an entropy production term in the corresponding equations. However, at the usually employed cooling rates, q, this entropy production is of a value, negligible in a first approximation, when compared with ΔS_g. Thus, Simon's assumptions turned out to be of exceptional significance and accuracy in creating a general concept of the glass transition and for more than 70 years they determined the development of ideas in vitrification. Simon's first two assumptions gave, moreover, the impetus to use thermodynamics of irreversible processes as the phenomenological basis of glass science. This is clearly seen in the subsequent treatment of the glass transition in the framework of irreversible thermodynamics by Prigogine and Defay [14] and by Davies and Jones [19, 20]. These authors used in fact Simon's two approximations, also neglecting with him the possibility of entropy production; however, the latter two authors in general anticipated entropy production as a possibility.

Using more or less complicated structural molecular models, the values of ΔS_g of various glasses can be calculated accepting Simon's first assumption. In this respect besides the already mentioned paper by Gutzow [191] in which a generalized lattice-hole model was used to estimate ΔS_g of various glasses, particular attention should be paid to models for calculating ΔS_g in silicate glasses and in vitreous SiO_2 by using the arrangement principles of silicate bonding. These models are discussed in more detail in Chapter 5 of the monograph [1].

Following the same line of argumentation as in deriving Eq. (5.8) and again accounting for $\Delta C_p(T) = 0$ at $T < T_g$ it turns out that the enthalpy difference

$$\Delta H_g = \Delta H(T)|_{T<T_g} \approx \Delta H_m - \int_{T_g}^{T_m} \Delta C_p(T) dT \tag{5.9}$$

is frozen-in below T_g to the value $\Delta H_g > \Delta H_0$, where ΔH_0 is the zero-point enthalpy, corresponding to the fictive equilibrium state of the undercooled liquid, if it could be realized at temperatures approaching absolute zero. This frozen-in enthalpy, ΔH_g, is in fact another contradiction to the third principle of thermodynamics which requires the approach of the equilibrium value of ΔH_0 at $T = 0$.

Introducing the thus obtained constant values of ΔH_g and ΔS_g into a well-known thermodynamic dependence,

$$\Delta G = \Delta H - T\Delta S , \tag{5.10}$$

derived in Chapter 2, another violation of the third law of thermodynamics becomes evident. For systems in equilibrium, obeying classical thermodynamics, the dependence

$$\left[\frac{d\Delta G(T)}{dT}\right]_{T \to 0} = 0 \tag{5.11}$$

has to be fulfilled as predicted by Nernst's theorem. However, with the results from above we have to write Eq. (5.10) for $T < T_g$ as (see also Section 5.5):

$$\Delta G_g(T) = \Delta H_g - T\Delta S_g . \tag{5.12}$$

Thus it becomes evident that the thermodynamic potential of the glass approaches zero-point temperatures not as predicted by Eq. (5.11) but with a constant nonzero slope:

$$\left[\frac{d\Delta G_g(T)}{dT}\right]_{T<T_g} = -\Delta S_g . \tag{5.13}$$

Thus, by accepting Simon's approximation it turns out that the glass formed at the temperature quench from the temperature $T \approx T_g$ remains at room temperature T_r frozen-in in a state of increased disorder (manifested by the increased entropy, ΔS_g), of higher energy (indicated the increased enthalpy, ΔH_g), both of the values ΔS_g and ΔH_g corresponding to the higher temperature $T = \tilde{T}_g > T_r$.

The third principle also requires (again in the form of Nernst's heat theorem) that when approaching the absolute zero temperature, besides Eq. (5.11), the relation

$$\Delta H(T)|_{T\to 0} = \Delta G(T)|_{T\to 0} = \Delta H(0) \tag{5.14}$$

must also be fulfilled (see Eq. (5.10)). For the frozen-in system, Eq. (5.12) (written again for $T \to 0$) is analogous to Eq. (5.14):

$$\Delta H_g\big|_{T\to 0} = \Delta G_g\big|_{T\to 0} . \tag{5.15}$$

However, here $\Delta H_g > \Delta H_0$, where ΔH_0 corresponds to the fictive equilibrium undercooled liquid at $T \to T_0$. These two deviations from the third principle of classical thermodynamics are best illustrated by constructing Nernst's $\Delta H(T)/\Delta G(T)$ diagram and comparing it with the diagram as it corresponds to Eqs. (5.14) and (5.15) (see Figure 2.29).

The above discussed values of ΔS_g and ΔH_g are obtained via $C_p(T)$ measurements according to Eqs. (5.8) and (5.9). This is a tedious experiment, requiring expensive calorimetric apparatus, covering (as done in Simon's classical investigations [21, 59]) the whole temperature interval from the respective melting temperature, T_m, down to temperatures, approaching the zero-point of temperature (Witzel [163] with SiO_2 and Gibson and Giauque [160] with glycerol experimented only down to hydrogen boiling temperatures, i.e., $T \sim 20$ K). Simon brought the experiments to helium temperatures, that is, close to $T \approx 4$ K. Below the mentioned low temperatures even significant changes in the $C_p(T)$ course cannot affect the values of the thermodynamic functions (see [1, 163] and the discussion given in [1] on contemporary $C_p(T)$ measurements with glasses at ultra-low temperatures, i.e., for $T < 1$ K).

In present day experiments, the convenient method of differential scanning microcalorimetry (DSM) is mostly used. However, with this technique measurements cannot be extended to ranges lower than nitrogen boiling temperatures ($T \approx 80$ K). In most cases authors, analyzing the kinetics of vitrification by DSM-methods, extend their measurements to several degrees below the corresponding T_g temperature, more or less correctly assuming that the foregoing summarized classical results allow one to extrapolate ΔS_g=const. down to $T \to 0$ K. The application of DSM measurements is illustrated in Figures 5.5 and 5.6. In Figure 5.6, the $C_p(T)$ course of a vitrifying metallic alloy is also given. These cases of undercooled metallic melts require fast experimentation, due to the high crystallization tendency of vitrifying metallic alloy melts.

There exist also alternative methods for determining the caloric properties of glass-forming melts and glasses. Most of them are usually applied to the direct determination of the frozen-in enthalpy, ΔH_g. They are discussed in Section 2.5.3.

The simplest possible method in this direction is the determination of the frozen-in value of the enthalpy difference, ΔH_g, that is, to measure the reaction heat, Q, for a chemical change, in which the substance under consideration is reacting once as a glass and a second time as a crystal. A well-known example in this respect gives the determination of the reaction heats of quartz crystals and of silica glass in hydrofluoric acid according to the equation

$$\text{SiO}_2 + 4\text{HF} \rightarrow \text{SiF}_4 + 2\text{H}_2\text{O} + Q \tag{5.16}$$

thus ΔH_g can be determined as

$$\Delta H_g = Q_{\text{glass}} - Q_{\text{crystal}}, \tag{5.17}$$

where Q_{glass} and Q_{crystal} are the respective reaction heats measured.

5.3
Consequences of Simon's Classical Approximation: the $\Delta G(T)$ Course

As already mentioned, Simon [23, 59] assumed that vitrification takes place not in a temperature *region*, but at the temperature $T = T_g$. According to this approximation the change in the $C_p(T)$ course takes place not along the corresponding sigmoidal curve, but as a simple step-like jump. This step-like change of $C_p(T)$ at $T = T_g$, as seen from Figure 5.1, is only a first approximation and a very substantial but realistic idealization. Simon's approximation introduces a vertical straight line, going through the turning point of the real, sigmoidally changing curve. At heating run experimentation, when the reverse transition glass $\rightarrow$ undercooled melt takes place, the sigmoidal (i.e., s-shaped) $C_p(T)$ curves appear again, however, with the mentioned peculiar "nose," its height increasing with increasing heating rates. Several examples in this respect are given in Chapter 2 of the present book.

A thorough analysis of the real $C_p(T)$ course upon glass transition is possible, as shown in Chapter 3, on the basis of the thermodynamics of irreversible processes and of one of its basic dependencies: on its phenomenological law and on De Donder's and Bragg–Williams' equations derived from it. The empirical approach discussed here in terms of Simon's classical assumption gives not only a safe basis in defining T_g (as the temperature of the stepwise $C_p(T)$ jump through the inflection point of the real sigmoidal $C_p(T)$ curve) but also of calculating the temperature course of the thermodynamic functions and the thermodynamic potential upon vitrification.

Let us now define the course of the Gibbs potential difference glass/crystal in terms of Simon's approximation, as it is indicated with Eqs. (5.6) and (5.7). It becomes evident, recalling the definition of a straight line tangent to the point $(x_0; y_0)$ of the plane curve $f(x)$

$$y = (x - x_0)\left[\frac{d f(x)}{dx}\right]_{x \to x_0} + y_0 \tag{5.18}$$

and the thermodynamic significance of the respective derivative in Eq. (5.18): here $f(x)$ has the meaning of the Gibbs potential difference, $\Delta G(T)$, between the metastable undercooled liquid and the respective crystal and thus

$$\frac{df(x)}{dx} = \left[\frac{\partial \Delta G(T)}{\partial T}\right]_p = -\Delta S(T). \tag{5.19}$$

For the point $(T_g; \Delta G(T_g))$ we have thus

$$\left[\frac{\partial \Delta G(T)}{\partial T}\right]_{T \to T_g} = -\Delta S(T)\big|_{T \to T_g} = -\Delta S_g. \tag{5.20}$$

Writing Eq. (5.10) for the metastable undercooled liquid (i.e., for an equilibrium system) for $T = T_g$ as

$$\Delta G(T_g) = \Delta H_g - T_g \Delta S_g, \tag{5.21}$$

it follows with Eq. (5.9) that Eq. (5.21) corresponds in fact also to the same point $(T_g; \Delta G(T_g))$ of the straight line y from Eq. (5.18) now written as

$$\Delta G_g(T) = (T - T_g)\left[\frac{\partial \Delta G(T)}{\partial T}\right]_{T \to T_g} + \Delta G(T_g). \tag{5.22}$$

This straight line, describing the temperature dependence of the thermodynamic potential of the glass, is tangenting the $\Delta G(T)$-line of the undercooled liquid at $T \to T_g$. This result is graphically illustrated in Figure 2.29b. A more general formulation of above considerations in terms of two tangental (G, T, p) planes (i.e., in (G, p, T) coordinates) was given by Gupta and Moynihan ([365], see also [366]).

In Chapters 3 and 8 of the present book, the problem mentioned above is reconsidered in terms of the thermodynamics of irreversible processes and the generic approach to glass transition, derived by two of the present authors. The essential point to be mentioned in this respect is that in all these more elaborate approaches the results obtained by the simple geometric considerations from above are fully confirmed. The geometric approach employed here opens also a simple way of analyzing the thermodynamic properties of other amorphous solids, of "glassy" crystals and of any solid with frozen-in defect structure, including the classical defect of molecular crystals, like CO, and of liquid crystals. Examples in this respect are given in two of the following sections of this chapter, as well as in Chapter 8. First, however, we have to continue the analysis of the kinetics of glass transition as discussed in the present section.

5.4
Change of Kinetic Properties at T_g and the Course of the Vitrification Kinetics

The kinetic properties of liquids – their *kinetic coefficients* (viscosity, $\eta(T)$, time of molecular relaxation, $\tau(T)$, coefficient of diffusion, $D(T)$, etc.) – are at equilibrium (i.e., for the melt at temperatures, $T > T_m$, and for the undercooled metastable liquid at $T < T_m$) unambiguous functions of the state parameters (pressure, temperature, electric and magnetic field strength, etc.) in the same way, as this applies for thermodynamic functions of equilibrium systems. Moreover, as far as the kinetic properties of melts and liquids are determined by the thermodynamic functions of the system (see [1]) the break-point in the temperature dependence of the thermodynamic functions is reflected also in a break-point of the $\eta(T)$ curves at $T = T_g$ as seen from Figures 3.21, 3.22, and 5.4.

Figure 3.21 is a remarkable generalization of the results of measurements on the temperature dependence of the viscosity obtained by several authors with a series of classical soda-lime silicate glasses with nearly equal composition performed years ago by Winter-Klein [289, 367]: there is still no other example in glass-science literature, where the measurements cover such a wide range of $\log[\eta(T)]$ values. From the above cited figures, as well as from other similar examples, it is evident that in the temperature dependence of viscosity the break point upon glass transition appears in the logarithmic representation of the viscosity. According to the well-known exponential formula for the temperature dependence of the time of molecular relaxation introduced in Chapter 2

$$\tau(T) = \tau_0 \exp\left(\frac{U(T)}{RT}\right) \tag{5.23}$$

and taking into account the linear dependence $\eta(T)$ vs. $\tau(T)$ also given (see Section 2.4.3), these results indicate that this break-point appears in fact in the temperature dependence of the activation energy, $U(T)$, of molecular flow. In a similar way the equilibrium–nonequilibrium transition upon vitrification is also manifested with a break-point at a temperature, corresponding to T_g, in the temperature dependence of the other kinetic coefficients of the glass-forming system. In other cases (as with the coefficient of heat conductivity, $\lambda(T)$), when the respective kinetic coefficient depends both on molecular flow, via $\tau(T)$, and additionally and more significantly on a thermodynamic coefficient (e.g., on C_p as it is the case with $\lambda(T)$), a step-like change is observed, as seen in Figure 2.36.

Another crucial experimental finding, concerning the kinetic characteristics of vitrification, is that the value of the glass transition temperature, T_g, depends on the cooling rate, $q = (-dT/dt)$. Thus follows, as already discussed in Chapter 3, a course established for the first time by Bartenev [169, 272] and Ritland [194] as

$$\frac{1}{T_g} = C_1 - C_2 \log |q|, \tag{5.24}$$

where C_1 and C_2 are constants and q is a constant cooling rate. Upon heating runs, in experimentation starting with already formed glasses a similar dependence is observed determining T_g in the glass to undercooled melt transition (see [272]). The kinetic overshot at T_g leads to somewhat higher T_g values and to the formation of the already discussed "nose" in the $C_p(T)$ curve and in the temperature dependence of any other thermodynamic or kinetic coefficients. At cooling run experiments, this "nose" has never been observed.

5.4 Change of Kinetic Properties at T_g and the Course of the Vitrification Kinetics

An explanation of these experimental findings can be given in terms of the already mentioned theoretical approach, based on the Bragg–Williams kinetic equation, as discussed in [11, 12, 17, 170, 319] and here in Chapters 3 and 8. In essence, as first observed already by Volkenstein and Ptizyn [170], this explanation stems from the Bragg–Williams kinetic equation, written in its simplest form as

$$\frac{d\xi(T)}{dT} = -\frac{[\xi(T) - \xi_e(T)]}{q\tau(T)}, \quad q = \frac{dT}{dt}. \tag{5.25}$$

This equation gives the change of the generalized internal structural parameter, ξ (see, again, Chapter 3), when the temperature, T, of the system is changed with a rate $q = dT/dt$. It is shown in [170] that by introducing appropriate temperature dependencies of both the time of molecular relaxation, $\tau(T)$ (see Eq. (5.23)), and of the temperature dependence of the equilibrium structure, $\xi_e(T)$, of the liquid into Eq. (5.25) (discussed in Chapters 3 and 8), the temperature course of $(d\xi/dT)$ displays *one inflection point* and no extremal points on cooling. The $(d\xi/dT)$ curves according to Eq. (5.25) have to show, however, an *inflection point* plus a *maximum* after that at heating runs. As far as it can be shown, that $C_p \sim d\xi/dT$, this also implies a corresponding difference in the course of the specific heats of the system when a glass transition is observed either at cooling or at heating run experimentation: in the $C_p(T)$ course upon heating, a rate-dependent maximum is always registered. In [17] it is also shown, that the solutions of dependencies like Eq. (5.25) are such that a difference up to several Kelvin are observed between the T_g values at cooling and of heating, as this has also been observed experimentally.

As shown in Chapter 3, the Bartenev–Ritland formula (Eq. (5.24)) can be easily derived, employing a dependence introduced years ago (in the 1950s) into glass science as a *postulate* by Bartenev [169, 272] attributed by him to Frenkel and Kobeko. It connects the time of molecular relaxation, $\tau(T)$, at $T \to T_g$ and the (constant) cooling rate, q, in the form

$$\tau(T)|q||_{T \to T_g} \cong \text{const.}, \tag{5.26}$$

where the constant usually has values of several Kelvin. The Bartenev–Ritland formula, Eq. (5.24), follows directly from Eq. (5.26), assuming that the $\tau(T)$ or $\eta(T)$ dependences can be written as an exponential function of the activation energy, $U(T)$ with the approximation that $U(T) = U_0 = \text{const.}$

A statement equivalent to Eq. (5.26) has been proposed (in fact *postulated*) by Reiner [368, 369] in a somewhat peculiar way: via the introduction of the number Dh, bearing the name of a Biblical character (of the prophetess Deborah). It is the ratio of the time of molecular relaxation, $\tau_R(T)$, of the system to the time, Δt_g, of observation (or the stay time of the system at $T \approx T_g$ in the transformation range)

$$Dh|_{T \to T_g} = \left(\frac{\tau_R(T)}{\Delta t_g}\right)_{T \to T_g} = 1. \tag{5.27}$$

For undercooled liquids near the melting point $Dh \ll 1$ holds, for glasses $Dh \to \infty$ and for $T \to T_g$, according to the above proposal, $Dh \approx 1$. It is obvious that

Reiner's stay time, Δt_g, is not sufficiently defined and that Eq. (5.27) is introduced in a *Deus ex machina* manner. Only recently Gutzow *et al.* [11, 12] could show that both Eqs. (5.26) and (5.27) can be in fact *derived* from general dependencies following from the already outlined generic approach developed in Chapter 3 in the framework of the thermodynamics of irreversible processes. In this way, in [11, 12], both the kinetics and the thermodynamics of the glass transition become, as was to be expected, a consequence from the same general premise: the phenomenological law of irreversible thermodynamics in a more general, nonisothermal formulation, first employed by Bragg and Williams (see [17, 319] and the literature cited there) and later by Volkenstein and Ptizyn [170].

In the following section it is shown that from this general phenomenological approach, and in particular from Eq. (5.23), at least in principle, the so-called *glass transition invariants* can be derived, which have been up to now postulated in glass science in the long history of its development by several authors. A first introduction and an outline of these invariants, connecting glass transition temperature, T_g, and the values of the thermodynamic functions frozen-in there, with the values of the same quantities at the melting point, T_m, is given in several occasions in the present book, and especially in Chapters 2 and 3. In Chapter 8 these invariants are qualitatively considered in the framework of the general conclusions of the theory of physicochemical similarity and the theorems of corresponding states. Here, as said, an attempt is made to derive them in the framework of generic approach, developed in Chapter 3.

Let us first observe that in introducing the thermodynamic significance, which the linear formulations of this thermodynamic approach attributes to both $(d\xi/dT)$ and the difference $(\xi - \xi_e)$, the Bragg–Williams dependence, given with Eq. (5.25), can be written in its general thermodynamic form:

$$\Delta C_p(T, \xi) \cong \frac{\Delta H(T, \xi)}{q\tau(T, \xi)} . \tag{5.28}$$

In order to perform the following derivations we must, however, specify in a particularly precise way the significance of the quantities, appearing in above dependence, beginning with the value $\Delta C_p(T, \xi)$ on the left hand side. This is done in the following section.

5.5
The Frenkel–Kobeko Postulate in Terms of the Generic Phenomenological Approach and the Derivation of Kinetic and Thermodynamic Invariants

The specification analyzed here stems from Prigogine's definition of the specific heats of the undercooled liquid and their separation into two parts according to

$$\frac{dH(T, \xi)}{dT} = \left[\frac{\partial H(T, \xi)}{\partial \xi}\right]_{P,T} \frac{d\xi}{dT} + \left[\frac{\partial H(T, \xi)}{\partial T}\right]_{p,\xi} . \tag{5.29}$$

Here the configurational part of the specific heats of the undercooled vitrifying liquid is defined as:

$$\Delta C_p(T, \xi) = C_p^{\text{conf}}(T, \xi) \cong \left[\frac{\partial H(T, \xi)}{\partial \xi}\right]_{P,T} \frac{d\xi}{dT} \cong h_0 \frac{d\xi}{dT}. \quad (5.30)$$

This definition of $\Delta C_p(T, \xi)$ is used throughout the present book; it includes the configurational part of the specific heats of *both* equilibrium and nonequilibrium liquids. In the right hand side of above equation another of Prigogine's proposals is used: to connect linearly the internal structural parameter, ξ, of the system with the configurational part of its thermodynamic function, for example, as

$$\Delta H(T, \xi) \cong h_0 \left[\xi - \xi_e(T)\right]. \quad (5.31)$$

Here h_0 is a constant factor, which in the case of an undercooled liquid corresponds approximately to its enthalpy of melting, $h_0 \approx \Delta H_m$. The second additive member in the right hand side of Eq. (5.29) denotes the phonon part of the $\Delta H(T, \xi)$ function of the liquid

$$\Delta C_p(T) = \left[\frac{\partial H(T, \xi)}{\partial T}\right]_{p,\xi} \cong C_p \cong \left(\frac{dH^{\text{cryst}}(T)}{dT}\right)_{p,\xi=0} \cong C_p^{\text{cryst}}(T), \quad (5.32)$$

which according to above proposal is taken to be equal to the specific heats of the corresponding crystal.

The Bragg–Williams equation, above written as usual in the form of Eq. (5.25), or more generally as

$$\frac{d}{dT}\left[\xi - \xi_e(T)\right] = -\frac{\left[\xi - \xi_e(T)\right]}{q\tau(T, \xi)} \quad (5.33)$$

does not explicitly specify the nature of the equilibrium reference function, $\xi_e(T)$. In its original derivation (e.g., via De Donder's dependence, given below) Eq. (5.33) refers $\xi_e(T)$ to the equilibrium structure of the undercooled liquid. However, especially Eq. (5.29) also gives the possibility to connect $\xi_e(T)$ with the structure of the respective crystal as the corresponding equilibrium reference state. This double possibility is open also in the specification of De Donder's equation,

$$\frac{d}{dt}\left[\xi - \xi_e(T)\right] = -\frac{\left[\xi - \xi_e(T)\right]}{\tau(T, \xi)} \quad (5.34)$$

describing isothermal relaxation, where $\xi_e(T)$ for any form of this function *is a constant*. The definition of $\xi_e(T)$, as referred to the crystalline state, gives the possibility to introduce into Eqs. (5.33) and (5.34) the quantities $\Delta C_p(T, \xi)$ and $\Delta H(T, \xi)$, and $\Delta C_p(T)$ and $\Delta H(T)$ as values, taken from experimental data. In the framework of Simon's approximation, we have to use dependencies like

$$\Delta H(T) = \Delta H_0 + \int_0^T \Delta C_p(T) dT = \Delta H_m - \int_T^{T_m} \Delta C_p(T) dT, \quad (5.35)$$

where

$$(\Delta H_m - \Delta H_0) = \int_0^{T_m} \Delta C_p(T) dT \qquad (5.36)$$

holds, to determine the values of $\Delta H(T, \xi)$ especially in the case, when we are concerned with the the value of $\Delta H(T_g, \xi) = \Delta H_g$, that is, with the enthalpy, frozen-in in the glass. Here with ΔH_0 in the above dependencies, the zero-point enthalpy of the *fictive* undercooled liquid is indicated, if we could realize it at $T \to 0$.

In the present section, however, we are concerned with another problem: how to estimate the value of the product $q\tau(T, \xi)$ at vitrification (i.e., of $q\tau(T, \xi)|_{T=T_g}$), as it follows from Eq. (5.34) in the framework of the generic phenomenological approach employed throughout the present book. We must compare this value then with empirical postulates, as those given by Reiner or Frenkel and Kobeko, introduced in Chapter 3 and here with Eqs. (5.26) and (5.27). Such an estimate and the indicated comparison show again the additional possibilities opened up by the generic phenomenological theory summarized and discussed in Chapters 3 and 8.

To perform such a comparison, it turns out, we have to use both cases: to express $\xi_e(T)$ once as representing the equilibrium structure of the real or fictive undercooled liquid in equilibrium, and a second time in terms of the structure of the crystal (assumed here to be always in equilibrium). In the first case, the equilibrium value of the structural order parameter, ξ, will be indicated in this section by $\xi_e(T)$ to avoid confusion with $\xi_e^*(T)$, which we reserve here for the crystal.

In order to facilitate the understanding and the direct connection between the nonisothermal Eq. (5.33) and its isothermal form Eq. (5.34) we will use in both cases linear dependencies for both $\xi_e(T)$ functions. They allow us, according to simple mathematics, the direct change of argument in both equations, when in accordance with the definition $(dT/dt) = q_0$ the cooling and heating rate of the vitrifying system also is a linear dependence $(T \propto q_0 t)$.

In Chapters 3 and 8 the reader finds the formal derivation of De Donder's equation, Eq. (5.34), from the phenomenological law of irreversible thermodynamics and its transition into the generalized Bragg–Williams equation, Eq. (5.33). The two necessary linear $\xi_e(T)$ dependencies employed here can be written in the form

$$\xi_e(T) = \begin{cases} b_0(T - T_0) + \xi_0 & \text{for } T_m \geq T \geq T_0 \\ \xi_0 & \text{for } T_0 > T \geq 0 \end{cases} \qquad (5.37)$$

with

$$b_0 = \left.\frac{d\xi_e(T)}{dT}\right|_{T=T_g} = \text{const.} \qquad (5.38)$$

as referred above to the case, when we denote the temperature course of ξ in the equilibrium liquid with $\xi_e(T)$. For the case, that we refer the state of a nonequilibrium liquid with the crystal we have to write the above equation with $\xi_e^*(T)$, with

$$b_0^* \equiv \left(\frac{d\xi_e^*(T)}{dT}\right)\bigg|_{T=T_g} < b_0 \tag{5.39}$$

and with $\xi_0^* \equiv 0$.

We further on introduce (for temperatures $T_m > T > T_0$) the notations

$$\frac{d\Delta\xi_e(T)}{dT} = \frac{d\xi_e(T)}{dT} - \frac{d\xi_e^*(T)}{dT} = (b_0 - b_0^*) = \Delta b_0 \tag{5.40}$$

corresponding to the function

$$\Delta\xi_e(T) = \xi_e(T) - \xi_e^*(T) = \Delta b_0(T - T_0) + \xi_0 . \tag{5.41}$$

Now, analogous to the above dependencies and to Eq. (5.30) we can also introduce the difference in the heat capacities difference equilibrium liquid/crystal as:

$$\Delta C_p(T)_{l/c} = C_p(T)_{liq} - C_p^{cryst}(T) \cong h_0 \Delta b_0 . \tag{5.42}$$

Such a provision corresponds in fact to the frequently used approximation, employed for the temperature dependence of the specific heat difference undercooled liquid/crystal in the glass transition region of both the real and (below T_g^-) of the fictive undercooled melt

$$\frac{\Delta C_p(T)_{l/c}}{\Delta S_m} = \text{const.} = a_0 \quad \text{for} \quad T_m > T \geq T_0 , \tag{5.43}$$

$$\frac{\Delta C_p(T)_{l/c}}{\Delta S_m} = 0 \quad \text{for} \quad T_0 > T \geq 0 , \tag{5.44}$$

discussed here in its thermodynamic significance in Chapter 8. Note that above dependence refers to the equilibrium values of the specific heats of both phases (liquid and crystal) of the sigmoid course of the specific heats of the real, increasingly nonequilibrium melt being described by the $\Delta C_p(T, \xi)$ function, as indicated in Figure 5.7. It is also evident that, in general, defining the glass transition temperature as the inflection point of the $\Delta C_p(T, \xi)$ curve as

$$h_0 \frac{d^3\xi}{dT^3}\bigg|_{T=T_g} \equiv \frac{d^2\Delta C_p(T, \xi)}{dT^2}\bigg|_{T=T_g} = 0 , \tag{5.45}$$

we also expect that at $T = T_g$

$$\Delta C_p(T, \xi)\big|_{T=T_g} = \Delta C_p(T_g, \xi) \approx \frac{1}{2}\Delta C_p(T_g)_{l/c} . \tag{5.46}$$

From Figure 5.7 it is seen that, for temperatures from T_m to T_g^+, we can also write

$$\Delta C_p(T, \xi) - C_p(T)_{liq} \cong 0 \tag{5.47}$$

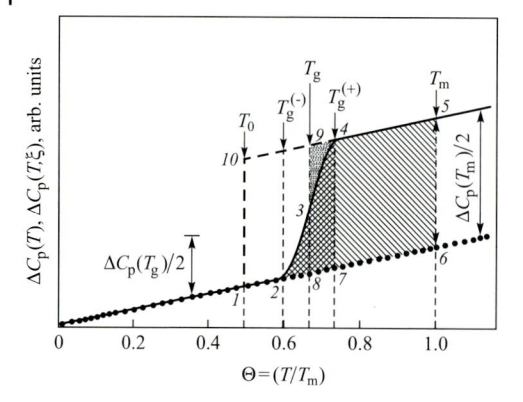

5 Thermodynamics of Amorphous Solids, Glasses, and Disordered Crystals

Figure 5.7 Temperature dependence of the specific heats, $C_p(T)$, of the equilibrium liquid, of the crystal and of the specific heats, $\Delta C_p(T, \xi)$, of the vitrifying liquid in the glass transition region (schematically). The triangle 3, 9, 4, *dashed areas* and the area of the figure marked by 1, 2, 8, 7, 6, 5, 4, are used in the text to determine the values of the enthalpy difference with respect to both the equilibrium liquid and the crystal.

and that for $T < T_g^+$

$$\Delta C_p(T, \xi) - C_p(T)_{\text{cryst}} \cong 0 \,. \tag{5.48}$$

With these relations, we can write the Bragg–Williams equation as:

$$\left.\frac{d\Delta\xi(T)}{dT}\right|_{T=T_g} = -\left.\frac{\Delta\xi(T)}{q_0\tau(T)}\right|_{T=T_g} \cong -\frac{\int_{T_g}^{T_m} \frac{d\xi\, dT}{dT} - \int_{T_g}^{T_m} \Delta b_0\, dT}{q_0\tau(T)|_{T=T_g}} \,. \tag{5.49}$$

In using Eqs. (5.30) and (5.31) we can now construct the dependencies Eqs. (5.37)–(5.42) in their thermodynamic form, according to which we have denoted

$$h_0\left(\frac{d\xi_e(T)}{dT}\right) \equiv \Delta C_p(T)_{l/c}, \quad h_0\left(\frac{d\xi(T)}{dT}\right) \cong \Delta C_p(T, \xi), \tag{5.50}$$

as well as

$$\Delta H(T)_{l/c} = h_0 \Delta \xi_e(T), \quad \Delta H(T, \xi) = h_0 \xi(T) \,. \tag{5.51}$$

This leads to the specific heat dependencies given in Figure 5.7, the corresponding enthalpy functions have to be schematically drawn as given in Figure 5.8.

Now, in a virtual cooling run experiment, performed in the framework of the schematically drawn, enlarged Simon-like approximation, we can determine with sufficient accuracy the value of the product $q_0 \tau(T, \xi)|_{T=T_g}$ both according to the thermodynamic form of the Bragg–Williams dependence as well as in its classical formulation, indicated with Eqs. (5.25) and (5.33).

From the $C_p(T)$ construction in Figure 5.7 and above derivations we can write that in Eq. (5.28), when we are expressing the deviation from equilibrium in terms

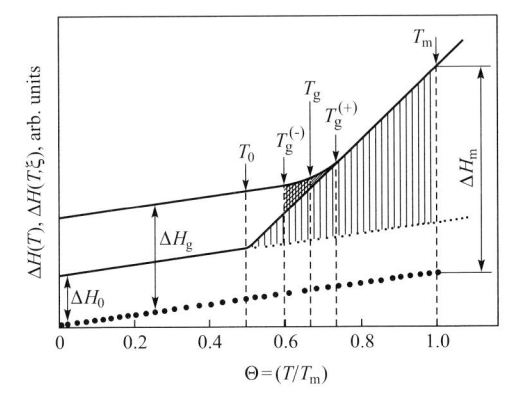

Figure 5.8 Temperature course of the enthalpy of liquid, crystal and glass, obtained by the approximated integration of the specific heat contents according to Figure 5.7. Note that here the double shaded, nearly triangular area in the temperature interval $(T_g^+ - T_g)$ corresponds to the triangle 3, 9, 4 in Figure 5.7.

of $(\xi - \xi_e(T))$, that is, in terms of deviation from the equilibrium state of the undercooled liquid, we have

$$\Delta H(T_g, \xi) \approx \int_{T_g}^{T_m} [\Delta C p(T, \xi)] dT - \int_{T_g}^{T_m} [C p(T)_{l/c}] dT \tag{5.52}$$

$$\approx \int_{T_g}^{T_m} [\Delta C p(T, \xi)] dT - \int_{T_g^+}^{T_m} h_0 \Delta b_0 dT$$

$$- \frac{1}{2} \int_{T_g}^{T_g^+} h_0 \Delta b_0 dT - \frac{1}{4} \int_{T_g}^{T_m} h_0 \Delta b_0 dT$$

$$\approx \int_{T_g}^{T_m} \Delta C_p(T, \xi) dT - h_0 \Delta b_0 \left[\left(T_m - T_g^+ \right) + \frac{3}{4} \left(T_g^+ - T_g \right) \right].$$

Thus the $\Delta H(T, \xi)$ value in the whole interval (T_m, T_g) is approximately given with the dotted-shaded area of the triangle 3, 4, 9 (see Figure 5.7 and also Figure 5.9 for $T < T_g$). With the approximation for $\Delta C_p(T_g, \xi)$ according to Eq. (5.46) we can now also write

$$\Delta H(T_g, \xi) \approx \int_{T_g}^{T_m} \Delta C_p(T, \xi) dT - \int_{T_g}^{T_m} \Delta C_p(T)_{l/c} dT \tag{5.53}$$

$$\cong \frac{1}{4} \Delta C_p(T)_{l/c} \left(T_g^{(+)} - T_g \right).$$

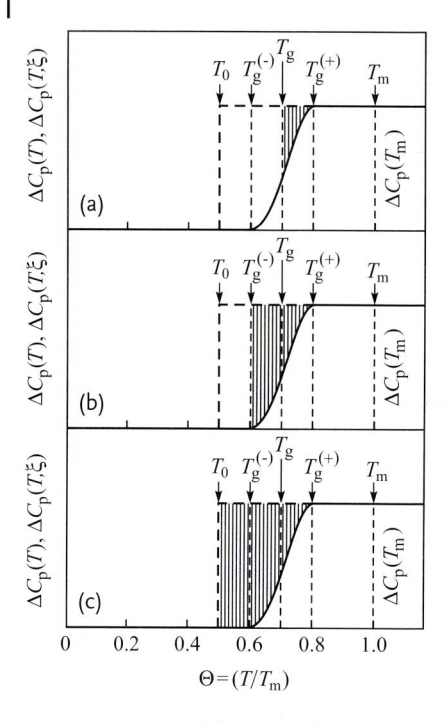

Figure 5.9 Change of the specific heats, $\Delta C_p(T, \xi)$, of a vitrifying liquid (the *solid sigmoid line*) in the glass transition interval and the enthalpy difference, $\Delta H(T, \xi)$, glass equilibrium liquid at different reduced temperatures, θ, at $T < T_g^+$. In the three pictures, this enthalpy difference is given by the shaded area above the respective $\Delta C_p(T, \xi)$ lines. Note that at decreasing temperature (as indicated in pictures (a–c)) the shaded area, and thus the value of $\Delta H(T, \xi)$ are increased. At (a) the temperature reached is $\theta = T_g/T_m$, at (b) the temperature reached is T_g^-, and at (c) the temperature reached is $T = T_0$, where the highest possible deviation from equilibrium is approached.

Thus we can represent Eq. (5.28) at $T = T_g$ as:

$$\frac{1}{2}\Delta C_p(T)_{l/c} \approx \frac{1}{4}\left(\frac{\Delta C_p(T)_{l/c}\left[T_g^{(+)} - T_g\right]}{q_0 \tau(T, \xi)\big|_{T=T_g}}\right). \tag{5.54}$$

As far as T_g is symmetrically situated in the $T_g^{(+)} - T_g^{(-)}$ interval,

$$\left(T_g^{(+)} - T_g\right) \approx \frac{1}{2}\left(T_g^{(+)} - T_g^{(-)}\right), \tag{5.55}$$

and, since we know from experimental data that for typical glass formers

$$\left(T_g^{(+)} - T_g^{(-)}\right) \cong \left(\frac{1}{10} \div \frac{1}{20}\right) T_g, \tag{5.56}$$

from the three previous dependencies it follows that for typical glasses

$$q_0 \tau(T, \xi)\big|_{T=T_g} \cong \left(\frac{1}{40} \div \frac{1}{80}\right) T_g \approx \text{const}_0 \approx 5\,\text{K}. \tag{5.57}$$

The above evaluation corresponds approximately to the Reiner–Frenkel–Kobeko postulate, as it was expressed with Eq. (5.24) by Bartenev, when we take into account that in general the T_g values of the glasses we consider here are located in the range from 350 to 550 K.

It was already mentioned that this postulate can be derived in the framework of the generic phenomenological theory of vitrification, summarized here in Chapters 3 and 8. In fact, with a linear dependence for the $\xi(T)$-function of the type, given with Eq. (5.37), and with a simple exponential dependence ($U(T) = U_0 =$ const.) for the time of molecular relaxation $\tau(T, \xi) \approx \tau(T)$ in Eq. (5.23), Gutzow and Babalievski [320, 370] (see also [11, 12]) obtained the result

$$q_0 \tau(T)|_{T=T_g} = 1.67 \left[\frac{d \ln \tau(T)}{dT} \right]^{-1} \bigg|_{T=T_g} \cong 1.67 \frac{RT_g^2}{U(T_g)}, \tag{5.58}$$

where $U(T_g)$ stands in Eq. (5.23) for the constant value of the activation energy of relaxation at $T \approx T_g$. Accounting for the circumstance that from Bartenev's times it is well-known from many experiments that for most glass-forming melts

$$\frac{RT_g^2}{U(T_g)} \approx \text{const.} \approx (10 \div 5) \text{ K}, \tag{5.59}$$

it becomes evident that the classical empirical postulate Eq. (5.26) has to be considered as a particular case of Eq. (5.58). Moreover, in [17] it is shown that Eq. (5.58) can be considered as the simplified expression of more complex dependencies, which can be obtained at more realistic $\tau(T, \xi)$-, $\Delta \xi(T, \xi)$- and $\Delta H(T, \xi)$-dependencies, via the condition given with Eq. (5.45), defining T_g as the inflection point of the $\Delta C_p(T, \xi)$ curve. From this condition, as already mentioned, follows directly both Eq. (5.58) and the Bartenev–Ritland dependence, Eq. (5.24), connecting glass transition temperature, T_g, with the cooling rate, q.

A classical example concerning the experimental proof of the Bartenev–Ritland dependence can be seen here in Figure 5.10 (according to Bartenev's own measurements). Bartenev and Lukyanov [371] have observed further on that in general

$$\frac{U(T_g)}{2.3 R T_g} \approx \text{const.} \approx (30 \div 35) \tag{5.60}$$

for the glass-forming substances they investigated (see also further data, compiled by Avramov [372, 373]). In fact, considering the above relations, this result follows also in the framework of the simple analysis performed here.

Now again, employing Eq. (5.49), we obtain the results from above in more general terms of this classical equation. Introducing there the dependencies, connecting the above expressions with the already discussed thermodynamic formalism, we can also introduce the enthalpy value and the specific heat differences in the way as they are obtained in caloric experimentation: as the differences against the respective values of the corresponding crystal. Thus we can write

$$\Delta C_p(T_g) \approx \frac{\Delta H_g}{q_0 \, \tau(T)|_{T=T_g}} \approx \text{const.}^* \tag{5.61}$$

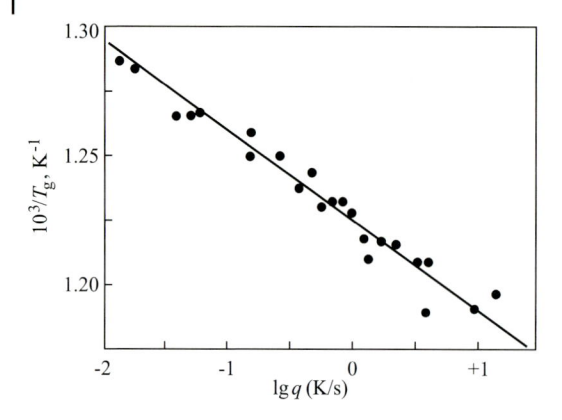

Figure 5.10 First proof of the kinetic dependence of T_g on cooling or heating rate q (for an alkali-silicate glass, after Bartenev [169, 272]): experimental proof of Eq. (5.24), giving for this case an activation energy of $U(T_g) \approx 267$ J/mol.

Dividing both sides of Eq. (5.61) with the melting enthalpy ($\Delta H_m = T_m \Delta S_m$), we can write it also in the invariant form:

$$\frac{\Delta C_p(T_g)}{\Delta S_m} \frac{\Delta H_m}{\Delta H_g} \approx \text{const.} * \frac{U(T_g)}{R T_g} \frac{T_m}{T_g}. \qquad (5.62)$$

We know already from Chapters 2 and 3 and from above considerations that any of the four ratios entering Eq. (5.62) have for many years been considered as invariants of glass transition, empirically established by well-known authors in the field of glass science and polymer physics: by Wunderlich, Tammann, Bartenev, Kauzmann and Beaman [24, 109, 110, 169, 173, 374]. The respective publications are cited in the literature, given at the end of the present book. Here we derived in the framework of the adopted generic phenomenological approach an expression, Eq. (5.62), showing that there has to be a distinct connection between the thermodynamic (on the left hand side of this equation) and the kinetic invariants (on the right hand side of the same dependence) of glass transition. We have made here the mentioned derivations. Performing the respective computations, we here retain the limitations of the Simon enlarged model of vitrification: it is used in calculating both the thermodynamic and kinetic data, necessary for the computation.

In terms of the schematic construction, given in Figures 5.7 and 5.8, the value of the melting enthalpy is determined by $\Delta H_m \approx C_p(\Delta T_g)_{l/c}(T_m - T_0)$ and according to Eq. (5.35). In the same construction ΔH_g is given by the area 1, 2, 8, 7, 6, 5, 4 in the mentioned figures. In determining the value of the product $q_0\tau(T_g)$ according to Eq. (5.62), we thus operate in the glass/crystal case with the ratio $(\Delta H_g/C_p(T_g, \xi))$, which is quite different from the equivalent ratio, determined from the glass/equilibrium liquid case. In doing this comparison, we have to account for the circumstance, that the value of specific heat $\Delta C_p(T_g, \xi)$ of the glass is the same in both cases, and given by Eq. (5.46). Thus a different value is obtained for const$_0$ in Eq. (5.57) and for the constant indicated by (*) in Eqs. (5.61) and (5.62). From a general point of view the "true" value of the product $q_0\tau(T_g)$

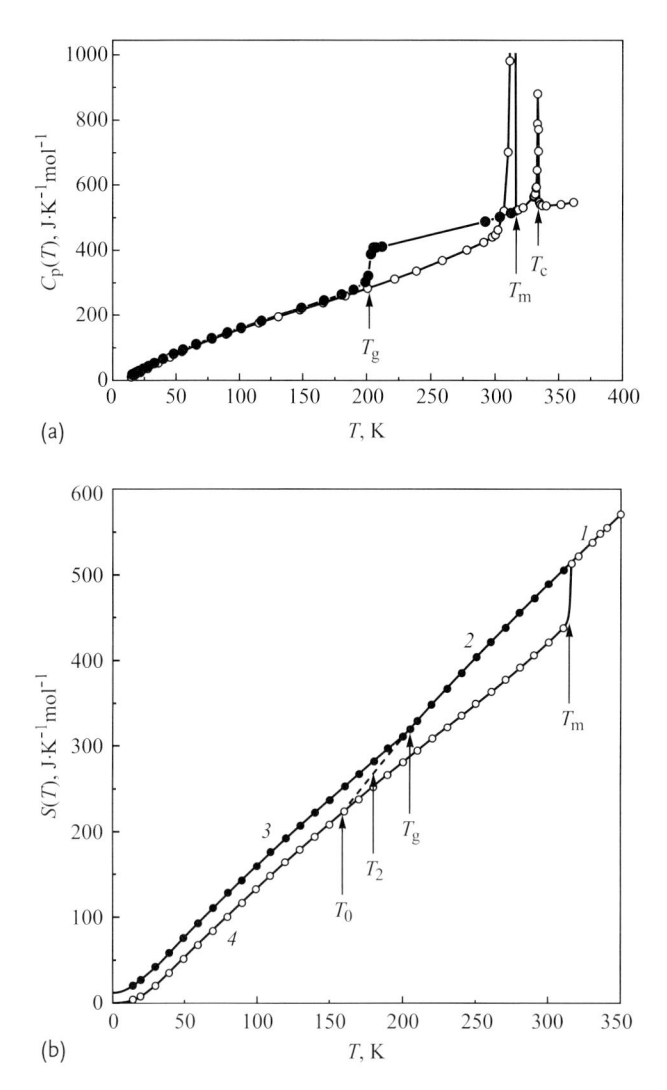

Figure 5.11 Temperature course of (a) the specific heats, $C_p(T)$, and (b) the entropy, $S(T)$ of a typical liquid crystal denoted as (OHMBBA) according to Sorai and Seki [347–349]. *Open circles* in both (a) and (b) represent the isotropic, nematic and crystalline phases of the same substance. *Black circles* in both (a) and (b) represent the glassy liquid crystal and the super-cooled liquid crystal of the nematic phase. T_c indicated the so-called clearing point at which a second-order phase transition takes place between the isotropic and nematic liquid. T_m indicates the melting point between the molecular crystal and the nematic liquid, T_g indicates the glass transition point and T_0 the respective Kauzmann temperature, where $S(T_0)_{\text{nematic liquid}} - S(T_0)_{\text{crystal}} = 0$. Note that at T_0 the frozen-in nematic liquid has a zero point entropy, $S_0 > 0$, corresponding to a glass.

is obtained from the glass/liquid case formulation as it is given with Eqs. (5.53) and (5.57): it corresponds to the basic idea of the whole theory, operating with the deviation from equilibrium in the system under consideration itself.

Nevertheless, the main result formulated here with Eq. (5.62) is obtained in the framework of the glass/crystal deviation from the equilibrium model. It is more convenient, as already mentioned, for comparison with experimental caloric data. This result can be formulated in the sense that the generic theory of glass transition developed here, having already given a correct temperature course of the kinetic and thermodynamic properties of the systems at glass transition, which permitted a new and more correct interpretation of the Prigogine–Defay ratio, of fictive temperature and pressure change at vitrification, can also forecast the existence of the long discussed and often employed invariants of vitrification. In Section 5.9 it is shown that from the four invariants, as they are connected in Eq. (5.62), all other invariants also follow. Figure 5.9 also shows another result of principal significance obtained in the framework of the derivations made in this section: the decrease of the entropy difference glass/crystal with decreasing temperature, T, when $T < T_g^+$. This picture needs no additional interpretation: it shows that the maximal deviation from the equilibrium liquid is obtained at $T = T_g^+$.

The above discussed derivations are confirmed for a great variety of *typical* glass-forming melts. Now we consider glass transitions in two particular but nontypical groups of representatives of glass-forming systems. Systems belonging to one of these groups, liquid crystals, have significant technical applications.

5.6
Glass Transitions in Liquid Crystals and Frozen-in Orientational Modes in Crystals

As liquid crystals are usually denoted those liquids (mostly constituted of organic compounds with complicated chain-like structures) in which a temperature change (*thermotropic* liquid crystals) or a decrease of the percentage of solvent employed (*liotropic* liquid crystals) cause different forms of orientation (nematic, smectic or combined) and/or structural change in the essentially liquid system – either a melt, or a solution [375]. The oriented liquid state is easily recognized optically under polarized light by the typical birefringence, characteristically changing for different types of orientational order, either externally induced (e.g., by applied electric fields) or inherent in the liquid in a given temperature interval. In thermotropic systems the liquid can exist as a smectic or a nematic-oriented liquid body in one or more temperature intervals. With appropriate cooling procedures both smectic and nematic liquids can be frozen-in into the respective solid glass, in which the optical birefringence of the initial liquid is preserved [346]. The process of liquid crystal vitrification as far as it is investigated up to now (see [346–349, 375, 376]) shows all the typical features of the glass transition in "normal" or typical molecular glass-forming liquids already discussed in the foregoing paragraphs (s-shaped decrease of $C_p(T)$ in the glass transition region, a salient point in the $S(T)$ and the $H(T)$ curves at $T = T_g$).

However, the glass transition in liquid-crystal systems is usually preceded by typical changes in the caloric properties of the liquid, when it undergoes the *isotropic liquid ↔ oriented state* (smectic or nematic) transition, which usually shows all the features of a second-order phase transformation (or as stated in [35]) in some cases: so-called "weak" first-order phase transitions. The phenomenology of typical second-order phase transitions is discussed in Chapter 2. Its manifestation in a liquid crystal is clearly seen in the $\Delta C_p(T)$ and $\Delta S(T)$ curves of a typical liquid crystal substance (OHMBBA, Figure 5.11) as they are constructed first by Sorai et al. [347]. However, in going into the quantitative analysis of glass transitions in liquid crystals peculiar details can be observed, when (as done in the following sections) the main vitrification characteristics in liquid crystals are compared with those of normal glass formers, vitrifying without previous orientational transitions.

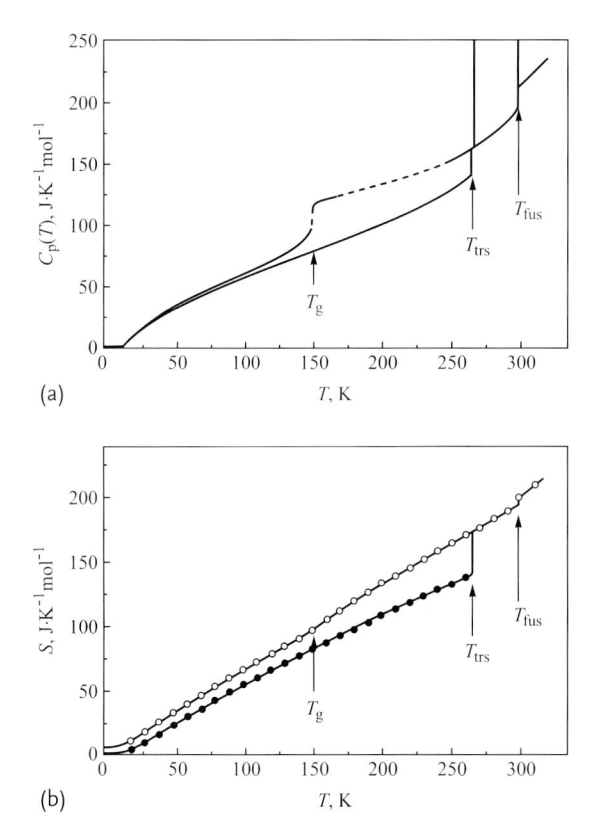

Figure 5.12 Temperature course (a) of the specific heats, $C_p(T)$, and (b) of the molar entropy, $S(T)$, of a typical plastic crystal, cyclohexanol, according to Suga [377]. T_m indicates the melting point of the high temperature crystalline modification of this substance and, by T_{trs}, the transition point between the two crystalline modifications. With T_g the temperature is denoted at which the entropy difference ΔS_0 at $T \to 0$ between the two crystalline modifications is frozen-in like in a typical glass transition.

Some of the dependencies outlined in the previous sections are observed not only in the glassy state of typical glass-forming melts and in the glass transition in the oriented liquids (somewhat irrelevantly called liquid crystals), but also in crystalline solids with different degrees of orientational structural disorder. A typical example in this respect is given in Figure 5.12 where as a result of $C_p(T)$ measurements and calculations similar to the those discussed in Eqs. (5.7) and (5.8), the entropy of molten and of crystalline samples of cyclohexanol in its crystalline modifications is constructed [377]. This substance does not give an undercooled melt able to vitrify. However, a break-point in the $\Delta S(T)$ curve analogous to that at T_g upon typical melt vitrification is clearly observable in the temperature dependence of the entropy

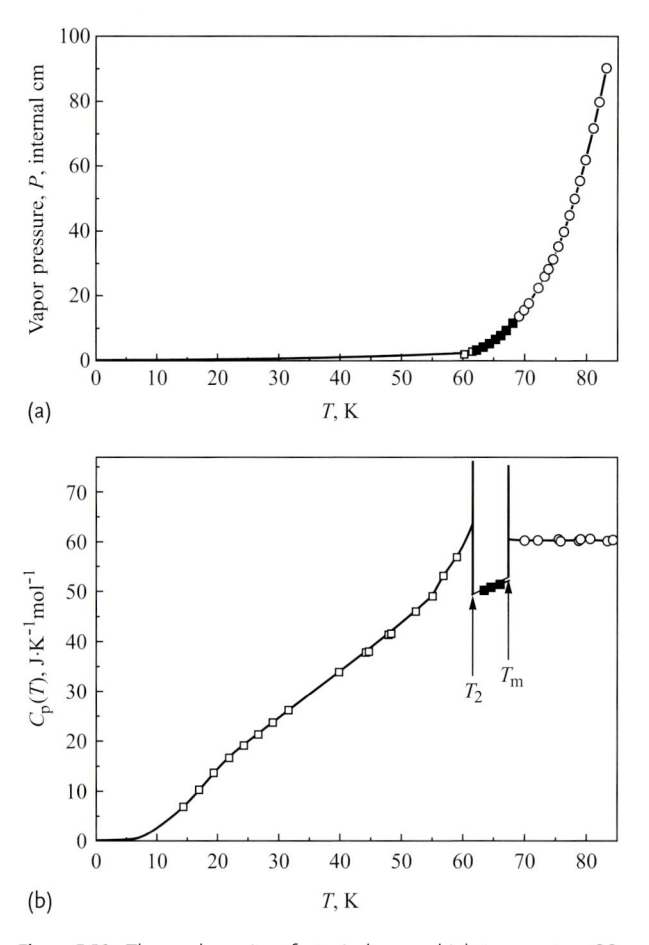

Figure 5.13 Thermodynamics of a typical disordered molecular crystal, CO (according to Clusius and Teske [352] and Clayton and Giauque [353]): (a) the temperature course of the vapor pressure, $p(T)$, of the second, low-temperature phase (open squares), of the high-temperature CO-crystal (black squares) and (open circles) of the liquid CO; (b) temperature dependence of the specific heats, $C_p(T)$, of the above indicated modifications of CO. The same symbols are used as in the $p(T)$ and $C_p(T)$ representations.

of the high-temperature crystalline modification of this substance. The same can be also seen in the $C_p(T)$ curves of cyclohexanol shown in Figure 5.12a, where the λ-type transitions between the crystalline phases is visible together with the step-like change of $\Delta C_p(T)$ at $T = T_g$, the freezing-in temperature of the disordered crystalline phase.

Cyclohexanol is only one of the representatives of a whole class of crystalline solids, sometimes somewhat misleading called glassy crystals (see [1, 184, 237, 341, 377]) in which different forms of orientational disorder can be frozen-in. Because of its particular mechanical properties more exactly cyclohexanol, cyclohexene, cycloheptanol and similar crystals are called *plastic crystals*. Further examples and experimental evidence on ΔS_g^0 values in this respect may be found in [237, 341, 377].

It is of significance to note that the freezing-in temperature in plastic crystals depends on the cooling rate in a way similar to Bartenev's dependence, Eq. (5.24). This result was manifested in an exceptionally instructive way by Suga [377] by analyzing the T_g-dependent change of the dielectric constant $\varepsilon(T)$ in cyclohexanol "crystalline glasses" at different frequencies of temperature alteration.

Certain thermodynamic properties of other classes of crystals with frozen-in structural disorder are in some ways similar, but also very different with respect to the caloric effects observed upon the vitrification of typical glass-forming melts. These are the classical "glassy" molecular crystals [168, 184, 352–355]. In them, by an as yet not clarified mechanism, disorder is frozen-in at the process of crystallization itself as observed by Kaischew [355] in detailed experimental investigations. No vitrification step or break-point is observed in the temperature course of their $C_p(T)$- or $\Delta S(T)$-functions (see Figures 5.13 and 5.14). Nevertheless the $S(T)$ curves of all these disordered molecular crystals show a definite zero-point entropy, $\Delta S^0 > 0$. The thermodynamic proof of this statement was given by a method, dif-

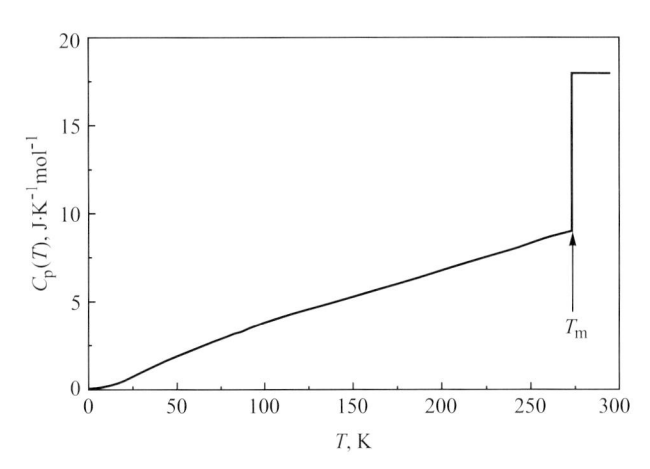

Figure 5.14 Specific heats, $C_p(T)$, of ice (I)-crystals according to the measurements by Giauque and Stout [340]. Note that no singular point is seen on the $C_p(T)$ curve from nitrogen temperatures down to the melting point, T_m.

ferent from the approach indicated with Eqs. (5.7) and (5.8) used in glass transition thermodynamics as it is developed in the previous sections.

5.7
Spectroscopic Determination of Zero-Point Entropies in Molecular Disordered Crystals

In determining the zero-point entropies, $\Delta S^0 > 0$, of the afore-mentioned simple molecular crystals (CO, H_2O, N_2O_2, etc.) the entropy of the vapor phase in equilibrium with the respective liquid state (at a temperature $T > T_m$) is calculated once by using the vapor pressure formulas, derived from the third principle of thermodynamic (see [55, 60]) and secondly, from the statistical partition function of the same vapors. The two values are then compared.

According to the third principle of thermodynamics the vapor pressure, p, of a solid or liquid is given by

$$\log p(T) = -\frac{L(T)}{2.3RT} + \frac{1}{2.3R} \int_0^T \frac{dT}{T^2} \int_0^T \left[C_p^{\text{vap}}(T) - C_p^{\text{cond}}(T) \right] dT + I_0(p).$$

(5.63)

Here $L(T)$ is the enthalpy of evaporation (sublimation), and the terms $C_p^{\text{vap}}(T)$ and $C_p^{\text{cond}}(T)$ are the specific heats of the gaseous and the respective condensed phase (liquid, crystalline) in the whole temperature range, beginning above T_m and ending as close as possible to the absolute zero of temperatures [55, 60]. The measurements of both $p(T)$ and $C_p(T)$ of these substances in Clusius and Giauque's laboratories in the 1930s (see [352, 353]) were performed down to 10–20 K. The necessary caloric analysis had to include besides the specific heat determinations and tensimetric measurements (i.e., vapor pressure measurements) also exact determinations of the enthalpies of melting, ΔH_m, of sublimation, $L(T)$, and evaporation, ΔH_b (at the melting, T_m, and boiling points, T_b), and the temperature dependence of $L(T)$, $H(T)$ and of both $C_p^{\text{vap}}(T)$ and $C_p^{\text{cond}}(T)$ in an as broad as possible temperature interval. The term $I_0(T)$ in Eq. (5.63) is the so-called chemical constant of the respective gas. For relatively simple molecules like those of the discussed molecular substances, the value and the temperature dependence of $C_p^{\text{vap}}(T)$ can be calculated, using the respective formulas of statistical thermodynamics, giving the connection with the thermodynamic functions. These calculations have to be supported by spectroscopic measurements (hence the name of the method of calculation) in order to determine from the band spectra of the vapor the exact values of the molecular constants, determining the partition function and thus $C_p^{\text{vap}}(T)$, $C_p^{\text{vap}}(T = 0)$, and $I_0(p)$.

The chemical constant $I_0(p)$ is determined in the framework of Nernst's theorem as

$$I_0(p) = \left(\Delta S^0 - C_p^{0,\text{vap}}(T) \right) \frac{1}{2.3R},$$

(5.64)

where $\Delta S^0 - C_p^{0,\mathrm{vap}}(T)$ is the specific heat of the gaseous phase at the lowest temperatures, where only the known structure-dependent rotational contribution to $C_p(T)$ has to be accounted for. In determining the caloric value of $I_0(T)$ via Eq. (5.63) by combined tensimetric and calorimetric data and comparing it with the statistically determined value of the chemical constant from Eq. (5.64) (when $\Delta S^0 \equiv 0$ is assumed) the true value of the zero-point entropy, ΔS^0, for the given crystal is determined. In the subsequent literature [168, 184, 352, 353], usually the values of $S(T)_{\mathrm{vap}}$ at the "normal" thermodynamic reference temperature ($T = 298$ K) are given as determined from caloric measurements (with the real value of ΔS^0). They are compared with the (naturally higher) $\Delta S(T)_{\mathrm{vap}}$ value, determined "spectroscopically."

The briefly summarized procedure, which gave strong evidence for the existence of frozen-in disordered crystals and for another exception from Nernst's theorem, is thus based on two calculations of the entropy: once employing caloric measurements, and in a second approach on its direct determination as

$$S = k_\mathrm{B} \left[\ln Z + T \left(\frac{\partial \ln Z}{\partial T} \right)_V \right] \tag{5.65}$$

from the partition function, Z, of the respective gas. In order to calculate $Z(T)$ the mentioned spectroscopic data have to be known. In calculating zero-point entropies of glasses according to Eq. (5.8), introduced at the beginning of this chapter (and derived in Chapter 2 in the framework of Simon's approximation, which in fact is also used here), the respective "ideal" crystal is considered as the reference state to both the undercooled liquid and the frozen-in defect solid. Moreover, it is known that at T_g the specific heats of the liquid decrease to their value in the crystal: thus ΔS_g can be calculated via Eq. (5.8) with a sufficient accuracy.

In the discussed defect molecular crystals, as demonstrated by Figures 5.13 and 5.14, no change, indicating a freezing-in process, is observed: disorder is frozen-in upon the formation of the defect crystal, giving as it seems, two (for H_2O and D_2O) or three (for CO, N_2O_2, and N_2O) energetically nearly equivalent configurations, as discussed in detail in [168, 184, 352–355].

5.8
Entropy of Mixing in Disordered Crystals, in Spin Glasses and in Simple Oxide Glasses

In assuming that the zero-point entropy of disordered crystals can be considered as an ideal entropy of mixing of two structural units arranged in distinct orientations (see [55]) we have to write

$$\Delta S^0 \approx -R \left[x \ln x + (1-x) \ln(1-x) \right]. \tag{5.66}$$

Here x and $(1-x)$ denote the molar fractions of molecules in the two orientations. As far as these two positions are energetically nearly equal (as assumed by Pauling [354, 378], Giauque [353] and Kaischew [355]) we have also to assume $x \approx 0.5$

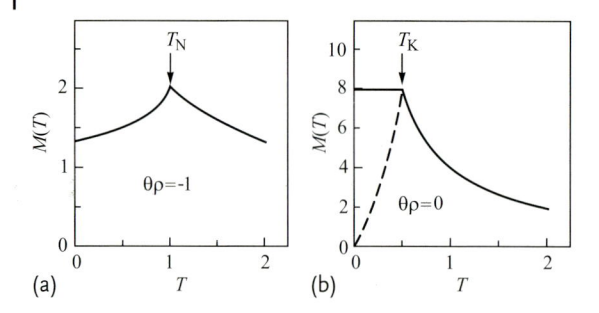

Figure 5.15 Thermodynamics of an antiferromagnetic crystal (a) and of a spin glass (b) in terms of the dependence of magnetization, $M(T)$, on temperature, T (given in relative units with respect to Neel's critical temperature, T_N). At $T = T_N$, the disordered paramagnetic crystal changes into an antiferromagnet (a). In part (b) of the figure, the magnetic spin-disorder is frozen-in at $T = T_K$ (*solid line*) or relaxes to $M = 0$ along the dashed line. Note that here $M(T)$ corresponds in its temperature behavior, determined by the magnetic susceptibility and the applied low magnetic field strength, to a thermodynamic function. In this sense, the process shown in (b) corresponds roughly to the freezing-in of either enthalpy or entropy in common glasses (schematic representation according to Moorgani and Coey [338]).

and thus $\Delta S^0 \approx R \ln 2$ is to be expected, as it is in fact observed in CO, N_2O_2, N_2O and also in cyclohexanol. The structure of ice (and of crystalline D_2O) according to Pauling [354] indicates the possible existence of three energetically nearly equal structural positions of the structural bonds and thus, via Eq. (5.66), an analogous entropy expression for the three structural units 1, 2, and 3 with $x_1 \approx x_2 \approx x_3$ as $\Delta S^0 \approx R \ln(3/2)$ is obtained and in fact also verified experimentally in the framework of the formalism derived in the previous section.

By changing the magnetic field strength, X, the resulting changes in magnetic properties and in freezing-in of magnetic orientational disorder in crystalline and amorphous alloys are of very particular nature. In many ways also similar to the caloric effects observed upon vitrification of typical glass-forming melts (see [338] and Figure 5.15). In these cases (at $T = $ const.) changing the magnetic field strength yields, as was demonstrated by Dobreva and Gutzow [337], the dependence similar to Eq. (5.26):

$$\tau(T, X)\omega_0 \cong \text{const.} \,. \tag{5.67}$$

Here $\omega_0 = (dX/dt)$ is the rate of change of magnetic field strength and $\tau(T, X)$ is the characteristic time of molecular relaxation, determining the process of magnetization.

A detailed description of the properties of magnetic spin glasses may be found in [338] and in the additional literature given in the introduction to the present chapter. Here of essence is the analogy of the "magnetic" freezing-in process with the freezing-in in molecular processes, as this is indicated by Eqs. (5.67) and (5.26). In assuming that, in spin glasses, orientations (e.g., parallel and antiparallel) are frozen-in, in terms of Eq. (5.66) again the ΔS^0 value can be estimated and a result nearly equal to $R \ln 2$ should be expected (see Jäckle [343, 379]). However, no direct caloric confirmation of this expectation is given in the cited literature, nor is it known to the present authors.

In 1962, Gutzow [191] proposed to consider and calculate the frozen-in topological disorder in glasses in terms of an entropy of mixing: either of free and occupied volume or later-on of differently coordinated structural units in accordance with Bernal's [380] and Scott's [188] models of random packing of equal spheres. A detailed description of these calculations is given in [1].

It was proposed many years ago (see [1] and the literature cited therein) to consider glasses as systems with frozen-in free volume (the so-called free volume concept of vitrification). In fact it was found that for most organic polymer and molecular glasses vitrification takes place at a relative free volume $v = 0.1$, where v is defined as

$$v = \frac{\rho_{\text{glass}} - \rho_{\text{cryst}}}{\rho_{\text{cryst}}} \tag{5.68}$$

by the respective densities ρ (or molecular volumes) of glass and crystal. For simple network glasses (like SiO_2, BeF_2, GeO_2, etc.; see [1]) the thus defined free volume (referred to the respective most stable crystalline modification, e.g., quartz for SiO_2) is considerably higher, that is, $v \approx 0.2$. By employing Eq. (5.66) the relative contribution of topological disorder to the frozen-in zero-point entropy of glasses turns out to be $\Delta S_g \approx R$ (when $v = 0.2$ as in the network glasses) or $\Delta S_g \approx (1/2) R$ as in organic molecular substances vitrified at $v = 0.1$ (see [1]). A thorough discussion on this subject was given by Petroff et al. [364], where an estimate of the contribution of various forms of disorder to the total entropy of undercooled liquids is given in the framework of appropriate lattice-hole models of simple and polymer glasses.

Thus it follows that the frozen-in topological disorder may be nearly a constant, as this is the case with the already discussed orientational disorder: at least for simple isostructural glasses (like network glass-forming oxides SiO_2, BeF_2, GeO_2), where it gives the main contribution. Computer results also give an additional indication in this respect with different models of disorder for frozen-in liquids constituted of equally sized spheres. In such calculations [381], $\Delta S^0 \approx R$ is also obtained as in Gutzow's first estimates [191] and in the already mentioned more exact models of silicate structures given by Beal and Dean [382].

5.9
Generalized Experimental Evidence on the Caloric Properties of Typical Glass-Forming Systems

In Chapter 2 and here in Sections 5.4–5.6, results are summarized on the caloric behavior of typical glass-forming systems with quite different structures: inorganic, inorganic-polymer, or molecular, metal-like. The obvious question to be answered in considering these and similar results of more than one hundred glassy systems so far investigated is the following: is there sufficient evidence in the behavior of glass-forming substances permitting, or even requiring, a generalized phenomenological treatment? The positive answer to this question is to some extend guarantied by the existence of several general empirical rules, connecting the caloric and kinetic properties of glasses. A derivation of the necessity of the existence of such invariants was given here in Section 5.5, in the framework of the phenomenological approach employed throughout the present book. Here we connect in more details some of these empirical results with the approach developed in Section 5.5.

From an empirical point of view, the first attempt for a generalized description of the thermodynamic properties of glass-forming melts was given by Kauzmann in 1948 [109]. Summarizing the experimental evidence available then concerning the temperature dependence of thermodynamic functions (of about 15 typical glass-formers), he has found that the temperature dependence of entropy $\Delta S(T)$, enthalpy $\Delta H(T)$, molar volumes, $\Delta v(T)$, of undercooled liquids in reduced coordinates (e.g., $\Delta S(T)/\Delta S_m$ vs. reduced temperature T/T_m) is similar, although not equal, for all substances analyzed. The absolute value of the melting temperature, T_m, turned out to be a sufficiently reproducible correlating factor for all substances analyzed (see Figure 2.40). The latter circumstance is not surprising, accounting for the well-known fact that at least for simple, nonassociating liquids, T_m is directly connected with the critical temperature, T_c, via linear dependencies, for example, $T_m \cong (3/5) T_c$.

Chapter 8 provides a more detailed discussion of this part of the invariant problem. In Section 2.5.2, the invariants of glass transition are discussed both in their historical and general empirical significance, especially in connection wit the Kauzmann paradox. Here we have only to add, that we need three of these results in order to attempt an additional analysis of the dependencies, derived in Section 5.5 (in particular, Eqs. (5.54)–(5.62)).

The first invariant, mentioned already several times in this book, is that, the glass transition temperature obtained at normal cooling rates (e.g., at $q = 10$ to 10^2 K/s, which were employed for the glasses as analyzed by Kauzmann) is for typical glass formers given by

$$T_g = \frac{2}{3} T_m . \tag{5.69}$$

For multi-component glass-forming melts instead of T_m the corresponding liquidus temperature T_l can be used [111]. The expression given with Eq. (5.69) is known as the (2/3)-Kauzmann–Beaman rule (see [1]).

The subsequent analysis of experimental data of about 40 mostly organic polymer glass-forming substances performed by Wunderlich [173] has shown (see also [1, 112]) that in general

$$\frac{\Delta C_p (T_g)}{\Delta S_m} \approx \frac{3}{2} \tag{5.70}$$

holds. Two other simple dependencies

$$\frac{\Delta S_g}{\Delta S_m} \approx \frac{1}{3}, \quad \frac{\Delta H_g}{\Delta H_m} \approx \frac{1}{2} \tag{5.71}$$

were also established by Gutzow years ago (in 1971 [1, 110], the second one by Tammann in 1933 [24]).

The degree of fulfillment of all the dependencies mentioned above is to a great extent evident from Figures 2.14 and 2.26, where the results of a survey of caloric results of about 110 glass-forming substances (organic-molecular and polymer, inorganic, etc., see the figure captions) are summarized (see [112]). At present it is not possible to say whether the deviation from the above mean values (or more characteristically, from the respective median values) is systematic or accidental (see the discussion in [1, 112]). It is also essential to note that of the above dependencies only Eqs. (5.71) are also applicable to metallic glasses (see [1, 112, 342]). However, it may be of particular significance that in all cases of vitrifying liquid crystals analyzed, it turned out that, although the dependencies given with Eqs. (5.69), (5.70), and (5.71) are fulfilled for them (as far as the only two or three existing measurements are characteristic data), the first of Eqs. (5.71) is not obeyed for vitrified liquid crystals (Figure 5.16; data are taken from [1, 237, 341, 347–349, 376] for Figure 5.16a; from [168, 184, 237, 352–354, 376] for Figure 5.16b).

By introducing the values of these three invariants into Eq. (5.62) the value of the constant there is readily determined. In comparing it with the respective constant from Eq. (5.58), it gives the ratio between these two constants as corresponding to the ratio between areas 1, 2, 8, 9, 10 and the triangle 3, 4, 9 in Figure 5.7, determining, according to the discussion in Section 5.5, the enthalpy difference glass/crystal and glass/equilibrium liquid at $T = T_g$. In the ratio $a_0 = (\Delta C_p(T_g))/\Delta S_m$, moreover, two other very essential thermodynamic properties of the undercooled liquids are connected. First this is $C_p(T)$, the susceptibility of the melt with respect to temperature changes. It determines the way temperature rises when an amount of heat is transferred to the melt. In a_0, this quantity is divided by the measure of structural change (measured by the value of $\Delta S(T)$ at $T = T_m$) caused in the liquid by the corresponding temperature rise. This is the reason why in earlier publications of one of the present authors [1, 17, 177] a_0 is called the *thermodynamic structural factor*: it determines even in simplest *thermodynamic models* (see [1, 17] and here Chapter 8) the temperature course of all thermodynamic functions.

We know that the configurational part of the thermodynamic functions of the liquid, according to several models (especially those of Adams and Gibbs [148] and of Avramov and Milchev [141]) exponentially determines the temperature dependence of the viscosity, $\eta(T)$ of glass-forming melts. This same factor, a_0, is also essential in order to derive the viscosity of glass-forming melts. In order to derive the kinetic invariants, in a next step we have to introduce an appropriate viscosity equation. One of the most universally accepted $\eta(T)$ dependencies, which is based on a sound statistical foundation, has been proposed by Adams and Gibbs [148] in the form

$$\log \eta(T) = A_0 + \frac{B}{2.3T\Delta S(T)} . \qquad (5.72)$$

Here A_0 is a constant (see [1]) and B, according to the original derivation of this viscosity equation, is dependent both on its complexity, expressed by ΔS_m, and its

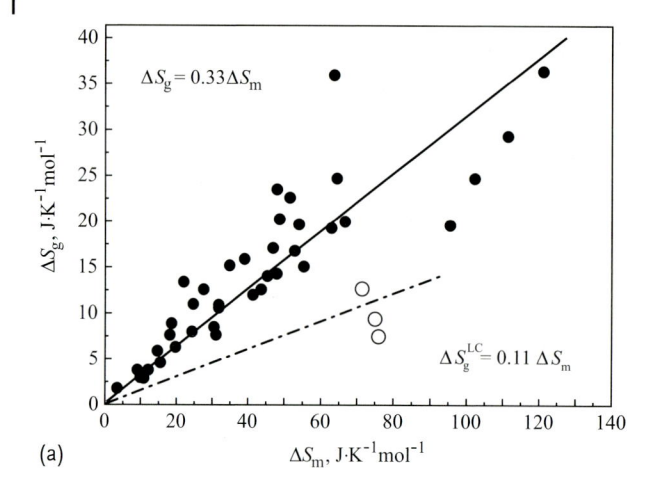

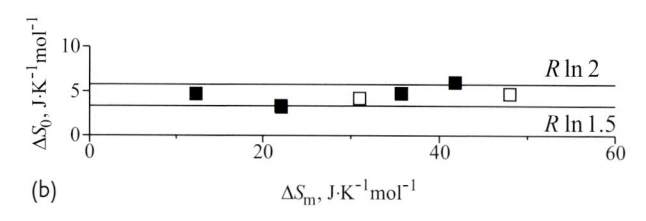

Figure 5.16 Zero-point entropies of typical glasses and of defect crystals. Melt-dependent and melt-independent thermodynamic invariants for disordered solids with frozen-in zero-point entropy (ΔS_g, ΔS_g^{lc}, ΔS_0) vs. melting entropy ΔS_m: (a) data for 36 typical glass-formers (*black circles*) approximately confirming Eq. (5.71) as $\Delta S_g \sim (1/3)\Delta S_m$. The substances are given here in the order of increasing ΔS_m values, they are: (poly)ethylene, SiO_2, Se, GeO_2, rubber, BeF_2, methanol, (poly)dimethylsiloxane, $Au_{0.077}Ge_{0.136}Si_{0.094}$, H_2O, poly(propylene), $NaPO_3$, transpolypentanamere, lactic acid, poly(oxacyclobutane), B_2O_3, ethanol, polytetrahydrofurane, n-propanol, isopropylbenzene, butene-1, isopentane, polyglycolide, poly(ethylene-therephtalate), poly-ε-caprolactane, As_2S_3, orthoterphenyle, 2-methylpentane, betol, benzophenone, glycerol, $As_2S_{/3}$, diethylphtalate, phenolphtaleine, $H_2SO_4 H_2O$, $H_2O4B_2O_3$, $Na_2O4B_2O_3$. For polymeric substances the gram formula value of the repeatable structural unit is taken; open circles correspond to three liquid crystals mentioned (b) data for the zero-point entropy of the known disordered molecular crystals (*black squares*), in order of increasing ΔS_m values (CO, H_2O, D_2O, N_2O, N_2O_2) and of two plastic crystals (ethanol and cyclohexanol: *open squares*), all of them are located in between $\Delta S_0 = R \ln 2$ and $\Delta S_0 = R \ln 1.5$ (see text) and are not proportional to ΔS_m, as this is the case for the previously discussed glass formers.

bonding strength, determined by T_m,

$$B \approx b_0^* \Delta S_m T_m . \tag{5.73}$$

Introducing, again, the reduced temperature, $\theta = (T/T_m)$, we can write Eq. (5.72) in the form

$$\log \eta(T) \cong A_0 + \frac{b_0^*}{2.3\theta(1 + a_0 \ln \theta)} , \tag{5.74}$$

and approximately the temperature function of the configurational part of the entropy can be given as

$$\Delta S(T) = \Delta S_m (1 + a_0 \ln \theta) . \tag{5.75}$$

The activation energy for viscous flow corresponding to Eq. (5.74) is now

$$U(T) = \frac{Rb^*}{1 + a_0 \ln \theta} \tag{5.76}$$

and at $T = T_g$ the ratio $(U(T_g)/(2.3RT_g))$ is found to be in accordance with Eq. (5.59) an invariant, as far as θ_g and a_0 in the above equations are invariants. Thus invariants are also the ratio $(U(T_g)/(2.3RT_g^2))$, which according to Eq. (5.24) determines substantially Bartenev's equation via the values of C_1, C_2 and also the expression for $(d/dT)[U(T)/RT]_{T=T_g}$, giving the slope of the activation energy at $T \to T_g$. This slope and the value of the constant in Eq. (5.26) are directly given by this ratio according to theoretical concepts, developed by Gutzow et al. [11, 12]. Thus, it determines to a great extent the vitrification possibilities of liquids. In this respect, the simple derivations from above give an additional support to the more general analysis made in Section 5.5.

5.10
General Conclusions

Considering the available information, summarized in the preceding sections, the following conclusions of more general nature can be drawn:

1. The experimental evidence on the thermodynamic properties and the kinetics of vitrification of typical glass-forming systems, on the glass transition in liquid crystals, on the freezing-in of "glassy" plastic crystals (like cyclohexanol) and most probably also of spin glasses shows that the respective nonequilibrium frozen-in glasses or glass-like states are formed in a very similar process. In any of these cases, dependencies equivalent to Eqs. (5.26) and (5.67) are obtained, which can be written as:

$$\tau(T, p, X) w|_{T_g, P_g, X_g} \approx \text{const.} \tag{5.77}$$

They determine at a given rate of change, w, of some external parameter (T, p or X) a singular point (e.g., $T = T_g$ at $p = $ const., $X = 0$) or more precisely a T, p or X-range at which vitrification or a vitrification-like process takes place. At such conditions, the characteristic relaxation time, $\tau(T, p, X)$, at a given rate, w, of change of the external parameters, the relaxation time of the system then becomes nearly equal to the stay time of the system in the critical region. In any of these processes a dependence, similar to Bartenev's formula Eq. (5.24), determines the kinetics of relaxation.

In liquid systems (glass-forming melts, liquid crystals) the time of molecular relaxation, $\tau(T, p, X)$, is determined directly by the viscosity of the corresponding fluid. In "glassy" plastic crystals $\tau(T, p, X)$ has to be determined by the relaxation time of the structural units in the respective "plastic" crystalline phase. In both crystalline and in amorphous spin glass systems, more particular considerations have to be added to determine the true nature of the time $\tau(T, p, X)$ (see [339]). However, in all these cases vitrification can and must be considered as a process taking place at a steadily increasing relaxation time; and when $\tau(T, p, X)$ has reached the w-determined stay time of the system, it is frozen-in to a glass-like structure. In [11, 12], the phenomenology of such a non-isothermal process of relaxation leading to glass transition is quantitatively established, when w is given by the cooling rate q at $p, X = $ const. Similar dependencies have been established also for increasing pressure (when $w = dp/dt$) or at changing magnetic field strength, X, etc. [337, 383, 384].

2. The thermodynamic properties of glasses (and especially of ΔS_g^0) formed in melt cooling (or quenching) processes are determined by the properties of the corresponding undercooled liquids, and as far as the melt at $T = T_m$ is taken as the reference state, the value of ΔS_g (and of any other thermodynamic function (enthalpy, ΔH_g, free volume, ΔV_g)) is for all substances a nearly invariant part of the value of the corresponding property at melting temperature, T_m. This result is a consequence of the fact that the thermodynamic structural factor, $a_0 = \Delta C_p / \Delta S_m$, varies in relatively close limits between 1 and 2 (see also the analysis, given in [1] and in Chapter 8). The highest values of a_0 are observed according to [177] for organic molecular or polymeric glass-formers while for simple oxide glasses (SiO_2, BeF_2) and for metallic glass-forming alloys [342] the relation $a_0 \cong 1$ holds. Thus both the relative invariance of the several ratios (Eqs. (5.70)), mentioned in the preceding two sections, finds its explanation, together with the fan-like spread of the temperature course of the respective ratios ($\Delta S(T)/\Delta S_m$, see Figure 2.40) as first observed by Kauzmann [109]. In this way, the ratio a_0 determines via the discussed appropriate equations both the individuality of the ($\Delta S(T)/\Delta S_m$) course of differently structured liquids ("fragile" at $a_0 = 2$, "strong" at $a_0 \cong 1$) for different substances and also its relative invariance at $T = T_g$. This is the reason why the significance of a thermodynamic structural factor must be attributed to the ratio $a_0 = (\Delta C_p(T_g))/\Delta S_m$.

3. However, when kinetic properties like the course of the viscosity, $\eta(T)$, for different glass-forming melts are compared, this individuality is brought into the exponent, for example, of the Adams–Gibbs equation (see Eq. (5.72)) and the fan-like divergence into "strong" and "fragile" temperature courses of viscosity becomes more pronounced. A remarkable confirmation in this respect gives Angell's $\eta(T)$ vs. (T_g/T) construction [5], shown in Figure 2.41. It must also be mentioned (see [10]) that those glass formers, which are thermodynamically "fragile" or "strong" are also more "fragile" or more "strong" in their viscosity behavior. This is to be expected by comparing Eqs. (5.72) and (5.75).

4. No distinct mechanism can be proposed for the formation of defect structures in molecular crystals, like CO, H_2O, and so on. Here the common point with

typical glass-formers seems to be only the possibility to describe the zero-point entropy in terms of both simple and more complicated forms and expressions of entropies of mixing. In [364] we have in fact shown that all three forms of configurational disorder existing in liquids – topological, orientational and configurational – can be described as different, generalized forms of entropy of mixing. In defect molecular crystals (such as CO) only the orientation determined structural disorder is expressed in the form of the entropy of mixing, discussed in one of the preceding sections. No process analogous to vitrification has been observed in these cases.

The derivations, given in Section 5.5, show that the generic phenomenological approach, derived by two of the present authors in the last several years, and summarized here in Chapters 3, 8 and 9, can give a sound theoretical basis for treating a number of problems, considered up to now in glass science only in a purely empirical approach. Here we have to mention, particularly in relation to the wide range of applications discussed in Chapter 3, that both the postulate of vitrification and the invariants of glass transition are now derived, instead of being postulated or extracted from experiment.

6
Principles and Methods of Collection of Glass Property Data and Analysis of Data Reliability
Oleg V. Mazurin

6.1
Introduction

The main chapters of this book show the importance of theoretical approaches to the analysis of the essence of a glassy state, specific features of glass structure and the possible influence of this structure on glass behavior. Existing methods of prediction of the dependence of various glass properties on glass compositions permit us to develop new particularly efficient commercial compositions of glasses. However, every scientist (belonging to both basic and applied glass science) should understand that any achievement in glass science depends quite considerably (and in many cases totally) on the amount and quality of experimental studies of glasses and glass-forming melts. Among these studies the measurements of the influence of composition and temperature on physical and chemical properties of glasses play an important role.

Systematic measurements of some physical properties of glasses began in Germany at the end of the nineteenth century [385]. Since then many thousands of scientists around the world have performed many hundreds of thousands of glass property measurements. These results were published in thousands of journals, reports, books, patent applications, theses, and other kinds of sources. Unfortunately, for several reasons quite a considerable amount of data has not been published anywhere and has already been lost or will be lost in the near future. It should be noted that the quality of a part of these unpublished works was much higher than the quality of some of the data published in our best scientific journals.

Regretfully, we are unable to prevent the loss of valuable unpublished information on glass properties. However, we should at least try to make accessible to any scientist information on glass properties that is already published. Without outside help, trying to find original publications containing information on properties of a substance can be difficult. The only rational way to do this is to organize a team of specialists with the main objective to search, collect and systematically present in a handbook or electronic database as much glass property data as possible. If such a team works for dozens of years, there is a chance that the majority of the published data on this particular subject will be collected.

The author of the present chapter, together with T.P. Shvaiko-Shvaikovskaya and M.V. Streltsina, started their work at the end of the 1960s in this area. At the beginning, we published the compiled data in handbooks [2, 3]. However, from the middle of the 1990s we were able to transfer all our data to an electronic database SciGlass [116]. For many years we worked in the Laboratory of Glass Properties of the Institute of Silicate Chemistry in Leningrad, belonging to the Academy of Sciences of the USSR/Russia. The main objective of this laboratory was to develop precise methods to measure properties of glasses and glass-forming melts, to study the dependence of these properties on glass compositions and temperature, and to develop new commercial glasses with improved technological characteristics. Thus, in the same laboratory we had not only developers of the glass property reference books and database but also their active users. This cooperation ensured the most advantageous conditions for development of the systems of collection, analysis, and presentation of the glass property data, which was obtained worldwide. Some of the results of this work and the problems connected with it will be presented in this chapter.

We should also mention the particular attention given at that time to the problems of correct measurements of properties of glasses and glass-forming melts by glass specialists in the USSR, in general, and in Leningrad, in particular. Such attention was stimulated by the scientists working at the State Optical Institute. They worked in close enough cooperation with the scientists of the Institute of Silicate Chemistry and Lensoviet Technological Institute also located in Leningrad. In terms of their achievements in the quality of glass property measurements, these scientists could be viewed as the second top-ranked scientific community in the world at that time (the first being the group of American glass specialists working in the National Bureau of Standards and Corning Glass Works company).

We, in Leningrad, worked in accordance with the following principle. If a materials scientist does not pay proper attention to the reliability of results he/she presents to be used by colleagues, this scientist not only spends his/her time and funding for nothing but he/she can also cause serious damage to people who trusted these results and used them for their work. That is the reason why my colleagues and I pay special attention to the ways of determining the reliability of published data, as well as to the problem of protecting scientists who are in need of such data from using obviously wrong values of glass properties.

The point is not only for finding quickly and easily a great amount of data obtained by scientific communities during long enough periods of time that any large material property database permits, but this work can also be used as an instrument for determination of the reliability of published data. Accordingly, this paper consists of two parts: in Section 6.2, the main principles of collection and presentation of glass property data are described. It seems that thanks to our specific fortunate situation (see above), our team was able to develop a unique system of collection and presentation of glass property data. And, time has come for an attempt to describe this system in some detail in a publication that will be accessible to all concerned in a near enough future, when the leading members of our team have stopped their active work on the glass property database. Sections 6.3 and 6.4 describe the possibility of using any large glass property database in order to determine reliable dependencies of properties on glass compositions. These sections also examine the possibility of using the given results in order to evaluate the quality of measurements reported in papers of interest. A general decrease of the reliability of an impressive number of published experimental results in materials science is one of the serious challenges of the twenty-first century. Thanks to some specific factors, glass science is one of few areas of materials science where the quality of published experimental data can be evaluated and where, in principle, efficient measures can be undertaken to improve this quality. Unfortunately, at present only a small fraction of glass scientists realizes the importance of this problem. I am using the present opportunity for a new attempt to attract the attention of the glass community to the indicated problem.

6.2
Principles of Data Collection and Presentation

6.2.1
Main Principles of Data Collection

Our database is formed on the basis of the following set of principles:

1. All published data independently of the source of their publication and supposed quality of the data should be included into the SciGlass database. The word "published" means here patents, reports, theses and other sources, even if the source existed in only a few copies. We have also considered information presented on the Internet during the last several years as published. In a few cases when the source of measurements was especially well-known and particularly reliable we have included the data given to us by scientists who had not published these findings at all.
2. It is well-known that in various kinds of scientific literature one can find numerous data on properties of glasses whose compositions are given only approximately (for example, only the total amount of RO is reported without information on the content of every oxide of a corresponding metal) or not given at all. The first case can be found in many publications in technical journals and is connected with the fact that exact compositions of many glasses possessing useful characteristics have been classified. The second case is the widespread practice employed for publications of property data in catalogs of glass companies. Most companies inform potential users of a glass code along with a long list of various properties of this glass. It is clear that such information can be of interest for a considerable number of glass specialists, especially if they want to select commercial glasses for some special purposes. Such information

can be found in the INTERGLAD glass property database[1]. However, in the SciGlass database only the data for glasses with known compositions are included. Among such glasses there are also a considerable number of commercial glasses, if their compositions could have been found in the nonclassified literature.

3. Although the authors of the SciGlass database are qualified enough to estimate the probable reliability of many published data, no attempts have been made to propose classifications of the presented data by the level of their reliability. We limit ourselves to only two ways of warning a user about the necessity to use some data with a certain caution:
   - Some suspicious (from our point of view) experimental values we mark with asterisks and note: "as printed in the original source," which means that if a user is surprised by a certain value, he/she will know that it was not a misprint of the authors of the database.
   - Sometimes we make a note that some data presented in the database differ considerably from the data of some other authors who studied the same property at the same compositions.

4. All data are included in the database in the same units, with the same number of significant decimals (even if the number of these decimals is obviously too large), and with the same way of composition description (wt., mole, or atom %, wt. or molar parts). At the same time a user can easily transform these values into the most convenient format for his/her way of composition presentation.

5. Every table containing experimental data is accompanied by a header where not only bibliographic information (including the title of the publication and the number of the last page of this publication) is given but also a concise description of the synthesis and measurement procedures for the studied glasses is presented.

6. Sometimes the authors publish their experimental results in several publications. References to all these publications are given in the corresponding header to the table.

7. We regret mentioning that some authors forget to inform readers what kind of percentage they used to describe compositions of studied glasses. Also when authors used certain additives to the initial compositions they often enough missed mentioning whether the content of these additives was included into 100% of the final glass compositions, or the content of the additives was given as "above 100%." In all these cases we request additional information from the authors. Several decades ago we received the answers from nearly all authors to whom we had applied. Now the number of answers to our queries is considerably below 50%. If the authors do not supply us with the necessary information we start analyzing other publications of the same authors, compare some of their data with similar data by other authors, and try analyzing some other features of the paper (for example, sometimes authors' discussion on the structural interpretation of their data can lead to a definite enough conclusion

---

[1] International Glass Database System INTERGLAD 6.2 (6.2.1), http://61.194.5.20/interglad6/index.html (accessed 28 January 2011).

about their approach to composition presentations). Then we write in headers that "there are reasons to suppose that the authors have used such and such kinds of percents."

8. It is well-known that the term "glass" can cover a large variety of amorphous substances. In the database developed by us and described here, only property data for inorganic nonmetallic glasses are included. The most important groups of these glasses are oxide, chalcogenide, and halide glasses. To the glasses collected in the database also belong glasses of mixed groups and salt glasses. The main reason why we did not include into the database any other kinds of glasses was the fact that the whole team of the SciGlass database developers consists of specialists of the previously mentioned kinds of glasses. We are sure that a reasonably good material property database can be developed only by people having special knowledge in the selected types of materials. At the same time we do not restrict the collected data by temperature ranges where the corresponding materials are in the glassy state. In the database, the properties of glass-forming substances are collected for any temperatures at which these properties were reported in publications. Note that only properties of glasses and melts without any inclusions of crystals are selected for the database. Exceptions are made only for the studies of the areas of glass formation, as well as determination of liquidus temperatures. At the same time the rates of crystallization of glass-forming melts are considered as a property of glass and all data on crystallization are also collected in the database.

9. As to the properties, only physical and chemical properties that could be of certain interest to glass technology and practical application of glasses are collected in the SciGlass database. This spectrum includes density, electrical, elastic, optical (including optical spectra), magnetic properties, viscosity, surface tension, chemical durability, strength, microhardness, thermal expansion, heat capacity, thermal conductivity, crystallization characteristics. We do not include into the SciGlass database the data describing the influence of radiation on glass properties. In the already mentioned INTERGLAD database the spectrum of properties included is several times broader. It includes thermodynamic characteristics, various spectral characteristics, results of X-ray investigations, and so on. We have decided from the very beginning to limit the number of properties in order to be able to concentrate on the search for the maximum amount of information on the selected compositions and properties.

10. For SciGlass, regular (every half-year) updates are released. The main information for every update is collected from publications that have appeared during the last two or three years. However, the search for publications that appeared in earlier time periods continues. As is seen from Table 6.1, even the number of the found sources published before the year 2000 increases steadily. It is due to the fact that in addition to examinations of dozens of journals from cover to cover, studies of reference journals, patent databases, as well as systematic use of an Internet search, we can regularly inspect references in survey papers and books. This approach helps us to find information in the sources that we have missed to find earlier in spite of a very thorough search for appropriate

Table 6.1 Numbers of sources of the data which were used for compilation of the SciGlass database: the data are given for six consecutive versions of SciGlass with the respective year of release. Both the total number of sources is given as well as a specification for different periods of time of publication of the sources included.

Number of sources	7.0 2007	7.1 2008	7.2 2008	7.3 2009	7.4 2009	7.5 2010
Before 2000	22 473	23 684	23 988	24 223	24 496	24 661
2000–2005	4106	4679	4797	4962	5124	5240
2006	615	754	791	838	885	938
2007	215	684	767	852	907	959
2008	None	62	242	650	781	877
2009	None	None	None	84	362	790
2010	None	None	None	None	None	82
Total	27 409	29 763	30 585	31 609	32 551	33 547

publications carried out during several decades. It therefore means that it is absolutely impossible for anybody to find all publications worldwide containing glass property data within the described area of substances and their properties. At the same time it seems obvious that the SciGlass database is nowadays the biggest possible collection of the corresponding information.

11. All information in the database is presented in English independently of the original language. As to the references, the names of books and journal titles are given in original languages if originals are written in the Latin alphabet, or in English transliteration of Cyrillic or Japanese. It is necessary for a search for publication originals in libraries. The titles of papers are given translated into English, if the language of publication is Russian, Bulgarian, or Japanese.

6.2.2
Reasons to Use the Stated Principles of Data Collection

It is probably clear without further explanation why it is advantageous to use some of the principles from above given list. On the other hand, the reasons for using some of the other principles listed above may be not so clear. These reasons will be described in some detail in this section.

A considerable amount of data presented in the database is impossible to find in most libraries of the world. These types of sources are theses, various kinds of reports, proceedings of some conferences, and collections of papers published by various research institutes or universities. However, even when the list of publications of interest contains, say, hundreds of papers the copies of which can be found in a library or bought via the Internet, it will take a considerable amount of time to select or purchase all these copies, and inspect the corresponding texts. That is why we tried to supply the presented data by a header containing a certain set of information that in most cases should be sufficient for understanding the ways of glass synthesis, sample preparation, methods of measurements and (if it was reported) temperature schedules of measurements. If authors report measurement errors, these data are also included in table headers. It is known that for lists of references some journals require titles of papers and not only the first, but also the last page of each paper. As is mentioned above, this information is also included in the header.

The next point that requires some explanation is the presentation in a header of all references to the publications, where the corresponding data were reported by their author(s). There are two reasons for this decision. First, if a user decides that he/she needs to inspect the original paper containing these data (for example, in order to find some additional information on the details of an experiment or discussion on structural interpretation of the obtained data), it may be important for him/her to know about all the possible ways of finding these data, because for practically every scientist the amount of time necessary to find different kinds of sources can be quite different. Second, if a user finds in a certain paper a reference to a publication where he/she may expect to find some data on properties of glass compositions of his/her interest, it may be reasonable for this user to begin with the search for such data in the SciGlass database. If this source can be found in the database, the corresponding data can be copied from there. If not, one will have to find this publication in a library or purchase an electronic copy of the whole paper from the corresponding publisher. It should be noted that for some data in table headers there is no information except for a reference. It means that such information was entirely absent in the original source.

Now let us consider the last, but not least, problem. We have entered into the database absolutely all data that we have been able to find in all kinds of literature. Sometimes even the first reading of the description of an experimental technique shows clearly that the author of the corresponding paper is not qualified enough to obtain reliable results. More often the same conclusion follows from the comparison of data taken from a certain publication with the data of many other authors who studied the same glass-forming system (see, for example, Section 6.3.2). However, we have worked in accordance with the following principle. One should try to compile in the universal glass property database absolutely all data published in the scientific or technical literature (including theses, reports, patents, etc.) worldwide from the end of the nineteenth century to recent years, irrespective of the quality of these data. There are several reasons for this decision. First, it seems wrong for any specialist (or a group of specialists) to take the liberty to decide what kind of data are worth including into the database and what kind of data are not worthy. Second, if a user of the database finds a reference to a certain publication where the properties of glasses of his/her interest are described, he/she should be able to find these data in the database. After that the user will be able to make a decision whether it is reasonable to use these data or not. Third, only statistical evaluation of all published data permits selecting the most reliable authors or groups of authors whose results could be used with confidence in their reliability. In some cases the authors of the database can help users in this respect by some notes in headers to glass property tables or by publications on this subject. However, the final decision should always be taken by a user and to do this he/she should possess all available information without any exceptions.

6.2.3
Problems in Collecting the Largest Possible Amounts of Glass Property Data

It is clear that two of the most important characteristics of a database similar to that described here are as follows: the level of completeness of the collected data and the activity in updating the corresponding database.

During several decades my colleagues and I tried to find in the existing literature all data on glass properties. To do this we have used the following means. First we selected dozens of journals where papers on glass properties were published regularly and we inspected these journals from cover to cover, beginning from the first volume of every journal. Then we inspected the corresponding parts of some reference journals. We also used library subject catalogs. In addition, we tried to use all our professional connections in order to find theses, reports, and other kinds of documents containing information of interest. A considerable source of the required information was (and still is) the inspection of references in survey papers and papers containing experimental data. We intensively inspected databases containing information on patents. Recently, the Internet has become a very useful source of various kinds of information on glass properties. All these efforts made it possible for us to compile the greatest body of experimental data on glass properties collected so far. We have approached a considerable number of scientists with the request to provide us with full lists of their publications and sometimes with copies of papers from publications which were particularly difficult to access. In many cases the corresponding response was quite positive.

Nevertheless, my colleagues and I are sure that there is a great amount of existing glass property data that has not yet been included in the database. First of all it is the information belonging to large glass companies. In many of them all results of various studies performed for many decades, including those which have no practical importance, are considered classified. Then there is a lot of information in various theses written worldwide that we were unable to find. Sometimes it is very difficult to find conference proceedings containing information of interest. And, finally, it appears that for various reasons, results in various journal publications were not timely detected by our team.

Actually, one of the best ways to make the SciGlass database even considerably more complete than it is today would be the help of the numerous users of this database. Every user can find out whether all of his/her publications are included in the database and, if not, he/she can send us copies of the absent publications. It seems obvious that all data included in this database will be saved for future generations of materials scientists. Even if further development of the SciGlass is stopped for some time, the existing numerous copies of the database throughout the world will be kept by various institutions for practically an unlimited amount of time, and thus these data will in one way or another be available to the public.

At the same time a single publication in a not quite popular (for glass scientists) journal may be easily overlooked by most people. Thus, if one wants the results of his/her work to be useful for other people, he/she should be interested in sending results of the corresponding publication to the SciGlass. Regretfully, this open exchange happens much more seldom than it should be (at least from our point of view).

6.2.4
Main Principles of Data Presentation

As was pointed out in Section 6.2.1, we try to present in the database any table in a form as near to the original as possible. That is, a user can see the data in the form where compositions and properties are presented in exactly the same units as they were given in the original. Here however it is necessary to mention the following fact. If in the original the data are given in the table form, we try to reproduce them as near to the original as possible. However, it is well-known that in many papers and books experimental data are presented in the form of figures. In these cases we do not reproduce the corresponding figures in the database. All graphic information is transformed by a specially developed digitizing program into numerical information that is presented in the database in the form of tables.

We consider the principle of digitizing all graphical information as one of the most important for any electronic materials science database. It helps to compile all existing data for a selected property of the composition of interest. It also provides the possibility to present data in any form in a way most convenient for a user: to shift from one kind of percentage to another, to use any units, and so on. This is particularly important in the cases when a graphic presentation is the main method of analysis of experimental data, namely, for the study of areas of glass formations and in the case of absorption spectra. To compare areas of glass formation published by different authors it is important to be able to select similar percentage and similar arrangements of apexes of triangles. To compare absorption spectra taken from different publications it is also necessary to select similar axes for curves taken from all publications of interest. In this case we have about one hundred possible versions of coordinates. The same dependence in different coordinates can look quite different. Thus the ability to transform a spectrum presented in coordinates selected by the authors of a publication into a curve presented in another coordinate system is actually an indispensable condition of any serious comparison of results by different authors.

A user of the database can see any information not only in units used by the authors of a publication, but also in any units selected by the user. It enables a user not only to see the results in the form that may be the most convenient for him/her, but also to compare the data taken from different sources. To make this comparison as universal as possible special "standardized" values are determined on the basis of temperature dependencies of some properties reported in various publications. Standardized values are the values of some properties at certain fixed temperatures or sometimes the values of temperatures corresponding to certain fixed values of a property (mainly viscosity). This permits one comparing any number of results published by different authors. Examples of such comparisons are presented in Section 6.3 of this chapter.

6.3
Analysis of Existing Data

6.3.1
About the Reliability of Experimental Data

Among other factors, the efficient use of experimental data both for theoretical and practical purposes depends to a great extent on data correctness. When one finds in the literature a paper containing experimental data of interest, it is in general difficult enough to determine the level of accuracy of these data. Even if an author describes an experimental procedure in some detail and presents an estimation of measurement errors, it should not always be considered as a confirmation of the reliability of presented data. The origins of possible errors are numerous enough.

One of the most important factors are the errors in estimation of glass compositions. In general, only about 10% of all studied glasses are analyzed in this respect. One should also take into account that really reliable results of analysis are only the ones of the so-called wet chemical analysis. Modern instrumental methods of the analysis can lead to considerable errors, if they are used for studying glasses with compositions essentially different from the standard ones. Extreme caution should be exercised in the cases when an author used alumina, silica or porcelain crucibles instead of platinum or gold ones. Unfortunately, recently the use of such cheap crucibles has become more and more popular.

Another factor is a wide use of sophisticated automatic equipment for some of the property measurements. This equipment makes the measurements much quicker and cheaper. However, there is a tendency of lending too much credence to this equipment. Thus an experimenter often tends not to pay enough attention to the influence of various factors on the results of measurements (thermal history and other specific characteristics of specimens, correspondence between the mean temperature of a specimen and the temperature recorded by an instrument, etc.). The third factor is the tendency of the authors of papers to pass the preparation of glasses and their measurements to technicians, often without a proper control over their activity. Possible consequences of such practices are obvious enough. And the last factor is the tendency of shifting the maximum activity in glass studies to the countries where only the first generation of materials scientists work. These young scientists start organizing their investigations from the very beginning, and in such situations, an increase of erroneous results is inevitable.

The obvious question arises: how can one select the most reliable values of glass properties? The answer to this question is also quite obvious: one should apply the statistical analysis of similar data reported by a considerable number of different authors. It is the only reliable way of any reasonable approach to solving this problem. It is also one of the reasons why the compilation of the maximum amount of data in one database is so important. The main specific features of such an approach and main regularities of statistical evaluation of experimental data will be demonstrated here by the example of binary systems.

6.3.2
Analysis of Data on Properties of Binary Systems

6.3.2.1 General Features of the Analysis

As was shown in Section 6.3.1, there are a considerable number of reasons for various deviations of experimental data from the correct ones. Certainly, the values of experimental errors depend greatly on the qualification of an experimenter, quality of equipment, and extent of an experimenter's desire to do his/her best in the course of property measurements. However, in general, certain errors in the course of any measurements are inevitable.

It is well-known that such errors can be divided into two main groups: random errors and systematic errors. The theory of random errors has been well-developed [386]. Among other factors these errors depend on the number of similar experiments conducted. By increasing this number, one can widely decrease the values of random errors. By making several measurements under similar conditions and using well-known equations a scientist can easily enough determine the confidence interval of the measured value for a selected confidence probability. On the other hand, predictions of systematic errors are much more difficult and much more uncertain. These errors depend on some factors that can only be partly known by an experimenter. Usually, the exact values of the influence of these factors is unknown. Sometimes even the sign of the influence of some factors is unknown to an experimenter. Obviously, this is the reason why some of the authors of experimental papers report the values of random errors (although often enough these values are too optimistic), but practically never discuss the problems of systematic errors and ways of decreasing them.

If it is therefore difficult for the authors of publications to determine the total errors of their measurements, then it is nearly impossible to do this for users of the results of such measurements. The only way to determine reliable values of glass properties is to use the statistical analysis of data published by a considerable number of independent groups. One can assume that in most cases (some possible exceptions will be discussed below) for any kind of measurement, the probability of positive and negative signs of a certain kind of systematic errors for every selected group of scientists should be nearly equal. If this assumption is correct, the statistical analysis of many publications should lead to randomization of systematic errors, which permits in solving the above-formulated problem.

In the classical approach to the theories of errors an experimenter performs all experiments using strictly similar procedures. In the presented approach we have used the results of studies of a certain property of glasses belonging to a certain

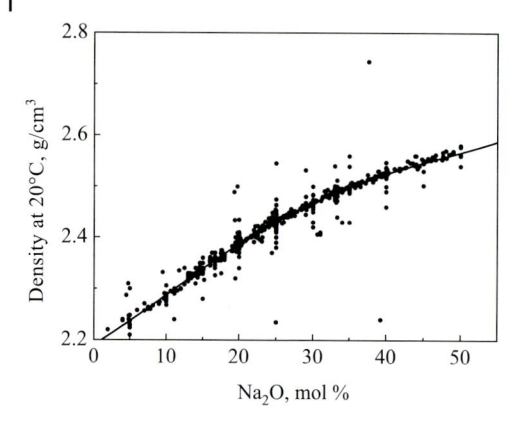

Figure 6.1 Room temperature densities of binary sodium silicate glasses containing 2–50% Na_2O. The 736 data points were taken from 202 different sources. The *solid line* shows the approximating polynomial.

binary system reported by any author, in any source, at any period of time. The reasonableness of such an approach can be shown by the statistical analysis of the obtained results. All graphs below were constructed by using version 7.5 of Sci-Glass, released in May 2010. To begin with let us select the case with the maximum number of experimental data. It is a study of density of sodium silicate glasses.

In the latest version of the SciGlass database within the composition range from 2 to 50 mol% of Na_2O and with other components not exceeding 0.2 mol% the density values for 736 glasses were found. These data were published in 202 sources between 1889 and 2010. All data points are presented in Figure 6.1. The solid line describes an approximating fourth-order polynomial. It is seen that most of the experimental points are positioned in the vicinity of the approximating curve. On removal of the data points with the highest values of residuals (the differences between experimental values and values calculated by the approximating curve) the standard deviation for 570 remaining data points proved to be only ± 0.005 g/cm^3.

It is seen that most of the data points are positioned near enough to the approximating curve. However, the number of these points is so high that it is difficult to see the exact distribution of points around this curve. This distribution becomes much clearer if we use a graph in coordinates frequently applied in the theory of errors, namely, the dependence of the number of these points on their distance from the most probable value of the property. In our case the most probable value of density for each content of Na_2O is described by the approximating curve.

The corresponding dependence is shown in Figure 6.2. We see that the distribution of most data points is described very well by a Gaussian curve. It means that the theory of errors can be used with a reasonable approximation in the case in question, even though our case differs in several aspects from the classical cases. Firstly, we do not have numerous samples of the same composition and the same thermal history. Just the opposite, there is a wide range of compositions and various kinds of sample preparations. For the classical case the procedure of measurements

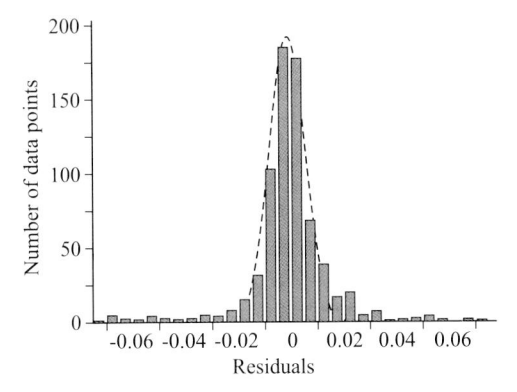

Figure 6.2 Distributions of residuals (in g/cm^3) for densities of sodium silicate glasses (cf. Figure 6.1). The *dashed line* describes the best-fit Gaussian curve. 723 data points were processed. 13 data points with especially high values of residuals were not included into the figure.

should be the same for all samples. In our case there is a really enormous diversity of the procedures of measurements, levels of qualifications of people measuring density, levels of attention to the quality of the measurements, and so on. And nevertheless, when the number of data points and number of sources of publications of these data points are high enough, the theory of errors can be applied.

Accordingly, there is good reason to state that the property values corresponding to residuals equal to zero should closely approximate the true values of the properties. These values are described by the corresponding approximation polynomial. An additional confirmation of this statement is the fact that in cases when the list of data for a selected property of a selected system includes data reported by particularly reliable sources (that is, well-known scientists who have worked in the world's best scientific centers) these data always prove to be closely approximating the center of residual distributions. Thus we may call the values described by an approximation polynomial as "the most probable property values."

It is well-known that even in the classical evaluations of the measurement results the existence of the values of residuals that prove to be outside the Gaussian distribution (the so-called gross errors) is practically inevitable. In the cases discussed here gross errors are considerably more often met than in classical cases. They are mainly connected with the so-called human factor. Their probability increased during the last decades, which is demonstrated below (cf. Figures 6.3 and 6.4). However, in the cases when the number of sources of the values of interest is high enough, this factor has only a minor influence on the final results (at least, so far). At the same time anybody who is interested in finding reliable experimental data in the literature should take into account the fact that the general quality of glass property measurements has deteriorated during the last few decades. Various aspects of this factor were described, illustrated, and discussed elsewhere [387–389]. Here, we illustrate this factor in one more form.

Let us divide all the existing data on densities of sodium silicate glasses into two nearly equal parts: the data published between 1889 and 1971 (361 data points) and

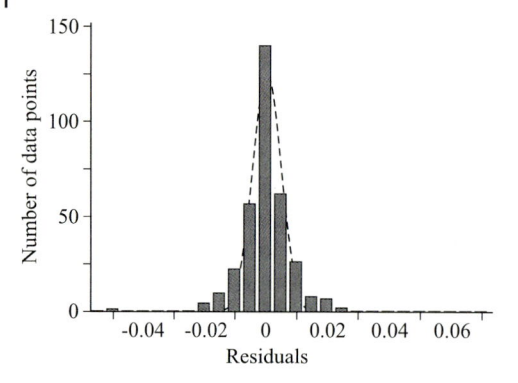

Figure 6.3 Distribution of residuals for values of densities of sodium silicate glasses published in the period from 1889 to 1971 (359 data points after removal of two data points). The approximating polynomial is the same as in Figures 6.1 and 6.2.

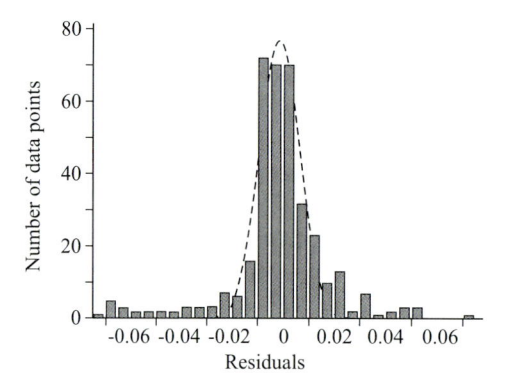

Figure 6.4 Distribution of residuals for values of densities of sodium silicate glasses published in the period from 1972 to 2010 (364 data points after removal of 11 data points). The approximating polynomial was the same as in Figures 6.1 and 6.2.

the data published between 1972 and early 2010 (376 data points). In the first part there are two particularly large gross errors. On their removal we obtain the results presented in Figure 6.3. Here we see a nearly ideal Gaussian distribution. Except for a small enough number of gross errors the only deviation from this distribution is a too large number of data points with minimal residuals. By the way, this feature of residual distributions is found in old publications often enough.

In the second part of the data much more gross errors were found. Thus here I had to remove 11 data points from the whole selection. The results of processing of the remaining 364 data points are presented in Figure 6.4. To my mind, the result is impressive enough and there are no reasons to discuss it here in detail. At the same time it is important to stress that the maximum of the Gaussian curve is positioned very near to the zero value of residuals. Note that for all four figures I have used the same approximation curve. Thus the results of approximations should be practically the same for processing of the first and second parts of the data in question. Nevertheless, it is clear, that if the total number of data points for some other properties and compositions decreases considerably, the danger of obtaining unreliable results for the recently reported data should be much higher than for the older ones.

6.3.2.2 Some Factors Leading to Gross Errors

When some authors manage to publish the results of their experiments that differ from the most probable ones by values exceeding the usual error of measurements by one or even two orders of magnitude, the question arises as to what can be the reason for such enormous errors. Note that in the standard measurement procedure any gross error appearing in a series of measurements is immediately discovered by an experimenter and removed from further processing of the data. In our case we have come across systematic gross errors that were not revealed by the author of the publication. Very often a detailed analysis of the corresponding publications does not make it possible to understand what particular factor could lead to such errors. However, sometimes it can be done. One such case can be illustrated by the analysis of the property data for alkali borate glasses.

An example of such a group of dependencies is presented in Figure 6.5. The data points for glasses with the maximum concentration of Na_2O equal to 40 mol% were selected. We can see that in this case the number of data points corresponding to the too-small densities is considerably higher than the number of the too-high values. However, before I explain this fact I shall dwell on some general conclusions concerning the scatter of the presented data. In the figure, the data points whose positions deviate particularly strongly from the most probable ones are marked by the numbers that refer to citations of the corresponding references. If one looks through these references, he/she will see that all of them belong to the sources published between 1975 and 2008. In 26 sources published between 1934 and 1974, and containing about one third of all existing data, there were no results with particularly large errors. It is a widespread enough situation. What was the reason for this?

If one finds time and looks through the publications from the 1940–1960s, he/she will see how attentive most of the authors were to the comparison of their glass property data with similar data published earlier. At that time there were no proper handbooks (and certainly no databases) and every author searched for the required data on their own, often using abstract journals. Certainly, scientists working in the 1930s often were unable to find anything in the literature: they were the first who studied the corresponding dependence. However these scientists usually paid a special attention to the reliability of the results of their measurement. Most of them made chemical analysis of the studied glasses, and they obtained the data that proved to be at the very center of the distributions of hundreds of various data reported on during the following decades.

Certainly, these esteemed pioneers of glass property studies spent a lot of time in their investigations. Today one wants to do the measurements as quickly as possible. However, one has to think about the reliability of his/her data as well. The only way to combine both requirements is to compare the results of new experiments with those already published. If the main objective of a new study is the investigation of multi-component glasses or glasses belonging to systems not studied before, then it is necessary to test the selected procedure of the study by performing some preliminary measurements of glasses and properties in the areas where a large enough number of scientists have studied and already reported similar data. However, at present it is a rather rare occasion. One of the examples of the consequences of such a regrettable lack of attention to the reliability of published data can be seen in Figure 6.5.

Now let us discuss the possible reason for the huge errors shown in Figure 6.5. We can see that unlike the data for sodium silicate glasses, the distribution of data points with increased values of residuals for sodium-borate glasses is obviously asymmetrical. Thus there should be a certain general factor leading to an increased probability of obtaining values of densities much lower than the correct ones. It seems quite probable that this factor is the presence of quite considerable amounts of water in all glasses whose residuals are highly negative. There are two possibilities in which the synthesis could have influenced the composition of the obtained glasses and correspondingly their density. One is the great difference between the volatility of pure B_2O_3 and sodium diborate or metaborate melts. According to various sources the intensity of volatilization of sodium-rich borate melts is from 5 to 10 times higher than that of boron oxide melts. Thus we can expect that in the course of melting for long enough times the content of Na_2O in borate glasses can essentially decrease, which should lead to the decrease in their density. The other possibility is the influence of the presence of H_2O in these highly hygroscopic glasses which should also lead to a decrease in their density (see for example [400]).

However, one can wonder why for these particular glasses some scientists managed to obtain densities that are much higher than the most probable ones. In these cases we can also expect that the real compositions of the studied glasses could dif-

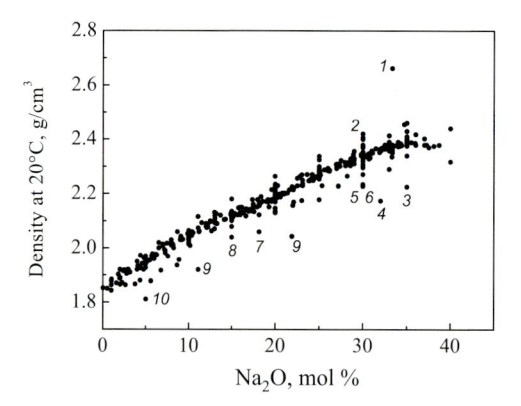

Figure 6.5 Room temperature densities of binary sodium borate glasses. The 526 data points were taken from 104 different sources. Specially marked data points were taken from the following sources: (1) [390]; (2) [391]; (3) [392]; (4) [393]; (5) [394]; (6) [395]; (7) [396]; (8) [397]; (9) [398]; (10) [399].

fer quite considerably from the reported ones. This is particularly clear in the case of the paper by Kutub [390] (see mark 1 in Figure 6.5). Not only was the sodium borate glass melted in an alumina crucible but it was also stirred by an alumina rod. It should have led to intensive dissolution of Al_2O_3 in a rather aggressive glass. It is to be noted, however, that according to the ternary diagram constructed by the corresponding option of SciGlass on the basis of the existing data the addition of Al_2O_3 to sodium borate glasses practically does not influence the density of sodium borate glasses. Thus the reason for such enormous error made by Kutub remains unknown.

It is clear that, when a scientist measures density of the glass belonging to the system in which this property was reported in many previous publications, any errors made by this scientist cannot influence our knowledge of the composition dependencies of this property. And the objective of this scientist is not to find the correct value of density. Along with this he/she usually studies some other properties of the same glasses or various structural characteristics of these glasses. Thus in many cases the only reason for density measurements is actually to test the correspondence of actual glass compositions to the expected ones. Due to the quite rare application of wet chemical analysis, such tests should be considered as very important. If the absolute values of residuals for density measurements prove to be unreasonably high, all other results of the corresponding publications should be considered as misleading and accordingly harmful for glass science.

6.3.2.3 Some Specific Examples of the Statistical Analysis of Experimental Data

It is clear that in most cases the number of experimental data used to describe a certain property of glasses in a certain binary system is much lower than in the cases described above. In this section, I will try to show that the statistical approach is quite useful in such cases as well. It is seen from Figures 6.6 and 6.7 that in the case of viscosity of lead silicate glasses the distribution of residuals differs appreciably from those described by the Gaussian curve. Certainly, it can be expected because the numbers of points and sources is comparatively small. However, even in this case the symmetry of the distribution and position of the Gaussian curve are quite admissible and the center of the Gaussian curve is positioned very near to the approximating values (these values correspond to the position "0" on the residuals scale).

At the same time the comparison of some experimental data with the most probable ones leads to completely puzzling results. An example of such results is presented in Figure 6.8. However, at first let us discuss the distribution of the main amount of data points in the figure.

In Figure 6.8 the existing data for T_g values of sodium tellurite glasses are presented. It should be noted that from the point of the statistical analysis of glass property data, the results of T_g measurements have certain specific features. This is connected with the fact that the procedure of T_g measurements applied by different scientists can be essentially different (see, for example [401]). Firstly, T_g values can be determined by both DTA/DSC and dilatometric measurements. The results for these two types of measurements can be substantially different. Secondly, even a greater difference can be generated due to the fact that various authors determine the positions of T_g in different parts of DTA/DSC curves (often without mentioning this information in the corresponding paper). Thirdly, the values of T_g depend quite appreciably on the rate of heating in the course of measurements. Fourthly, these values depend on the thermal history of the used samples. One can assume that in this case any kind of the usual distribution of residuals cannot occur. Nevertheless, we also have the usual kind of distribution (see Figure 6.9). Definitely, it is not an ideal one, but quite acceptable. One can conclude that if the number of various groups of authors is large enough, all the above-mentioned factors are also included in the probabilistic distribution of the resulted data.

Now let us analyze particular results presented by seven data points described by squares that were taken from the paper by El-Damrawi [402]. This dependence has nothing to do with the other results. In the paper no comparison of the obtained values with those published earlier (16 publications appeared in the literature before 2000) was made and accordingly no explanations of possible reasons of such differences were proposed. A reader of the paper by El-Damrawi might suppose that the author was the first person who studied T_g values of binary sodium tellurite glasses. The simplest explanation of the appearance of such impressive discrepancy between the experiments by El-Damrawi and the most probable values of T_g in the studied system is the suggestion that a technician providing the author with the experimental data merely mixed up weight and mole percentage. Let us suppose the following. The initial compositions of the studied glasses were written in mol%. However, somebody decided that these values describe compositions in wt% and recalculated them once more in mol%. The difference between molecular weights of Na_2O and TeO_2 is high enough and such recalculations would lead to enormous errors in the final results. There are at least two facts supporting this supposition. First, if we accept that compositions of the glasses measured by El-Damrawi are in wt% and recalculate them to mol%, we will obtain the results that will not dif-

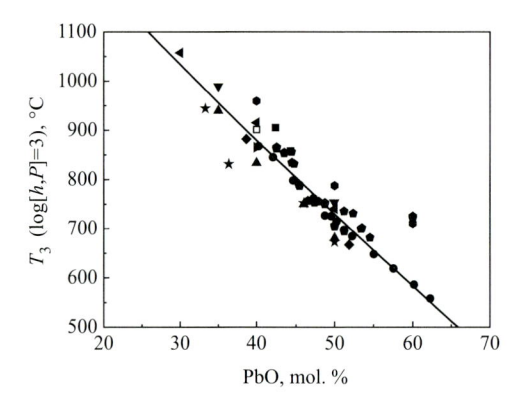

Figure 6.6 Values of T_3 (i.e., the temperatures at which the viscosity of a glass is equal to 10^3 Poise) for binary lead silicate glasses. The 62 data points were taken from 11 sources. Data points taken from each source are marked by unique symbols.

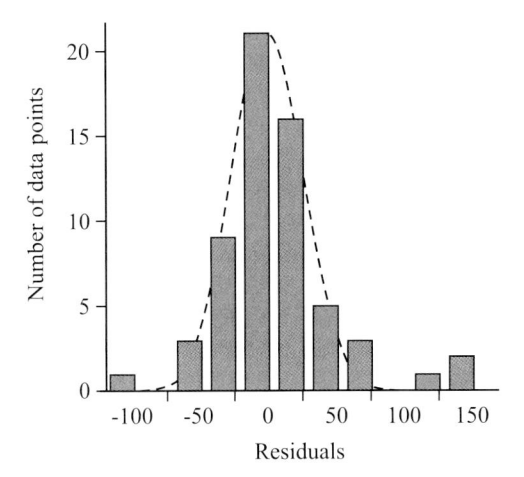

Figure 6.7 Distribution of residuals (in °C) for the data points presented in Figure 6.6.

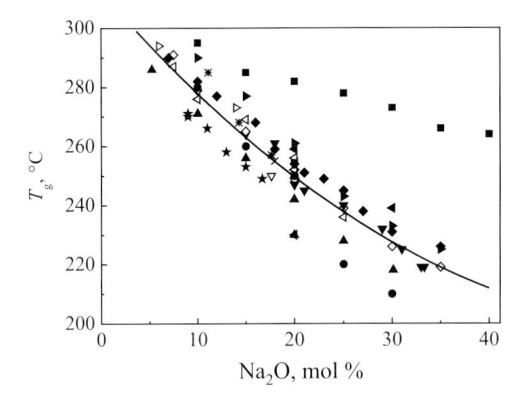

Figure 6.8 Values of T_g for binary sodium tellurite glasses. In the figure there are 81 data points taken from 26 sources. The approximating curve was formed on the basis of 74 data points taken from 25 sources (see explanations in the text).

fer considerably from the other data points presented in Figure 6.8. Second, in the paper by El-Damrawi the values of T_g are presented for glasses containing 40 and 50 mol% of Na_2O (the last data point is not shown in Figure 6.8). At the same time, according to several studies whose results are presented in the SciGlass database, the upper border of glass formation for Na_2O in the system Na_2O-TeO_2 is equal to 35–38 mol% of Na_2O (depending on the cooling rates of the samples).

Thus, probably, the measurements made by El-Damrawi or his technicians were not as bad as it might be assumed by comparison of his data with data of all other scientists. The main errors might be connected with the fact that El-Damrawi and/or his technicians have confused recalculations in various kinds of percentages. Everybody who has carried out experimental work knows that some or other kinds of errors may appear at various, sometimes absolutely unexpected steps of

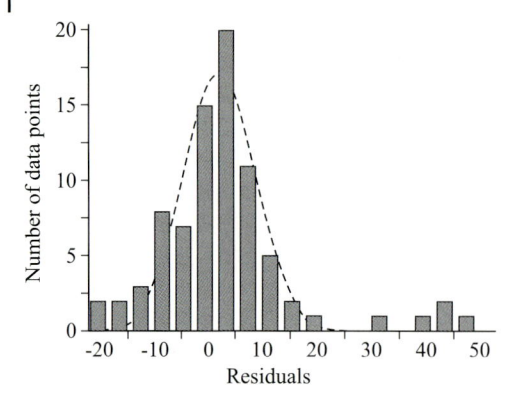

Figure 6.9 Distribution of residuals (in °C) for data points presented in Figure 6.8.

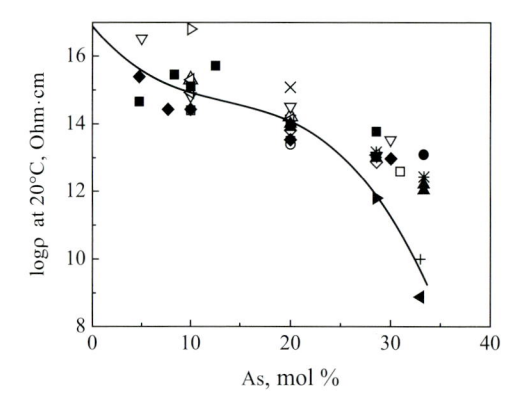

Figure 6.10 Values of the logarithm of electrical resistivity for As–Se glasses. The 40 data points were taken from 18 sources.

scientific work. That is why it is necessary to be very careful and check and recheck one's results. However, to be 100% safe against any kind of errors one should compare one's own results with results of previous studies. It is so easy to do this. It is so easy to understand the necessity of this action. And, I cannot apprehend the psychology of people who ignore these obvious ways to ensure reasonable quality of results of laborious studies.

In Figure 6.10 there is one more specific example of the distribution of property data. It should be noted that the scatter of data for electrical conductivity is usually particularly large. This is connected with a number of specific problems arising in the course of measurements of this property. It is not reasonable to consider this subject in any detail. It is only worth noting that often even the error in half the order of magnitude of the logarithm of electrical resistivity should not be considered as too high. However, as it is seen from Figures 6.10 and 6.11, in such cases the statistical approach also permits estimating the composition dependencies of such properties with a reasonable accuracy.

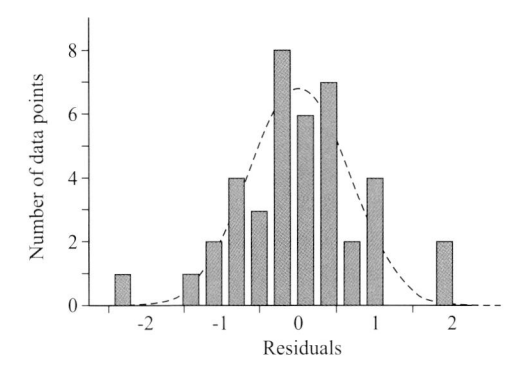

Figure 6.11 Distribution of residuals (logarithm of electrical resistivity) for the data points presented in Figure 6.10.

6.3.2.4 What is to Do if the Number of Sources Is Too Small?

Situations similar to those described above are specific for measurements of many hundreds of property-composition dependencies. At the same time there are even more cases when the number of sources containing experimental data is too small. In this case it is necessary to analyze each case separately. Below I present several specific examples of such cases.

The values of thermal expansion coefficients of glasses in the system Bi_2O_3–SiO_2 were reported only in three publications. However, fortunately, all results correspond to each other quite reasonably. This fact is shown in Figure 6.12. It is important that the three groups of authors worked absolutely independently of each other (which is quite clear from the corresponding references). Thus we can practically be sure that the dependence presented in the figure is reliable within $\pm 3 \cdot 10^{-7}$ K^{-1}, which is a very good accuracy for measurements of thermal expansion.

Figure 6.13 illustrates a somewhat different case. Here we have an excellent correspondence between the results of two publications and quite different data taken

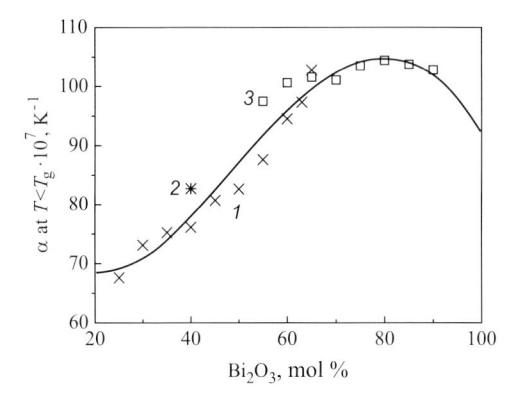

Figure 6.12 Values of thermal expansion coefficients of Bi_2O_3–SiO_2 glasses: (1) [403]; (2) [404]; (3) [405].

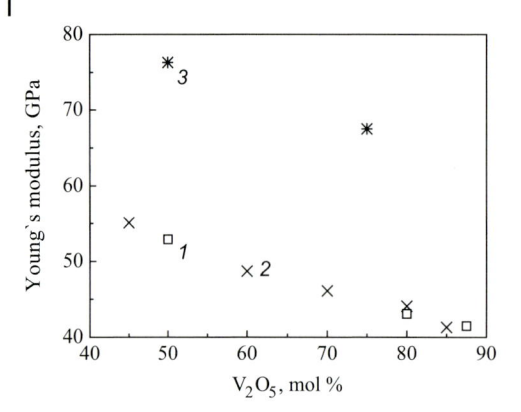

Figure 6.13 Values of elastic moduli of V_2O_5–P_2O_5 glasses: (1) [406]; (2) [407]; (3) [408].

from the third source. Certainly, there are reasons to trust these two independent studies yielding the same results. Nevertheless, it is reasonable to check the validity of the results by Drake et al. [408]. Luckily, in all three sources the data on densities of the studied glasses were also given. The comparison of these data with the data of many other authors showed that the density values reported in the papers by Field and by Farley and Saunders are very near to the approximating curve, while the density value of one of the two glasses studied by Drake et al. differs from the most probable density value by $0.1\,g/cm^3$, which is a large error for this property (cf. Figure 6.2, for example) [408].

An example of an unfortunately not so seldom case is shown in Figure 6.14. Neither of the authors published any other papers with microhardness data. Thus it was not possible to assess the quality of these particular measurements by comparison of the results of such measurements with the results of other authors for the systems which were studied by a larger number of authors. The analysis of the density studies of V_2O_5–TeO_2 glasses did not lead to any clear results. Which of these two sources is more reliable? Certainly, a great scatter of the data presented in [410] casts some doubt upon the quality of the measurements. At the same time this is a poor enough reason for drawing a definite conclusion on the reliability of the data presented in [409].

In this case, I decided to combine all data for microhardness of binary glasses of the systems R_mO_n–TeO_2. The result of this procedure is shown in Figure 6.15. It is clear that for glasses with a high percentage of TeO_2 microhardness has values within the range of 2–3.5 GPa. The lowest microhardness was found for a glass containing about 20% of Tl_2O. No other sources reported abrupt changes in microhardness with small changes in composition (similar to the results reported by Sidkey et al.). All these factors allow us to conclude that the data for microhardness of V_2O_5–TeO_2 by Chopra et al. are much more reliable than those by Sidkey et al. and exclude the latter data from any further analysis of the dependence in question.

The latter example shows that in the cases when the number of existing sources is small the determination of the most reliable data sometimes requires apprecia-

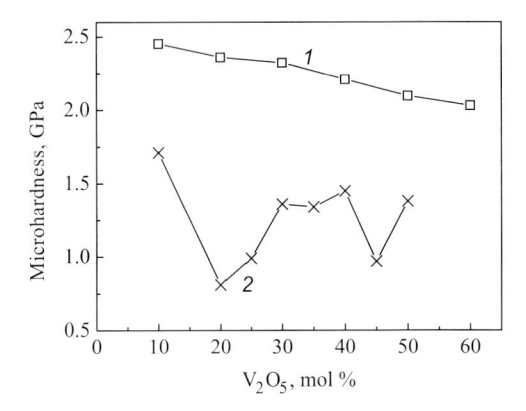

Figure 6.14 Microhardness of V_2O_5–TeO_2 glasses according to two different sources: (1) [409] and (2) [410].

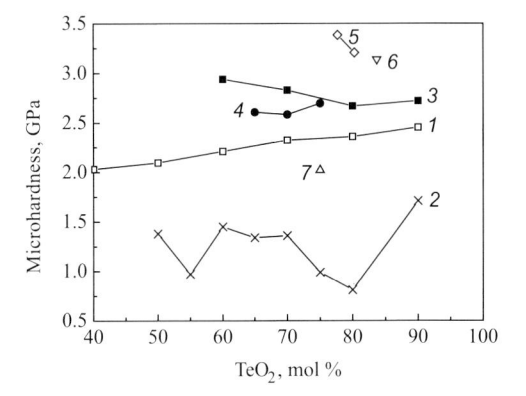

Figure 6.15 Microhardness of binary tellurite glasses according to different sources: (1) [409] (V_2O_5–TeO_2); (2) [410] (V_2O_5–TeO_2); (3) [411] (ZnO–TeO_2); (4) [412] (ZnO–TeO_2); (5) [413] (B_2O_3–TeO_2); (6) [414] (Ta_2O_5–TeO_2); (7) [415] (Tl_2O–TeO_2).

ble efforts. However, it is definitely reasonable to spend some additional time for the assurance that you have managed to select the best of the existing results. Even more difficult can be the situation when one finds only one source where the dependence of interest is reported. One encounters such situations particularly often in the case of multi-component glasses. If you want to grasp an idea how reliable the data may be, you should try to form your own opinion about the reliability of data presented by the same group of authors in other publications, where such results could be compared with similar studies performed by larger sets of scientific groups.

As follows from the above examples, if the number of independent authors studying the property of interest of the corresponding binary system or of the pseudo-binary section in the ternary or quaternary system is numerous enough (say, six or more), there are very good chances that the most probable composition dependence of the property will prove to be quite reliable. If the number of independent authors is smaller and at the same time the results of different authors differ from each other quite considerably, one needs to undertake a special investigation in order to find the most reliable data. These investigations can be quite diverse, but it could take no more than half an hour to conduct them. Quite often they lead to positive results. If we have results presented only by one author (or one group of authors), it is necessary to analyze some other publications of the same author(s) in order to determine the level of reliability of these results.

6.4
About the Reliability of the Authors of Publications

6.4.1
The Moral Aspect of the Problem

Every specialist in any branch of science knows that there is a great difference in the qualities of various experiments described in scientific journals. Sometimes a qualified reader can clearly see this quality in the course of reading a paper. Sometimes it is difficult to determine this without an attempt to repeat a certain experiment. For young scientists the possibility to select reliable experimental data from unreliable ones is even more difficult. Presumably, in materials science this problem is particularly important due to a large number of new experimental data appearing every month in thousands of sources (including now numerous Internet sites). Some of these data are totally wrong. It is impossible to calculate the losses of the world economy connected with this factor, however, they are definitely high enough. Accordingly, the attempts to answer the question, "What is to be done?" seems inevitable.

There are some fields in materials science where it is very difficult to propose anything efficient in order to decrease the above-mentioned losses. This is caused by a considerable dependence of properties of organic polymers, metals, ceramics, and so on, on characteristics of raw materials, various details of technology, small concentrations of impurities, and so on. Fortunately, most properties of glasses depend mainly on their composition and temperature. The third factor is thermal history. However, its influence is much smaller than the influence of the first two factors. Thanks to this specific feature of glasses and glass-forming melts we can efficiently carry out the analysis of experimental data published by various authors. There are at least two possibilities to considerably diminish the damage impaired to glass science and industry by publications of unreliable data.

One possibility to act in this direction is quite obvious. It is the necessity to increase the attention of referees of scientific journals to the quality of experimental data presented in papers submitted for publications. I believe that everybody interested in using reliable glass property data will agree with me. How to implement this is another question and it should be a special subject of a detailed discussion. The other possibility also seems obvious, but at the same time doubts are cast upon its implementation. This concerns selection of a few groups of authors who have published a considerable number of papers containing obviously wrong data, as well as information of the glass community about such groups of authors. I do understand that such a suggestion may look shocking for many bona fide scientists. This is something like an attempt to put a group of scientists onto a certain black list and impede their scientific careers. However, let us look at this problem from another point of view. For many years these scientists have been using one or another financial funds, scientific journals and scientific conferences for publications or reporting rubbish, and, what is even worse, they have been misleading people who were so naive as to consider the corresponding data as true and valid. Are we not entitled to attempt to warn scientists against using these data? I can agree that the general system of selecting such groups of authors should be discussed in detail and implemented with utmost care. Probably, this can be done under the leadership of the International Commission on Glass.

Certainly, the best way in the discussed direction would be the development of a special program of analyzing the consistency of the data of every author who published the values of properties of various binary glasses with the most probable data for these glasses. This program should automatically determine the "reliability rating" of experimental data of every author and present the complete list of such ratings in a particular site. This list should be regularly updated and accordingly every author will be able to improve it by paying more attention to the quality of his/her experimental work. Unfortunately, the development of such a program will require an ample amount of resources, because any program of this kind will need a long meticulous testing procedure of revealing and removing any possible defects in the discussed evaluation of data quality. Without this procedure, a systematic application of such program can lead to some quite harmful results.

Then what is reasonable to do now? To let any ignorant, unscrupulous or just light-minded author publish any number of totally erroneous data in high ranking journals to earn their regular grants? Such attitudes may result in a real disaster in glass science. One of the possibilities of doing something to save the situation is to select some examples of systematic publications of obviously incorrect data by a certain author or by a certain team of authors. Certainly, it could be considered as somewhat unjust because some authors who may probably publish even less reliable data will appear free of the corresponding accusations. However, I think this is better than doing nothing. And if the proposed analysis of experimental errors of some authors is absolutely correct, it may not only help these authors to improve their further experimental performance but can also serve as a warning to other light-minded authors. As a preliminary attempt of such discussion, I will permit myself to give an example of this approach in the next section of this paper.

6.4.2
An Example of Systematically Unreliable Experimental Data

As the basis for illustrating the proposed approach to the problem I have selected the data presented in Figure 6.8. In the paper by El-Damrawi, from where these data were taken [402], the results of measurements of several glass properties were

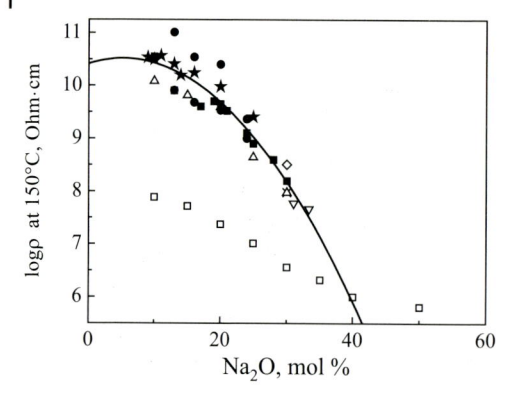

Figure 6.16 Electrical resistivity of Na_2O–TeO_2 at 150 °C. The 43 data points were taken from 8 sources. *Squares* (□) specify the results from [402].

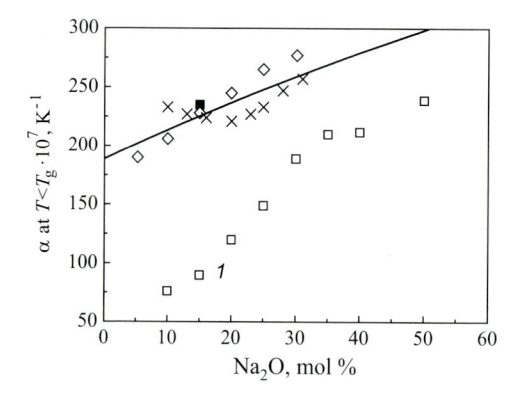

Figure 6.17 Thermal expansion of Na_2O–TeO_2 glasses. The 23 data points were taken from 4 sources. *Squares* (1) specify the results from [402].

presented. The comparison of these results with similar data reported by other authors are shown in Figures 6.16 and 6.17. It should be pointed out that in the case of resistivity only one result was reported after the publication of the paper by El-Damrawi. In the case of thermal expansion all values were published before 2000. However, the author has simply ignored the existence of any data except his own.

It is also to be noted that (as discussed above) in the case of T_g measurements it was reasonable to suggest that the values of this property were measured correctly and the found discrepancy between the data of [402] and the most probable values of T_g could be attributed to errors in calculations of glass compositions, whereas in the case of the data presented in Figures 6.16 and 6.17 the situation is quite different. The largest errors appear for glasses with the smallest concentrations of Na_2O and no changes in percentage would decrease such errors. Two orders of magnitude for electrical resistivity exceed considerably any reasonable standard deviations in such kinds of measurements. However, particularly impressive are

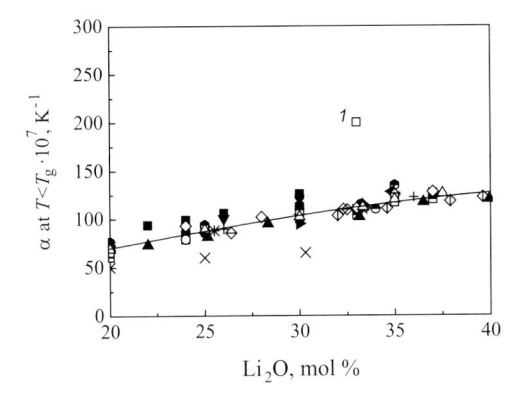

[Figure: plot of α at T<Tg·10⁷, K⁻¹ vs Li₂O, mol %, ranging 20 to 40 mol%, with data scattered around 100, point labeled "1" near 200]

Figure 6.18 Thermal expansion of Li$_2$O–SiO$_2$ glasses. The 112 data points were taken from 34 sources: (1) specifies the results from [416].

errors in measurements of thermal expansion. It is funny (I cannot find any other word in this case) that according to [402] "the accuracy of the measured values of α was ±2%," that is, for glasses with the lowest concentrations of Na$_2$O the errors should be less than $2 \cdot 10^{-7}$ 1/K.

One could eventually assume that the sketched above deviations may have been a certain exception in connection with this particular system caused by some unfortunate circumstances. For example, a person whom the author of the paper had trusted supplied the author with a series of glasses having compositions that had nothing in common with the expected ones. Therefore I have checked the reliability of the results of some other publications by the same author.

Figures 6.18–6.20 illustrate the results of studies of some properties of binary lithium silicate and lithium borate glasses. Again we can see a large difference between the results of the experiments by El-Damrawi and Doweidar and the data taken from a large number of other publications. Note that the sign of the obvious error in the thermal expansion measurements of a silicate glass (Figure 6.18) is opposite to the sign of errors of the thermal expansion measurements of tellurite glasses in Figure 6.17 although in both cases these errors are unbelievably high.

It is important to mention one more aspect of the discussed problem. Quite often glass scientists study the influence of a third component on a property by preparing a series of glasses that include an initial binary glass. It is an absolutely correct approach. However, if we find that the value of the property of the initial, binary glass is incorrect, the results of all other glasses become questionable. This is illustrated in Figure 6.19. According to El-Damrawi and Doweidar the magnitude in the decrease of thermal expansion in the course of replacement of Li$_2$O by Al$_2$O$_3$ in silicate glasses is two or three (for various parts of the dependence) times higher than the real one. It should be noted that not all property values presented in the cited papers for lithium silicate and lithium borate glasses are erroneous. For example, the values for specific resistivity of these glasses reported by El-Damrawi

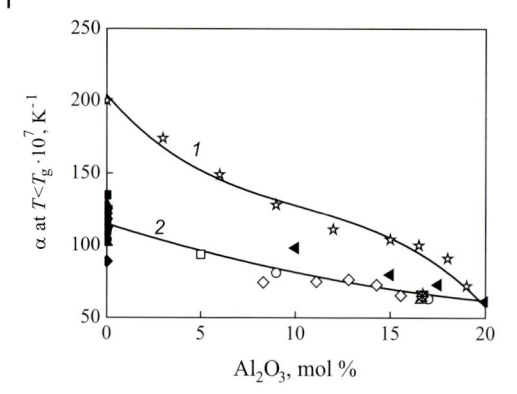

Figure 6.19 Expansion coefficients of glasses in the system $Li_2O-Al_2O_3-SiO_2$ for the section with content of SiO_2 equal to 65–67 mol% of SiO_2. The 51 data points were taken from 26 sources: (1) specifies the results from [416], and (2) specifies the results from all other data sources.

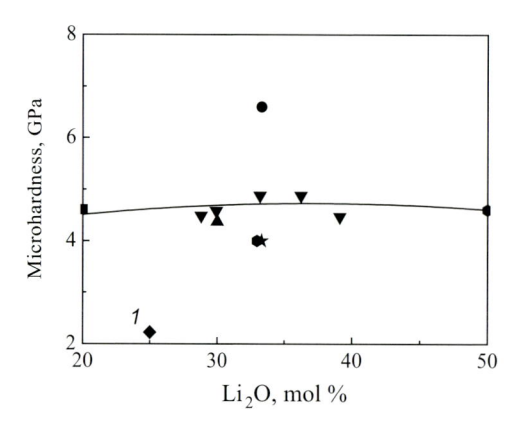

Figure 6.20 Microhardness of $Li_2O-B_2O_3$ glasses. The 13 data points were taken from 9 sources: (1) specifies the results from [417].

and Doweidar are reasonably reliable. However, it is impossible to predict which particular result is reliable and which one is not.

As it is seen from the above-cited references, often enough El-Damrawi published his papers together with Doweidar. Thus, I have selected a few publications by Doweidar written without El-Damrawi where the properties of some binary glasses were investigated and have compared the corresponding data with those of other authors. The results of several comparisons proved to be acceptable. Some others presented fully unreliable data (see Figures 6.21 and 6.22). It is worth noting that in Figures 6.21 and 6.22 we can see the results taken from two publications by Doweidar and/or El-Damrawi. And in both cases the results taken from two publications presenting the same kind of data are quite different. In one publication the reported results are quite (or more or less) reliable, and in the other – appallingly wrong. At the same time the descriptions of the measurement procedures are

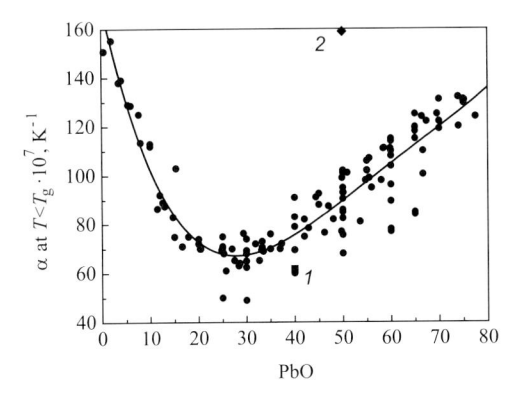

Figure 6.21 Thermal expansion of PbO–B$_2$O$_3$ glasses. The 131 data points were taken from 35 sources: (1) specifies the result from [418], and (2) from [419].

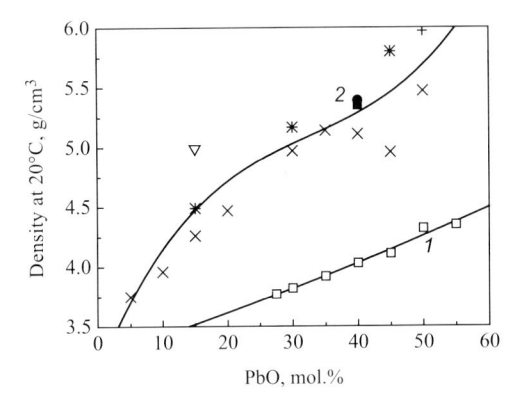

Figure 6.22 System PbO–ZnO–B$_2$O$_3$; the content of B$_2$O$_3$ is 40 mol%. The 23 data points were taken from 7 sources: (1) specifies the results from [420], and (2) from [421].

identical in all publications (it seems that the authors just copied certain standard phrases). All experiments were done in the same place, the Department of Physics, Faculty of Science, Mansoura University, Egypt. And in the papers in question, there are no corrections, no comparisons, no explanations.

6.4.3
Concluding Remarks

I have cited as an example for the illustration of a general problem some publications by El-Damrawi and Doweidar taken actually at random. Among the recent scientific publications it is possible to find publications by an appreciable number of scientific teams that regularly publish unreliable results.

Every person may commit errors from time to time. Some errors may appear as the results of misprints somewhere on the way between obtaining initial results and final publications of data. However, if a person values his/her good name and if he/she is properly qualified, these errors are rare enough. When an author finds his/her own error he/she either sends a special correction note to a journal or pays particular attention to the explanation and correction of this error in his/her further publications. One of the most efficient (and at present very simple) way of preventing various kinds of errors is the comparison of new data with the already published ones. If the composition area of a new investigation is outside the compositions studied so far, it is necessary to additionally prepare and measure a few glasses that have already been studied and whose compositions are at the same time as near to the compositions of the main group of glasses as possible. In principle, it is possible that certain differences in properties can arise from special ways of glass synthesis or thermal histories of the samples. This is now a rare possibility, however its probability is far from being zero. Nevertheless, in this case an author, who finds a clear difference between his/her data and similar data reported by other scientists, should pay special attention to the study of this possible phenomenon and prepare samples both for the standard and special procedures to study this in some detail.

However, in this case and many similar cases, the situation is quite different. It looks like that some authors were never aware that there is somebody else studying the same property of the same glass compositions. As follows from the data presented in Figures 6.21 and 6.22, these authors even did not try to compare similar data from other publications written by their own team. By the way, the publications by El-Damrawi and Doweidar contain some interesting analysis of their data (which was definitely the reason why their papers were published in high-ranking scientific journals). These publications show that the authors are knowledgeable and qualified enough. Thus, I think, we should assume that these scientists were not interested in such a routine work as synthesis of glasses and measurements of their properties. Probably they have just used the data given to them by their technicians. Referees seeing that the presented results are discussed at the appropriate level are inclined to consider the experimental data as fully reliable. Both the authors and referees are so busy that they are unable to spend their precious time for testing the quality of the data. And, as a result, we have what we have.

Problems of such kind are one of the serious aspects of the actual situation in glass science at the beginning of the twenty-first century. And the question remains: Is it permissible (or probably is it desirable) to do in a much wider scale what I have done in Section 6.4.2? That is, to warn all people who may be interested in using the data published by certain teams of authors that it is necessary to be very cautious when using these data for any scientific or practical purpose.

At the end, I would like to point out that all I wrote does not mean that further publications by the scientists' team including El-Damrawi and Doweidar should be not allowed. I only suggest that the editors and referees of scientific journals should pay special attention to the descriptions of experimental procedures given in submitted papers, as well as to proofs of the reliability of the presented data. It seems evident that the only totally safe evidence of such reliability is the reasonable consistency of new experimental results with the results of a considerable number of data taken from earlier publications. It is much safer than the use of the so-called standard samples. By using these samples you can test the reliability of your measurement procedure. But as to the correctness of your synthesis (together with the correctness of measurements), you can test it only by comparison of your data with the published ones (see the demonstration of this statement in Figure 6.5).

6.5
General Conclusion

The glass property database SciGlass is not the only database of this kind. However, it is based on the system of compilation and presentation of glass property data that was developed, tested and improved during about a period of 50 years by a team of specialists in glass science. Thus it is possible to assume that the main principles of collection and presentation of experimental data may be of some interest not only to users of this database but also for people who would like to develop some other databases in various areas of materials science.

Although the first version of the SciGlass database was released more than fifteen years ago, many specialists on glass science and technology have not used this database efficiently up to now. One of the very important (although far from being the only one) ways of using this database is to find the most reliable glass property data for binary and ternary glasses. In this paper mostly the statistical evaluation of the properties of binary glasses has been considered. For ternary glasses the approach to find the most reliable data is the same. Some options of SciGlass permit one to determine the most reliable data for multi-component glasses as well. However, this is quite a different story. It is worth pointing out that the possibility to determine the most reliable experimental data is important not only for using these data for the solution of various problems of basic and applied glass science but also for the development of efficient methods of predictions of glass property values (see, for example, Chapter 7 of this book).

This database permits not only the determination of the most reliable property, composition dependencies, but also permits finding publications where highly erroneous data on certain properties of certain glasses are presented. In principle it reveals the groups of authors who systematically publish totally unreliable data. It is worth discussing this problem in a broad scale: should the glass community try to defend itself from such authors or will any such attempt violate the main principles of political correctness? And if such defense may be considered possible, how could it possibly be organized?

7
Methods of Prediction of Glass Properties from Chemical Compositions

Alexander I. Priven

7.1
Introduction: 120 Years in Search of a Silver Bullet

The metaphor of a "silver bullet" referring to the folkloric belief that such bullets are the only weapons which can kill a vampire or werewolf is widely used in science, technology and engineering. This metaphor means a simple guaranteed solution for a difficult problem which remained unresolved for a long time. One of such problems is the development of new compositions of glasses with prescribed combinations of physical properties. Up to now, such development remains a very laborious and expensive work: each time one has to synthesize and study hundreds and thousands "trial" glasses until a single composition with desirable properties is found – if found at all.

Mathematical models allowing one the prediction of physical properties of glasses from their chemical compositions are obvious candidates for the role of the mentioned "silver bullet." Inorganic glasses and glass-forming liquids (melts) are unique materials which physical properties are mostly determined by only two factors: chemical composition and temperature.

There are few exceptions of this rule. First, some glasses are phase-separated: liquation or particular crystallization can considerably change physical properties. Second, physical properties of glasses somewhat depend on the preliminary time-temperature regime of heat treatment, the so-called thermal history. Third, properties of any system are more or less pressure-dependent. Fourth, physical properties of glasses can be modified by some treating their surfaces (e.g., ion exchange); such treatment is widely used for changing optical and some other properties of glass articles, such as wave guides and low-transmitting glasses. However, all of them can be considered as special cases. The common case concerning the majority of scientific papers and practical applications is the consideration of single-phase oxide glasses and glass-forming melts at atmospheric pressure, with the assumption that glasses are being synthesized by cooling the melts with moderate cooling rates having an order of few or few dozen K/min. Only this case will be considered below. With the above-made stipulation, we come to the conclusion that, at least in theory, knowing chemical composition of a particular glass or melt is sufficient to be able to predict its multiple properties at an arbitrary temperature.

It is worth to note that there are two considerably different practical tasks that can be resolved with the use of models that predict physical properties of glasses from their chemical compositions, and the requirements to models used in these two cases are considerably different. The first of these tasks consists in the development of new compositions of glasses having prescribed combinations of properties. The required properties are determined by the usage of glass: for instance, window, chemical ware, optical lens and LCD panel require glasses with very different properties.

To find a glass with a new, specific combination of properties one has to look at various composition areas. If we try to solve the problem by using a model that predicts glass properties from chemical compositions we need a model (or a set of models) that allows one the prediction of multiple properties in wide composition areas. The most important characteristic for such model (or models) is *generality*, that is, ability to calculate multiple properties and applicability to as wide composition areas (and, sometimes, temperature ranges) as possible.

The second task is modification of an existing glass composition to improve one or more properties without worsening others. For example, one may need increasing thermal shock resistance of a glass for chemical ware without noticeable changing the chemical durability, viscosity and thermal conductivity. In this case, one does not need to look at a wide composition area but needs to accurately calculate multiple properties in a local system within narrow enough concentration ranges of components. For this purpose, the main criterion of model quality is accuracy of calculations. Combination of generality and accuracy in one model is, as a matter of fact, the above-mentioned "silver bullet" which glass producers need and glass scientists are searching for.

More than a centenary history of trials of searching for these functions is full of discoveries, insights, hopes and disappointments. Probably the only one that is still not available is a final solution. Nonetheless, the progress in this direction is impressive and demonstrative. It demonstrates multiple traps, hidden dangers and nonobvious ways of solving difficult problems that initially seemed very simple. In 1988, an excellent monograph was published by Volf [422]. In this monograph almost all of more or less significant findings concerning the practical calculation of glass properties from chemical composition and temperature were considered in fine detail. During the next two decades, several new approaches to glass property simulation were suggested, some of them seeming promising. Below, the history and contemporary state of glass property simulations are considered and some prognoses for foreseeable future are made. The author does not aim at a detailed characterization of particular models as it was done in [422] but mainly tries to focus the attention of the reader on the basic ideas and approaches of different authors to glass property simulation.

When preparing the present chapter the global glass property information system SciGlass [423] was widely used. This system contains the largest public database of glass properties containing more than a million data points concerning about 350 000 glass compositions. Also, SciGlass allows one the calculation of multiple properties by using more than 100 models. The property calculation models incorporated into SciGlass [423] are disclosed in [290, 307, 385, 424–531] and a few private communications; the most important of which are briefly characterized below.

7.2 Principle of Additivity of Glass Properties

7.2.1 Simple Additive Formulae

At the end of the nineteenth century, the German chemist, glass technologist, and, in particular, inventor of borosilicate glasses, Otto Schott, founded a glass laboratory (Glastechnisches Laboratorium Schott & Genossen (Schott & Associates Glass Technology Laboratory)) developing later into a world-wide known company aimed to produce optical glasses with a desired wide spectrum of technological applications. One of the challenges which he had to account for was the broadening of the nomenclature of optical glasses required for producing the lenses for microscopes, telescopes and other devices. A good lens consists of several glass elements with different optical characteristics determined by destination and construction of the lens. Thus, to satisfy the needs of optical industry a variety of glass compositions were required; each of them had to have prescribed values of several properties: refractive index and its partial dispersions, thermal expansion coefficient, elastic properties, and so on.

As it was stated above, to fit all of these requirements in a single glass composition a lot of glasses had to be synthesized and investigated in detail; for that, very laborious and expensive research was required. To save money and time Schott asked A. Winkelmann to develop some mathematical models describing the relationships between chemical compositions and physical properties of glasses. At that time (1890s), about a dozen of oxides were commonly used for production of optical glasses: SiO_2 (as the main glass-forming component), Li_2O, Na_2O, K_2O, MgO, CaO, ZnO, BaO, PbO, B_2O_3, Al_2O_3, and P_2O_5; each glass contained from 5 to all 12 different oxides in specific proportions, which gave the required combination of properties. Thus, Winkelmann and Schott had to develop a model (or a series of models) describing dependencies of glass properties on the concentrations of these 12 oxides.

It is worth to note that at this time no systematic data about concentration dependencies of glass properties were published and no theoretical models were suggested. Most of available data concerned few multi-component glasses; for many of them exact compositions were not reported (e.g., Na_2O and K_2O were specified together as "R_2O" or "alkalis"). Thus, Winkelmann and Schott needed an approach allowing one the development of rather accurate mathematical models from minimum available data, most of them being obtained by themselves. As a solution to this problem, Winkelmann used the idea of "additivity" which implies that contributions of particular oxides to physical properties of glasses are proportional to their concentrations,

$$P = \sum_{i=1}^{n} c_i p_i, \quad (7.1)$$

where P is a property to be calculated, p_i stands for "partial coefficients," or "weight factors," or "specific contributions" of the components (these synonymous terms are used by different authors), c_i is the concentration of the ith component, and n is the total number of components in the studied glasses.

In [528], Winkelmann suggested to calculate the specific heat capacity (C_p) of glasses near to the room temperature as an additive sum of weight percent (w_i) of the above-mentioned oxides characterized by constant partial specific heat capacities ($C_{p,i}$) as:

$$C_p = \sum_i w_i C_{p,i}. \quad (7.2)$$

In the following papers [385, 529, 530], Winkelmann and Schott extended the same principle to other properties important for optical glasses: density, coefficient of thermal expansion, Young's modulus and refractive index. The partial coefficients for 12 oxides were determined from the experimental data on several dozen "test" glasses which they melted and investigated in their laboratory.

It should be noted that the additive approach was not suggested first in theory by Winkelmann. In his first paper [528], he referred to a similar relationship for the specific heat of simple one-component solids that was discovered several decades before; now it is known as the Neumann–Kopp rule. According to this rule, atomic contributions to the molar heat of a solid only slightly depend on the chemical nature of this solid; in other words, atomic heats in solids can be considered as having nearly constant values. Then, Kopp extended his approach to molar volume and, thus, made it possible to calculate the densities of some organic substances from their chemical formulas. The approach of Kopp was described in [468] and other papers published from 1850 to the 1860s. In the papers of Kopp and other scientists (e.g., Loschmidt [532]), additive models were suggested for the prediction of physical properties of various inorganic and organic substances. Winkelmann and Schott only applied the additive approach to the calculation of physical properties of silicate glasses. Nonetheless, the practical importance of the considered simulations is difficult to overestimate. Having very limited experimental data, Winkelmann and Schott could develop a set of models which allowed them the prediction of multiple properties of silicate glasses in a wide-range of concentrations.

7.2.2
Additivity and Linearity

The simplest and the most popular statistical procedure of fitting various experimental dependencies including multi-factor relationships (such as concentration dependencies of glass properties) is linear regression. In terms of the designations used in Eq. (7.1), the regression equation for a glass property, P, as a function of chemical composition in an n-component glass-forming system can be written as

$$P = k_0 + k_1 c_1 + k_2 c_2 + \ldots + k_{n-1} c_{n-1} = k_0 + \sum_{i=1}^{n-1} k_i c_i , \qquad (7.3)$$

where k_i ($i = 0, 1, \ldots, n-1$) are fitting parameters, or regression coefficients. As far as the sum of concentrations of all glass components is constant ($\sum c_i = 100\%$), the concentration of the last (nth) component (most often, SiO_2 is selected for this role) is not considered as an independent variable; indeed, it can be written as:

$$c_n = 100 - \sum_{1}^{n-1} c_i . \qquad (7.4)$$

That is why in an n-component system only ($n - 1$) components are incorporated into a regression equation of the type of Eq. (7.3).

If we substitute Eq. (7.4) into Eq. (7.1) we come to the following result:

$$P = \sum_{i=1}^{n-1} c_i p_i + \left(100 - \sum_{i=1}^{n-1} c_i\right) p_n = 100 p_n + \sum_{i=1}^{n-1} c_i (p_i - p_n) . \qquad (7.5)$$

The latter equation becomes identical to Eq. (7.3) if we substitute k_0 for $100 p_n$ and the other k_i for $(p_i - p_n)$. Thus, the additive equation is mathematically identical to a trivial linear equation. The conversion of these equations into each other is just a mathematical trick. In other words, any essential statement concerning an additive formula can be equally applied to linear regression and vice versa.

7.2.3
Deviations from Linearity

If the concentration dependencies of glass properties really were linear the approach of Winkelmann and Schott (or the essentially identical procedure of linear regression) would be exactly what the scientists needed. Unfortunately, reality is not so simple.

In Figure 7.1, some concentration dependencies of glass densities in binary systems containing sodium oxide are shown. As we can see from the figure, in more or less wide concentration ranges the concentration dependencies of glass density can really be considered as linear. However the figure clearly shows that this is definitely not the general case. In some cases the concentration dependencies of glass properties are much more complicated – see the examples in Figures 7.2–7.5 below. Actually, nearly all ideas suggested by the authors of multiple models developed in the twentieth century were aimed at an accurate description of nonlinear effects. Below, the most well-known of them will be briefly characterized.

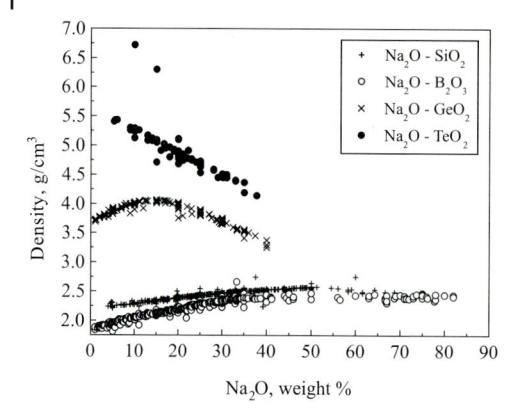

Figure 7.1 Concentration dependencies of density of binary glasses containing Na_2O at room temperature, according to the experimental data of multiple authors presented in the SciGlass database [423]. Systems are specified in the legend.

7.3
First Attempts of Simulation of Nonlinear Effects

7.3.1
Winkelmann and Schott: Different Partial Coefficients for Different Composition Areas

In the above-mentioned models of Winkelmann and Schott [385, 528–530], the first attempt of considering nonlinear effects was made. The authors have ascribed different "partial coefficients" to two oxides, B_2O_3 and PbO, in different composition areas. In principle, such an approach could work; however, the authors had too few experimental data to determine the coefficients as well as concentration boundaries of their application. As a result, their model has nearly no "predictive power" when calculating the property of lead- and boron-contained glasses. Nonetheless, in later years, this idea was successfully used by other authors who had more data.

7.3.2
Gehlhoff and Thomas: Simulation of Small Effects

The fact that concentration dependencies of glass and melt properties are, in the general case, not linear was well-known to glass scientists from the very beginning. This statement especially refers to the viscosity. This property is definitely the most important for practice, and its systematic studies have been started in the beginning of the twentieth century. In the 1920s, the nonlinearity of concentration dependencies of viscosity even in rather narrow concentration ranges became obvious to glass scientists. However, the nature of nonlinearity was unclear.

The first (known to the author) attempt to predict the viscosity of glass-forming melts in rather large concentration and temperature ranges was made by Gehlhoff and Thomas [545]. They suggested a very simple but quite promising approach.

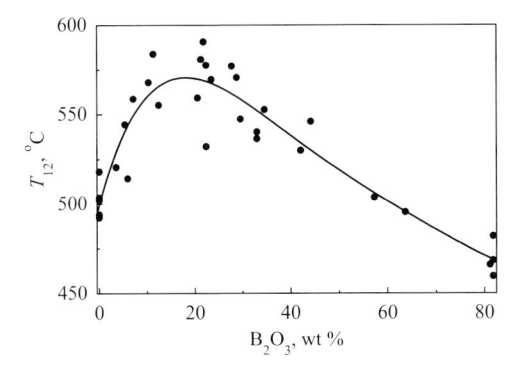

Figure 7.2 Effect of replacement of SiO_2 for the same weight portion of B_2O_3 on the temperature T_{12} corresponding to the viscosity 10^{12} Poise in ternary sodium borosilicate melts containing 20 ± 2 wt % Na_2O. Points correspond to experimental data of different authors presented in the SciGlass database [423]. The line shows the results of the polynomial approximation.

This approach was based on the assumption that general trends of influence of particular oxides on the viscosity characteristics are similar in a rather broad composition area. In this case, replacing one component by another in the same proportion should cause more or less similar changes of a given property in different concentration ranges.

Based on this assumption, Gehlhoff and Thomas suggested three sets of parameters describing changes of three viscosity characteristics (the temperatures corresponding to three fixed viscosity values) caused by replacing one weight percent of silica for particular oxides (Na_2O, CaO, Al_2O_3, etc.). To calculate absolute values of these characteristics in some or other concentration ranges they suggested to experimentally measure the viscosity of one glass belonging to this range and then to use the measured data as "reference points" for further calculations. For that, one only needed to present an arbitrary composition as a result of subsequent (rather small) replacements of silica by other components that were very easy to calculate. The idea of Gehlhoff and Thomas promised reducing the amount of experimental studies in new composition areas by orders of magnitude – if it would have been correct. Unfortunately, this idea turned out to be wrong.

Figure 7.2 presents the results of subsequent replacement of SiO_2 for the same weight fraction of B_2O_3 in sodium borosilicate glasses. As we can see from the figure, an addition of the first portions of B_2O_3 increases the viscosity whereas the next portions decrease it. In many other cases, replacement of one component by another leads to quite different changes of viscosity depending on a particular composition. As a result, any attempt to employ the mentioned approach in really wide composition areas seems inevitably to fail. A "silver bullet" was not found. Nonetheless, for the calculation of other properties such as density or refractive index (which Gehlhoff and Thomas also suggested to calculate in a similar manner), the described idea can be rather useful.

7.3.3
Gilard and Dubrul: Polynomial Models

If a simulated dependence is not linear but smooth the idea of polynomial (first of all, quadratic) approximation offers itself as the simplest approach. This idea was suggested by Gilard and Dubrul [447–449] to predict several properties of silicate glasses in wide concentration ranges.

It has to be noted, however, that the mentioned authors did not consider their approach in terms of polynomials. They only suggested considering the partial coefficients in the additive formula Eq. (7.1) as linear functions of concentrations:

$$p_i = k_{0,i} + k_{1,i} c_i . \tag{7.6}$$

In other words, they only tried to "improve" the additive formulas to consider the deviations of actual concentration dependencies of glass properties from these formulas. However, it is obvious that mathematically this modification means the same as a quadratic approximation without cross-terms:

$$P = \sum_{i=1}^{n} c_i (k_{0,i} + k_{1,i} c_i) = \sum_{i=1}^{n} k_{0,i} c_i + \sum_{i=1}^{n} k_{1,i} c_i^2 . \tag{7.7}$$

For comparison, the "classical" quadratic formulas can be presented as:

$$P = \sum_{i=1}^{n} c_i (k_{0,i} + k_{1,i} c_i) = \sum_{i=1}^{n} k_{0,i} c_i + \sum_{i=1}^{n} k_{1,i} c_i^2 + \sum_{i=1}^{n} \sum_{j=i+1}^{n} k_{i,j} c_i c_j . \tag{7.8}$$

The possible origin for ignoring just cross-terms in the formulas of Gilard and Dubrul is unclear. There are no physical or chemical reasons to assume that cross-terms $k_{i,j} c_i c_j$ in the latter formula are less important than the quadratic terms $k_{1,i} c_i^2$. Certainly this approach simplifies the final formulas but the reason of accepting *this* particular simplification is questionable.

Another feature of the model of Gilard and Dubrul seems to be more important. As far as the author knows, they were the first who specified concentration limits where the model was recommended to use. In principle, this innovation could be very important: it allowed one a distinction of the concentration range where the model can be used without serious risk of obtaining absolutely erroneous results. Unfortunately, for multiple reasons the authors of the model could not properly determine these limits. As we can see from Figure 7.3, within the specified concentration range the model in many cases results in huge errors of predictions. In fact, a specialist who would like to use this model for an arbitrary composition gets nothing useful from such specification.

After Gilard and Dubrul, the opinions of the authors of glass property models about specification of their applicability limits had varied. Some authors, for example, Appen [424], paid serious attention to a specification of these limits; others, for example, Demkina [440], did not consider it as reasonable to specify them at all. In some models (e.g., [443, 546, 547]) the limitations are specified not only

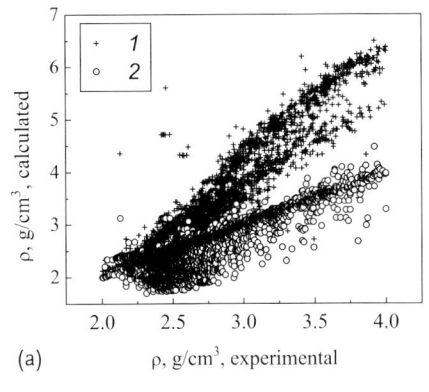

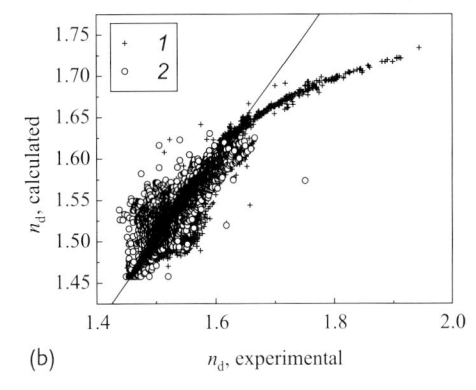

Figure 7.3 Comparison of predictions of the models by Winkelmann and Schott [385, 528–530] (1) and Gilard and Dubrul [447–449] (2) with experimental data from multiple authors presented in the SciGlass database [423] for glasses containing 50 and more mol% SiO_2. (a) Density at room temperature; (b) refractive index (n_d). The straight lines correspond to coincidence of experimental and calculated property values.

for concentration limits but also for their products and other derived characteristics.

Which of these or other existing approaches (if any) is correct? The question is not as simple as it might seem at a first glance. A more or less detailed consideration of this problem would require a special paper. Here the author can only state that any model has some applicability range, and for empirical models (including the models discussed in this paper) these ranges are usually more or less close to the boundaries of the data area in which the model parameters were determined. However, for complex multi-factor dependencies such as concentration dependencies of glass properties in multi-component systems this criterion cannot be considered as a general rule. The only reliable way of estimation of the prediction errors of one or another model consists in the comparison of its predictions with experimental data (desirably being obtained by several independent authors) in a corresponding composition area (see the contribution of O. Mazurin in Chapter 6 of this book for more details). The limits specified by the author(s) of a model can only give a general idea (which, as we saw, might not be correct) but cannot guarantee safe application of a particular model to a particular glass composition.

In general, the formulas of Gilard and Dubrul give lower systematic deviations than the formulas of Winkelmann and Schott do; nonetheless, the accuracy of predictions by both models is similar (see Figure 7.3). On the whole, the discussed approach did not allow one considerably improving the generality and accuracy of predictions compared to the "pure" additive approach.

7.4
Structural and Chemical Approaches

7.4.1
Nonlinear Effects and Glass Structure

The most known appearance of nonlinearity in concentration dependencies of glass properties is denoted as a boron (or boric) oxide anomaly. This term is widely used for description of multiple nonlinear effects in concentration dependencies of properties of glasses containing boron oxide. Historically this term was used for the first time by Biscoe and Warren [548] in 1938 for a description of the concentration dependence of the coefficient of thermal expansion in a sodium borate system. In Figure 7.4 we can see this dependence according to experimental data of multiple authors presented in the SciGlass system. After that, a lot of similar effects were found in concentration dependencies of various properties of borate and borosilicate glasses. Thus, the term boron oxide anomaly is now often used as a common name for multiple effects of this kind.

The authors of the mentioned paper [548] assumed that the boron oxide anomaly is caused by a change of the coordination transformation of a boron atom that can be threefold or fourfold coordinated with oxygen in oxide glasses. So, if one assumes that this transformation completes or changes the sign of about 20 mol% Na_2O then it is possible to explain the form of the dependence in Figure 7.4 as a result of a different expansion of $[BO_3]$ and $[BO_4]$ groups.

Further studies ([534, 541], etc.) disproved the above given explanation. It was found that the $[BO_4]/[BO_3]$ ratio really has a maximum in the sodium-borate system but this maximum corresponds to ~33 mol% Na_2O, not ~20 mol% as seemingly follows from the above plot. Nonetheless, the general idea of structural changes as a main reason of nonlinear concentration dependencies of glass properties was not

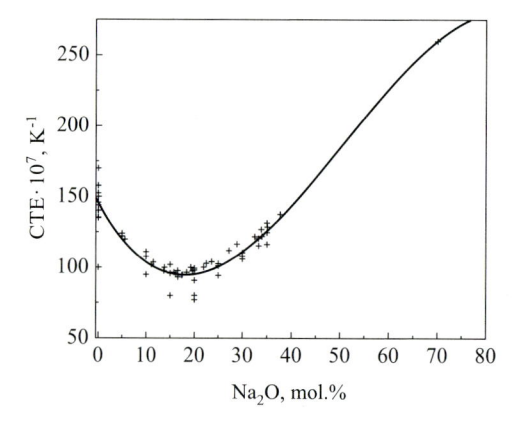

Figure 7.4 Concentration dependence of the linear coefficient of thermal expansion of glasses in the system Na_2O–B_2O_3 according to the data by multiple authors presented in the SciGlass Information System [423].

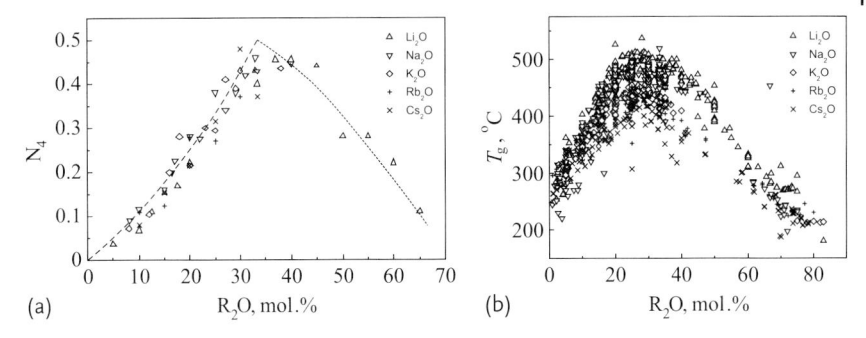

Figure 7.5 Concentration dependencies of T_g and N_4 in the systems R_2O–B_2O_3: (a) dependence of N_4 by using the data from [533–544]; (b) concentration dependence of T_g by using the data presented in SciGlass [423].

erroneous. A strong correlation of the coordination state of glass-forming cations with the form of concentration dependencies of glass properties was found for a lot of glass properties in multiple systems. As an example, Figure 7.5 demonstrates concentration dependencies of the glass transition temperature, T_g, in binary alkali-borate systems together with the available results of NMR studies. In the latter studies the authors measured the value of N_4:

$$N_4 = \frac{[BO_4]}{([BO_3] + [BO_4])}, \tag{7.9}$$

where $[BO_3]$ and $[BO_4]$ are the mean concentrations of the corresponding structural groups in the glass. As we can see from the figure, concentration dependencies of these characteristics (T_g and N_4) are very similar.

As for thermal expansion of sodium-borate glasses, the form of its concentration dependence shown in Figure 7.4 is caused by the interaction of several structural factors, and that is why the minimum of the expansion coefficient does not correlate with the maximum of N_4. A structural explanation of the dependence plotted in Figure 7.4 that is consistent with available structural data can be found in multiple papers, for example, in [548]. For us, the most important is the fact that the specific features (like extremes and breaks) in concentration dependencies of glass properties can be explained in terms of structural changes, in general, and coordination changes, in particular. This idea was widely used by the authors of multiple models developed for prediction of glass properties from chemical compositions.

It is important, however, to underline that a deviation from linearity is not always structurally caused. Indeed, in the general case, there is no physical reason to consider the concentration dependence of "a glass property" linear even in a system without any structural changes. For example, the density of an ideal mechanical mixture is not a linear function of molecular or weight percentages of components but only of volume percentages; if we use other kinds of percentages the dependence can be more or less close to a linear one but cannot be exactly linear (this question will be discussed below in more detail). If we consider a property such as viscosity or its logarithm, there is no physical reason at all to consider its concentration dependencies linear in any system by using any kind of percentage. Strictly speaking, there is only one case when the effect of structural (or chemical) factors on physical properties of glasses (not concerning the liquidus temperature) is obvious: if we have extremes or breaks in concentration dependencies of properties. However, in very many practical cases the "structural" viewpoint provides a clear explanation of observed nonlinear effects and, what is more important, simplifies the mathematical description of these effects.

7.4.2
Specifics of the Structural Approach to Glass Property Prediction

Prior to a more detailed discussion of the structural approach the author has to stipulate what this means because the studies discussed in this chapter have a significant difference from typical structural studies.

Typically, the term structural approach is applied to a model that is based on some theoretical concepts and/or experimental data about structure, and the main destination of such a model consists in a qualitative or quantitative description of some structural characteristics such as the distribution of structural groups, coordination numbers, angles, bond lengths and energies, and so on.

Sometimes such models contain some equations for the calculation of some physical properties as well, for example, glass densities [549, 550] or melt viscosities [551]. However, regardless of the authors' words about such calculations, these equations are not developed for practical prediction of glass or melt properties but for *verification of structural models* by using external data. For this purpose, it is generally quite sufficient to compare the model predictions with property data for glasses or melts of the simplest compositions, that is, binary or ternary (very rarely, quaternary) glasses. The extension of such models to more complicated systems is not considered in these studies, and in fact – for multiple reasons – the simplest glass-forming systems remain the only object of application of reported "property models."

At the same time, any practical application of models which claims the prediction of glass properties assumes the applicability of the model to multi-component glasses, that is, glasses containing 5–10–15 and more components *together*. Glasses of simpler compositions mostly (with only few exceptions) have no practical applications. In other words, the structural models (in general understanding of this term) cannot be directly applied to the practical prediction of glass properties from their chemical compositions and, thus, are outside the scope of our consideration; some of these concepts and models will be mentioned below but not analyzed in detail.

In this chapter, we discuss the models that do allow practical predictions. These models are, strictly speaking, not structural models. Their authors only tried to implement some structural concepts into their empirical formulas describing relationships between chemical compositions and physical properties of glasses.

7.4.3
First Trials of Application of Structural and Chemical Ideas to the Analysis of Glass Property Data

The observation of the boron oxide anomaly in [429] was not the first study where the glass structure was discussed. As far as the author knows, the idea to treat glasses as an under-cooled liquids was developed already by Mendeleev (see [552, 553]). This idea was further developed by Tool [69] who proposed a first quantitative model for calculation of changes of glass properties when heat-treating within the glass transition range, based on the concept that glass is a substance whose structure continuously deviates from equilibrium when cooling. The idea of chemical equilibria in glass-forming melts was then used in multiple papers; for example, in 1947, Yesin [554] considered some particular chemical equilibria as major factors determining structure and viscosity of glass-forming melts.

At the beginning of the last century, more detailed assumptions about glass structure and its constituents were made. In 1921, Lebedev [477] suggested to consider the glass structure as consisting of "crystallites," very small multi-atomic groups similar to the lattice units. An alternative hypothesis was suggested in 1932 by Zachariasen [555] who considered the structure of silicate glasses as a continuous random network composed of $[SiO_4]$ tetrahedra. Both of these hypotheses were discussed individually and opposed each other in many publications over the course of the twentieth century. Another view to glass structure was anticipated in 1930 by Schtschukarew and Müller [556] who considered glass as consisting of "polar" and "nonpolar" parts and tried to use this concept for an explanation of the observed concentration dependencies of electrical conductivity of glasses. Later by using this concept Müller suggested a model for calculation of the heat capacity of glasses [488, 489].

The ideas suggested in these and other papers were used later when developing various models for prediction of glass properties from chemical compositions. However, these studies themselves (except for [489]) as well as many other studies published later (e.g., [557–561]) were not aimed at predicting multiple glass properties of multi-component glasses in wide concentration ranges. The suggested models could be directly applied only to the simplest glass-forming systems. In the models designed for practical application in glass industry, structural and chemical approaches were developed in a somewhat different way. Below we consider the most important of these models.

7.4.4
Evaluation of the Contribution of Boron Oxide to Glass Properties

The coordination number of boron atoms is the most important structural factor that affects glass properties. The extraordinary importance of this factor is caused by specific influence of boron oxide on glass properties. Addition of this component allows for the improvement of multiple characteristics of oxide glasses: it increases chemical durability, prevents crystallization, increases thermal shock resistance, improves insulating properties, and so on. Also, addition of boron oxide changes multiple glass properties to a large extent, often in opposite directions depending on glass composition: it increases or decreases viscosity, refractive index, density, and so on. Due to these specific features, boron oxide is widely used in multiple kinds of commercial glasses: optical glasses, insulators, glassware, chemical ware, medical equipment, and so on.

All of the above-mentioned effects are mostly caused by changing the coordination number of boron atoms. For example, Appen [424] determined the partial molar volume of B_2O_3 to be equal to $\sim$38 and $\sim$18 cm^3/mol for threefold-coordinated and fourfold-coordinated boron, respectively; that is, change of coordination number from 4 to 3 *more than doubles* the partial density of boron oxide in glass. Thus, the evaluation of the coordination state of boron atoms in oxide glasses is a key point for any accurate prediction of multiple properties of various commercial glasses.

7.4.4.1 Model by Huggins and Sun

As far as the author knows, the first people who attempted to use the idea of boron coordination for the prediction of properties of multi-component glasses were Huggins and Sun [463, 464, 526]. Their model estimates the N_4 value from a linear (i.e., additive) formula with implementing a natural limitation $0 \leq N_4 \leq 1$ (i.e., negative values are considered as zeros and the values larger than one as unity). Their model then calculates glass properties ascribing different partial coefficients to B_2O_3 with threefold and fourfold coordination of boron.

Unfortunately the mentioned model could not properly predict the properties of glasses containing boron oxide. Figure 7.6 compares model predictions with experimental data for hundreds of glasses containing a considerable amount of B_2O_3, reported by many dozen authors. From the figure we can see that the results of predictions systematically differ from the experimental data. Figure 7.7 compares the results of model prediction with experimental N_4 values [535, 538–540, 543, 544] measured by the NMR method. As we can see, the calculated concentration dependence of N_4 does not coincide with the experimental data (we will return to this figure below).

The contradictions between model predictions and experimental data are caused by multiple factors. Probably, the most important of them is the lack of available experimental data for both N_4 values and simulated properties (density, refractive index and mean dispersion).

7.4.4.2 Models by Appen and Demkina

The next major step in the prediction of glass properties from chemical compositions was made by the Soviet scientists Appen and Demkina. The first publications of their models in open literature appeared in the 1940s [562, 563], and final variants of models were published in the 1970s [424, 440]. The model of Appen was somewhat further improved and extended by his follower Gan Fuxi [446] and Demkina's model was further developed by Shchavelev [521–524].

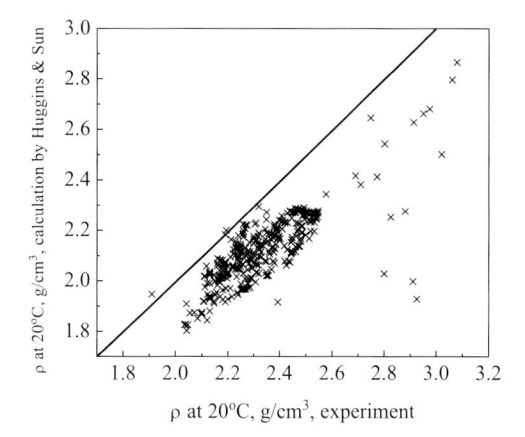

Figure 7.6 Comparison of results of prediction of glass properties by the Huggins and Sun model [463, 464] with experimental data for density of glasses containing ≥ 20 mol% B_2O_3. Points correspond to experimental data presented in the SciGlass database [423] for glasses whose compositions are located inside the concentration limits specified by the authors of the model. The *solid line* shows where the calculated and experimental property values would be identical ($Y = X$).

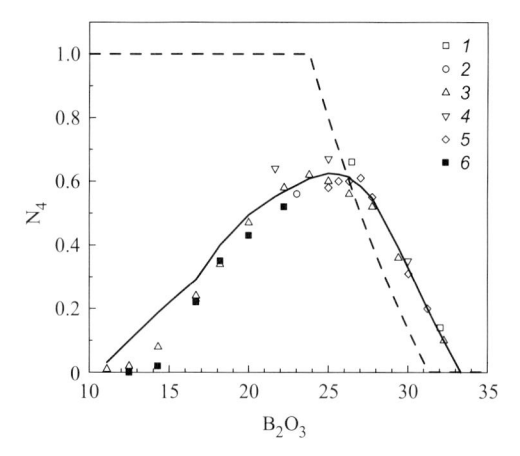

Figure 7.7 Concentration dependence of N_4 for glasses of composition (mol%) $(100 - 3x)Na_2O \cdot xB_2O_3 \cdot 2xSiO_2$. Points correspond to experimental data taken from [544] (1), [538] (2), [539] (3), [540] (4), [543] (5), [535] (6). *Dashed and solid lines* correspond to the results of prediction by the models by Huggins and Sun [463, 464] and Priven [503]. At the abscissa, the molecular percent of B_2O_3 is given.

The approaches used by Appen and Demkina were somewhat different in detail but the basic idea was the same. They considered the chemical nature of interaction of boron oxides with other components (glass modifiers) causing a change of boron coordination. Both of them considered the transformation B^{III} to B^{IV} and vice versa (where the Roman numbers mean coordination numbers of boron atoms) as a result of chemical interaction of boron oxide with the network-modifying oxides

R_2O and RO that can be (in the simplest form) presented as

$$2[BO_{3/2}] + MO \Rightarrow 2[BO_{4/2}]^- + M^{2+}, \qquad (7.10)$$

where $[BO_{3/2}]$ and $[BO_{4/2}]^-$ mean boron-oxygen triangles and tetrahedrons (fractional indices mean that each oxygen atom belongs to two net-former polyhedra) and M means one bivalent or two monovalent cations.

Appen and Demkina also considered similar reactions of modifying oxides with Al_2O_3 which transform octahedra $[AlO_{6/4}]$ to tetrahedra $[AlO_{4/2}]^-$:

$$2[AlO_{6/4}] + MO \Rightarrow 2[AlO_{4/2}]^- + M^{2+}. \qquad (7.11)$$

From multiple source data they came to the conclusion that in the case of the lack of modifying oxides, Al_2O_3 always "wins" the "fighting" with B_2O_3 for the modifier's oxygen. In other words, aluminum oxide always bounds as much oxygen as possible, and only the remaining can be used for building $[BO_{4/2}]$ tetrahedra.

To incorporate this kind of understanding in practical models, Appen, Gan Fuxi and Demkina used the so-called oxygen number that can be presented as

$$\psi = \frac{\sum_i m_i \psi_i - m_{Al_2O_3}}{m_{B_2O_3}}, \qquad (7.12)$$

where m_i are the mole fractions of modifying oxides (R_2O and RO), $m_{Al_2O_3}$ and $m_{B_2O_3}$ are the mole fractions of Al_2O_3 and B_2O_3 respectively, and ψ_i are coefficients determining the "power" of the corresponding modifier oxides which react with aluminum and boron oxides. After calculating the ψ value Appen and Demkina calculated the partial values determining the contributions of boron oxide to physical properties by using empirical (nonlinear) formulas. Although these formulas and their parameters in these two models are somewhat different the final results of calculations are generally very similar (see Figure 7.8 as an example). From the same figures we can see that both of these models predict the properties of borosilicate glasses much better than the above-mentioned model of Huggins and Sun [463, 464].

The possibility of an accurate description of the influence of boron oxide on multiple glass properties by using a very simple general approach (and rather simple calculation formulas) promised a quick resolution of a variety of problems concerning the prediction of glass properties from compositions. The most attractive was the idea of using this approach for the prediction of the viscosity and some other properties (so-called transport properties, mainly electrical conductivity and diffusivity) by employing a more or less similar formalism. Unfortunately, this idea became unproductive. Appen, Demkina and their followers tried to apply their ideas to the viscosity prediction, but the only results of their work were new experimental data. A model that could be applied to a more or less wide range of concentrations was not suggested. The "silver bullet" was not found again.

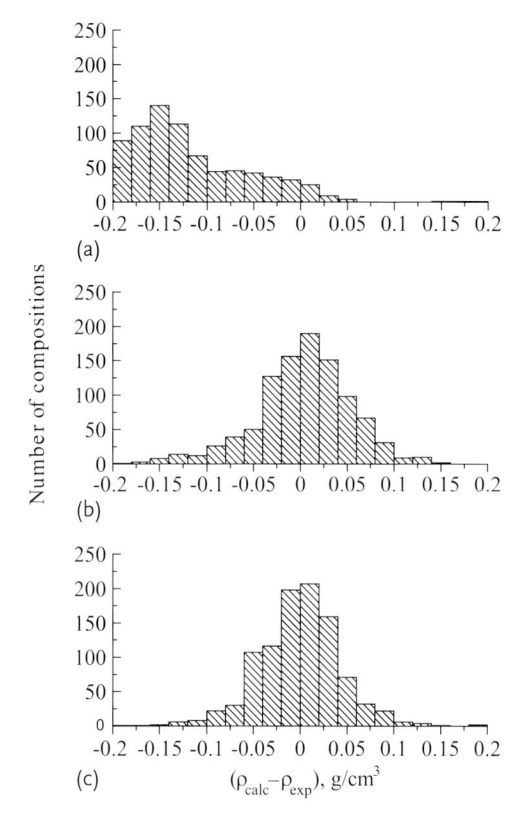

Figure 7.8 Comparison of predictions of models by Appen [424] (a); Demkina [440] (b) and Huggins and Sun [463, 464] (c) with the same array of experimental data about glass densities, that is, 1005 data points measured by multiple authors presented in the SciGlass [423] database. Each glass composition lies within the concentration limits specified by the authors of all models. The concentration of SiO_2 varies in all glasses from 45 to 80 mol% and B_2O_3 from 10 to 40 mol%.

X-axis: $(\rho_{calc} - \rho_{exp})$, g/cm^3
Y-axis: Number of compositions

7.4.5
Use of Other Structural Characteristics in Appen's and Demkina's Models

The authors of the above-mentioned models tried to use other structural characteristics to improve the accuracy of predictions and widen the applicability ranges of their models. In particular, Appen [424] analyzed in this respect the fraction of bridging oxygen in the silicon–oxygen glass-forming network and the mean number of bridging oxygen atoms in a [SiO$_4$] tetrahedron. Demkina ([438–440, 563], etc.) tried to evaluate the compositions and concentrations of multi-atomic structural groups which were assumed to exist in glasses. However, contrary to the change of boron coordination, the structural meaning of other "structural" characteristics used in the calculation formulas is disputable. For example, Appen estimated the connectivity of the glass-forming network as a function of molecular percent of SiO_2, without consideration of obvious (and easy-to-estimate within Appen's approach) contributions of Al_2O_3, B_2O_3 and other network-forming oxides. This simplification is difficult to explain from structural positions.

In the author's opinion, the only ingredient in Appen's and Demkina's formulas that has some relation to structure is considering the interaction of network modifiers with aluminum and boron oxides. As to other elements of their algorithms, they look "purely empirical" more than structural. However, two mentioned kinds of interactions are critical for property simulation, and their proper consideration (even approximate) allowed the mentioned authors to seriously improve the generality of the models and the accuracy of the predictions.

7.4.6
Recalculation of the Chemical Compositions of Glasses

There is one more distinctive feature that has some relation to structural factors: the kind of percentage used in calculations by additive formulas.

It is worth noting that the simulations considered in this chapter were aimed from the very beginning at a practical use of the models. Glass technologists used to express glass compositions in weight percentage; therefore, during the decades after Winkelmann and Schott this kind of percentage was nearly the only one that was used in additive formulas. At the same time, it is clear from very general considerations that the deviations of the concentration dependencies of glass properties from straight lines significantly depend on the kind of percentages used.

Obviously, the concentration dependence of the same property cannot be equally linear when expressing the compositions in different kinds of percentages. Figure 7.9 well confirms this statement, for lead-silicate glasses the components (PbO and SiO_2) of which have very different molecular weights. The dependence is rather close to a straight line (although not exactly linear) when using molecular percentages but has considerable curvature when using weight percentages.

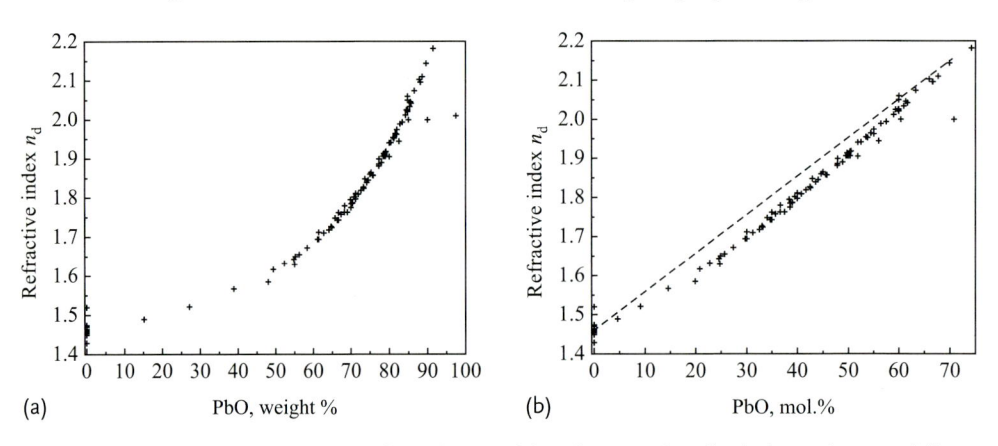

Figure 7.9 Concentration dependencies of the refractive index of lead-silicate glasses in different kinds of percentages: weight % (a) and molecular % (b). The *dashed line* is drawn to show the curvature.

Appen and Demkina were the first two scientists who suggested using other measures of composition than weight percentages. Their choices were, however, somewhat different. Appen used molecular percentage as a very clear and easily understandable parameter that often makes concentration dependencies of glass properties rather close to linear ones (like the one shown in Figure 7.9b). The remaining curvature was taken into account by introducing nonlinear terms, in most cases, restricting the relations to a quadratic approximation. Demkina was tending initially to use volume percentage. She was primarily interested in the prediction of refractive indices, and knew that in an ideal solution the refractive index is a linear function of concentrations expressed in volume percent. So, she expected that if one expresses the concentrations of oxides in volume percentages then the concentration dependencies of the refractive index of glass would be linear.

In theory, the formulas for conversion from weight and molecular percentages to volume percentages are simple and well-known:

$$C_{i,\text{vol}} = 100 \frac{C_{i,\text{wt}}/\rho_i}{\sum_{k=1}^{n} C_{k,\text{wt}}/\rho_k} = 100 \frac{C_{i,\text{mol}} v_{m,i}}{\sum_{k=1}^{n} C_{k,\text{wt}} v_{m,k}}, \qquad (7.13)$$

where i is the index of a component the percentage of which we need to convert, k are indices of all glass components (including the ith component), n is the number of components in a glass, C_{vol}, C_{mol} and C_{wt} are volume, molecular and weight percentage, respectively, ρ is the partial density, and v_m is the molecular volume of a component. So, if we know the partial densities or molecular volumes of glass components then we can calculate their volume percentage from chemical composition. The problem is however that it is impossible to experimentally determine the partial densities or molecular volumes of a particular glass component. So, to go this way we need to make some assumptions having no direct confirmation.

Demkina's assumption was as follows: *if* one finds such a set of coefficients that when substituted for molecular volumes, they make the concentration dependencies of properties virtually linear, *then* these coefficients *are* (with an accuracy to a constant multiplier) "true" molecular volumes. Thus, she found a series of empirical parameters that made concentration dependencies of glass properties (in the composition range she was interested in) nearly linear and *named* the results of conversion "volume percentage."

From a physical point of view, this was a serious mistake. Indeed, her calculations had no relation to the determination of partial volumes or densities of glass components and, strictly speaking, these volume percentages could not be verified by any experiment at all. Moreover, when trying to give theoretically a foundation for these calculations Demkina made several curious mistakes. In particular, she replaced molecular masses in the calculation formulas by some "structural masses of silicates" in order to consider the "actual compositions of structural units existing in glasses," but "forgot" to subtract the corresponding amount of "bounded" silica from the total percentage of SiO_2 substituted in additive formulas and, thus, considered the contribution of silica twice. Moreover, even for glasses containing no silica at all, she still considered "silicates" as structural units. These and other

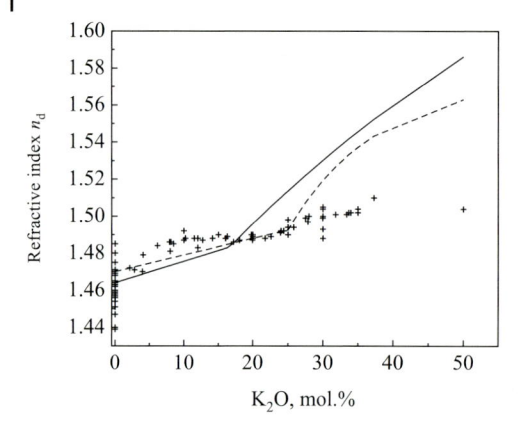

Figure 7.10 Concentration dependence of the refractive index in the system $K_2O-B_2O_3$. Points correspond to experimental data of multiple authors presented in the SciGlass database [423]; the *solid line* shows the predictions of the Demkina [440] model, and the *dashed line* shows the predictions of Appen's model [424]. Note that Appen recommended the use of his model only in a limited concentration range of silicate systems; borate glasses presented in the figure contain no silica and, thus, are far outside of this range.

examples of confusion in Demkina's theoretical reasoning and calculation formulas were discussed by Appen. In fact, as Appen stated in [424], Demkina simply introduced one more set of empirical parameters having no other meaning than fitting experimental concentration dependencies of glass properties.

However, let us put aside physical theories and focus on practice. Indeed, the main goal of Demkina was the accurate prediction of optical properties of glasses, not theoretical findings. So, if the parameters of an empirical model are properly determined (regardless of their meaning – the parameters of empirical formulas often have no inherent physical meaning at all) then we could expect accurate predictions. In many practical cases, especially considering borosilicate glasses, Demkina's model actually gives very accurate results. So, can we say then that finally a "silver bullet" was created now in spite of all the "physical" data that disprove Demkina's model? The answer is no again. This is shown in Figure 7.10 where the refractive index of binary potassium borate glasses is presented.

As we can see from the figure, Demkina's predictions have little in common with experimental data. For glassy potassium metaborate (50 mol% K_2O), the error of prediction reaches about 0.1, which is a catastrophic error for this property. However, why do we consider this system at all? Indeed, Demkina was focused on silicate glasses and did not try to fit borate systems. So, this result seems to be located far away from the applicability area of her model. But, this is not true. Such conclusion would contradict Demkina's approach as a whole. Demkina insisted that she used "true volume percentages" which are equally applicable to any glass-forming system including nonsilicate systems as a particular case. That is why she paid no attention to the specification of the applicability range of her model: indeed, why should one restrict artificially the application of a "universal" model? Thus, the concentration dependence plotted in Figure 7.10 should be considered as a serious error *within* the applicability range specified by the author of the model. In fact, paying no attention to the specification of the concentration limits Demkina only said that she *had no idea* about the real applicability range of her model.

In this connection, it is worth noting that Appen tried to carefully specify the applicability range of his model. Not all of his specifications are exact: for example, some parameters for MgO are recommended for use in "some practical area of glass compositions containing sodium oxide." Formulas for alkali oxides, lead oxide and even silica also contain some (not critical) uncertainties. However, basic concentration limits were defined clearly. Numerous comparisons of predictions of his model with experimental data (totally, the author of this paper verified Appen's model by using property data for *several dozen thousand glasses* presented in the SciGlass database [423]) show that within the concentration range specified by Appen, his model nearly everywhere gives accurate predictions. In other words, when using this model within this concentration range we are almost sure in correct results. In many other cases the results of Appen's predictions are more or less accurate even far outside of this (rather broad) applicability range; Figure 7.10 clearly demonstrates this feature. Below we will return to the problem of concentration limits of calculation models.

Now let us come back to the use of volume percentages in calculation formulas. In 1965, Yakhkind [557] suggested a very simple solution of the problem. He substituted Eq. (7.13) into the volume-additive formula for a property, P,

$$P = \sum_{i=1}^{n} C_{i,\text{vol}} P_i = \frac{\sum_{i=1}^{n} (C_{i,\text{wt}}/\rho_i) P_i}{\sum_{i=1}^{n} (C_{i,\text{wt}}/\rho_i)} . \tag{7.14}$$

Here $C_{\text{vol},i}$ is concentration of the ith component in volume percent and ρ_i is its partial density. The denominator of the right part of Eq. (7.14) gives the reciprocal density of glass, or specific volume, $1/\rho$, which in an ideal mixture is a weight-additive property. Thus, we can rewrite Eq. (7.14) as:

$$P = \frac{\sum_{i=1}^{n} C_{i,\text{wt}}(P_i/\rho_i)}{1/\rho} = \rho \sum_{i=1}^{n} C_{i,\text{wt}} \left(\frac{P_i}{\rho_i}\right). \tag{7.15}$$

After dividing both parts by the density, ρ, we finally obtain the equation:

$$\frac{P}{\rho} = \sum_{i=1}^{n} C_{i,\text{wt}} \left(\frac{P_i}{\rho_i}\right). \tag{7.16}$$

This very simple equation looks impressive. Indeed, after an identical mathematical transformation the volume-additive formula with partial volumes, which are problematic to estimate from experiment even approximately, is converted into a traditional weight-additive formula. Indeed, if we *only* know the density of glass, ρ (determined by any number of ways including direct experiment) then we can

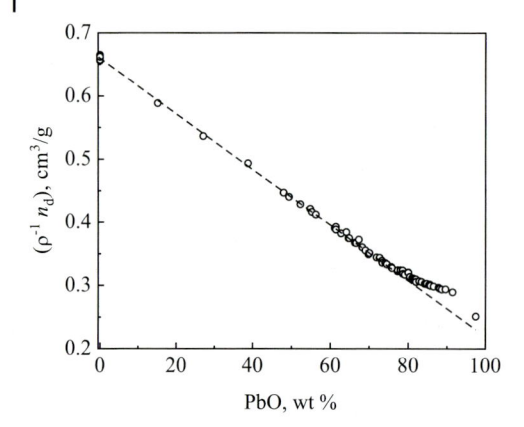

Figure 7.11 Concentration dependence of refractive index n_d divided by density ρ in a lead-silicate system. Points correspond to experimental data of multiple authors presented in the SciGlass database [423]. The *dashed line* shows a linear approximation in the concentration range 0–80 wt% PbO.

calculate the ratio (p/ρ) by using a weight-additive formula. We do not need to know partial volumes or partial densities of components but need only partial coefficients, $(p/\rho)_i$, which, at least for simple systems, can be determined from clear single-meaning experiments.

Figure 7.11 demonstrates this possibility in the lead-silicate system. In the concentration range from 0 to 80 weight% (∼0–52 mol%) PbO, the dependence of the ratio (n_d/ρ) on weight percentage of PbO is nearly an ideal straight line (cf. Figure 7.9b). The plot shown in Figure 7.11 looks like the "lead anomaly" (indeed, contributions of PbO to most physical properties of glasses are usually denoted as "difficult to consider") is just a result of the use of an improperly chosen coordinate system.

In 1988, the author of the present chapter tried to develop a method of calculation of the refractive index in multi-component glasses [511] based on Yakhkind's formula, Eq. (7.16). This attempt was rather successful. However, comparison with multiple experimental data presented in the SciGlass database [423] has shown that the accuracy of prediction by this method is generally the same as by Appen's method [424], which uses traditional molecular percentages.

In fact, the idea of the use of special coordinates making concentration dependencies of glass properties simpler was suggested long ago. This is the idea of "refraction," a function of the refractive index that can be considered as a sum of "partial refractions" of components. In particular, this idea was employed by Young and Finn [531] in 1940: these authors used the refraction expressed as $(n-1)/\rho$ that was considered as an atomic-additive function of chemical composition. They did not suggest the model for a calculation of density, and thus it is impossible to directly compare the values of refractive index predicted by their model with experimental data. However, they also suggested the refractions for the mean dispersion $(n_F - n_C)$ that allows one to directly calculate Abbe's number, $\nu_d =$

$(n_d - 1)/(n_f - n_c)$:

$$v_d = \frac{n_d - 1}{n_F - n_C} = \frac{(n_d - 1)/\rho}{(n_F - n_C)/\rho} . \qquad (7.17)$$

This value can be compared with numerous experimental data and, thus, allows one the evaluation of this idea in practice. A comparison with the experimental data presented in SciGlass has shown that the accuracy of predictions by the model of Young and Finn is considerably worse than by the models of Appen and Demkina.

So, why does the idea of linearization of concentration dependencies of physical properties by using optimal units of chemical compositions and/or property values (volume percentage, refraction, etc.) not work? If it nevertheless works, why does it not help to increase the accuracy of property predictions as compared to Appen's model which uses improper molecular percentages? Indeed, although no oxide glass consists of molecules of SiO_2, PbO, P_2O_5, and so on, Appen's model based on molecular additivity gives the most accurate property predictions – the question is why? Also, why do multiple mistakes and internal inconsistencies in Demkina's model not prevent the successful use of this model for practical calculations, at least for traditional optical glasses? And, finally, which kind of concentration units can be recommended for future simulations – and is this factor not significant at all?

It is quite unlikely that somebody could give a simple, single and solely correct answer to these questions. The problem is not trivial, and the solution probably as well.

Which kind of percent is optimal? In general, the best choice is the kind that linearizes concentration dependencies of a given property in an ideal mixture. This kind depends on the physical nature of the calculated property and, in some cases, on the unit in which this property is expressed: for instance, the mean molar volume is a mole-additive function, specific volume is weight-additive, and density is volume-additive, although all of them present the same physical characteristic, the ratio between weight and volume of a given substance.

In this connection, the author believes that such frequently used units as molecular percentage of oxides can be considered as optimal units for simulation of only those properties which are directly related to the amount of substance expressed in moles of oxides, for example, the mean molar volume. For any other properties, molecular percentages are not the optimal concentration unit. Indeed, glasses are not molecular substances. Compositions of their structural units, whatever they would be, are not identical to the molecular formulas of oxides. Indeed, why should we use, for example, a mole of SiO_2 and P_2O_5 but not Si_2O_4 and $PO_{2.5}$? There are no physical or chemical reasons for any of these choices, each of them is not more than a convention and, thus, cannot have essential advantages to alternative choices. The only advantage of molecular percentage is that the use of these concentration units, in general, simplifies the consideration of structural and chemical factors as compared to weight or volume percent. Indeed, chemical reactions are much easier to calculate in molecular units than in weight or volume ones.

Before the invention of the computer, this factor could have played a significant role in a proper simulation of concentration dependencies of glass properties, because it was more convenient and much less laborious to express the concentrations in the same units when considering structural factors and when calculating properties themselves. However, the author of the present chapter believes that nowadays any simulation of this kind must be oriented to the use of computers. In this case, any transformation between different kinds of percentages can be programmed in a minute and performed in a microsecond. Thus, there no longer exists any reason for using the same kind of concentration units in all cases. The authors of new models should base their choices on those units which are preferable for a particular set of properties calculated, instead of making the same choice for all possible computations.

At the same time, in the author's opinion, simplification of concentration dependencies of glass properties by using corresponding transformations for both chemical compositions and physical properties is not a panacea. Glasses are not ideal mixtures; so, in the general case, the concentration dependencies of their properties would not become linear when using *any* coordinate system. To gain an advantage from the use of an optimal coordinate system one has to properly consider more important factors such as the difference of structural roles of particular components in glasses in different composition areas. Otherwise, the use of an optimal kind of percentage in additive formulas might not yield the desired effect.

7.4.7
Use of Atomic Characteristics in Glass and Melt Property Prediction Models

In Appen's papers (see [424] and references therein), the idea of similarity of contributions of chemically similar oxides (like Na_2O and K_2O) to glass properties is widely used to explain various facts. However, Appen did not use this idea for determining the numerical values of parameters of his model: all of them were derived from the experimental concentration dependencies of glass properties.

Other authors ([451, 456, 483, 490, 519, 520, 564–569], etc.) tried to use the characteristics of particular atoms forming glass structure (atomic radii, lengths and energies of bonds, coordination numbers, fractions of bridging and nonbridging oxygen atoms, etc.) for calculation of model parameters. The models were developed for the determination of Young's modulus [483], the coefficient of thermal expansion [490], the viscosity [451, 564–566] and some other properties.

Comparison of the results of predictions by these models with experimental data stored in the SciGlass database [423] shows that usually (although not always) general trends of concentration dependencies of simulated properties are reflected properly; however, the accuracy of predictions is generally much worse than that for the empirical models described above. In the author's opinion, it is caused by too much serious simplification of the effects of particular components on glass and melt properties. At the same time, the models of this type, with no or very few fitting parameters, can be used for a rough estimation of property values in very broad composition areas including those parts that are not well-studied yet. In other words, this approach has good chances to be general, but it is unlikely that it will be as accurate as the traditional empirical approach.

7.4.8
Ab Initio and Other Direct Methods of Simulation of Glass Structure and Properties

The idea of calculating physical properties of substances by using models directly derived from "first principles" seems to be very attractive. With some more or less verisimilar assumptions, it is possible to simulate atomic arrangements and motion and, thus, to get an idea about the probable glass structure. Properties such as density, viscosity, glass transition temperature, expansion coefficient, heat capacity and others can be directly calculated from this information. The only information required for such simulations is a rather short set of atomic characteristics.

The *ab initio* approach is (with some stipulations) used in molecular dynamics and Monte-Carlo methods. Some authors ([570–575], etc.) tried to employ this method for prediction of properties of glasses and glass-forming melts. In these studies, the accuracy of property predictions and the covered concentration ranges were not as good as for the above-considered empirical models, and the author has serious doubts that in the foreseeable future it would be possible to reach the practical value comparable to these models. However, these simulations might give an idea about the *main structural factors* that could affect the physical properties of glasses in a new composition area that was not studied yet. In particular, the molecular dynamic simulations might help to determine the compositions of the most probable structural groups, most probable coordination numbers of different atoms, and so on.

Another approach was suggested by Shakhmatkin and Vedishcheva ([576, 577], etc.) who tried to derive the properties of glasses and glass-forming melts from thermodynamic simulations. They considered a glass-forming melt as a mixture of structural units that are in chemical equilibrium with each other; correspondingly, the equilibrium concentrations of these structural units can be calculated from their thermodynamic potentials. Then Shakhmatkin and Vedishcheva made the assumption that the mentioned structural units are identical to the existing crystalline phases with respect to their thermodynamic potentials and physical properties. Finally, they assumed that in a solid glass each kind of structural units is presented in a concentration corresponding to thermodynamic equilibrium at the glass transition temperature, T_g.

Provided the basic concept is correct and above-mentioned simplifications and assumptions can be accepted, then the only requirement for the prediction of glass properties would consist of the knowledge of the thermodynamic potentials and physical properties of the corresponding crystalline phases. Thus, *in theory* it is possible to calculate the properties of an *infinite number* of glasses from the properties of a *finite number* of crystalline phases. However, currently this approach can be applied only to some binary and a limited number of ternary systems for which the required thermodynamic information is available. In nearly all multi-component systems, the available thermodynamic data are not complete and thus the application of this approach is not possible.

7.4.9
Conclusion

Among the models developed in the twentieth century for the calculation of most of the glass properties from chemical compositions, structural models were surely the best predictive tools. The use of rather simple empirical equations having an approximate structural background allowed one a proper description of the general trends of changing multiple physical properties depending on chemical composition as well as rather accurate prediction of properties of particular glasses from their compositions. An advantage of this approach is the use of the same or very similar formulas for considering the influence of structural factors on multiple properties. Indeed, within the framework of the above-discussed structural approach, different properties are considered simply as different appearances of the same major factor making concentration dependencies of glass properties nonlinear: this factor is glass structure.

Probably in the twentieth century this approach was applied most consecutively in the models of Demkina [440]. However, some methodological mistakes in her approach (see [424, 578]) reduced the value of her efforts considerably. As a result, these models are usually a little bit worse than the models of Appen [424]. Anyway, the models of Appen, Demkina and their followers [444–446, 521–524] are the most general and, in most cases, the most accurate among the models developed in the last century. The most serious drawback of their approaches is the impossibility to predict the transport properties, the most important of which is the viscosity of glass-forming melts. The formalism of Demkina's and Appen's models was too simplified to describe concentration dependencies of viscosity characteristics which are generally much more complicated than those for density, refractive index and other properties calculated by their models. So, let us now discuss approaches that simulate the viscosity of glass-forming melts.

7.5
Simulation of Viscosity of Oxide Glass-Forming Melts in the Twentieth Century

7.5.1
Simulation of Viscosity as a Function of Chemical Composition and Temperature

When predicting the viscosity of glass-forming melts, it is equally important to consider concentration and temperature dependencies. Both chemical composition and temperature greatly affect viscosity changing it up to 15–20 orders of magnitude.

The general approach to the simulation of viscosity is the same in all known models. This approach assumes that the temperature dependencies of viscosity are similar for all glass-forming substances (more exactly, temperature dependencies of viscosity are similar for nearly all known substances; however, in this paper only glass-forming substances are considered). Based on this similarity, it can be approximated in a very wide viscosity range (from $\sim 10^0$ to $\sim 10^{13}$ Poise and even more) within $\pm 0.03\ldots 0.06$ units of decimal logarithm (the accuracy of a good experiment) by the Vogel–Fulcher–Tammann (VFT) equation,

$$\log \eta = A + B/(T - T_0) . \tag{7.18}$$

Here η is the viscosity, T is temperature, and A, B and T_0 are empirical constants. Thus, if one determines these three constants for some temperature or viscosity range he/she can calculate the viscosity at any particular temperature in this range. Above $\sim 10^{13}$ Poise the viscosity of glass-forming substances considerably depends on "thermal history," that is, preliminary heat treatment. This dependence is caused by a deviation of the thermodynamic state from equilibrium when heat-treated with the usual rates (0.1 to 10 K/min). Using a special regime of heat treatment, so-called stabilization, allows one to achieve the equilibrium state where the temperature dependence of viscosity matches the Vogel–Fulcher–Tammann equation up to the highest values that are possible to measure ($\sim 10^{16}$ Poise and even more). However, such measurements are technically very difficult and performed very rarely (cf. also Chapter 3).

In theory, the constants A, B and T_0 are functions of chemical composition and their values can be calculated similarly to any other glass properties. This approach is employed in several models [428, 472, 474, 579]. However, this approach is rarely used. The main reason is that it is possible to describe the same experimental data within their experimental errors by using quite different combinations of constants. An example is presented in Figure 7.12. As we can see from the figure, curves 1 and 2 calculated by using quite different values of all three constants coincide with each other within the usual experimental error of $\pm 0.05 \log \eta$ units; for example, the values of the constant B for these curves differ from each other by 2000 K. At the same time, a change of only one of these constants for only *a quarter* for this difference (curve 3) changes the results of calculation by *an order of magnitude* that exceeds the experimental error level 20 times. In fact, this means that nobody can determine the values of the *particular* constants more or less accurately. In other words, the exact values of these constants themselves, taken separately from one another, cannot be verified by any available experiment. The determination of these values is always performed by using some *voluntary rule*. This is very inconvenient.

In most of the known models, the VFT constants are calculated indirectly: they are derived from three data points (i.e., the viscosities at three different temperatures) which are calculated from chemical composition. There are two commonly used variants of selection of these data points: (a) viscosities at three fixed temperatures and (b) temperatures corresponding to three fixed viscosity values. In theory, there are no principal differences between these two variants, and the simulation technique in both cases is very similar. For practice, however, it is usually more convenient to fix the values of viscosity because in most of practical applications it is more important to know the viscosity – temperature curve in a given range of viscosities than in a given range of temperatures. That is why in most practical

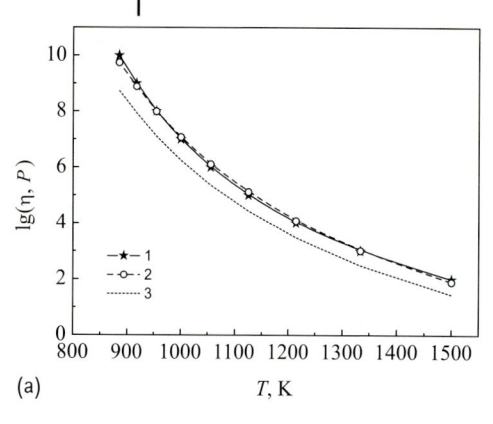

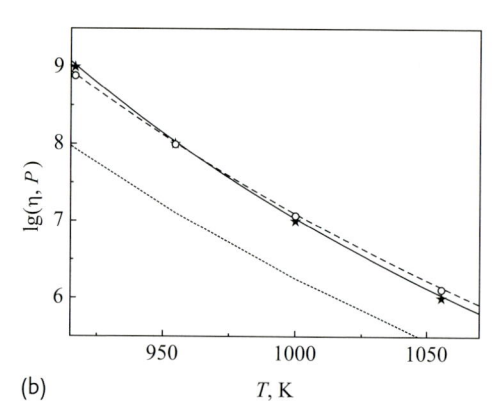

Figure 7.12 Comparison of the results of calculation by the VFT equation for different combinations of constants (a): (1) $A = -3$, $B = 5000\,K$, $T_0 = 500\,K$; (2) $A = -4.4$, $B = 7000\,K$, $T_0 = 390\,K$; (3) $A = -4.4$, $B = 6500\,K$, $T_0 = 390\,K$, in a wide range of viscosity values (a) and in a larger scale (b).

models, the values calculated from chemical composition are the "characteristic" (or "standard") temperatures corresponding to three viscosity values.

7.5.2
Approaches to Simulation of Concentration Dependencies of Viscosity Characteristics

7.5.2.1 Linear Approach

It is well-known that the concentration dependencies of viscosity characteristics (both viscosities at fixed temperatures and temperatures corresponding to fixed viscosity values) are considerably nonlinear in almost any more or less wide concentration range. Nonetheless, a linear approach to their calculation became quite productive for particular types of practical glasses. The most well-known and the most important of them is certainly soda lime glass. This type covers window and automotive glasses, mirror glass, some kinds of glassware and special glasses. In a wide interpretation, the term *soda lime glass* can be applied to compositions containing ~10–25 weight percent of alkali oxides (mostly, Na_2O), up to 15–20 wt% of RO (mainly, $CaO + MgO$), up to 5–10 wt% of $Al_2O_3 + B_2O_3$, and the rest of SiO_2. In a narrow interpretation, the soda lime glasses can be considered as glasses whose composition is close to $Na_2O \cdot CaO \cdot 6SiO_2$, with possible small additions of MgO and Al_2O_3.

Since this type of glass is used especially widely, a lot of scientists tried to simulate concentration and temperature dependencies of their viscosity. The task was not too difficult as long as concentration limits (see above) are narrow and no complicated structural effects are found. Thus, the linear approach was quite sufficient to rather accurately describe the available experimental data and predict viscosities of new glasses having similar compositions. The author's analysis [496] has shown that among the known models the most accurate is the model of Okhotin and Kim Eun San [491, 492], but other models also give quite accurate predictions. For wide concentration ranges, such an approach is obviously ineffective.

7.5.2.2 Approach of Mazurin: Summarizing of Effects

In the 1940–1960s, Mazurin and colleagues suggested a new approach to the prediction of "transport properties," that is, of viscosity and electrical conductivity, from chemical composition and temperature [484–486]. The idea of Mazurin was to summarize the "effects" instead of components. The effects were considered as the results of interactions between particular components. For example, if we replace Na_2O by K_2O in silicate glasses or melts we should consider two effects: effect of addition of K_2O and "mixed alkali effect." The resulting change of a property is the sum of these two effects.

The authors incorporated each effect by using empirical formulas that include the concentrations of components responsible for this effect. For instance, to consider the mixed-alkali effect they suggested a formula containing concentrations of alkalis as variables. In general, Mazurin's approach can be presented by the formula

$$P = \sum_{i=1}^{n} c_i p_i + \sum_{k=1}^{n} \sum_{l=1}^{n} f(c_k, c_l), \tag{7.19}$$

where $f(c_k, c_l)$ stands for the effect of interaction of the kth and lth components. According to Mazurin, the form of the function f can be different for different pairs of components even when calculating the effect of the same pair of components on similar characteristics. For example, when calculating the temperature T_2 corresponding to the viscosity 10^2 Poise the effect of interaction between Na_2O and BaO is described as $7[Na_2O]/[BaO]$ where $[Na_2O]$ and $[BaO]$ are molar percent of the corresponding oxides; the effect of the same interaction on the value of T_3 corresponding to the viscosity 10^3 Poise is described as $4[Na_2O] \cdot [BaO]$, and so on.

The obvious advantage of this approach is its flexibility. Indeed, for each couple of oxides the authors can derive a specific formula that accurately fits the experimental data. However, this flexibility makes it practically impossible to extend the approach to complicated systems with multiple interactions between components. Indeed, if we consider a glass-forming system containing, say, 20 components and add only one new component to it then we have to study 20 *only binary* interactions of this component with others, that is, 20 new effects. If we consider ternary effects (indeed, ternary interactions between glass components take place) we need to study several hundred other effects as well. Each of these effects requires the study of at least 10–15 new glasses. In fact, this means that describing the effect of *one* new component with 20 others requires experimental research with a scope comparable to *all* experimental data about viscosity of glass-forming melts available worldwide. In practice, the authors of the model used simple quadratic terms and cross-terms (like $4[Na_2O] \cdot [BaO]$) for almost all effects – this was the only way to determine the model parameters from a very limited amount of available experimental data. In fact, the described model exhibits only a minor difference as compared to other polynomial models. Nonetheless, this model was actually the first tool allowing one the prediction of viscosity of silicate melts in wide ranges of concentrations and temperatures.

It is worth noting that by using the mentioned approach Mazurin suggested a model for calculation of another transport property, the electrical conductivity of glasses, in a system containing 13 components (including two alkali oxides, Na_2O and K_2O) in rather wide concentration ranges. As far as the author knows, for this property nobody else could suggest something considerably better till now.

7.5.2.3 Approach of Lakatos: Redefinition of Variables

In the 1970s, Lakatos and colleagues suggested a series of models for the prediction of high-temperature viscosity of silicate glass-forming melts in several concentration ranges. The approach of Lakatos has two specific features as compared to previous (and past) models. First, the authors suggested to re-define "chemical variables:" instead of molecular percentage of oxides they used molecular *ratios* to SiO_2. For instance, instead of [MgO] they used the ratio $[MgO]/[SiO_2]$. According to the authors of the model, this re-definition made the described concentration dependencies simpler. Second, in some papers (e.g., [472, 474]) they did not calculate particular (measured) viscosity characteristics but directly calculated the constants A, B and T_0 of the Vogel–Fulcher–Tammann (VFT) equation (see above) as linear functions of the mentioned ratios.

The idea to use molecular ratios, R_mO_n/SiO_2, can be considered as quite reasonable if we remember that in the glass-forming systems R_2O–RO–SiO_2 (where R means any monovalent or bivalent cation) for which the models were mainly developed, the ratio $\{(\sum[R_2O]+\sum[RO])/[SiO_2]\}$ exactly determines the mean number of the "bridging oxygen" atoms in a [SiO_4] tetrahedron n_{br} [424]:

$$n_{br} = 4\left(1 - \left(\sum[R_2O] + \sum[RO]\right)/[SiO_2]\right). \tag{7.20}$$

Correlation between the viscosity of silicate melts and the fraction of bridging (and nonbridging) oxygen atoms is well-known and widely used, especially in geochemical studies ([432–435, 450, 451, 456, 519, 520, 565–569, 580, 581], etc.). Thus, this ratio can be considered as a structural characteristic affecting the viscosity. Another eventually possible alternative approach, that is, the direct calculation of the VFT constants, seems not to be optimal (see above).

7.5.2.4 Polynomial Models

In 1974, Lyon [482] suggested a new viscosity model for glasses of soda lime type. This model covered a rather narrow concentration range but this was the range which is especially important for practice. On the basis of available experimental data the author derived a series of polynomials describing concentration dependencies of viscosity at several fixed temperatures (from 600 to 1300 °C). The results of simulation can easily be interpolated for intermediate temperatures and, thus, the model covers wide temperature and viscosity ranges.

In this paper, Lyon tried to give a foundation of his selection of particular terms of polynomials by chemical reasoning. However, this reasoning looks too general, and it is not clear *from a chemical viewpoint* why the interactions between particular pairs of components are more important than others. The chemical meaning of square terms (e.g., [MgO]^2) is also unclear. In fact, the author used a polynomial model similar to those of Gilard and Dubrul [447–449] with the only difference: he used not only quadratic terms but also cross-terms. Essentially, Lyon's model is very similar to Mazurin's one; these authors only used different words for an explanation of similar concepts.

In the 1990s, the interest in polynomial models greatly increased due to the widespread availability of powerful personal computers. Many specialists had optimistic expectations concerning the statistical treatment of large data arrays by using computer technologies. As examples, the author could refer to the papers [428, 469, 579]. The authors of these papers suggested "first drafts" of models based on the polynomial regression method. The particular calculation algorithms suggested in these papers were different in details but the idea was the same: this method has no principal limitations, and in the near future the use of this technique should allow one to develop accurate viscosity models for a wide range of concentrations. However, the only task that the authors could accomplish in this direction were these "drafts" that were applicable only to specific binary and ternary systems.

7.5.3
Conclusion

In the twentieth century, some models for prediction of viscosity characteristics were suggested in some very important composition areas, first of all, for the glasses of the soda lime type. The generality of these models is not comparable with that for other properties such as density, optical or mechanical characteristics. This feature is caused by the much more complicated form of the simulated dependencies. The approaches to simulation of viscosity suggested in the twentieth century were considerably more diverse than those for other properties. This is natural: more complicated dependencies required more efforts for simulation, and different authors tried to resolve the presented problems in different ways.

It is worth mentioning that nobody could successfully use the "structural approach" similar to that suggested for other properties (see above). At the same time, it was obvious from the very beginning that the structure is the key factor affecting the viscosity of glass-forming melts, at least, in silicate systems. If we read nearly any paper where the concentration dependencies of viscosity of silicate glass-forming melts are discussed we could find some phrases related to structural factors, such as "strength" of the glass-forming network. Indeed, it is very well-known that molten silica has the strongest structural network and the highest viscosity among all known glass-forming melts. However, this structural knowledge did not allow one to develop a more or less general model for prediction of viscosity characteristics.

Some attempts to apply structural concepts to viscosity simulations were made but they did not result in increasing the generality and/or accuracy of models. Mostly, structural reasoning was used only as scientific background for usual polynomial models. It is not surprising that the success of these models had little relation to the presence or absence of such a background.

7.6
Simulation of Concentration Dependencies of Glass and Melt Properties at the Beginning of the Twenty-First Century

7.6.1
Global Glass Property Databases as a Catalyst for Development of Glass Property Models

At the end of the last century two global databases, Interglad [582] and SciGlass [423] appeared in the world market. Their first versions contained chemical compositions and property data on about several dozen thousand glasses and melts; now, on about 300 000 compositions in Interglad and more than 350 000 in SciGlass (note however that in Interglad a considerable part of glass compositions is not disclosed). These databases allow one to practically instantly search for a particular property in a required concentration range. The SciGlass database also contains more than 100 built-in models allowing one to calculate multiple properties for a user-specified glass composition and temperature within the applicability ranges of these models. So large collections of data "all-in-one" made it possible to instantly (in seconds) get a sample of data for any property and any concentration range where this property was studied by anybody. Within the last decade, several specialists have suggested new models, developed with the wide use of the mentioned electronic databases. Below, the results achieved in this direction are characterized.

7.6.2
Linear and Polynomial Models

A new attempt of the use of the polynomial regression approach was made by Fluegel and colleagues [443, 547]. In multiple publications these authors say that their models are global, that is, very general, and very accurate at the same time. However, the analysis of the real ranges of applicability of these models (specified in details in the software tools developed by the authors but not in the publications themselves) shows that these ranges are even narrower than that for the above-mentioned Lyon's model [482]. Moreover, comparison of the predictions of these models with experimental data available in the SciGlass database shows that in the same concentration ranges both models supply us with results of practically the same accuracy (see Figure 7.13). In other words, use of powerful computer programs and huge data arrays did not help Fluegel *et al.* to considerably improve the accuracy of predictions as compared with a very similar model developed manually more than two decades before; the concentration range specified by Fluegel *et al.* (considering internal limitations) is considerably narrower than that for the Lyon's model. The author also knows a few linear models developed by using the SciGlass database [470, 480]. However, these models also did not considerably increase the generality or improve the accuracy of predictions as compared to previous models.

The following question arises: Why has the use of powerful computers and huge data arrays stored in the electronic databases caused so modest results? The author

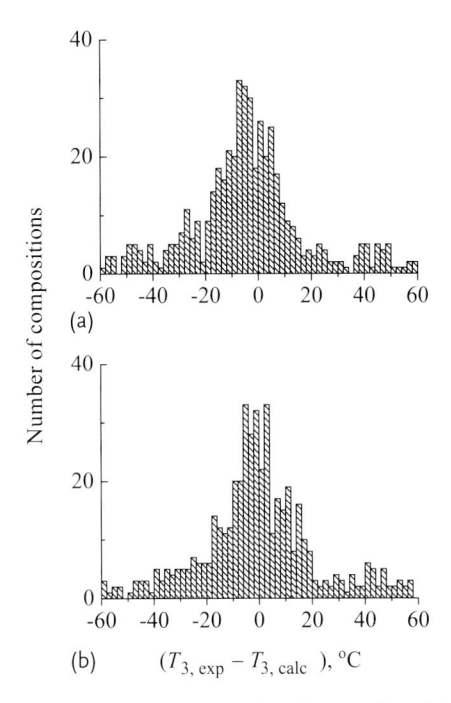

Figure 7.13 Comparison of predictions of models by Lyon [482] (a) and Fluegel et al. [443] (b) with experimental data contained in the SciGlass database [423] within the same concentration range that fits all limitations specified by the authors of both models (515 data points).

sees several reasons for such unexpected result. The most important one, in the author's opinion, is the improper understanding of possibilities and limitations of the polynomial equations as such.

The *first very serious problem* of any polynomial approximation is the number of fitting parameters employed. In polynomial equations for multi-factor dependencies, the number of parameters increases in geometric progression with increasing of *both* the number of considered independent factors (in our case, these factors are glass components) and the power of the polynomial. For the determination of these parameters, huge arrays of the experimental data are required. In fact, even cubic regression for, say, 10 components in wide concentration ranges (measured by several dozens of molecular percent for each of them) is far outside of practical possibilities. In all literature available worldwide we do not have enough data.

The *second problem* is the high sensitivity of the results of simulation to experimental errors in the source data. A series of data points with large errors in a little investigated composition area might take a model far away from an actual dependence and introduces various artifacts into it. The danger is caused by the fact that large errors in the reported glass property data are usually *systematic errors* and thus might appear in quite a considerable part of reported data (see examples in [578] and Chapter 6 of this book). When developing most of standard statistical procedures including polynomial regressions their authors proceeded from the assumption that the experimental errors are random and normally distributed. Processing the data with considerable (and generally unpredictable) systematic errors is not a simple task, and it requires serious efforts to resolve the respective problems. Use of standard statistical software packages can help to do this but cannot replace a human when performing this task.

The *third problem* is the essential limitation of the polynomial regression itself. This limitation is never considered in statistical handbooks – but it exists. Let us demonstrate this limitation by using the example of interaction of a glass modifier (for example, Na_2O) with aluminum and boron oxides causing a change of boron coordination and, as a result, the effect of boron oxide on glass properties. These interactions (in very simplified form) can be presented as

$$Na_2O + Al_2^{VI}O_3 \leftrightarrow 2Na^+[Al^{IV}O_{4/2}]^- , \qquad (7.21)$$

$$Na_2O + B_2^{III}O_3 \leftrightarrow 2Na^+[B^{IV}O_{4/2}]^- , \qquad (7.22)$$

where Roman superscripts indicate coordination numbers. As follows from various data (structural studies [533, 534, 536, 563], etc., thermodynamic simulations [576, 577], analysis of concentration dependencies of glass properties [424]), reaction with B_2O_3 can take place only in the case when there is some excessive amount of Na_2O remaining after the reaction with Al_2O_3. In other words, the interaction between Na_2O and B_2O_3 *might be blocked* by Al_2O_3. However, in the case of an excess of Na_2O this "blocking effect" disappears.

Now let us assume that we have to predict properties of glasses containing all four components (Na_2O, Al_2O_3, B_2O_3 and SiO_2) based on the experimental data from ternary aluminosilicate and borosilicate systems. The polynomial regression approach has no formal limitations for predictions of this kind. In polynomial regression models, interaction between factors is considered by using cross-terms for corresponding variables. In particular, interaction between Na_2O and B_2O_3 can be considered by using the product of their concentrations $[Na_2O] \cdot [B_2O_3]$. Interaction between Na_2O and Al_2O_3 can be considered in a similar manner. It is even possible to indirectly describe the blocking effect of Al_2O_3 for the reaction given by Eq. (7.21) by introducing a cross-term responsible for (imaginary) "interaction" between Al_2O_3 and B_2O_3. However, within a polynomial model there is no way to account for the possibility that one interaction *might block one and might not block another one*. The situations when an interaction between two components can be blocked by a third component are very typical for oxide glasses. For example, in the system $Na_2O–CaO–Al_2O_3–SiO_2$ the interaction between CaO and Al_2O_3 can be blocked by Na_2O.

The above-mentioned reasoning does not mean that it is impossible to describe the concentration dependencies of glass properties in aluminoborosilicate systems by using polynomials. If we have enough data we can approximate them by polynomials very well. However, *this polynomial would not be predictive* for any combination of components which is not presented (or presented insufficiently) in the source data.

Fluegel [442, 546] tried to incorporate this specific feature of polynomial approximation in his models. He suggested to implement the limitations not only for concentrations of particular components but also for their products (e.g., [Na_2O] · [CaO]). This approach, at least in the opinion of the author of the model, preserves the improper use of models (is this opinion correct or not – this is a subject of special consideration) but it often makes them impractical if we try to predict the properties of *new* glasses which have not yet been studied. The use of computers and electronic database can help to properly consider this limitation, but cannot exclude it.

Certainly, the viscosity is one of the most complicated (for simulation) glass properties. Concentration dependencies of many other properties are much simpler. Nonetheless, starting from Gilard and Dubrul [447–449], nobody could suggest a more or less general model for any glass property on the basis of polynomial regression. So, the author considers the polynomial regression approach as having no visible prospects in the simulation of concentration dependencies of glass properties in wide concentration ranges.

7.6.3
Calculation of Liquidus Temperature: Neural Network Simulation

The liquidus temperature is the most difficult for prediction quantity among commonly used properties of glasses and glass-forming melts. The origin of these difficulties is obvious for anybody who has an idea of typical concentration dependencies of this property: they are not smooth but generally have sharp minima at eutectic points. In Figure 7.14 we can see a typical dependence of this type.

Some authors [462, 473, 583] tried to calculate the liquidus temperatures of multi-component silicate glasses without a consideration of the above-mentioned specific features. As a result, the models often fail when compared with the experimental data. For example, comparison of predictions of the model [473] with

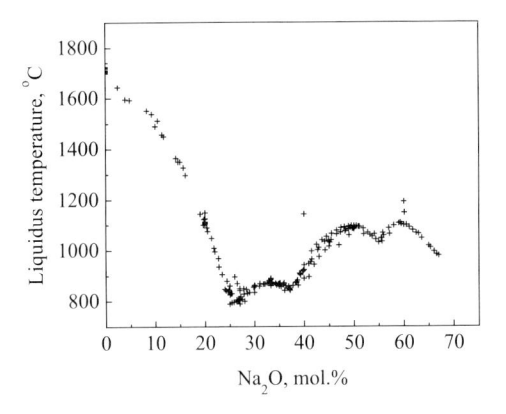

Figure 7.14 Experimental concentration dependence of liquidus temperature in the system Na_2O–SiO_2 according to the data of multiple authors presented in SciGlass [423] (outliers are not removed).

experimental data available from the SciGlass system [423] shows that the prediction errors might reach 200 °C and more, instead of 5–10 °C claimed by the authors of the model.

The specific form of concentration dependencies of liquidus temperature is well-known from thermodynamics. Thermodynamic calculations *in theory* allow one an exact approximation and even prediction of this characteristic for any chemical system. In practice, however, the thermodynamic approach somehow fails even in the simplest, binary systems. For example, for the composition (in mol %) of $50SiO_2 \cdot 50B_2O_3$, the publications [584–586] predict liquidus temperatures of 530, 790 and 870 °C, respectively – it is unlikely that such results, each of them obtained by using the thermodynamic approach, can be considered satisfactory for any purposes.

Some authors [512, 546, 587] tried to modify the polynomial regression approach by considering the primary crystalline phases. It is well-known that within a concentration range where the same primary phase is crystallized when cooling a melt, the concentration dependence of liquidus temperature is always smooth. Thus, if we properly determine the boundaries of this concentration range we could use simple statistical methods (like polynomial regression) for approximation of the experimental data. The main problem is how to properly prescribe these boundaries. In simple systems with many experimental data available to determine these boundaries, it is not a very difficult task. However, for multi-component systems having few (or a few dozen) experimental data points, use of these approaches is also unlikely.

In 2007, C. Dreyfus and G. Dreyfus [588] suggested a new approach to the prediction of the liquidus temperature in multi-component systems. Their approach was used (as far as the author knows) in cybernetics and is known as a neural network simulation. This approach is completely empirical but it is specially developed for the prediction of complicated multi-parameter dependencies with multiple singular points such as extremes and breaks. In neural network models, a dependent variable, Y, is presented as a linear function of nonlinear "responses" of "neurons"

$$Y = p_0 + \sum_{m=1}^{M} p_m y_m, \tag{7.23}$$

$$y_m = \tanh\left(\sum_{n=1}^{N} w_{nm} x_n + w_0\right), \tag{7.24}$$

where Y is a predicted property, y_m is a "neuron;" p_m, p_0, w_{nm}, w_0 are empirical parameters, and x_n are input variables (in our case, they can be considered as concentrations of components or their functions). Finally, the expression for a predicted property can be written as:

$$Y = p_0 + \sum_{m=1}^{M} p_m y_m = p_0 + \sum_{m=1}^{M} p_m \tanh\left(\sum_{n=1}^{N} w_{nm} x_n + w_0\right). \tag{7.25}$$

The specific feature of the neuron function is its unique possibility to accurately approximate both smooth and broken functions by using the same formalism.

Another specific feature of the neural network approach (as stated in [588]) is the use of the "training procedure" for calculation of the model parameters by using a "training set." However, in the author's opinion, this feature is not specific for this particular approach: it can be applied to a number of other approaches as well. Below, another procedure that uses this approach will be considered.

The advantage of this approach is the parsimonious use of fitting parameters. The flexibility of the neural network approach allows one a justification of the form of particular neuron functions directly when training a model, without prescribing its structure in advance, with the use of a minimum possible number of parameters. For the functions like concentration dependencies of liquidus temperature, this factor (the number of fitting parameters) is critical: otherwise, we might not have enough data even in very simple systems as it was clearly shown in [588]. However, the same feature (use of the same basic function for datasets of any nature) can become a drawback as far as some (in principle, known in advance) peculiarities of the described dependencies might not be considered properly.

In particular, in the case of liquidus temperature, it is well-known that *only minima can be broken (and are usually broken) but maxima are always smooth* like the ones shown in Figure 7.14. This feature is principal and can be derived from very general thermodynamic considerations. Obviously, a direct implementation of this feature into a model could help (with other model characteristics being similar) to simplify its structure and decrease the number of fitting parameters. By analogy, no other function (e.g., polynomial, Fourier, etc.) can approximate an exponential function as parsimoniously as the exponent itself. Anyway, the basic approach (not to use a "universal" form of the simulated function but to justify it for a particular data-set) seems very promising, and most probably this approach can be applied not only to the liquidus temperature but to other properties as well.

The author has to note, however, that the basic idea of the neural network approach (the use of a linear sum of nonlinear functions of linear inputs) is not something cardinally new in the simulation of glass properties. In the final analysis, this is the same idea that Appen and Demkina used when implementing the "oxygen number" (see above) for evaluation of the "partial coefficients" of boron oxide. In Demkina's model [440], even the particular expression for N_4 uses the same basic function as the neural network model [588] employs: the hyperbolic tangent. In fact, the N_4 parameter in Demkina's model can be considered as a "hidden neuron" without great reserve. In other words, in the model [588] the *technology* of simulation is new for glass property simulations but the basic idea is not.

7.6.4
Approach of the Author

7.6.4.1 **Background**
In [307, 495, 497–505] and references cited therein, the author of this chapter described his own approach to the simulation of concentration and temperature dependencies of properties of oxide glasses and glass-forming melts in wide ranges of compositions and temperatures. The model allows one the prediction of mul-

tiple (viscosity, density, optical, mechanical, thermal and some other) properties, many of them being possible to calculate in wide temperature ranges, for 64 oxides allowed in any combinations within the concentration ranges from 0 to mostly 50–100 mol% for each component (see [307] for details).

The described approach is based on the assumption that properties of glasses and glass-forming melts can be presented as functions of concentrations of structural groups each of them containing one or more coordination polyhedra of network-forming cations. These polyhedra can also be bounded with modifying cations (e.g., $[AlO_{4/2}]^-Na^+$). Structural groups are formed in chemical reactions taking place in a melt, for example (in very simplified form), via the reactions

$$0.5Al_2O_3 + 0.5Na_2O \leftrightarrow [AlO_{4/2}]Na , \qquad (7.26)$$

$$[AlO_{4/2}]Na + 2[SiO_{4/2}] \leftrightarrow ([AlO_{4/2}]Na \cdot 2[SiO_{4/2}]) , \qquad (7.27)$$

and so on. These and other reactions between structural groups (finally formed from oxides) are reversible. The author assumed that each of these reactions in the melt approaches equilibrium. All of these assumptions are widely used in glass science and have theoretical and experimental foundations. The only problem is the impossibility (for multiple reasons) of prediction of exact compositions and, moreover, concentrations of structural groups in more or less wide composition areas. Thus, the author suggested an empirical approach to the estimation of these characteristics based on experimental data about glass structure and glass properties.

7.6.4.2 Model

The mathematical model for calculation of concentrations of structural groups is similar to thermodynamic models but uses empirical coefficients instead of equilibrium constants and activity coefficients. Thus, the author considers his model as being essentially empirical but using some elements of the structural approach.

The structural data were used for simulation of the most important characteristic, the coordination state of boron atoms. The model [503] based on available NMR data postulates the compositions of structural groups and calculates their concentrations. The N_4 value is calculated from the obtained concentrations of structural groups containing threefold and fourfold coordinated boron. A comparison with the source data shows that the model calculates the N_4 value (see above) with the standard error of about 0.05 that is close to the experimental errors of measuring this characteristic. One example of comparison of model predictions with experimental data is shown in Figure 7.6 above; many other examples are presented in [503].

Assumptions about compositions and concentrations of other groups were mostly made on the basis of the analysis of concentration dependencies of glass properties. All model parameters were derived from the data about the simplest (binary and ternary, very rarely quaternary) glass-forming systems. The data about most quaternary and all more complex systems were used only for an estimation of the accuracy of the predictions.

For calculations of some properties, especially those that were measured for very few compositions (note that for the 64-component system even several hundred studied glasses are very few), the author used their relationships with other properties which were studied in more detail. For example, for heat capacity of melts the SciGlass system contained the data on about only a few hundred compositions. Considering the rather complicated form of its concentration dependencies this is surely not enough to develop a more or less accurate model. Thus, in [507] the author suggested an indirect calculation of this property from other characteristics: heat capacity of glasses (measured in much more detail), glass transition temperature, T_g, and temperature coefficient of the viscosity logarithm in the glass transition range. The latter characteristic has a strong correlation with the jump of the heat capacity, ΔC_p, that is, the difference between heat capacity of melt and glass in the glass transition range.

As a result, it was possible to develop the model for heat capacity of melts without the use of the experimental data about this characteristic at all. As far as developing the model did not require experimental data its applicability range is not restricted by the composition area where this property was studied, and the model can predict heat capacity of melts even in the systems where the properties of the melts were not measured at all. Comparison of model predictions with available experimental data (none of which were used for determination of model parameters) shows that the standard error of model predictions is only about twice as large as for the local models that were directly derived from these data. The author considers this property as a very good result for a model covering an incomparably wide composition area. Other examples of the use of inter-property relationships are presented in [307, 495].

Finally, to be able to extend the model to wide temperature ranges the author suggested new empirical equations for the temperature dependencies of some properties. In particular, two equations were suggested for viscosity [502]. One of them contains only two fitting parameters, β and θ,

$$\log \eta = -3.5 + \frac{\beta}{\log T - \log \theta}, \qquad (7.28)$$

where η is the viscosity in units of Poise and T is temperature in degrees Kelvin. This equation approximates the temperature dependencies of viscosity of melts of various chemical nature in a very wide range (from ~ 1 to $\sim 10^{13}$ Poise) with the mean error close to 0.1–0.2 units of decimal logarithm. This value is not as low as the typical experimental error but quite satisfactory for prediction of viscosity from chemical composition and temperature. Reducing the number of fitting parameters has greatly simplified the determination of their values from chemical composition. The three-parameter extension of the equation allows one an approximation of the experimental data better than the VFT equation; this extended equation is also presented in [502].

In this connection, however, it should be noted that the term "experimental error," even if we talk about the same experiment, can have a considerably different meaning *and different numerical value* depending on the subject of consideration.

This aspect of the problem is usually ignored in experimental papers; moreover, very often their authors have no idea about the magnitude of the real errors of the reported data. Usually, they only report the random errors that show the level of reproducibility of property values measured on the same sample by the same person using the same device in a very limited period of time. This approach ignores all kinds of systematic errors: differences in sample geometry, errors of measuring the temperature (indeed, all glass properties are more or less temperature-dependent), effects of annealing, and so on.

If we consider the *concentration* dependencies of glass properties we should consider one more factor: the *inconsistency of the reported glass composition with the actual one*. In fact, nobody can exactly determine the chemical composition of a glass sample being studied. Usually the authors of the papers report only the "batch compositions" of glasses, that is, the compositions derived from the batch materials. This composition does not account for volatilization of some components (alkalis, B_2O_3, PbO, etc.), dissolution of crucible materials (SiO_2 from quartz crucible, SiO_2 and Al_2O_3 from chamotte or porcelain), redox reactions with oxygen and other gases present in the furnace atmosphere (FeO/Fe_2O_3, SnO/SnO_2, etc.) and other changes of glass composition in the course of melting. As a result, the content of some components in molten glasses can deviate from the prescription for up to 10–15 weight percent and sometimes even more (some results of such "studies" are demonstrated in Chapter 6 of this book). The only reliable way to determine the actual glass composition is wet analysis – but even in this case we generally know the concentration of the major components with an accuracy no better than ±0.5...1 wt%. Usually the wet analysis is not performed because of its costs, and the deviation of the reported glass compositions from actual ones is about 1–2 weight% for the major components.

The typical effect of 1 weight% of alkalis and some other components on the viscosity at a fixed temperature has an order of several dozen percent that corresponds to 0.1–0.2 units of decimal logarithm (sometimes this effect can reach even several orders of magnitude – see, for example [478] – but such cases are not typical). Thus, when studying the concentration dependencies of the viscosity the errors of $0.1-0.2 \log \eta$ are unavoidable. Correspondingly, if we are aimed at calculation of viscosity as a function of chemical composition and temperature there is no sense to approximate its temperature dependencies with very high accuracy. In this situation, the application of Eq. (7.28) with only two fitting parameters becomes very useful.

7.6.4.3 Comparison with Previous Models

In general, the author's approach allowed one to develop a model that is applicable to very wide concentration and temperature ranges. In fact, the applicability range of this model includes the applicability ranges of nearly all existing models for prediction of properties of oxide glasses and melts and also a very wide composition area where none of the alternative available models is applicable. In the composition areas where other models are applicable, the suggested model is about as accurate as the best of them, somewhere being even better (see numerous

Table 7.1 Comparison of the method proposed by the author of this chapter and the method by Appen, within the limits of application of Appen's method according to the data presented in the SciGlass database (cited from [307]). Density ρ, thermal expansion coefficient α, refractive index n_d, mean dispersion $(n_F - n_C)$, Abbe's number ν_d, Young's modulus E, shear modulus G, surface tension σ, the number of compositions n, and ΔP_{avg} stands for the absolute root-mean-square deviation.

Property	n	ΔP_{avg}	
		Proposed method	Method by Appen
ρ, g/cm^3 at 20 °C	6879	0.043	0.038
$\alpha \cdot 10^7$, K^{-1} at 210 °C	1583	6.49	6.44
n_d at 20 °C	6063	0.0076	0.0068
$(n_F - n_C) \cdot 10^4$	1571	4.3	5.2
ν_d at 20 °C	1579	1.5	3.0
E in GPa at 20 °C	1155	4.3	6.4
G in GPa at 20 °C	452	1.8	3.0
σ in mN/m at 1300 °C	253	17	13

examples in previous publications of the author [307, 495–509]). So, the suggested approach could extend the applicability ranges of glass property prediction models very much without significant loss in the accuracy of predictions. The general results of comparison with Appen's method, the best of earlier developed methods for multiple properties, are presented in Table 7.1.

It is also interesting to compare the present approach with the neural network simulation. Unfortunately, the results of this approach do not extend to properties other than the liquidus temperature; therefore, let us compare the approaches, not the results of their application. As stated above, the mentioned approach has some common features with the structural approach of Demkina who calculated the N_4 value in a manner very similar to the "hidden neurons" in neural network models. The author also calculates the N_4 value and other characteristics by using some structural assumptions that are, ultimately, analogs of the hidden neurons.

Another similar feature with the neural network simulation is paying great attention to the parsimony of fitting parameters. However, the author's approach to this frugality is different. The author has the opinion that the use of the same mathematical function for all cases is not the optimal way. That is, if we know that a simulated dependence has some specifics, it is better to use the equations that consider these specifics immanently. For example, in the concentration dependencies of the characteristic temperatures corresponding to fixed viscosity values, many researchers found sharp maxima (e.g., [589, 590]) but no one observed sharp minima. In the author's approach (see [497–500], etc.), such behavior is implicit in the formalism.

As a result, to describe both concentration and temperature dependencies of viscosity in a 20-component system where all components are available in high concentrations and any combinations (including Al_2O_3 from 0 to ~90 mol% and B_2O_3

from 0 to ~50 mol%) [500] the author used only ~40 parameters, which corresponds to a one-neuron model in the neural network approach. The author hesitates that a one-neuron model could accurately describe both concentration and temperature dependencies of viscosity in such a complicated system. Anyway, the author does not consider the neural network simulation as somewhat alien to the chemical or structural approaches. It seems rather a possible "technological" realization of this approach than something principally different.

7.6.4.4 Conclusion

As follows from the above-mentioned consideration, in general, the author's approach looks much better than the earlier approaches (except for the neural network for which we have not enough data to compare). But have we finally sought a "silver bullet" mentioned in the introduction? Surely the answer is no. This model is very general as compared to others but not always very accurate, especially when predicting properties of glasses in composition areas which have not yet been studied. Certainly, a rough estimation is much better than nothing, but sometimes specialists need more.

In addition, although the basic approach is universal the particular relationships used in the model are often "oxide-oriented:" in particular, it is rather difficult to extend the model to mixed glasses containing both oxide and nonoxide components. These (e.g., oxyfluoride) glasses are widely used in industry, and prediction of their properties is an important practical task – but probably for that other approaches need to be suggested. In other words, the suggested model is generally better than others and has nearly no compositional limitations for oxide systems; at the same time, it also has limitations that should be considered when using the model in practice: this is a good instrument but not a universal instrument by any means.

7.6.5
Fluegel: a Global Model as a Combination of Local Models

In [547], Fluegel suggested a new approach based on the wide use of glass property databases. The basic assumption is very clear: in a narrow enough concentration range concentration dependencies of properties are close to linear ones. This assumption is correct for almost all cases except for the vicinity of sharp breaks in concentration dependencies of properties, but (except for liquidus temperature, see above) these breaks are observed very rarely. Thus, if we have a large data array it is possible to find the property data for glasses whose compositions are close to the composition of interest and then to make a linear model for these particular compositions. Then, to estimate the error of prediction Fluegel suggests generating several models basing on different datasets; this technique is essentially similar to the well-known "cross-validation" method. The model that has the lowest error of prediction is used for calculation.

In principle, there is nothing that could invalidate this approach, at least in theory. In practice, however, it is also of limited use. The limitation is caused by the lack of required data in very many concentration areas where only few (or even no) data points are available from the worldwide literature. Only one example is the following: in the SciGlass database, among 350 000 glass compositions studied over more than 150 years, there is *nothing* containing Li_2O and CdO together for which high viscosities (greater than $10^{7.6}$ Poise) were measured. Note that both of these oxides belong to the top 20 glass components studied in utmost detail. It is quite unlikely that any automatically generated linear model could predict the viscosity of such glasses.

In fact, the mentioned approach can be very useful only for the local concentration ranges when there is a sufficient number of data points measured by different authors (if all or almost all data points were measured by the same authors the model would reproduce their systematic errors that might be high enough but would be "hidden" and unable to be estimated). These concentration ranges surely exist, and many of them are of great practical importance. But this approach ultimately cannot be considered as universal.

7.6.6
Integrated Approach: Evaluation of the Most Probable Property Values and Their Errors by Using all Available Models and Large Arrays of Data

In SciGlass [423], the following option suggested by Mazurin is available. A user specifies a concentration range and finds all data about a property calculated within this range. Then the program calculates this property for all found compositions by using all available models. After that, the far outliers (i.e., glass compositions with largest differences between calculated and experimental property values) are removed, and for the remaining data the program calculates the statistical characteristics; one of them is the systematic deviation, that is, the average value of difference between calculated and experimental property values for each particular model. After that, based on the statistical characteristics of different models, a user can select one or more best models to be utilized in further calculations.

Finally, a user specifies an arbitrary glass composition (within the specified concentration range), and the program calculates the property value by using the specified models. For each model, the program extracts the systematic deviation from the result of the calculation. The obtained results are averaged, and the average value is considered as a final result of the calculations. Thus, the program helps the user to determine the best models in a given concentration range and compensates the systematic deviation of predictions made by each of these models from experimental data. The result of such calculations is, in general, more reliable and more accurate than the results of prediction by any particular model.

In essence, this approach can be considered as a further development of the above-described idea of Gehlhoff and Thomas [545, 591, 592] to calculate small changes of properties after small compositional changes. Indeed, extraction of systematic errors of model predictions from the calculated property values gives essentially the same result as addition of calculated property changes to an experimental value. The only difference is that Mazurin suggests using multiple models and multiple experimental data instead of only one model and one data point as

Gehlhoff and Thomas did. However, the above-described difference is very important: in fact, it becomes possible to calculate all properties in any concentration range where at least one model is available. In the cases when multiple models are available a user can select the best.

In [593], the author of the present chapter together with Fluegel improved this approach. In the new algorithm, all calculations are completely automated. In particular, the "reference concentration range" is calculated programmatically, separately for each particular model. Also, the compensation of systematic deviations is performed only in the case when *model predictions have similar deviations from the data reported in several independent publications.* This approach prevents reproducing the systematic errors of property measurements in the results of calculations. Next, for each model a "weight factor" is calculated as a function of the standard deviation of model predictions from the experimental data, and the final result of calculation is simply a weighted sum of predictions made by particular models. Thus, all models contribute to the final result, but the best models make greater contributions, so that the final result of calculations becomes more reliable than each of the particular results. Finally, the program estimates the confidence limits of prediction on the basis of analysis of the deviations of predicted property values from the experimental data. As a consequence, it is possible not only to get a result of prediction itself (as it was possible in previous models) but also to estimate how accurate it is.

It is important to note that within this approach any model can be used for calculations. In particular, it might be a linear model generated by the program by Fluegel's algorithm (see above), or a model built by using the "neural network" approach. As a result, this approach allows one the prediction of property values and estimation of their errors nearly for any glass composition for which at least one model is available or can be generated.

This approach has practical advantages in comparison to any of the above discussed particular approaches. In the composition areas where accurate models are available they are primarily used, and the final result of calculation simply reproduces their results accomplishing them with the estimates of the confidence limits of predictions. In other composition areas, the best of available (or programmatically generated) models are also used; the results of calculations might not be very accurate but it is as much accurate as possible. In other words, this approach allows one the most accurate property prediction among all possible ones for each property and each particular composition as well as to estimate the error of such prediction – a characteristic that is important itself.

So, did we really invent a "silver bullet?" Unfortunately, no, again. Indeed, if an accurate enough prediction is available by using at least one model then the result would be accurate as well (or even a little bit better if several accurate models are available simultaneously). However, this approach cannot give an accurate prediction if no particular model gives it. This approach really integrates the advantages of particular models and reduces their disadvantages – but *it cannot give more than the particular models could give.* In the case when no particular approach works, the integrated approach does not work as well. In other words, this is not a magic tool but only the best possible one at the conditions given.

7.7
Simulation of Concentration Dependencies of Glass Properties in Nonoxide Systems

All models considered above were developed for oxide glasses. However, oxide glasses are not the only type of inorganic glasses. Starting from the middle of the twentieth century, other types of glasses were intensively studied; among the first were halide and chalcogenide glasses. These glasses have many serious advantages in some special applications, such as infrared optics.

Simulation of concentration dependencies of physical properties of these glasses is not as well-developed as for oxide glasses but several models are known. In [444], Gan Fuxi suggested an additive model for the calculation of density, refractive index and mean dispersion of fluoride glasses that can contain about 40 different fluorides. From Figure 7.15 we can see that in binary fluoro-beryllate systems the concentration dependencies of these properties are really very close to linear ones. This fact can be considered as an indication of only minor effects of structural factors on the mentioned properties of these glasses.

The situation with chalcogenide glasses is, however, greatly different. The author cannot discuss the specific features of these glasses in detail (referring the reader to [467] and other special literature) and will only characterize the relationships between their chemical compositions and physical properties. These relationships are generally rather complicated; they have multiple extremums and breaks being generally similar to concentration dependencies of liquidus temperatures shown in Figure 7.14. In Figure 7.16 we can see typical examples of these dependencies.

Also in this figure we see very many outliers – the points that are far outside the general trends. In oxide systems far outliers were also found (see [578]) but they are generally not so numerous. The most probable reason for this phenomenon is crystallization. In general, when cooling, chalcogenide melts are crystallized much

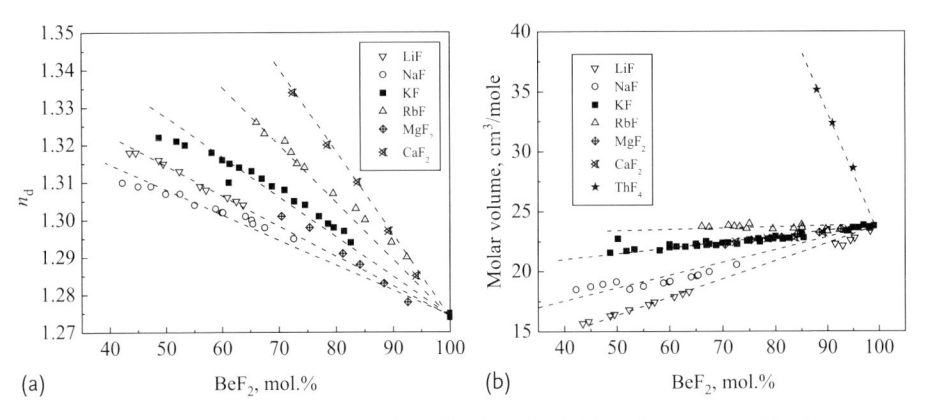

Figure 7.15 Concentration dependencies of the refractive index, n_d, (a) and molar volume (b) in binary fluoride systems with BeF_2 as a glass-forming component, according to the data of multiple authors presented in the SciGlass [423] database. The second component of the glasses is specified in the legend.

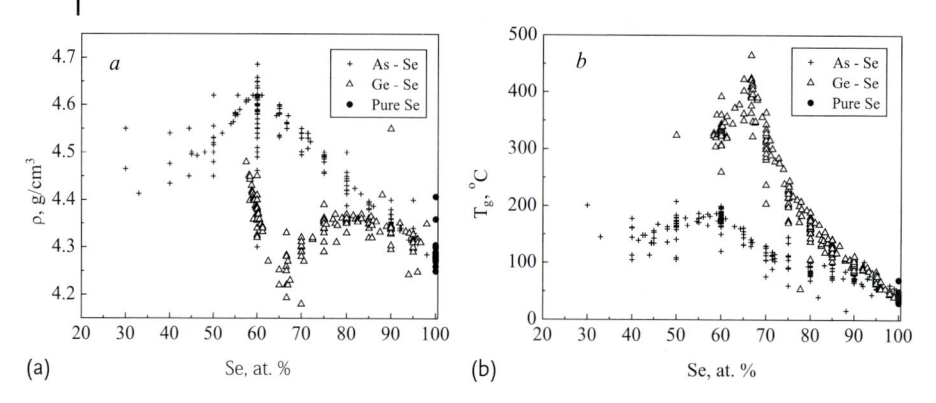

Figure 7.16 Concentration dependencies of density (a) and glass transition temperature (b) in binary systems As–Se and Ge–Se, according to data from multiple authors presented in the SciGlass database [423].

easier than oxide melts, and to obtain crystal-free glasses some special measures are required (see [467]). However, the authors of multiple publications did not pay very careful attention to the fine details of synthesis; that is why their glasses might contain considerable amounts of a crystal phase that could affect property values.

The calculation of physical properties of chalcogenide glasses is, therefore, a rather complicated task. Nonetheless, several models for these calculations were suggested [426, 427, 431, 467]. Among these models, the most general and in most cases the most accurate is the model of Kokorina [467]. The author has to stipulate that in the cited book no "model" in a strong meaning of this term was suggested, but it is possible to derive a model from the equations presented in the book. The model is based on the chemical approach. The reasonability of this approach is clear from Figure 7.16: indeed, the sharp extremes in the plots exactly coincide with the compositions that correspond to the known crystalline compounds $GeSe_2$ and As_2Se_3. Thus, the consideration of chemical interactions in these glasses offers itself. Unfortunately, Kokorina did not use computers in these studies and could not develop a model covering many components. She only suggested the equations and found the values of their parameters for six atomic components: As, Ge, Sb, Sn, S and Se.

The author tried to improve the chemical approach suggested by Kokorina. Unfortunately the paper describing the calculation algorithm has not been published yet, but the users of the SciGlass database [423] can employ this model (it is labeled as "Priven-Ch" where "Ch" means "chalcogenide") for the calculation of several properties (density, thermal expansion coefficient, refractive index, glass transition temperature, dilatometric softening temperature and microhardness) in the system containing 16 elements most commonly used in the chalcogenide glasses. In Figure 7.17 we can see the good correlation between experimental and calculated density values. Thus, the chemical approach demonstrated its applicability to this type of glass as well, similar to the oxide glasses.

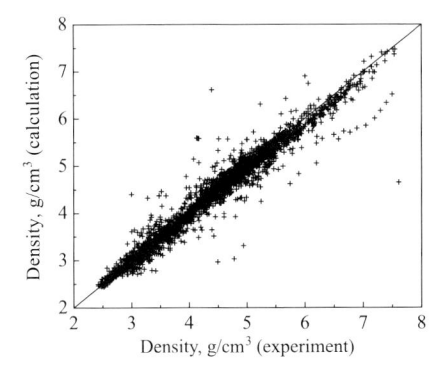

Figure 7.17 Comparison of results of prediction of glass densities by the author's model with experimental data for density of chalcogenide glasses. Points correspond to experimental data presented in SciGlass database [423] for glasses with compositions that fall inside the concentration limits specified by the author of the model. *Solid line* show the place of points for which calculated and experimental property values would be identical ($Y = X$).

7.8
Summary: Which Models Were Successful in the Past?

If we look at the whole history of simulation of concentration dependencies of glass properties we can distinguish three time periods when the generality and/or accuracy of models drastically increased. The first period was during the 1890s, the time when Winkelmann and Schott [385, 528–530] suggested the first set of models allowing for the prediction of multiple properties in a multi-component system on the basis of the additive approach. As mentioned above, they did not invent this approach (indeed, is it possible to invent a straight line?) and did not even apply it to glasses for the first time: Kopp [468] did this several decades before. However, they were the first who suggested this approach as a basis for simulation of multiple glass properties in a broad enough area of chemical compositions.

The second period was the middle of the twentieth century when Appen and Demkina [562, 563] suggested their models which incorporated some structural concepts. It is worth noting that Appen and Demkina were also not the people who invented the chemical or structural approach to glasses. The chemical analogy of glasses and solutions was suggested as early as in nineteenth century by Mendeleev [552], and the structural analogy between silicate glasses and crystals was known at the beginning of the last century after the papers of Zachariasen [555] and Lebedev [477] were published. However, these theoretical concepts by themselves provided nothing for the practical prediction of glass properties. Successful attempts of Appen and (to a somewhat less extent) Demkina were caused by their personal contributions much more than by the ideas of their famous precursors mentioned in their papers.

The third period started at the end of the last century after the global glass property databases Interglad [582] and SciGlass [423] appeared. At first glance, there is nothing common between the models which were suggested during this time, except for the general idea of the use of electronic databases. Indeed, the author's version of the chemical approach [307, 495], the approach of Dreyfus and Dreyfus [588] and the integrated approach in its different variants (see above) seem to have nothing in common. However, if we look deeper into all of the above-mentioned approaches that made great advances in simulations, we can observe one common peculiarity: *the authors of each of these models tried to properly guess general trends* of simulated dependencies more than to exactly consider their fine details. This conclusion requires some explanation to avoid misunderstandings.

It is well-known that the accuracy of an approximation of experimental dependencies by using empirical equations can be increased if we use more complicated models having more fitting parameters. For example, quadratic polynomials usually approximate measured dependencies better than linear equations; a cubic polynomial does the same approximation better than a quadratic one, and so on. However, does it mean that the more parameters we introduce the more accurate *predictions* we will have? No, it does not. If a model contains more parameters than necessary its predictions become *less* accurate. In statistics, this phenomenon is known as over-fitting. A detailed consideration of this phenomenon is far beyond the scope of this paper; the author can only refer to the excellent monograph of Esbensen [594] and other sources where this problem is considered in connection to simulation of multi-factor nonlinear dependencies. In a few words, for any sort of experimental data and any kind of simulated dependence, there is some level of complexity which an empirical model must not exceed in order to predict new data. Otherwise, one can only accurately approximate the source data points but cannot more or less accurately predict *new* data.

Addition of new terms to an empirical model often makes it too sensitive to erroneous data. In [594] it is shown that various statistical procedures for evaluation of the contributions of particular data points in a final model (e.g., "leverage") might not work properly in the case of multi-factor dependencies. This is especially dangerous when *extrapolating* a model outside of the data area where its parameters were determined. At first glance, this danger can be easily overcome: one only needs to specify the boundaries of the mentioned data area and not use a model outside. The problem, however, is that for such subject of simulation as concentration dependencies of glass properties in multi-component systems, a proper specification of the source data area is not a simple task.

Let us return now to the system $Na_2O–Al_2O_3–B_2O_3–SiO_2$ where the reactions described by Eqs. (7.26) and (7.27) take place. Let us assume that we have experimental data about glasses belonging to only three binary systems: $Na_2O–SiO_2$, $B_2O_3–SiO_2$ and $Na_2O–Al_2O_3$. In these systems we can easily synthesize the glasses containing from 0 to 100% SiO_2, from 0 to 100% B_2O_3, from 0 to ~50 mol% Na_2O and from 0 to ~55 mol% Al_2O_3. Does it mean that on the basis of these data we are able to predict the properties of any glasses in the specified concentration range? Definitely no: in the above-mentioned systems there are no glasses containing Na_2O and B_2O_3 together and, thus, there is no chance to properly consider the interaction between these components.

Let us add two ternary systems to our dataset: $Na_2O–Al_2O_3–SiO_2$ and $Na_2O–B_2O_3–SiO_2$. In this case, the participants of each reaction, Eqs. (7.26) and (7.27), will be present together at least in some glasses. So, we, in principle, can properly simulate the effect of each of these reactions. However, can we simulate the *blocking effect* of Al_2O_3 to the reaction Eq. (7.27)? Certainly not: for that, we need some glasses containing Na_2O, Al_2O_3 and B_2O_3 *altogether*. If we do not have the data for these glasses we are unable to simulate the mentioned effect whatever the concentration limits of each component in our glasses would be. Thus, when determining the boundaries of the studied data area we have to consider not only the overall concentration limits of each component but also its *combinations* with other components. This means that in the case of prediction of glass properties in multi-component systems there is no principal difference between interpolation and extrapolation: interpolation of concentration dependencies for *new combinations* of components is, at least, not a less difficult task than extrapolation for *new concentration ranges of the same components*. In fact, both cases can be considered as *extrapolations* from the known composition areas to unknown ones. Thus, the only possible way of simulation is the use of models that *can be used for some extrapolations*.

The basic requirements to such models can be formulated as follows:

1. A model should *properly account for the general trends* of simulated phenomena (otherwise any extrapolation is impossible by definition).
2. A model must *not implement artifacts* that do not follow from the source data, that is, must not predict complicated behavior where it does not take place (otherwise extrapolation becomes unreliable).

The characteristics of the "milestone" approaches to glass property prediction well satisfy both these requirements. Winkelmann and Schott actually could only guess the most general trends of influence of chemical compositions of glasses on their physical properties. Their model implements no artifacts to simulated dependencies as far as it is linear. A linear equation cannot be over-fitted by definition: it does not contain terms that might be considered as "excessive" ones.

The model of Appen satisfied the above-mentioned requirements at a new level of knowledge. The chemical approach allowed for the consideration and quantitative description of general trends of *nonlinear* relationships between chemical compositions and physical properties of glasses, especially in the silicate systems containing aluminum and boron oxides together with multiple modifiers (R_2O and RO). The oxygen number (see Eq. (7.12)) well describes the general trends of influence of multiple oxides taken in various combinations on the coordination states of network-former cations of aluminum and boron. At the same time, this equation (as well as other equations used in Appen's model) implements no artifacts into the results of simulation: the model can *overlook* actual complexity of simulated dependencies but cannot *introduce* excessive complexity into the results of simulation.

The use of the concept of oxygen number made it possible to *extrapolate* the behavior of multiple oxides from one composition area to another. Moreover, it allowed for the *prediction* of the effects that *were not experimentally measured at all*. Indeed, if we can calculate the effect of some oxide on the coordination number of boron then we can estimate the effect of interaction between this oxide and B_2O_3 on *any* property *without* experimental studying this effect: to evaluate its value we only need to calculate the oxygen number and substitute the result of calculation into an *already existing* formula for contribution of boron oxide. This procedure gives no chance to overcomplicate the description of the simulated dependencies.

In this connection, one more question arises: if the best models are those which allow extrapolations then why do we need to consider any limitations at all? Or was Demkina who considered her formula "universal" ultimately right – and if so, why do her formulas *allowing* for extrapolations often fail outside traditional silicate systems?

In the author's opinion, the answer is contained in the first of the above-presented conditions: to be suitable for extrapolations a model must properly describe general trends. This is possible only in the case if *a model includes the extreme cases* of the simulated phenomena. Pure silica, borate glasses with 70 and more mol% of Na_2O and other modifiers, aluminoborate glasses containing more than 50 mol% of Al_2O_3 are examples of the extreme cases for oxide glasses. Without proper consideration of these extreme cases, it is unlikely possible to predict the properties of such glasses. So, "overall" concentration limits of components are useful information that must be considered when determining the applicability limits of glass property models. If a model is developed on the basis of the property data for glasses containing, say, from 0 to 20 mol% Al_2O_3 then this model might not be applied to higher concentrations of this oxide. However, this fact says *nothing* about the applicability of a model *within* the mentioned range. The author of the present chapter has the opinion that, in the general case, the most important (and the most difficult to evaluate) are *internal* limitations of glass property models caused by improper considering the interactions between glass components.

Obviously, the chemical approach in general can help to properly consider the mentioned factors. However, the author has to underline that the "chemical approach" as such does not guarantee a proper simulation. Note that Huggins and Sun were the first who tried to consider the most important chemical factor, the change of the coordination number of boron atoms depending on the chemical composition of the glass. However, their formulas *did not properly reflect* the chemical nature of this effect and did not properly account for the factors affecting this process; as a result, these formulas failed when trying to perform calculations in new composition areas.

Another hidden danger becomes visible if we consider the thermodynamic approach. This approach considers chemical factors implicitly, and the potential ability of thermodynamic models to be used for extrapolations is obvious and unlikely can be disproved. However, to use any model in practice one has, at least, *to determine its parameters* first. Although the authors of [576, 577] declare the use of "no fitting parameters at all" this declaration is not consistent with reality. The authors of the model actually use a lot of fitting parameters, with the only stipulation that these parameters are not derived from the experimental data about simulated properties of glasses. From the "pure" theoretical viewpoint, this stipulation might be an advantage. However, for the practical use of a model it is not as important which data to use but is important *to have* these data and to be able to *accurately determine* the parameters from them. Thermodynamic calculations require such data as (a) an exact list of all phases possible in a simulated chemical system, (b) temperature dependencies of their chemical potentials, and (c) their activity coefficients in the liquid phase that, in their part, depend on chemical composition and temperature of the liquid. To obtain these data from experiment is, in general, a much more laborious and much more expensive procedure than to measure the desired glass properties, and even if all required data are available in principle they might not allow for an accurate determination of model parameters: the uncertainty of their values caused by various factors is sometimes considerably greater than the uncertainty of the glass property data. That is why the author of the present chapter does not see any great advantage of using this approach for the prediction of glass properties, at least, in the foreseeable future.

If we come back to the best models developed in the last century we can find one more common feature of these models: their authors tried to develop these models in such a way that *their parameters would have a clear sense*. The physical meaning of "partial coefficients" in the additive model is clear and visible as well as the sense of oxygen number (the amount of oxygen available for building [BO_4] tetrahedra), fraction of bridging oxygen and other particular parameters used in Appen's formulas. In contrast, the meaning of coefficients of polynomial and other models developed by using standard statistical procedures is usually unclear. In the latter case, the chances of bringing artifacts into a model and, thus, to make it unsuitable for extrapolations considerably increase.

One more factor that prevents introducing artifacts into a model is *minimization of the number of fitting parameters*. The greater the number of parameters, the higher the chance that some of them overcomplicate the model. The harm of the excessive parameters usually does not appear when approximating the source data points (i.e., the data from which the model parameters were determined) but becomes apparent when predicting new data – this is the phenomenon of "over-fitting" that was mentioned above. The use of a minimum possible number of parameters allows for reducing the risk of over-fitting and overcomplicating a model that can help to widen the applicability range and increase the reliability of predictions. On the other hand, minimization of the number of fitting parameters should not be considered in absolute terms. The models which were developed without or almost without the use of property data (*ab initio* simulations [570–575], models that use atomic characteristics as found in [483, 490], etc.) are inevitably simplified, and they might not reflect (or improperly account for) some important factors affecting physical properties of glasses and glass-forming melts. As a result, such models usually give considerably less accurate predictions than the empirical models.

In the author's opinion, at present, more or less accurate models can be developed only on an empirical basis. In other words, the aim should not be to avoid the use of empirical data at all but to use the *reasonable minimum* of this kind of data and, correspondingly, to develop the models with the *most reasonable minimum number of fitting parameters*. This means that *among various models giving similar accuracy of predictions, a model with a minimum number of fitting parameters would generally be preferable; in particular, it has a much better chance of becoming applicable to a wider composition area*.

7.9
Instead of a Conclusion: How to Catch a Bluebird

The above-made conclusions do not answer the main question: in which direction of glass property simulations could we expect considerable progress and how do we achieve this progress? Let us try to answer this question. Hereby the author does not have pretensions of exactness of his prognosis but only tries to foresee the future trends on the base of his understanding of the current situation.

If we cast one more glance at history we can come to the conclusion that the metaphor of the "silver bullet" used above is not exact. Indeed, this metaphor usually implies that all attempts to resolve a problem remain unsuccessful. But in the case of prediction of glass properties from chemical compositions, it is surely not the case. The models that are available nowadays do work, and they do allow predictions, often being not less accurate than ordinary experimental investigations. More likely, the problem is that the authors of models often expect (and the glass industry needs) more than the available models can give. The scientists make great progress – but the ultimate goal goes farther. In this situation, the better metaphor seems to be a "bluebird of happiness," a symbol of fortune that slips out of hand every time one seems to sneak up and finally catch it. It is worth noting another and much deeper idea that was introduced by Maurice Maeterlinck, the author of play "The Blue Bird:" in order to catch a bluebird one has *to see the invisible*.

In this connection, the author would like to go back to the discussion of the structural concept versus the concept of the "black box." Our consideration has shown that the quality of a model, that is, its generality and accuracy, does not directly depend on the type of its basic concept. Good glass property models were developed on the basis of the structural concept (starting from Appen [562] and Demkina [563]) as well as on the basis of the concept of the black box (e.g., Fluegel [547], Dreyfus and Dreyfus [588]). At the same time, other simulations based on structural or chemical concepts (such as thermodynamic simulations [576, 577]) as well as the concept of the black box (e.g., polynomial models [443, 546]) were not successful. In the author's opinion, the above-mentioned factors (proper consideration of general trends, minimum number of used parameters, etc.) are more important than the basic concept of a model. Nonetheless, the author believes that the understanding of the nature of the simulated dependencies can make a model more reliable and avoid undesirable artifacts. In other words, by using the same data it is usually possible to develop a more general and/or more accurate model if its author understands the (hidden) factors affecting the behavior of the studied objects.

For example, if we know exactly (from general thermodynamic reasoning) that sharp extremes in the concentration dependencies of liquidus temperature are *always* minima, that is, maxima are *always* smooth then we could restore an experimental dependence by using fewer data; in particular, we could successfully restore a dependence from the data that give no exact idea of the shape of the extremes. Otherwise, we need additional data to analyze the shape of the extremes. Thus, although the author considers the glass property models derived *only* from structural or chemical data having little (or even no) practical value, the understanding of the chemical and structural background could *help* to develop better models. In the author's opinion, his own model [307, 495] with the body of mathematics borrowed from thermodynamics and the most important structural parameter (N_4 value) calculated directly from structural data can be considered as a clear confirmation of this idea.

In this connection, the author considers any essential information about the structure of glasses and the nature of the simulated properties helpful, even if a model itself is developed on exactly an empirical base. In particular, the author believes that a simulation of glass structure *ab initio* could give very useful information for developing new glass property models. For the development of an empirical model, such as the author's [307, 495], this information could drastically reduce the required number of experimental data. If the direct information about glass structure is not available for some composition area, molecular dynamics and Monte-Carlo methods can provide a scientist with some idea about the probable structure. Thus, the author has the opinion that proper use of all available information (including glass property, structural, thermodynamic and other data as well as the results of various simulations) could help us to bring the simulation of concentration dependencies of glass properties up to a new level and, as a result, to make the models more general and more accurate. The keyword is the *proper* use of these data.

In conclusion, the author would like to say a few words about simulation of multi-factor dependencies. In the introduction, the author said that most of physical properties of inorganic glasses and glass-forming melts can be considered as functions of only two major factors: chemical composition and temperature, and above we considered only these kinds of properties. However, there are some properties for which we cannot assume this. The most practically important of these properties are chemical durability, strength, heat transfer, diffusivity and some optical properties (e.g., transparency). All of these properties depend on multiple factors which are individual for each particular property: chemical durability depends on the composition of the media and characteristics of the glass surface; mechanical strength depends on the surface defects and kind of deformation (e.g., strength to contraction greatly differs from strength to tension); heat transfer depends on the characteristics of the surrounding (convection, radiation); diffusivity depends on the media and penetrating compound (or ion); transparency depends on the wavelength.

Let us consider only one example: the effect of heat treatments in the glass transition range on physical properties (density, viscosity, etc.). Multiple variants of models based on the concept of structural relaxation allow one a prediction of these effects. However, each of these models contains some parameters which can be determined only from special experiments which most often are laborious and expensive. Essentially, we face the same problem as when simulating concentration dependencies of "usual" properties of glasses: some characteristics (so-called relaxation constants) depend on the chemical composition of glass but there is no model to predict them from composition. By mentioned reasons, the fewer the amount of experimental data is we need to determine the model parameters the greater is the chance to develop models for prediction of these parameters from chemical composition.

In [508] the author made the first attempt (as far as he knows) to evaluate the most important relaxation constant of the model of Mazurin *et al.* [290], the ratio of the viscosity to the Kohlrausch relaxation time (K_s), from chemical compositions of oxide glass-forming substances. By using the approach reported in [508], it became possible to drastically reduce the preliminary experimental research required for estimation of multiple characteristics (viscosity, relaxation times, elongations, linear expansion coefficients) in the glass transition range with the accuracy sufficient for multiple (although certainly not all) practical applications. Some important characteristics such as the boundaries of the annealing range (for arbitrary cooling rate) can be estimated even without any preliminary experiments at all. The author believes that further efforts in this direction could help to develop new models for accurate predictions of physical properties and other characteristics (such as internal stresses) when heat-treating glass-forming substances within the glass transition range without laborious and expensive preliminary experimental research of each particular glass.

To summarize, the author believes that for the simulation of concentration, temperature and other dependencies of glass properties the knowledge of any kind can be useful, be it a structural concept, the results of *ab initio* simulations, thermodynamic data, and so on. At the same time, it seems obvious that within a foreseeable period of time (at least, several next decades) the only way of building models allowing one practical predictions of multiple properties in wide concentration ranges would be empirical modeling. For that purpose, reliable experimental data are required. The amount of the required data considerably depends on the approach to property simulation but without at least a few reliable data points it seems unrealistic to accurately predict glass properties anywhere. Thus, one of the important tasks for the glass science should be obtaining reliable property data in new composition areas. In this connection, it is necessary to mention a dangerous trend: during the last decades, the quality of glass property data published in the available literature, unfortunately including even the leading journals, drastically decreased. This situation might be difficult to believe, but this is a fact: just look at the figures presented in Chapter 6 of the present book.

To be sure that experimental data are reliable there is only one way: compare them with independent data obtained by other researchers. However, even if such data already exist it is not a simple task to find them. Nowadays, glass property data are published in thousands of independent sources: the list of periodical journals which publish such data contains several hundred names. The only possible way is the use of global glass property databases which collect the information from the original sources, make this information easily searchable and present all results in a unified manner. Now there are two databases of this kind: SciGlass [423] and Interglad [582]; the largest collection of composition and property data is contained in the SciGlass database. In Chapter 6 of the present book, some theoretical and practical questions concerning the validation of experimental data by using this database are presented. Without the use of glass property databases, it seems problematic to support the experimental research of glass properties in various composition areas.

Thus, the development of new glass property models allowing for accurate predictions of multiple glass properties in wide composition areas requires united efforts of multiple specialists in multiple domain areas: property measurements, structural studies, fundamental investigations, and database development. The author believes that future decades will bring remarkable progress in this direction. However, it is unlikely that this progress will immediately make the earlier experience and knowledge obsolete and useless. The author believes that only the use of all available information could allow one reaching remarkable practical results.

8
Glasses as Accumulators of Free Energy and Other Unusual Applications of Glasses
Ivan S. Gutzow and Snejana V. Todorova

8.1
Introduction

In the first three chapters of the present book we described the thermodynamic properties of glasses, directing our efforts on their *typical* behavior as representatives, as the best known form of condensed matter with frozen-in, nonequilibrium amorphous structure and the corresponding constant values of the configurational contributions to the thermodynamic functions. This unusual structural and thermodynamic state of glasses determines also their particular behavior, when compared with the equilibrium forms of condensed states: with crystals and with stable or metastable liquids.

The behavior even of glasses, obtained at the "typical" conditions of glass transition, is in many respects particular, even strange. It becomes even more unexpected when we synthesize glasses at nontypical, even at extreme conditions, applying ultra-rapid cooling, the methods of nano-dispersion, vacuum quenching, as discussed in Chapter 2. In these unusual ways of synthesis, amorphous solids, that is, *glasses*, can be obtained which are exceptionally *strange* in both their behavior and thermodynamic properties. In most cases, the experimentalists and the technologists applying such unusual glass-forming methods of vitrification are somehow afraid from the sometimes unusual properties of the glasses thus synthesized, obtained at extreme cooling rates or other methods of rapid change of external parameters. Strangely enough the usual method, employed even in contemporary technologies, is to try to return glasses, obtained as solids by such unusual methods, to the "typical," usual properties of condensed matter. This attitude, as it is shown in the present chapter, may not be the best method of employing many of the glasses, formed by extreme techniques. On the contrary, there are unusual methods to employ amorphous solids and thus to use the particular properties of glasses as frozen-in, nonequilibrium systems.

It is well-known that common glasses in their different compositions and typical appearance have been synthesized and used for more than 3000 years. There is, however, in line with above statements a strange peculiarity in the way glasses and

Glasses and the Glass Transition, First Edition. Jürn W.P. Schmelzer and Ivan S. Gutzow.
© 2011 WILEY-VCH Verlag GmbH & Co. KGaA. Published 2011 by WILEY-VCH Verlag GmbH & Co. KGaA.

glass-like substances have been employed even at the first stages of their history: in most cases as common materials without using their particular glassy properties and possible applications as kinetically frozen-in, thermodynamically nonequilibrium amorphous solids. On the contrary, glasses were utilized over many years of applications even as substitutes of either crystals or of liquids, that is, of systems in thermodynamic and structural *equilibrium*. In ancient Egyptian times, primitive glass or glass-like beads were used to imitate gemstones. In their most striking classical "high-tech" applications – as optical glasses – vitreous materials have undergone, for more than a hundred years now, a particular technical treatment: prolonged annealing. This treatment brings them as close as possible to the initial thermodynamic equilibrium state of the fluid, from which they have been evolved, that is, to the corresponding undercooled metastable liquid. In most cases, this or similar treatments are a technological necessity: diminishing the deviation from equilibrium. Optically observable strains and inhomogeneities are thus removed, mechanical durability is improved and the optical characteristics of glassy materials brought to desired perfection. However, in such a process of annealing, of relaxation and thermodynamic stabilization glasses as materials loose to a great extent the particular properties, inherent in their very nature as amorphous, nonequilibrium physical states.

The particular thermodynamic properties of glasses can be, however, very useful. The discussion in the present chapter shows that distinct, very specific applications of glasses are possible, requiring on the contrary an increased extent of their deviation from equilibrium, an increase of their frozen-in disorder. It turns out that the very nature of frozen-in disorder, even of possible inhomogeneities, of anything increasing the thermodynamic potential of the system, may be of use. Thus glass can be easily transformed into accumulators of energy, of frozen-in reactivity, even in materials for a novel application: in batteries of energy, based on their frozen-in defect structure. Increased solubility, high reactivity, increased electrochemical potential: all these and further specific properties of glasses can or could be used.

Common silicate glasses are practically insoluble in water and aqueous solutions. However, it turns out that even the minimal solubility of "stable" silicate glasses in aqueous media can be of use or has to be accounted for (e.g., in the immobilization of radioactive waste products). On the other hand borate and phosphate glasses with controlled, especially with constant leaching rate in luminous acid aqueous solutions can be employed as microfertilizers in agriculture, and vitrified water-soluble organic glasses as drug solvents in modern medicine. The very possibility of frozen-in disorder may be of use: even as frozen-in increased chemical reactivity, which can be used when necessary or in biological applications such as in freezing-in life, and in the realization of absolute anabiosis in physiological aqueous solutions.

We have seen in Chapter 5 that molecular inorganic crystals with frozen-in defect structure and pseudo-plastic organic crystals with glass-like behavior and so-called liquid crystals can also accumulate disorder and potential energy. This is why such crystalline or semi-crystalline structures and materials are also discussed in brief in the present chapter. It turns out that extremely high melting defect crystals may also be of interest as possible accumulators of energy.

Let us first consider here the ways in which we can describe, in order to regulate or to increase the nonequilibrium state of glasses, their deviation from equilibrium. This will thus characterize them from a new standpoint and thus open new horizons of application of glasses.

8.2
Ways to Describe the Glass Transition, the Properties of Glasses and of Defect Crystals: a Recapitulation

In Chapter 3, we have given an outline of the way in which the generic phenomenological theory describes glass transitions. Let us reiterate that the starting point of this description is the generalized nonlinear form of the phenomenological law (Eq. (3.24) in Chapter 3). We repeat it here writing it again as:

$$\frac{d\xi}{dt} = -L f(X), \quad X = \frac{A}{RT}. \tag{8.1}$$

In the same way as in Chapter 3, with $d\xi/dt$ we specify here the rate of change of the structural order parameter, ξ, and X is the thermodynamic driving force of this change. We introduce again the two necessary conditions, determining the unknown $f(X)$ function: first, at $X = 0$ we expect:

$$f(X)|_{X=0} = 0. \tag{8.2}$$

Secondly, we require moreover that at small deviations from equilibrium (when $X \ll 1$, see Chapter 3)

$$f(X)|_{X \ll 1} = X, \tag{8.3}$$

that is, the fundamental law obtains at such conditions its classical linear form. The latter constraint can be also written as:

$$\left.\frac{d f(X)}{d X}\right|_{X \ll 1} = 1. \tag{8.4}$$

The function $f(X)$ is specified with the above two conditions only in its linear approximation. We need, however, at least another, a third condition to specify sufficiently accurately the $f(X)$ function as a *nonlinear* dependence. In Chapter 3, we determined $f(X)$ in three different ways, (i) via the Gibbs thermodynamic potential (Eq. (3.26)), (ii) employing in a general mathematical approach the second-order derivative in the respective Taylor expansion of $f(X)$ (Eq. (3.29)) or (iii) via the general dependence, Eq. (3.34), connecting $f(X)$ with the relaxation times of the system. Here we follow an alternative way of determination of $f(X)$ allowing us to utilize more directly our knowledge on the physics of the process of glass relaxation we have to analyze.

In general, the experience in studying isothermal relaxation in glasses shows [1] that this is a nonlinear process, depending on time, t, and more specifically, on the degree of completion of the process itself. Thus Eq. (8.4), proposed for small values of X only, could be applied for higher values of X also, if we assume that in the general, the nonlinear, case $df(X)/dX$ is a function of $f(X)$ itself. This is in fact so in most cases of growth and decay processes, observed in nature, in general, and in glass relaxation, in particular. Thus we have to add on the right hand side of Eq. (8.4) an additional term $\psi[f(X)]$ which indicates the following: the process is a function, ψ, of the value of $f(X)$ itself. In the simplest possible assumption, based on such an approach, we could write Eq. (8.4) with $\psi = 1$ as:

$$\frac{df(X)}{dX} = 1 + f(X). \tag{8.5}$$

In a previous publication [17] we have also discussed several more complicated forms of Eq. (8.5), leading to other dependencies of growth and development of a system brought out of equilibrium: nonrestricted increase, cyclic growth and decay, catastrophic development, etc. Let us consider here only Eq. (8.5), as it corresponds with $f(X)_{t\to\infty} = 0$ to relaxation as a process returning the system to equilibrium, where $X = 0$.

For a function, for which above Eq. (8.5) is fulfilled, we can write for any $n \geq 2$

$$\frac{d^n f(X)}{dX^n} = \frac{d^{n-1} f(X)}{dX^{n-1}} \tag{8.6}$$

as this corresponds to the definition of an exponential function. Indeed, after integrating Eq. (8.5) with the additional condition, Eq. (8.2), we arrive at

$$f(X) = \exp(X) - 1, \tag{8.7}$$

representing an appropriate choice for our unknown $f(X)$ function. This function is satisfying both Eqs. (8.2) and (8.4) and the requirements of the thermodynamics of irreversible processes in its linear formulations, according to which we should expect the fulfillment of Eq. (8.2).

Equation (8.7) as also discussed in the mentioned paper [17] corresponds in fact to the $f(X)$ function which has been used to describe in a generalized way in terms of Eq. (8.1) chemical reaction kinetics (leading at $X \ll 1$ to van't Hoff's simple reaction formalism). It is also used to describe in a generalized way the *rate of structural change* according to the derivations of the theory of physicochemical similarity [595].

As it is also discussed in Chapters 2 and 3, with known expressions for the connections between affinity A and thermodynamic potential $\Delta G(T, \xi) \approx G^{(2)}(\xi - \xi_e)^2$ and with $A = (d\Delta G(T, \xi))/d\xi|_{T,p}$ we obtain from Eq. (8.1) the classical De Donder equation, connecting the rate $d\xi/dt$ linearly with $(\xi - \xi_e)$ (cf. Eq. (3.31)):

$$\frac{d\xi}{dt} = -\frac{(\xi - \xi_e)}{\tau_e}. \tag{8.8}$$

This simple relation, where, τ_e, is the time scale of the process,

$$\tau_e = \tau_0 \exp\left[\frac{U(T)}{RT}\right] \tag{8.9}$$

fails to describe relaxation in glasses. In Chapter 3 and [179, 180] it is shown that with Eq. (8.7) a solution in terms of the semi-linear equation

$$\frac{d\xi}{dt} \cong -\frac{(\xi - \xi_e)}{\tau_e} \exp\left[\frac{\xi - \xi_e}{\xi_e}\right] \tag{8.10}$$

can be obtained in a simple rearrangement procedure. Above semi-linear equation follows from Eq. (8.7) when the affinity A is expressed after the respective truncated Taylor expansion as $A = G^{(2)}(\xi - \xi_e)$ and the approximation $G^{(2)} = (d^2 \Delta G(T, \xi))/d\xi^2 \approx (RT)/\xi_e$ is used in assessing the value of the thermodynamic driving force, $\Delta G(T, \xi)$.

Further on, by introducing a constant cooling rate

$$q = \frac{dT}{dt} = \text{const.} \tag{8.11}$$

and employing appropriate dependencies connecting the thermodynamic properties and the structural order parameter, ξ, for example,

$$\frac{d\Delta H(T, \xi)}{d\xi} = h_0 = \text{const.} \tag{8.12}$$

in the way described in Chapter 3 we obtain finally

$$\Delta C_p(T, \xi) = \frac{\Delta H(T, \xi)}{q\tau(T, \xi)}, \quad \tau = \tau_e \exp\left(-\frac{\xi - \xi_e}{\xi_e}\right). \tag{8.13}$$

This is one of the basic equations which gives the possibility to describe the thermodynamics of vitrification, as was already shown in Chapter 3 and [179, 180].

In deriving Eq. (8.13) we have moreover employed a simple $\xi_e(T)$-dependence (over a MFA model of undercooled liquids, described in [1] and in Chapter 3 or by using a Kauzmann-like linear $\xi_e(T)$-dependence, as done in [11]). We have indicated in Eqs. (8.10) and (8.13) with $\tau(T, \xi)$ the generalized $\xi(T)$-dependent value of the relaxation time in the vitrifying liquid given here by

$$\tau(T, \xi) = \tau_0 \exp\left[\frac{U(T)}{kT}\right] \exp\left[-\frac{(\xi - \xi_e)}{\xi_e}\right]. \tag{8.14}$$

Thus we can perform with Eq. (8.13) an approximated description of the change of thermodynamic functions upon vitrification.

The integration of Eq. (8.13) in terms of known thermodynamic dependencies discussed in Chapter 3 leads to distinct expressions for the temperature dependence of the thermodynamic functions like entropy $\Delta S(T, \xi)$ and enthalpy

$\Delta H(T,\xi)$ on temperature, T, and cooling rate, q. With another known classic thermodynamic dependence

$$\Delta G(T,\xi) = \Delta H(T,\xi) - T\Delta S(T,\xi) \tag{8.15}$$

we could also directly describe the change of the Gibbs thermodynamic potential upon vitrification.

However, in performing such a program, we should have in mind that in analyzing the behavior of thermodynamic functions and potentials in the glass transition region (T_g^+, T_g^-) we should also take into account the entropy production term, $d_i S(\xi)/dt$. This, as discussed in Chapter 3 and in the recent paper of one of the present authors [240], is a process which at least in principle has also to be taken into account in considering the thermodynamic properties of glasses obtained at various cooling rates, q.

In using the formalism, described in [240], we have in fact to add to the expression of $\Delta S(T,\xi)|_{T<T_g}$ for the frozen-in entropy also the value of the entropy $\Delta_i S(\xi)$ produced in the glass transition interval $(T_g^{(+)}, T_g^{(-)})$ via the entropy production term $d_i S(\xi)/dt$, which at nonisothermal conditions, as anticipated here, has to be defined via Eq. (8.11) in terms of $d_i S(\xi)/dT$. Thus we have to write in a better approximation that for temperatures $T < T_g^{(-)}$ the entropy of the glass is to be calculated via the dependence

$$\frac{dS(T,\xi)}{dT} = \frac{C_p(T,\xi)}{T}\left[1 + \frac{A(T)}{C_p(T,\xi)}\frac{d\xi}{dT}\right], \tag{8.16}$$

where the right-hand term in the square brackets stems from the entropy production term. This term, as given in more details in Chapter 3 and in [240, 316], can be expressed by the affinity $A(T)$ and the rate of structural change $d\xi/dT$ as:

$$\frac{d_i S(\xi)}{dT} = \frac{A(T)}{T}\frac{d\xi}{dT}. \tag{8.17}$$

With the above-mentioned approximations for the thermodynamic driving force $\Delta G(T,\xi)$, its second derivative $\Delta G^{(2)}$ and the already discussed connection between $A(T)$ and $\Delta G(T,\xi)$, it becomes obvious that in fact Eq. (8.16) can be expressed as:

$$\frac{dS(T,\xi)}{dT} \cong \frac{C_p(T,\xi)}{T}\left[1 - \frac{RT}{h_0}\frac{(\xi-\xi_e)}{\xi_e}\right]. \tag{8.18}$$

In the glass transition region $RT \cong RT_g$ holds, and using known approximations for the melting enthalpy, ΔH_m, introducing the Beaman–Kauzmann formula, connecting melting temperature, T_m, with T_g we have to expect that $h_0 \approx \Delta H_m \approx (3 \div 5) RT_m \approx (3/2)(3 \div 5) RT_g$. Thus in Eq. (8.18) we have $(RT/h_0) \approx (0.2 \div 0.1)$, and with $(1 > \xi > 0)$ and $\xi > \xi_e$ (for cooling), as defined in Chapter 3, it becomes obvious that the right-hand correction term in the square brackets of Eq. (8.18) can be in general neglected, and in calculating the course of $S(T,\xi)$ via the integration of a $C_p(T,\xi)$-curve, as it follows from the solution of Eq. (8.13) (as done in constructing Figure 3.10b) is quite permissible as a first approximation. In doing the respective construction, q has negative values in cooling and positive values in heating in Eq. (8.13), respectively.

In order to use the Gibbs fundamental equation in nonequilibrium thermodynamics (see again Chapter 3), which (for $p = $ const.) reads as

$$d\Delta G(T,\xi) = -\Delta S(T,\xi)dT - A(T)d\xi, \quad (8.19)$$

to construct in an analogous way the $\Delta G(T,\xi)$ course upon glass transition from the respective $\Delta S(T,\xi)$ curves via

$$\frac{d\Delta G(T,\xi)}{dT} = -\Delta S(T,\xi)\left\{1 + \frac{A(T)}{\Delta S(T,\xi)}\frac{d\xi}{dT}\right\}, \quad (8.20)$$

a similar correction has to be considered as indicated with Eqs. (8.16) and (8.18).

The ratio $a_0 = [\Delta C_p(T,\xi)/\Delta S_m]$, which we called in [1, 177] the *thermodynamic structural factor*, has for glass-forming liquids in the temperature range (T_m, T_g) values of the order $a_0 \approx (1 \div 2)$. As discussed below, these limits correspond to distinct structural types of glass-forming melts. Here it is essential only to observe that in this way a_0 determines in the glass transition region the absolute value of the nonequilibrium correction for $\Delta G(T,\xi)$ which is typically higher for the $\Delta S(T,\xi)$ curve. More significant in this respect is, as even the structure of Eq. (8.15) shows, that the changes in the course of $\Delta S(T,\xi)$ are multiplied by temperature (here by $T \approx T_g$) and thus more effectively change also the course of $\Delta G(T,\xi)$. A thorough analysis shows [180, 316] that with these corrections, the "bumps" in the $\Delta G(T,\xi)$-curves shown in Figure 3.10 are practically eliminated. Thus the "classical" course of the change of the thermodynamic potential, as we constructed it in [1, 596] and in Chapter 2 following Simon, can be used with sufficient accuracy at least as a first approximation. This is done in the following section to determine in a sufficiently general but simple way the energetics of vitrification and to estimate the value of the potential energy, accumulated in frozen-in in amorphous or disordered crystalline structures.

To try to solve this task, going the theoretically more substantiated route, using the "thermodynamic" solutions of the Bragg–Williams equation is at this stage of the theory difficult, because still the unknown constants either have to be introduced into the basic equations or have to be estimated in a process of optimization. Moreover, no theoretical equivalent to the generic thermodynamic and kinetic approach derived here and in Chapter 3 for processes of glass transition is known at present in describing the formation and the state of crystals with inherent defect structure. In this case, as done below, we have to employ semi-empirical simple models: the results obtained with them, it turns out, can be compared more or less directly with the predictions obtained in the framework of Simon's classical approximation and its geometric interpretation, as done here in the subsequent analysis.

8.3
Simon's Approximation, the Thermodynamic Structural Factor, the Kinetic Fragility of Liquids and the Thermodynamic Properties of Defect Crystals

Reducing again with Simon's approximation the glass transition region to a single temperature, $T = T_g$, thus neglecting entropy production and the influence of the corresponding term $d_i S(T, \xi)/dT$ upon vitrification we can write Eq. (8.15) for temperatures $T < T_g$ as:

$$\Delta G_g(T, \xi) \cong \Delta H_g - T \Delta S_g . \tag{8.21}$$

This equation, in which we have introduced the notations $\Delta H(T, \xi)|_{T<T_g} = \Delta H_g$, $\Delta S(T, \xi)|_{T<T_g} = \Delta S_g$, and $\Delta G(T, \xi)|_{T<T_g} = \Delta G_g(T, \xi)$ as already employed in Chapter 2, is a sufficiently correct approximation. Moreover, considering the thermodynamic potential of the glass, frozen-in at glass transition as the tangent at $T = T_g$ to the real (for $T > T_g$) or fictive part (at $T \leq T_g$) of the course of the thermodynamic potential, $\Delta G(T)$, of the metastable undercooled liquid we have to write:

$$\Delta G_g(T, \xi) = \left. \frac{d\Delta G(T)}{dT} \right|_{T=T_g} (T - T_g) + \Delta G(T_g) . \tag{8.22}$$

This construction, which is illustrated in Figure 8.1, where we denoted it as Simon's diagram, gives a simple way to calculate the thermodynamic functions, frozen-in a glass and according to Eq. (8.18) the corresponding value of the thermodynamic potential, $\Delta G_g(T, \xi)$, corresponding to it.

Employing the well-known thermodynamic dependence,

$$\left(\frac{d\Delta G(T)}{dT} \right) = -\Delta S(T) , \tag{8.23}$$

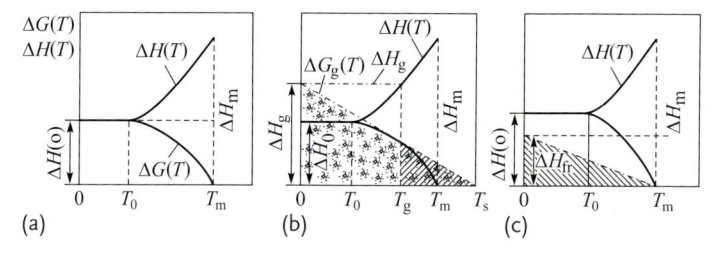

Figure 8.1 Schematic representation of Nernst's heat theorem (a), of Simon's glass transition diagram (b) and of the $\Delta G(T) - \Delta H(T)$ diagram of disordered crystals (c). In (a), the course and the transition liquid → crystal is given in accordance with the classical thermodynamic construction of Nernst; (b) illustrates Simon's approximate treatment of glass transition at $T = T_g$ in Nernst's terms; and (c) gives the $\Delta G(T)$, $\Delta H(T)$ course for the liquid → crystal transition in disordered crystals, in which the entropy $\Delta S_{mix} = R \ln 2$ is frozen-in. Note that $\Delta H_{frozen} < \Delta H_g$ according to Eqs. (8.54) and (8.55).

used already in Chapter 2, the slope of the straight line in Eq. (8.22) is given by $\Delta S(T)|_{T=T_g} \cong \Delta S_g$. The $\Delta G(T, \xi)$ course from Eq. (8.22) is illustrated here in Figure 8.1, where the construction we called Simon's $(\Delta G(T) \Delta H(T))$ diagram is given besides Nernst's classical $(\Delta G(T), \Delta H(T))$ course, illustrating the thermodynamic change melt/crystal in the corresponding equilibrium system. From Simon's construction (Figure 8.1b) and the corresponding triangle there (see the greater of the two triangles with differently shaded areas on Figure 8.1b) it is evident that the frozen-in enthalpy value, ΔH_g, in a glass and its frozen-in entropy, ΔS_g, are connected as:

$$\Delta H_g \approx T_s \Delta S_g . \tag{8.24}$$

In Eq. (8.24), T_s is the temperature (see [597]), where, if we should virtually prolong the Gibbs potential difference of the frozen-in glass *above* the temperature $T = T_g$, we should expect $\Delta G_g(T, \xi) = 0$. In reality, the frozen-in system could of course be sustained without measurable change at temperatures $T > T_g$ only for times, t, substantially smaller then the corresponding time of molecular relaxation, $\tau(T, \xi)$. It is also to be noted that, at $T > T_s$, the glass, virtually brought there, should be more stable than the respective undercooled melt.

Thus as far as Simon's approximated picture of the glass transition can be applied, via Eq. (8.24), we have connected in a general way frozen-in enthalpy and frozen-in entropy, provided T_s is known as a general or, at least, as a typical value. It has also to be noted that as far as $T_s > 0$, it is evident that in the case when ΔH_g is substantially different from ΔS_g (e.g., that at $\Delta H_g \gg 1$, we should expect $\Delta S_g \cong 0$) is precluded by the whole geometry of the thermodynamics of vitrification according to Simon and by Eq. (8.24), in particular. Such a case, derived in the framework of some recent molecular models (e.g., from the so-called energy landscape models of glass transition [252, 254]) is also impossible from more general viewpoints: for a structural parameter $\xi_e > 0$, any MFA-model of structural and configurational disorder provides an increased enthalpy only for an also increased value of the configurational entropy. Equation (8.24), can also be used to derive a generalized expression for the value of $\Delta G_g(T, \xi)$, the increased Gibbs energy of the frozen-in system at any temperature $T < T_g$, provided the course of $\Delta G(T)$ for the corresponding metastable melt is known.

For differently structured undercooled glass-forming melts, in general, different metastable liquid-crystal Gibbs potential courses of $\Delta G(T)$ should be expected. Any of these possible $\Delta G(T)$ courses has, however, to fulfill two general requirements [1, 112]: at the melting point

$$\Delta G(T_m) = 0 \tag{8.25}$$

has to hold and in the vicinity of the zero-point of absolute temperature

$$\Delta G(T = 0) = \text{const.} = \Delta H_0 \tag{8.26}$$

should be obeyed. Here ΔH_0 indicates the zero-point enthalpy of the respective fictive undercooled liquid (see Figure 8.1) at a temperature $T \to 0$ K.

Let us now expand the generally unknown $\Delta G(T)$-function in the vicinity of $T = T_m$ as a truncated Taylor series:

$$\Delta G(T) = \Delta G(T_m) + \frac{d\Delta G(T_m)}{dT}(T - T_m) + \frac{d^2\Delta G(T_m)}{2dT^2}(T - T_m)^2 . \quad (8.27)$$

Recalling another general thermodynamic dependence

$$\frac{dS(T)}{dT} = \frac{\Delta C_p(T)}{T} , \quad (8.28)$$

introducing the reduced temperature, $\theta = (T/T_m)$, and accounting for Eqs. (8.23) and (8.25) we can write Eq. (8.27) in the form

$$\Delta G(T) \cong \Delta S_m T_m \left[1 - \frac{\Delta C_p(T_m)}{2\Delta S_m}(1 - \theta)\right](1 - \theta) . \quad (8.29)$$

Here the ratio $a_0 = \Delta C_p(T_m)/\Delta S_m$ is the *thermodynamic structural factor* of the undercooled melts, introduced in 1981 by Gutzow [177] and by Gutzow and Dobreva ([10, 112], see also [1]) in describing the steepness of the temperature change of the thermodynamic properties of undercooled glass-forming liquids. Later on in analogy to a term proposed in 1985 by Angell [5, 362] to describe the temperature course of the viscosity, $\eta(T)$, of glass-forming liquids, this ratio we also called *thermodynamic fragility* [1, 10]. In general, as already said (see also [1, 10, 597]), the value of the parameter a_0 varies from $a_0 = 1$ for the simplest glass-forming structures (like SiO_2, GeO_2, BeF_2 or metallic alloy glasses) to $a_0 = 2$ for organic polymers, having a well-defined maximum of occurrence value at $a_0 \cong 3/2$ [1, 112] for the most typical glass formers. This most probable value of the ratio $[\Delta C_p(T_g)/\Delta S_m] \cong a_0$ was observed many years ago by Wunderlich [173] and is presently verified by ample experimental evidence summarized in [1, 112] and in the present book in Chapter 2.

With values of a_0, ranging from 1 through 1.5 to 2, it turns out we have in Eq. (8.29) a dependence, fairly well describing the temperature course of $\Delta G(T)$ of practically all typical classes of undercooled glass forming melts. In fact, introducing in Eq. (8.29) the three above mentioned a_0 values, several different $\Delta G(T)$ courses are obtained as approximations for the undercooled liquid/crystal Gibbs potential difference, frequently employed in crystallization equations as determining the respective driving force of crystallization [1, 112]. The respective potential difference courses are given below, beginning with the thermodynamically unrealistic, but mathematically also possible case $a_0 = 0$, leading to the convenient, often used classical expression for the thermodynamic driving force of crystallization

$$\Delta G(T) = T_m \Delta S_m(1 - \theta) , \quad (8.30)$$

proposed many years ago by J.J. Thomson and still employed in many publications in the current literature. For $a_0 = 1$, 1.5 and 2, Eq. (8.29) gives respectively (see

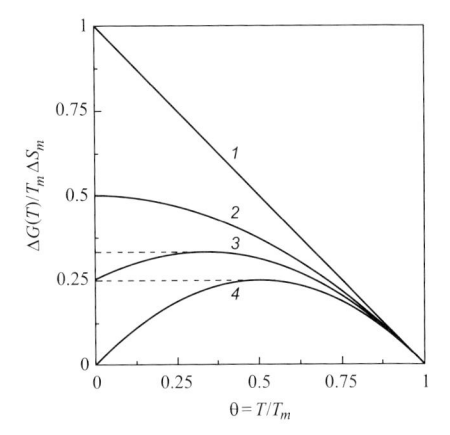

Figure 8.2 Temperature dependence, $\Delta G(T)$, of the Gibbs potential difference undercooled liquid/crystal for the four different solutions of Eq. (8.29): (1) according to Eq. (8.30) obtained for $a_0 = 0$; (2) according Eq. (8.31) with $a_0 = 1$; (3) according Eq. (8.32) with $a_0 = 1.5$; (4) according to Eq. (8.33) with $a_0 = 2$. A horizontal tangent indicates for $a_0 = 1.5$ and $a_0 = 2$ the continuation of $\Delta G(T)$ at $\theta < \theta_0$.

also [1])

$$\Delta G(T) = \left(\frac{1}{2}\right) T_m \Delta S_m (1 - \theta)(1 + \theta) , \qquad (8.31)$$

$$\Delta G(T) = \left(\frac{1}{4}\right) T_m \Delta S_m (1 - \theta)(1 + 3\theta) , \qquad (8.32)$$

$$\Delta G(T) = T_m \Delta S_m (1 - \theta)\theta . \qquad (8.33)$$

The dependence Eq. (8.31) was proposed by Spaepen for the temperature change of the thermodynamic potential in the crystallization of metallic liquids, while the relation Eq. (8.33), attributed to Hoffman, is usually employed in the analysis of crystallization of organic melts, and especially of polymer systems. Figure 8.2 illustrates the course of the four above discussed approximations for $\Delta G(T)$; the character of this change, when a_0 is increased from 0 to 2 is also clearly visible.

It is evident that the above dependencies define, with an increasing a_0 value, a change from a linearly *strong* (at $a_0 = 0$, $a_0 = 1$) to a substantially *fragile* (in Angell's terminology [1, 5]) $\Delta G(T)$-temperature course at $a_0 = 2$. In the above mentioned publications by Gutzow and Dobreva [10, 68, 112] it is shown, how a combination of a_0 values with kinetic factors and models of liquid flow leads to the formulation of a *kinetic fragility factor* similar to the one proposed by Angell to describe the temperature course of the viscosity of glass-forming liquids in generalized terms. Above derivations show also that the significance of the factor a_0 in the temperature dependence of all the thermodynamic functions comes in fact from its role in determining the temperature course of $\Delta G(T)$. Thus it enters in any possible approximation of the most substantial thermodynamic parameter of undercooled liquids: the temperature course of their Gibbs' potential.

According to known thermodynamic dependencies like Eq. (8.28), the Taylor expansion given with Eq. (8.29) and thus a_0 also determines the temperature course of any thermodynamic function of glass-forming liquids. In reduced coordinates, for example, in $\Delta S(T)/\Delta S_m$ vs. θ or $\Delta H(T)/\Delta H_m$ vs. θ terms, equal dependencies are thus expected for equal a_0 values. As far as $a_0 \cong 1.5$ is the most frequently observed value, the similar course of the reduced values of the thermodynamic functions first observed by Kauzmann [109] in 1948 thus finds its natural and simple *formal* explanation. In the classical theories of physicochemical similarity [595] the normalizing factors, bringing about corresponding states and similar dependencies (in appropriately reduced coordinates), stem from the values of the material constants, which the different substances have at the corresponding critical state (e.g., T_c, V_c, p_c in terms of the van der Waals equation). For most glass-forming substances the values of the critical constants are unknown and most of the corresponding substances of interest cannot be brought into this state. By this reason, an *alternative normalizing state* has to be employed, using the temperature of the melting point, T_m, and the values of the thermodynamic functions of (e.g., melting enthalpy, $\Delta H_m = T_m \Delta S_m$, and the entropy of melting, ΔS_m, itself) of the respective glass-forming substances. As far as the values of these functions are considered for the metastable state of equilibrium substances, the choice of glass transition point, T_g, as a reference state although often plasticized in glass-science literature (see [5]), is inadequate, as it refers the substances to the nonequilibrium state of the glass and to a time dependent, not thermodynamically defined but kinetically determined temperature, as T_g in fact is. The possibility nevertheless to use T_g as a reference temperature in characterizing glassy systems comes from the fact that at typical, or normal conditions of vitrification (e.g., at cooling rates $(q = (10^{-1} \div 10^{-3})$ K/s) the value of T_g is determined by the already discussed empirical Beaman–Kauzmann relation ($T_g \cong (2/3) T_m$) by the respective melting temperature.

More convenient, although less general in treating the thermodynamics of glass-forming liquids, is another approximation for $\Delta G(T)$, proposed in the monograph [1], according to which it is assumed that in the temperature interval (T_m, T_0), that is, from $\theta = 1$ to $\theta = \theta_0$ the value of $\Delta C_p(T)$ is a constant, while for $\theta \leq \theta_0$ in accordance with the requirements of the third principle of thermodynamics it states that $\Delta C_p(T) = 0$. With this simplified, but thermodynamically correct model, defining the temperature dependence of the specific heats of an undercooled glass-forming melt as

$$\frac{\Delta C_p(T)}{\Delta S_m} = \begin{cases} a_0 = \text{const.} & \text{for } \theta_0 \leq \theta \leq 1 \\ a_0 = 0 & \text{for } 0 \leq \theta \leq \theta_0 \end{cases} \quad (8.34)$$

the value of the Gibbs thermodynamic potential has to be written with Eq. (8.28) as

$$\frac{\Delta G(T)}{T_m \Delta S_m} = \begin{cases} (1 - a_0)(1 - \theta) - a_0 \theta \ln \theta & \text{for } \theta_0 \leq \theta \leq 1 \\ (1 - a_0)(1 - \theta_0) - a_0 \theta_0 \ln \theta_0 & \text{for } 0 \leq \theta \leq \theta_0 \end{cases}. \quad (8.35)$$

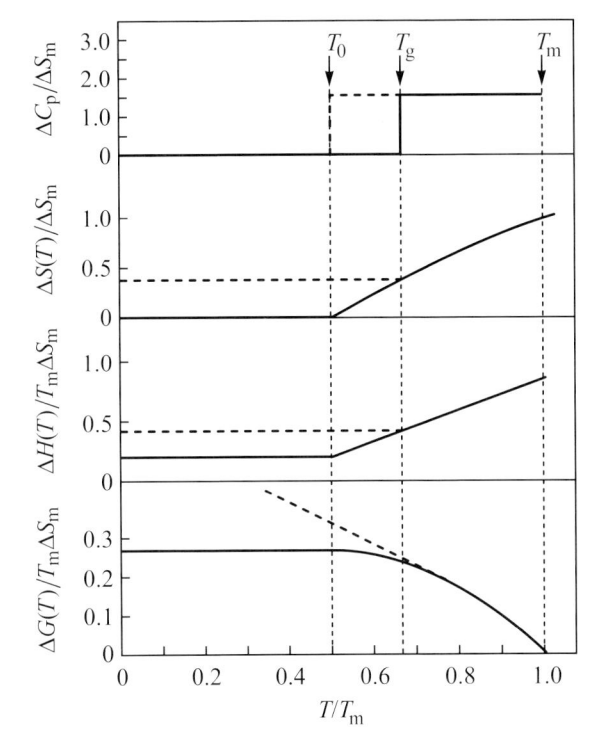

Figure 8.3 Temperature course of thermodynamic functions of the metastable undercooled melt calculated with the approximation Eq. (8.34) and the resulting dependencies (*solid lines*). Hereby the parameter a_0 was set equal to $a_0 = 1.5$, corresponding to the most probable value of this ratio for typical glass-formers. By *dashed lines* the respective dependencies for the frozen-in glass are given as they follow from this thermodynamic model.

In Figure 8.3 the temperature course corresponding to $\Delta C_p(T)$, the respective thermodynamic functions $\Delta H(T)$, $\Delta S(T)$ and $\Delta G(T)$ is illustrated for this thermodynamic model of an undercooled liquid. To a first approximation, as discussed in detail in [1], this approximation describes with sufficient accuracy the thermodynamic behavior of simple glass-forming liquids.

Following Eq. (8.23), we can also define by the differentiation of any of the above dependencies Eqs. (8.30)–(8.33) in the same general manner the temperature dependence of the configurational entropy of the undercooled melt, for example, as

$$\Delta S(T) = \Delta S_m [1 - a_0 (1 - \theta)] \approx \Delta S_m [1 + a_0 \ln \theta] \quad (8.36)$$

for any value of the reduced temperature, $\theta_0 \leq \theta \leq 1$. Here again the value of the parameter a_0 determines the course of the $\Delta S(T)$-dependence of a given type of glass forming melts. However, from the right hand dependence, indicated with Eq. (8.36), it becomes evident that, in general, the thermodynamic validity of any one of above indicated approximations for both $\Delta G(T)$ and $\Delta S(T)$ is restricted (as indicated above) to values $\theta \geq \theta_0$ where θ_0 is d determined by

$$\Delta S(\theta_0) = 0. \tag{8.37}$$

The logarithmic function on the right-hand side of Eq. (8.36) defines with above condition the value of a_0 by the expression

$$\theta_0 = \exp(-1/a_0). \tag{8.38}$$

The thermodynamic condition, indicated with Eq. (8.37), also defines the temperature limits of application (again in terms of the reduced temperature, θ) of the $\Delta G(T)$ course according to the approximations Eqs. (8.31)–(8.33), as this is seen in Figure 8.2.

The thermodynamic and structural significance of the ratio $a_0 = [\Delta C_p(T)/\Delta S(T)]$ comes from the circumstance, as we pointed out in [251], that it indicates (via the value of $\Delta S(T)$) the change of configurational structure, caused by the introduction of a unit of thermal energy into the system by the increase of the temperature of the system by one temperature unit. This quantity of heat absorbed is determined at any temperature, T, by the value of $\Delta C_p(T)$. In our definition of a_0, we characterize any substance by the value of the said ratio at the respective melting temperature, T_m. Thus, by introducing both $\Delta C_p(T)$ and $\Delta S(T)$ at $T = T_m$, the value of a_0 reaches the meaning of a characteristic material constant.

In the following section another characteristic, the entropy of melting, is introduced together with the enthalpy of melting to characterize the ability of a substance to accumulate energy. The value of a_0 it turns out is of significance also in this characterization of different glass-forming melts.

8.4
The Energy, Accumulated in Glasses and Defect Crystals: Simple Geometric Estimates of Frozen-in Entropy and Enthalpy

Let us now first determine the values of the thermodynamic functions, frozen-in in a glass upon vitrification at Simon's temperature, $T = T_g$. Then we will analyze in an analogous way this problem for crystalline systems with frozen-in defect structures.

8.4.1
Enthalpy Accumulated at the Glass Transitions

With Eq. (8.22), with the definition of ΔS_g according to Eq. (8.23) and the triangle indicated with a dotted-shaded area on Figure 8.1b, the temperature T_s, introduced in the previous section, can be determined as

$$(\theta_s - \theta_g) = \frac{\Delta G(T_g)}{T_m \Delta S_g}, \tag{8.39}$$

where $\theta_s = (T_s/T_m)$ and $\theta_g = (T_g/T_m)$. With Eqs. (8.24), (8.29), (8.39) and the first of the two right-hand expressions in Eq. (8.36) we have

$$T_s = T_g + \frac{\left[1 - \frac{a_0}{2}(1 - \theta_g)\right](1 - \theta_g)}{1 + a_0 \ln \theta_g} T_m \ . \tag{8.40}$$

Using again the left-hand expression from Eq. (8.36) we can also write

$$\Delta H_g = T_s \Delta S_m \left[1 - a_0 (1 - \theta_g)\right] . \tag{8.41}$$

Again recalling the above mentioned expression (Eq. (8.36)) we obtain as the direct connection between ΔH_g and ΔS_g the simple dependence, already indicated with Eq. (8.24), having estimated now, however, via Eq. (8.40) the value of T_s and its dependence on both structural characteristics (via a_0 and on the conditions of vitrification), which determine the value of the glass transition temperature, T_g. Thus, considering Eqs. (8.40) and (8.41), we have obtained in a simple geometric expression, accounting for both the structure of the vitrifying melt (via a_0) and for its change with the respective conditions of vitrification, indicated by the respective T_g value, determined according to Bartenev's equation (see [169] and its derivation in Chapter 3)

$$\frac{1}{T_g} = C_1 - C_2 \log |q| \tag{8.42}$$

by the cooling rate, q. Assuming for the activation energy, $U(T)$, in Eq. (8.9) for the time of molecular relaxation, τ, the simplest possible dependence ($U(T) = U_0 \approx$ const.) the constants in Bartenev's equation (see [1, 169, 371] and its derivation in Chapter 3) have the values

$$C_1 = \frac{2.3R}{U_0} \log \left(\frac{R T_g^2}{U_0} \frac{1}{\tau_0}\right), \quad C_2 = \frac{2.3R}{U_0} \ . \tag{8.43}$$

The term τ_0 as in Eq. (8.9) in the dependencies above is again identified with the characteristic time of eigen-frequency oscillations of the mean building units of the vitrifying liquid.

More generally instead of q we can introduce in Eq. (8.42) the rate of change, w, of any external parameter of state, determining the process of glass transition. In a series of publications by Gutzow and Dobreva parameters were considered, including cooling or heating rate, as well as the rate of pressure increase, (dp/dt) [383], and of the change of magnetic, M, or electric field strength, E [337, 383]. In doing so, the value of the activation energy, U, has to be considered as a function of T and also of p, M, or E, as in fact done in [337, 383, 384]. In this way a wide field of possibilities is indicated from the formalism above to obtain from the same substance (i.e., at the same T_m, a_0 values) glasses with a different amount of frozen-in disorder (measured, for example, in terms of the structural order parameter, ξ) and a corresponding frozen-in enthalpy and entropy, ΔH_g, and ΔS_g. It is essential to mention that, according to above derivations and Eq. (8.24), these two values are interconnected and with Eq. (8.21) they also determine the respective frozen-in value of the Gibbs free energy of the respective glass for temperatures $T < T_g$.

From Eq. (8.42) it is evident that at higher cooling rates, q, the value of T_g is increased (although in a logarithmic scale) and even T_g values approaching melting point T_m are to be expected. At $T_g = T_m$, according to Eq. (8.40), we would also have $T_s = T_m$ and according to Eq. (8.24) and also $(\Delta H_g/\Delta H_m) \approx (\Delta S_g/\Delta S_m)$: that is, at vitrification at melting point temperatures, the same relative part of both the enthalpy and the entropy of melting would be frozen-in. Both higher cooling rates, q, and the increase of the complexity of the glass-forming melt (at $a_0 \to 2$) enlarge according to Eq. (8.41) the value of ΔH_g. In general, systems with higher complexity (i.e., with higher values of a_0 and of ΔS_m) should be preferred as accumulators of frozen-in enthalpy, ΔH_g. Such an effect according to Eq. (8.42) is expected with higher melting systems, and this is natural if we recall that T_m (and even more, generally the enthalpy of melting, $\Delta H_m = T_m \Delta S_m$) is a simple measure of the bonding strength in a system.

In Section 3.3.2 of the present book and in Chapter 5 of the monograph [1] we introduced a simple lattice-hole MFA molecular model of a vitrifying liquid. According to this model the equilibrium value, $\xi = \xi_e$, of the generalized structural parameter, ξ, of the system is determined according to Eq. (3.50). In the vicinity of the respective glass transition temperature T_g (where $\xi \ll 1$ and thus also $(1-\xi)^2 \approx 1$) we get

$$\xi \approx \exp\left(-\chi \frac{\Delta H_{ev}}{RT}\right). \tag{8.44}$$

In the framework of the same MFA model the configurational enthalpy and entropy of the vitrifying model liquid system are defined as

$$\Delta H(T) = \chi \Delta H_{ev} \xi (1-\xi) \approx \chi \Delta H_{ev} \xi \tag{8.45}$$

and

$$\Delta S(T) = -R\left[\ln(1-\xi) + \frac{\xi}{1-\xi}\ln\xi\right] \approx 2.5 R\xi . \tag{8.46}$$

The right hand sides of both dependencies follow with the above indicated approximations considering the expected values of ξ in the glass transition region (see also Section 3.2 in Chapter 3).

Suppose now that in a cooling run performed with a cooling rate, q, we have frozen-in according to Bartenev's equation (8.42) at $T = T_g$ the value $\xi \approx \xi_g$, as it follows from Eq. (8.44). The frozen-in value of the configurational enthalpy and entropy according to above given latter two equations thus would be $\Delta H_g \approx \chi \Delta H_{ev} \xi_g$ and $\Delta S_g \approx 2.5 R \xi_g$. From the ratio of these two quantities we obtain according to Eq. (8.24) a very reasonable value of the temperature $T_s \approx ((\chi \Delta H_{ev})/(2.5R))$. Here as in Chapter 3 with ΔH_{ev} we indicated the evaporation enthalpy of the model liquid, and χ is a numerical factor, typically equal to ~ 3.8.

The significance of the aforementioned considerations is that they show that from a phenomenological as well as from molecular point of view between configurational enthalpy and entropy a nearly linear connection has to exist. Thus the possibility is precluded that, at a given value, ξ, the existence of a frozen-in state and a structure having both $\Delta H_g = \text{const.} > 0$ and $\Delta S_g = 0$ should be possible. It can be shown, as indicated in the above considerations, that to any value of $\xi_g > 0$ according to any molecular MFA-model these two configurational thermodynamic functions are interconnected in a way similar to the one indicated by Eqs. (8.45) and (8.46). In this way (at least, from the standpoint of more or less complicated molecular models), the real existence of frozen-in systems, which, having significant ξ_g and ΔH_g values, should correspond to a zero configurational entropy (as this is expected according to some recent models of glass transition [252–254]) has to be rejected.

8.4.2
Free Energy Accumulated at the Glass Transition and in Defect Crystals

The geometric approach described above in determining ΔH_g can also be applied to crystals with frozen-in configurational disorder, as another case of possible interest to accumulate not only disorder, but also the increased energy connected with it. Classical examples in this respect, already discussed in Chapter 5, are defect molecular crystals like CO, N_2O, H_2O (ice I), well-known from the efforts to verify experimentally the third principle of thermodynamics (see also Chapter 9 and [182, 598]) from glass transition like processes in so-called glassy crystals. We mentioned this second aspect in Chapter 5, with glass transition-like processes in plastic crystals like cyclohexanone [237, 377] and vitrification of so-called liquid crystals, which are in fact only liquids with an additionally imposed orientational order. In this sense, the simple geometric considerations and the respective derivations, made in the preceding section, can be more or less directly applied to both plastic and liquid crystals with frozen-in disorder: there both thermodynamic and kinetic measurements indicate the existence of a glass transition interval [377], and a glass transition temperature, T_{gc}, considerably lower than the melting point, T_m, can be defined as in typical glass-formers. This temperature, as demonstrated experimentally [377] depends typically on the cooling rate, q, in the way, as predicted by Bartenev's equation, Eq. (8.42).

It seems, however, that defect systems also exist, like the above mentioned molecular crystals, where the defect crystalline structure is formed in the immediate vicinity below the melting point, T_m, of the crystal, seemingly at T_m itself. It was also shown by experimental evidence, provided by Kaischew [355] many years ago for CO, that the formation of the defect structure in such crystals does *not* depend on the conditions of crystallization (e.g., on cooling rate), or even on the way of crystallization (from the melt, via vapor quenching, etc.). Usually it is assumed that the configurational disorder, frozen-in in defect molecular crystals of the mentioned type, is caused by the existence of two (in ice, there are three [354, 378]) configurational states of the building units of the crystal, characterized by a very low energy difference. Thus the molecular fractions x_1, x_2 of differently oriented molecules in both states 1 and 2 are nearly equal (i.e., $x_1 \approx x_2 = (1 - x_1)$). This determines a mixing entropy term of frozen-in configurational disorder

$$\Delta S_{mix} \approx \Delta S_{frozen} = -R\left[x_1 \ln x_1 + (1 - x_1) \ln (1 - x_1)\right], \tag{8.47}$$

approximately equal (at $x_1 = 0.5$) to $\Delta S_{frozen} \cong R \ln 2$ (in the particular case of ice with three different orientational states, $\Delta S_{frozen} \cong R \ln (3/2)$, see [354]). Thus ΔS_{frozen} is not a part of ΔS_m (see Chapter 5) and we have to expect a Nernst–Simon-like construction of the $\Delta G(T)$, $\Delta H(T)$ – diagram of a new type, as it is given on Figure 8.1c. In this case, the temperature T_s defined in the previous section with Eq. (8.24), is to be taken equal to the melting point (i.e., here $T_s \approx T_m$ as mentioned as a possibility with Eq. (8.40)). With the value of ΔS_{frozen}, calculated from Eq. (8.47), we have (see the corresponding shaded triangle on Figure 8.1c) to write in analogy with Eq. (8.24) that in a defect crystal of the considered type

$$\Delta H_{frozen} \cong T_m \Delta S_{frozen} \approx \Delta T_m R \ln 2 \approx T_m \left(\frac{\Delta S_m}{3}\right) \ln 2. \tag{8.48}$$

In writing the last right-hand part of the above expression we have expressed the entropy of melting of the considered crystal as

$$\Delta S_m = nR. \tag{8.49}$$

The value of the number n in the case of molecular crystals is usually found in the limits of $n = (2 \div 4)$. Thus $n \cong 3$ can be used as a rough approximation, leading to the right hand part of Eq. (8.48). Thus we can attempt a comparison of the value of enthalpy ΔH_{frozen} in the defect crystal with the enthalpy ΔH_g, typically frozen-in in a glass, for example, at "normal' cooling rates ($q \approx 10^1 \div 10^3$ K/s). In making this comparison we use the results, given in the previous sections, connecting the "typical" values of the frozen-in parameters of glass with the concepts of the thermodynamic and kinetic invariants of vitrification.

In order to estimate in the same rough approximated manner the values of the thermodynamic functions frozen-in in a glass at the mentioned typical cooling conditions, we have to recall existing approaches, connecting both via a kinetic derivation (given in Chapter 3) and also in a purely empirical manner ΔH_g, ΔS_g and T_g with the corresponding values of the same properties at the melting point, that is, with ΔH_m, ΔS_m and T_m in the form of the *glass transition invariants*. The first of these invariants, the Beaman–Kauzmann rule [1], determines the "normal" glass transition temperature as

$$\frac{T_g}{T_m} = \theta_g \cong \frac{2}{3}. \tag{8.50}$$

It was observed years ago by Gutzow [110] that, with Wunderlich's most probable mean value of the ratio

$$a_0 = \frac{\Delta C_p(T_m)}{\Delta S_m} \approx \frac{3}{2}, \tag{8.51}$$

it follows from the right hand side of the approximation of Eq. (8.36) that for the above Beaman–Kauzmann glass transition temperature

$$\Delta S_g \approx \frac{1}{3} \Delta S_m \tag{8.52}$$

has to be taken as a proper estimate. Thus with Eq. (8.51) and the above mentioned a_0 value we expect with Eq. (8.40) that the "normal" value of the temperature T_s is

$$\frac{T_s}{T_m} = \theta_s \approx 1.5 \,. \tag{8.53}$$

Thus Eq. (8.24) gives with Eq. (8.53) a possible approximate value of the frozen-in enthalpy in "typical" glasses, obtained at "normal" cooling rates

$$\Delta H_g \approx \frac{1}{2} \Delta H_m \,. \tag{8.54}$$

Experimental evidence, collected in the long years of thermodynamic glass science research, is summarized in the monograph [1] and elsewhere [68, 112], it gives a good coincidence with the invariant values according to Eqs. (8.50)–(8.54). It has also to be noted that at extremely high cooling rates (as realized in the vacuum quenching of Ar-vapors on He-cooled substrates in the experiments of Kouchi and Kuroda on the synthesis of Ar-glass [599, 600]) in fact T_g values approaching the melting point $T_s \approx T_m$ have been observed and also consequently the equality $\Delta H_g / \Delta H_m \approx \Delta S_g / \Delta S_m$ has been realized. In "normally" cooled typical glasses, however, typically only half of the enthalpy of melting, ΔH_m, is frozen-in.

A considerably lower estimate follows from Eq. (8.48) for the enthalpy accumulated in defect crystals with a disordered structure, formed at the melting point itself

$$\Delta H_{\text{frozen}} \cong \Delta H_m \left(\frac{\ln 2}{3} \right). \tag{8.55}$$

For crystals, in which a defect structure is frozen-in in the same way as in glasses (i.e., for "glassy crystals" [237]) with a freezing-in temperature, $T_{g,c}$, considerably below the respective melting temperature, T_m, follow ΔH_g values, lying in between the estimates indicated with Eqs. (8.54) and (8.55).

The value and the temperature dependence of the Gibbs free energy, $\Delta G_g(T, \xi)$, of the respective frozen-in systems (glasses, glassy crystals, vitrified "liquid crystals," defect crystals) determines, as shown below, their increased solubility, their high chemical reactivity, and their particular electrochemical properties. It is thus a measure of increase of the effective work, which could be done by a solid with frozen-in disorder. In writing Eq. (8.21) with the now known values of ΔH_g and ΔS_g, derived from above simple considerations, it follows that for normal cooling rates (i.e., for θ_g and θ_s according to Eqs. (8.50) and (8.53)) we could expect as a most-probable temperature dependence for the frozen-in Gibbs potential in "typical" glasses the expression

$$\frac{\Delta G_g(T, \xi)}{\Delta S_m T_m} \cong \frac{\Delta S_g}{\Delta S_m} (\theta_s - \theta) \approx \frac{1}{3} (\theta_s - \theta) \approx \frac{1}{2} - \frac{1}{3} \theta \,. \tag{8.56}$$

The right hand equality follows with the mentioned typical value of θ_s. From Eq. (8.21), accounting for the frozen-in entropy value $\Delta S_{\text{frozen}} \approx R \ln 2$, we obtain with Eqs. (8.55) and (8.49) that

$$\frac{\Delta G_{\text{frozen}}(T, \xi)}{\Delta S_m T_m} = (1 - \theta) \frac{R}{\Delta S_m} \ln 2 \cong \frac{1}{3}(1 - \theta) \ln 2 \tag{8.57}$$

holds. Thus a considerably lower value for the frozen-in Gibbs free energy is to be expected in typical defect crystals, where the defects are created in the vicinity of T_m.

Suppose, we would like to construct a device, an engine, based on the *glass* → *crystal* reaction. Using a well-known thermodynamic formalism [177] the thermodynamic coefficient of efficiency, Φ, employing such a device, should be given (at $T < T_g$) by

$$\Phi = \frac{\Delta G_g(T)}{\Delta H_g} \cong 1 - \frac{1}{6}\theta . \tag{8.58}$$

The right hand side approximation indicated here follows with Eqs. (8.54) and (8.56). It shows that with decreasing temperatures Φ approaches values close to one. However, as discussed in detail in Chapter 9, the absolute zero of temperatures cannot be reached by a *glass* → *crystal* reaction, and thus the value $\Phi = 1$ remains impossible to reach, as this is expected in any other case from the general formulations of the third law of thermodynamics. It is nevertheless to be noted, that according to Eq. (8.58) a device, based on any form of the *glass* → *crystal* reaction should be (from a thermodynamic point of view) more effective at *low* temperatures: its efficiency there will be however restricted by *kinetic* constraints, which become more important at low temperatures.

Let us now compare the possibilities, given by employing glassy materials, when compared with other traditional methods of activation of solids. A well-known possibility of increasing the amount of accumulated energy in solids is to disintegrate (e.g., by milling) the solid to an array of particles (e.g., spheres) of a mean radius R_s. Applying the Gibbs–Thomson equation (see, e.g., [1, 601]) the increased free energy of the system would be

$$\Delta G_s(T) = \frac{2\sigma V_m}{R_s} . \tag{8.59}$$

Here with σ is denoted the surface energy at the disintegrated solid (crystal or glass) to air (or solution) interface and V_m is the molar volume of the considered solid. We introduce as an estimate of σ the corresponding change of enthalpy, ΔH^* at the respective interface the Stefan–Skapski–Turnbull rule [1]

$$\sigma = \gamma_0 \frac{\Delta H^*}{N_A^{1/3} V_m^{2/3}} . \tag{8.60}$$

Here $\gamma_0 \cong 0.5$ is a steric coefficient and N_A stands for Avogadro's number. Thus Eq. (8.59) can be written (for temperatures $T \approx T_m$ and dividing it by $\Delta H_m =$

$T_m \Delta S_m$) in the form

$$\frac{\Delta G_s(T_m)}{\Delta S_m T_m} \cong \frac{2\gamma_0}{\mu}\left(\frac{\Delta H^*}{\Delta H_m}\right). \tag{8.61}$$

Here $\mu = (R_s/d_0)$ describes the dispersion of the system ($d_0 \approx (V_m/N_A)$) and the ratio ($\Delta H^*/\Delta H_m$) is, when ΔH^* indicates the evaporation heat of our solid, approximately equal to $8 \div 10$. In dissolution experiments, however, $\Delta H^* \approx \Delta H_m$ and in such a case this ratio is approximately equal to one.

Thus it turns out that even at disintegration to nano-sizes ($\mu \cong 5$), for example, in air, the increase of $\Delta G(T)$ to $[\Delta G(T) + \Delta G_s]$ is approximately equal to the effect of vitrification. Disintegration has thus to be mainly considered as a method of additionally increasing the $\Delta G_g(T)$ value of an already vitrified system to $[\Delta G_g(T,\xi) + \Delta G_s]$. In considering its applicability it has, however, to be accounted for, that Eq. (8.61) indicates that only disintegration to nano-sizes ($\mu < 10$) could guarantee measurable effects. Moreover, the effect of additional disintegration in solution (i.e., at $\Delta H^* \approx \Delta H_m$) is considerable lower.

Other methods of additional increase of the Gibbs free energy, $\Delta G(T)$, of a solid glass could include the possible influence of elastic strains (produced, for example, in quenching the glass sample down from the temperature of vitrification, T_g, to room temperatures, T_R). This case is discussed by the present authors in details in a previous publications [596, 602] with the result that in typical inorganic glasses vitrification strains, developed between T_g and T_R, could scarcely produce effects, leading to an overall additional increase of $\Delta G_g(T)$ more than $\Delta G_{add} \cong 1.02 \Delta G_g(T)$.

8.5
Three Direct Ways to Liberate the Energy, Frozen-in in Glasses: Crystallization, Dissolution and Chemical Reactions

The increased Gibbs free energy, $\Delta G_g(T,\xi)$, and the respective enthalpy, ΔH_g, frozen-in in a glass can be released, when the kinetic barrier, sustaining the nonequilibrium state of the system is eliminated. The first and simplest way to do this is to initiate (e.g., by a temperature rise to $T > T_g$) devitrification (i.e., crystallization) of the glass. Thus according to Eq. (8.54) half of the melting point enthalpy, ΔH_m, of the substance can be released and used. Of considerable help in initiating, conducting and directing this process could be different forms of nucleation catalysis: the preliminary introduction into the glass of a sufficiently high concentration, N^*, of active crystallization cores, initiating nucleation, or of soluble dopants, decreasing the crystal/melt interface energy, $\sigma_{c,l}$. By decreasing $\sigma_{c,l}$ again nucleation is enhanced. The possibilities in both ways of nucleation catalysis and their concrete applications in various technical glass-forming systems are summarized in Chapter 7 of the monograph [1] and also in the more recent publications [603, 604].

8.5.1
Solubility of Glasses and Its Significance in Crystal Synthesis and in the Thermodynamics of Vitreous States

The second way to free the frozen-in energy in a glass is to dissolve it in an appropriate solvent. Because the glass has a higher solubility than the crystal, in this way a metastable solution is formed, supersaturated with respect to the corresponding crystal. This peculiarity of glasses indicates possible applications in using glasses, because of their expected and experimentally verified high solubility, as sources of constant super-saturation in isothermal crystal growth syntheses from solution. The increased solubility of glass has also another very general significance: if properly measured, it gives a direct way of determining $\Delta G_g(T, \xi)$ and of demonstrating the particular properties of the vitreous state as a kinetically frozen-in thermodynamic system.

This general significance of the solubility and of the vapor pressure of glasses was first recognized by Simon [51], who however also pointed out the difficulties, connected with vapor pressure and solubility measurements of glasses: the possible relaxation of the bulk of the glass or of the solution/glass interface during the measurement itself. This is why most of existing measurements in this direction, known from the beginning of the 1930s, have in most cases to be considered only as more or less semi-quantitative determinations. They usually give a qualitatively true insight into the problem or have initiated technically important syntheses: in their physicochemical and thermodynamic details they are, however, mostly open to criticism.

In 1986, Grantcharova and Gutzow [185] (see also [172]) proposed to determine the solubility of glasses, $C_{gl}(T)$, by measuring (in accordance with Nernst's kinetic solubility model) the *rate of dissolution* of the glass, $dC(t)/dt$. This was done at constant temperature, T, by determining via direct *in situ* calorimetry the mass, $C(t)$, of the solute dissolved at time t in the solution. As the necessary dissolution model was used the dependence

$$\frac{dC(t)}{dt} = \frac{D_0 F^*}{\delta^* v^*} \left[C_{gl}(T) - C(t) \right]. \tag{8.62}$$

Here D_o is the diffusion coefficient of solute molecules in the solvent used, δ^* is is the thickness of Nernst's diffusion layer, F^* denotes the glass/solution interface area and v^* is the volume of the solution investigated. Denoting with $\tau^\# = (\delta^* v^*)/(D_0 F^*)$ the effective time-constant of the dissolution process, the solubility of the glass was obtained as the concentration $C_{gl}(T)/v^*$ either from the $C(t)$-measurements in the logarithmic terms of the integrated form of above equation, which reads

$$\ln \left\{ \frac{C_{gl}(T) - C(t)}{C_g(T)} \right\} = -\frac{t}{\tau^*}, \tag{8.63}$$

or directly from the initial slope of the dissolution curves according to

$$\left.\frac{d C(t)}{d t}\right|_{t \to 0} = \frac{C_{gl}(T)}{\tau^{\#}}. \tag{8.64}$$

As a model system in the investigation of Grantcharova and Gutzow [185] was chosen the dissolution of a low-melting organic substance (phenolphthalein) in alkalified aqueous solutions. In such solutions phenolphthalein is dissolved with an intensive red color enabling a direct colorimetric determination of $C(T)$ of both the vitreous and the crystalline samples. In order to preclude bulk relaxation of the glass the solubility measurements reported in [185] were performed at temperatures sufficiently below the glass transition temperature, T_g, of the chosen model glass, well-known in its rheological behavior from previous investigations. With the described dissolution way of direct $C(t)$-measurements Simon's re-condensation cycle and the possibility of relaxation processes in the glass/solution interface was avoided.

As seen from Figure 8.4, smooth dissolution curves from both the crystalline and the glassy samples were obtained, corresponding to the logarithmic form of Nernst's equation (8.63). At conditions where crystallization of the metastable solution, obtained by the dissolution of the glassy samples could not be avoided (i.e., when at higher temperatures $C_{gl}(T) \gg C_{cryst}(T)$), the solubility of the glass was determined via Eq. (8.64). The final results of the whole investigation of the solubility of both crystalline and vitreous phenolphthalein are given on Figure 8.5 in $\log C(T)$ vs. $1/T$ coordinates.

In general, for temperatures sufficiently below the vitrification temperature T_g it can be expected (see Gutzow [248] and [172, 185]) that the vapor pressure of the glass, P_{glass}, and the vapor pressure of the respective crystal, P_{cryst}, are connected

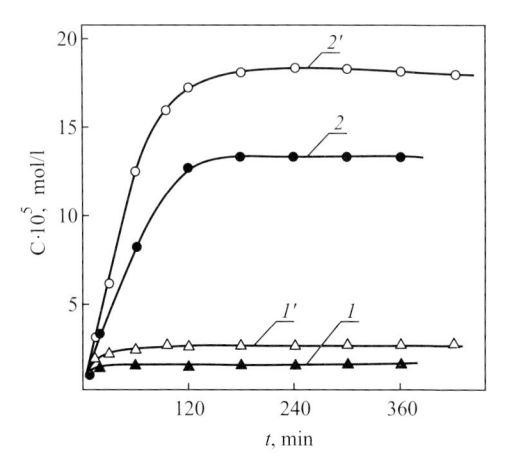

Figure 8.4 Kinetics of dissolution of phenolphthalein in water (in the absence of crystallite precipitation): curves (2) and (2') represent dissolution of the glass at temperatures 13 °C and 22 °C, respectively; curves (1) and (1') show dissolution of the crystalline phase at the same conditions (Grantcharova and Gutzow [185]).

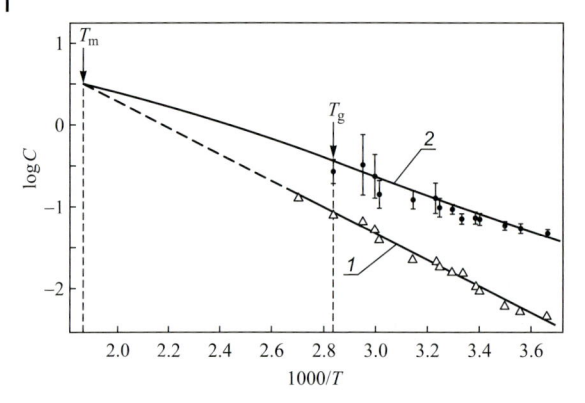

Figure 8.5 Temperature dependence of the solubility of vitreous and crystalline phenolphthalein in water according to Grantcharova and Gutzow [185] (*triangles*: solubility data for crystals; *circles*: solubility data for the glass). The straight line (1) through the crystal solubility data and its extension (*dashed curve*) is obtained by a least square fit. Curve (2) through the solubility data of the vitreous sample is drawn according to theoretical expectations (cf. Eqs. (8.65), (8.66)–(8.68)) with caloric data for ΔH_g and ΔS_g. With T_m and T_g, the melting point and glass transition temperature of phenolphthalein from caloric and rheologic data are indicated.

in the framework of Simon's approximation (which reduces the glass transition interval to the glass transition temperature T_g, only) by the dependence

$$\ln\left(\frac{P_{\text{glass}}}{P_{\text{cryst}}}\right) = \frac{\Delta G_g(T,\xi)}{RT} = \frac{\Delta H_g}{RT} - \frac{\Delta S_g}{R}. \tag{8.65}$$

Here ΔH_g and ΔS_g have the already discussed meaning and are connected with $\Delta G_g(T,\xi)$ according to Eq. (8.21). The above written formula, illustrated in Figure 8.6 for undercooled liquids, crystals and glasses describes also the solubility C_{glass} and C_{cryst} of sufficiently diluted solutions. Drawing this figure, it is assumed that in the considered temperature interval, where $\Delta G_{\text{subl}}(T) \approx \Delta G_{\text{evap}}(T) + \Delta G(T)$, the latter term refers via Eqs. (8.29) and (8.35) to the liquid/crystal change. Thus with Eq. (8.15) the vapor pressure of the crystal can be written according to

$$\ln p_{\text{cryst}} \approx -\frac{\Delta H_{\text{subl}}}{RT} + \frac{\Delta S_{\text{subl}}}{R}, \tag{8.66}$$

where ΔH_{subl} and ΔS_{subl} are approximately constant. It results as usual in a straight line in coordinates $\ln p$ vs. $1/T$.

However, for the undercooled liquid, accounting for the temperature dependence of the Gibbs potential difference liquid/crystal, given with $\Delta G(T)$ according to Eqs. (8.29) or (8.35), we have

$$\ln p_{\text{liq}} \approx -\frac{\Delta G_{\text{evap}}(T)}{RT} \approx \frac{\Delta G_{\text{subl}}(T) - \Delta G(T)}{RT}. \tag{8.67}$$

This dependence corresponds in $\ln p$ vs. $1/T$ coordinates to the respective curve as given in Figure 8.6. The expected vapor-pressure dependence of the glass is deter-

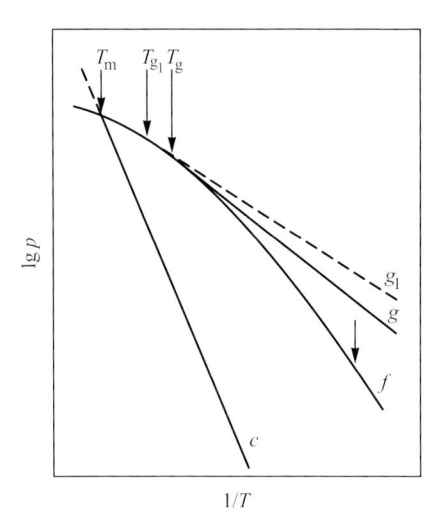

Figure 8.6 Temperature dependence of the vapor pressure of a crystal (specified by c), the undercooled melt (f) and of a glass (g), frozen-in at two different temperatures T_g and T_{g1} ($T_{g1} > T_g$). The expected change of the vapor pressure of the glass upon stabilization is indicated by an arrow.

mined by

$$\ln p_{gl} \approx -\frac{\Delta G_{\text{evap.glass}}}{RT} \approx \frac{\Delta G_{\text{subl}}(T) - \Delta G_g(T, \xi)}{RT} . \tag{8.68}$$

When we account for the already discussed temperature dependence of the Gibbs potential difference glass/crystal $\Delta G_g(T, \xi)$ (Eq. (8.56)), this has to be in $\ln p$ vs. $1/T$ coordinates a straight line tangential at $T = T_g$ to the $\ln p$ vs. $1/T$ line of the undercooled liquid, given with Eq. (8.67).

The above derived dependencies are schematically drawn in Figure 8.6. The significant point in this figure is [1, 248] that the vapor pressure dependencies of both the crystal and the glass have to be straight lines in $\ln p$ vs. $1/T$ coordinates, that the $\ln p_{\text{cryst}}$ and the $\ln p_{\text{liq}}$-lines are crossing each other at $T = T_m$ and that the $\ln p_{\text{glass}}$-line is tangenting the convexly bended $\ln p_{\text{liq}}$-line at $T = T_g$. In Figure 8.5, the experimentally obtained results on the solubility of phenolphthalein in water are drawn in such a way in $\log C$ vs. $1/T$ coordinates, in order to illustrate the above derived dependencies.

In Figure 8.7, we have compiled the results from quantitative or semi-quantitative $C_{\text{glass}}(T)$, $C_{\text{crystal}}(T)$-measurements of several authors with classically known systems: they all correspond in their temperature course to dependencies at least qualitatively described by Eqs. (8.66)–(8.68). It is of even greater significance to note that the values of ΔH_g and ΔS_g obtained for both phenolphthalein glass and for vitreous SiO_2 according to solubility measurements (see the data in Figures 8.5 and 8.7 and their detailed discussion in [172, 182, 185]) and from direct caloric determinations (i.e., from $\Delta C_p(T)$-measurements) give with sufficient accuracy the same result. Thus they confirm both above written solubility formulas and the

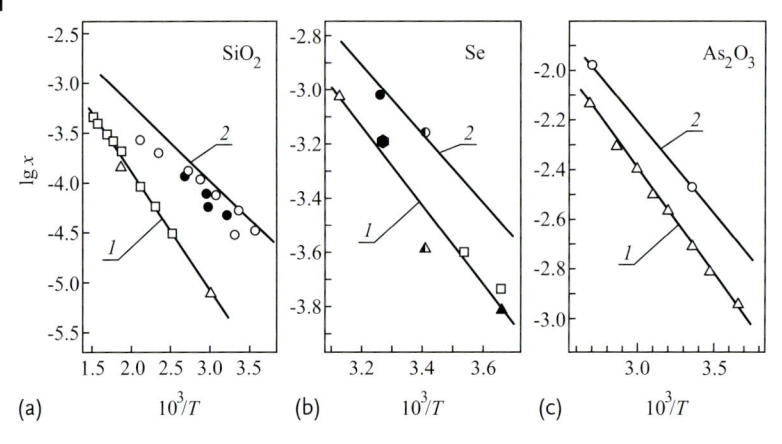

Figure 8.7 Experimental results on the solubility of simple inorganic glass-forming systems in coordinates corresponding to Eqs. (8.65) and (8.66)–(8.68). As a measure of the solubility the molar fraction χ of the solute in the solution is chosen. The curves refer to (a) SiO_2 in water (*triangles*) and *squares* to the solubility of quartz; *black* and *white dots* denote the solubility of amorphous samples, according to experimental data compiled by Iler [605]; (b) Se in CS_2 [*triangles*, *squares* and *hexagons* – solubility of crystalline monoclinic Se according to Mitcherlich et al. [606], Ringer [607] (see Gmelin [608, p. 257]); black and semi-black circles – amorphous Se according to Rammerlberger and Shidai (Gmelin [608, p. 257])]; (c) As_2O_3 in water (*triangles* – solubility of crystalline cubic As_2O_3 according to experimental data by Anderson (1923) (see also Gmelin [609, p. 278]), white circles – solubility of vitreous As_2O_3 according to a determination by Winkler (1885) (Gmelin [609, p. 278]), Landolt–Börnstein, Zahlenwerte und Funktionen, Part II, 2b, 6th edition, Springer, 1962). In all three cases, the straight line is a fit through the experimental data as it would have been expected from the outlined equations (see Eqs. (8.66)–(8.68)) by using the caloric data.

validity of Simon's approximation as an easy and convenient way of calculating the thermodynamic properties of glasses.

In introducing into Eq. (8.65) the results on the invariants of the thermodynamic properties of glasses for typical glass-formers (cf., Eqs. (8.50)–(8.54), obtained at "normal" cooling rates), it can be expected that at temperatures just below T_g the ratio $\log(C_{gl}/C_{cr})|_{T \leq T_g} \approx \Delta S_m/(5.5 R)$ should have, as derived in [182] the indicated constant value, determined by the complexity of the building units of the systems investigated. A measure of this complexity is the ratio $\Delta S_m/R$, which for simple systems (like metal melts or oxide glasses such as SiO_2, GeO_2) approaches unity and for organic liquids has values up to $3 \div 6$ and even more. In a detailed investigation, Parks et al. [610] determined in 1928 the solubility of a classical organic glass, glucose, as a crystal and as a glass – in four *different* organic solvents (see Table 8.1). An analysis of these results [182] shows that in fact in the four cases investigated nearly the same constant value of the above mentioned ratio is obtained. In good coincidence with the expectation for the complicated glucose molecules the invariant value of above ratio is found to be $\Delta S_m/R \approx (5.1 \pm 0.5)$ for all solvents used in the investigation, described in the cited paper [182].

Table 8.1 Solubility, C (g/100 g solvent) of glassy and crystalline glucose at 298 K in different organic solvents (according to Parks et al. [610]). Note that C does not depend substantially on the solvent but on the state (crystalline or vitreous) of the *solute*. According to the main value estimate made for $\log(C_{gl}/C_{cr})$ at 298 K $< T_g$ it follows that ΔS_m for glucose is of the order $\cong 5.5R$.

Solvent	Glassy solute	Crystalline solute	$\log(C_{glass}/C_{crystal})$ individual	$\log(C_{gl}/C_{cr})$ mean value
99% Ethyl alcohol	4.70	0.44	1.03	1.07
Pure ethyl alcohol	1.58	0.22	0.85	
Pure isopropyl alcohol	1.07	0.08	1.13	
Pure acetone	0.184	0.012	1.12	

Of particular technical significance is the formation of metastable solutions with increased concentration with respect to the corresponding crystal because of the possibility to use them in directed crystal growth processes. Such processes can be and have been employed in order to grow isothermally single crystals out of solutions, in which the super-saturation is determined by the respective *glass → crystal* potential difference at *constant* temperature. Figure 8.8 illustrates the possibilities opened up with our experiments [185] with the growth of phenolphthalein single crystal from aqueous solutions in which the super-saturation is sustained by glassy phenolphthalein. Also illustrated in Figure 8.8 is the growth of graphite crystals from vitreous carbon via appropriate gas transport reactions. Methods of isothermal solution growth were, it seems, first proposed for the crystallization of quartz or cristobalite crystals out of supercritical alkaline aqueous solutions, in which the supersaturation was sustained by samples of SiO_2-glass (see Nacken in [611, 612] and also Thomas and Wooster [613]). Also promising was the growth

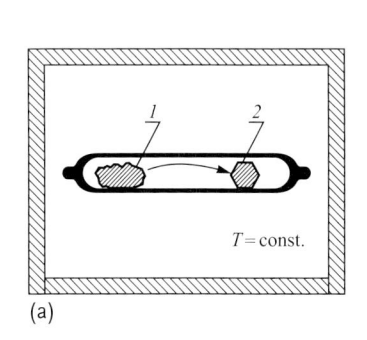

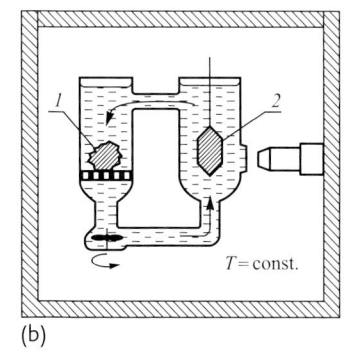

Figure 8.8 (a) Quartz ampoule experiment for gaseous transport reaction (via $C + S_2 \rightarrow CS_2$) for the realization of the *vitreous carbon → graphite* transition (1-vitreous carbon, 2-graphite); (b) illustration of the isothermal crystal growth experiments with vitreous phenolphthalein (1) as a source of supersaturation of the growth of crystalline phenolphtalein (2) in aqueous solution according to [185].

of hexagonal Se-crystals out of alkaline Na_2S-aqueous solutions, supersaturated with vitreous Se (experiments by Kolb [614]). There have also been several technically oriented proposals [601] to use the different solubility of vitreous carbon and of graphite and diamond in Ni/Cu-alloy melts to grow diamond or graphite isothermally. The different solubility of vitreous carbon and of graphite in Ni-melts is illustrated on Figure 8.9: it opens a new discussion of the thermodynamics of carbon (see [601]).

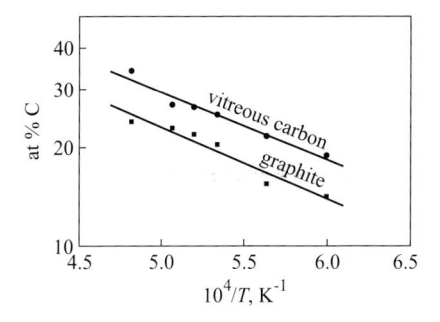

Figure 8.9 Solubility of graphite and of vitreous carbon in Ni-melts according to data reported by Weisweiler and Mahadevan from 1947, summarized in [601].

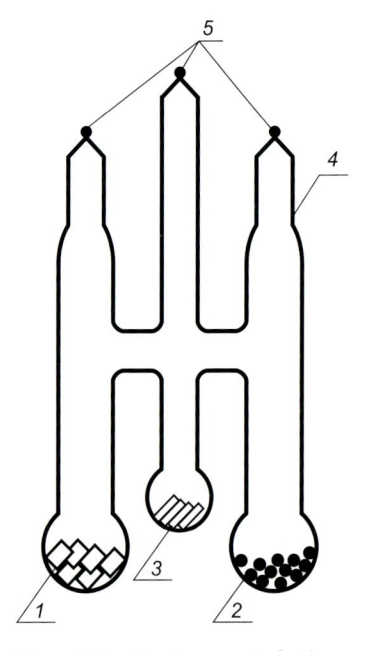

Figure 8.10 Quartz ampoule for the gaseous transport reaction (via reactions like $C + 2J_2 \rightarrow CJ_4$ or $C + S_2 \rightarrow CS_2$) for the realization of the *vitreous carbon* → *diamond* transition at isothermal conditions (1050 °C) via the metastable growth of diamond: (1) seed diamond crystals; (2) micro-sized vitreous carbon; (3) Pt catalyst in the transport gas filled ampoule; (4) quartz ampoule, (5) sealed ampoule inlets.

The second, more promising method to realize and even technically exploit the possibilities of the *vitreous carbon* → *graphite* or even of the *vitreous carbon* → *diamond* transition is to use gaseous transport reactions (e.g., the reaction $C + S_2 \to CS_2$ or $C + 2H_2 \to CH_4$ as done by Gutzow et al. in [601, 615]) to transport isothermally carbon from vitreous carbon to grow diamond single crystals at approximately 1100 °C (see Figure 8.10). This process involves in fact the broader question of the participation of vitreous substances in chemical reaction kinetics, a problem analyzed 1986 by Grantcharova and Gutzow also in the already cited paper [185] and also discussed below in some details.

8.5.2
The Increased Reactivity of Glasses and the Kinetics of Chemical Reactions Involving Vitreous Solids

Suppose that a heterogeneous chemical reaction

$$\gamma E + \sum_i a_i M_i = \sum_j b_j N_j \tag{8.69}$$

is taking place under definite conditions (at $p = $ const.) and the solid reagent, E, participates in it (at temperatures $T < T_g$) once as a crystal and a second time as a glass. Here with M_i and N_i are denoted the gaseous or liquid reactants and reaction products and γ, α and β are the stoichiometric coefficients of the reaction. The change of thermodynamic potential, $\Delta G_g^*(T)$, connected with above reaction is

$$\Delta G_g^*(T) = -RT \ln K_p(T) = -A(T). \tag{8.70}$$

Here $K_p(T)$ and $A(T)$ indicate respectively the constant of the mass action law

$$K_p = -\frac{\prod_i [M_i]^{\alpha_i}}{\prod_j [N_j]^{\beta_j}} \tag{8.71}$$

and $A(T)$ is the respective chemical reaction affinity at the considered temperature, T.

Employing the expression for the thermodynamic potential difference glass/crystal (Eq. (8.21)) we have shown above (see also [185]) that the ratio of glass to crystal vapor pressure can be expressed as indicated with Eq. (8.65). Thus with Eq. (8.70) it follows moreover that

$$K_{p_{\text{glass}}} = K_{p_{\text{cryst}}} \exp\left[\frac{\Delta G^*(T)}{RT}\right] = K_p \left(\frac{p_{\text{glass}}}{p_{\text{cryst}}}\right)^\gamma. \tag{8.72}$$

It is to be expected that, if the solid reagent E is placed in a closed volume in both its crystalline and glassy forms in contact with the reactants M_i, the $E_{\text{glass}} \to E_{\text{crystal}}$ transition will take place via above indicated gaseous reaction at a velocity proportional to the affinity $A(T)$ (see Eq. (8.70)). Since $(p_{\text{gl}}/p_{\text{cryst}}) > 1$, it follows from above equation also that $K_{p_{\text{glass}}} > K_{p_{\text{cryst}}}$ and that the above written heterogeneous reaction (Eq. (8.69)) will always be faster by the factor

$$f = \exp\left(\frac{\Delta G^*_{\text{glass/cryst}}}{kT}\right) = \left(\frac{p_{\text{glass}}}{p_{\text{cryst}}}\right)^\gamma \tag{8.73}$$

in the glass reagent case. This in principle is the third, the chemical reaction, way to free the potential energy, frozen-in in the glass: to bring it in contact with an appropriate gaseous or liquid reagent.

The above discussed problem is a possibility of great significance in any application of glasses as reagents, and not only as reagents. Equations (8.70), (8.72) and (8.73) are of particular significance in the mentioned cases of isothermal crystallization (e.g., in the considered case of diamond growth) and in analyzing the applicability of soluble glasses in agriculture (as so-called *agriglasses*, see [616] and below in this chapter), in medicine (*per os* application of medicaments in the form of fast dissolving glassy drugs, etc.). Equation (8.73) opens the problem of reactivity of vitreous solids in the same way as Eq. (8.65) gives the possibility of discussing the enhanced solubility of vitreous solids. However, it has to be taken into account that both solubility, C_{glass}, and chemical constant, $K_{p_{\text{glass}}}$, of vitreous solids have to be determined under conditions, where crystallization and even relaxation of the glass (to the respective metastable liquid) has to be excluded. The way such measurements are to be performed was already discussed and is described in details in [172, 185]: the essential moment in glass solubility measurements is to guarantee a constantly renewed glass to solution interface. Similar requirements have also to be fulfilled if it should be intended to determine true values of ΔH_g and ΔS_g via chemical reaction measurements.

A fourth method to release the energy, incorporated in glasses, is given by galvanic cells, constructed based on the possibility to perform the *glass* → *crystal* reaction under electrochemical conditions. This problem is discussed in the following section.

8.6
The Fourth Possibility to Release the Energy of Glass: the Glass/Crystal Galvanic Cell

This fourth possibility to use the energy, accumulated in glasses, first mentioned by Gutzow in 1981 [177], is based on the idea to construct a "physical" or an "allotropic" galvanic cell in which the transition *glass* → *crystal* is exploited to generate an electromotor driving force (EMF). The EMF should be expressed as:

$$\Delta E_g(T) \cong -\frac{1}{zF}\Delta G_g(T,\xi). \tag{8.74}$$

Here F is Faraday's constant and z is the change of electric charges in the electrochemical reaction employed. In considering the caloric equivalent of electric energy and the value of Faraday's constant ($9.65 \cdot 10^4$ C mol^{-1} = $9.65 \cdot 10^4$ As mol^{-1}) the EMF (in Volts) is directly obtained from Eq. (8.74) if we divide the value of

ΔG (expressed in J mol^{-1}) by the corresponding zF value (however, in employing thermodynamic data from the classical periods of electrochemistry, where ΔG and ΔH are expressed in cal mol^{-1} the corresponding value of zF is given by $zF = 2.31 \cdot 10^4$ to be used for the conversion factor). Figure 8.11 shows the simplest realization of this possibility using a metal glass and its metallic crystalline analog. In Figure 8.12 are displayed the EMF values obtained of the galvanic cell *vitreous carbon* → *graphite* as they are reported by Das and Hucke [359].

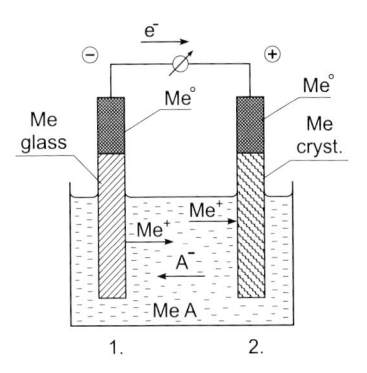

Figure 8.11 A possible realization of the galvanic cell glass/crystal in its normal variant: glass (1) and crystalline (2) metallic electrodes in an aqueous solution of the corresponding metal cation $M_e^{(-)}$. The vitreous cathode is dissolved into $M_e^{(-)}$-ions in the electrolyte and the crystalline anode grows, incorporating the metallic atoms $M_e^{(0)}$ reduced on its surface.

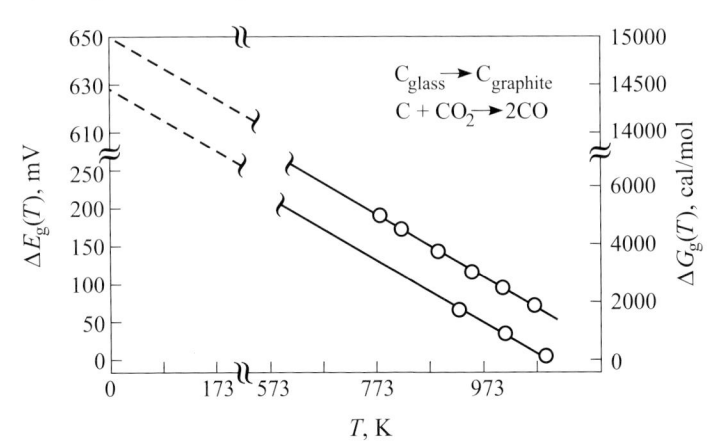

Figure 8.12 Temperature dependence of the EMF of the vitreous *carbon* → *graphite* galvanic cell, exploiting Boudouard's reaction ($C + CO_2 \rightarrow CO$) according to data by Das and Hucke [359] as they are summarized in [596] in terms of Eq. (5.74). Each straight line corresponds to a different vitreous carbon glass, obtained under distinct, but differing conditions, specified in [359]. Note the linear dependence of both $\Delta G_g(T, \xi)$ (on the left ordinate) and of $\Delta E_g(T, \xi)$ (on the right ordinate) with temperature, indicating nearly equal ΔS_g values.

The important feature in Figure 8.12 is the linear character of the $\Delta E_g(T)$-dependence, as it follows for the EMF of any *glass → crystal* system, where we have to expect according to Eq. (8.21) that (as in all previous derivations, at $p = $ const.)

$$\frac{d\Delta G_g(T)}{zF\,dT} = \frac{d\Delta E_g(T)}{dT} = -\frac{\Delta S_g}{zF_g} = \text{const.} \tag{8.75}$$

Such a course of the EMF could be of particular interest in many technical devices as a linearly temperature-dependent potential, $\Delta E_g(T)$. This dependence also follows from the general thermodynamics of electrochemistry and of the *glass → crystal* galvanic cell in particular via the Gibbs–Helmholtz equations. Thus writing (again for $p = $ const.)

$$\Delta G_g(T) = \Delta H_g(T) + T\left(\frac{\partial \Delta G_g(T)}{\partial T}\right)_p, \tag{8.76}$$

we come to the conclusion that

$$\Delta E_g(T) = -\frac{\Delta H_g(T)}{zF} - T\left(\frac{\partial \Delta G_g(T)}{zF\partial T}\right)_p = -\frac{1}{zF}\left(\Delta H_g - T\Delta S_g\right), \tag{8.77}$$

and thus with $\Delta H_g, \Delta S_g = $ const. to the expectation of an always linear $\Delta E_g(T)$-dependence. This conclusion follows, however, in assuming that Simon's approximation is correct: that is, here and in Eq. (8.21) we have neglected the "nonequilibrium" affinity correction term $(A(T)/\Delta S(T,\xi))d\xi/dT$, which takes into account that glass transition takes place not abruptly at $T = T_g$, but in a broader temperature interval, in which nonequilibrium effects may be of principal significance. Figure 8.12 shows that (as already discussed in considering above nonequilibrium correction) at least in a first approximation these effects can be in fact neglected. Equation (8.77) is also of additional significance, as it shows how the *glass → crystal* galvanic element would work, for example, in adiabatic isolation or in a reservoir at constant temperature: in doing the necessary predictions, we have again to keep in mind that we know from the thermodynamics of vitrification that $\Delta H_g, \Delta S_g$ are not only constants, but have also always *positive* values.

In the already cited previous publication of one of the present authors [177], several proposals can be found, describing the possibility to construct "normal" or more complicated forms of *glass → crystal* galvanic cells, for example, in terms of so-called bipolar cells. There also other convenient devices to use the EMF of the electrochemically performed *glass → crystal* reaction are discussed. These possibilities are inherent in the very nature of the vitreous state. They are most easily demonstrated by measuring the electrode potential or even the corrosion potential of glassy metallic alloy electrodes. Here differences of the order of $(10 \div 15)$ mV are to be expected, introducing into Eq. (8.77) the values of ΔH_g and ΔS_g, expected according to the previous derivations given in [177] in more details. The expected $\Delta E_g(T)$ differences ($5 \div 10$ mV) are also manifested in Figure 8.13 for the corrosion potential difference *normally quenched glass → crystal* and also for the *quenched glass vs. annealed glass* galvanic cell arrangement. As expected, not only the crystal,

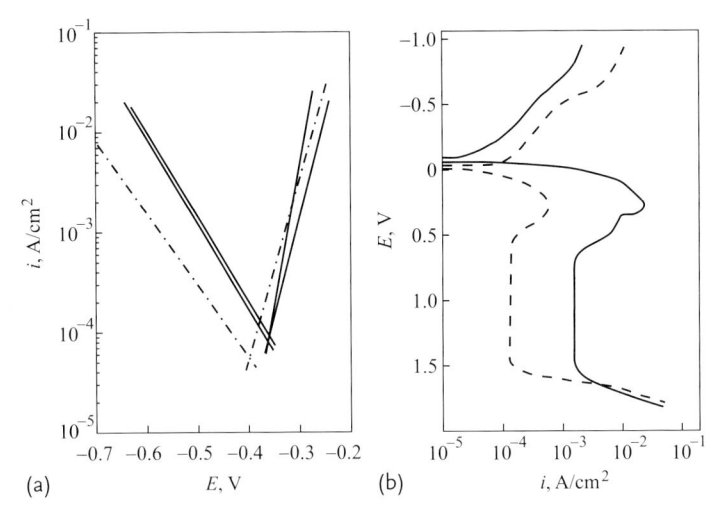

Figure 8.13 The corrosion potential at metallic alloy glasses and of the respective crystalline substance according to data by Heusler and Huerto [617] and by Zaprianova, Raicheff and Dimitrov [618]: (a) Corrosion potential of a Co_3B-alloy glass (*solid lines*) and of the respective crystal (*dotted lines*); (b) corrosion potential of the $Co_{70}Fe_5Si_{15}B_{10}$ alloy glasses: as quenched material (*solid lines*) and of the relaxed alloy. Note that in both cases an electrode potential difference of about $10 \div 12$ mV is observed.

but also the annealed metal glass has a lower potential than the quenched glass, according to the results seen on Figure 8.13 taken from [618].

It is also of interest to consider the construction of accumulator-like devices, in which the process *glass* → *crystal* (battery discharge) takes place spontaneously and the reverse process (*crystal* → *glass*, that is, battery charge) under externally applied electric potential. It could be also of technical interest to investigate it, for example, in terms of Eq. (8.77).

The fifth possibility to use the increased thermodynamic potential of the frozen-in glass is the existence of an EMF difference potential $\Delta E^{\#}$, stemming from the direct contact of a metallic or semi-conductor glass to the respective crystal at an appropriately constructed junction. There both the *Seebeck*-effect and the *Peltier*-effect, when applied to the frozen-in glassy system, should bring about interesting possibilities. The always present Seebeck-effect potential should influence the already discussed galvanic potential glass/crystal, as this is known from the general theory of the formation of the potential of every "normal" galvanic contact. This is evident from the considerations outlined in the subsequent section.

8.7
Thermoelectric Driving Force at Metallic Glass/Crystal Contacts: the Seebeck and the Peltier Effects

According to the general present day theory of the thermoelectric properties of solids, the thermoelectric driving force, $\Delta\varepsilon_{1,2}$, arising at the contact of two different metals or semiconductors 1 and 2 is determined according to

$$\Delta\varepsilon_{1,2} = \frac{S_1^* - S_2^*}{F} \tag{8.78}$$

by the difference in the values of the transport entropies S_1^* and S_2^* of the charge carriers in the two bodies in contact. Within the framework of the classical Drude–Lorentz concepts of a nondegenerate population of charge carriers (i.e., of an electron gas, following Boltzmann's statistics), occupying the volumes of the two different metals 1 and 2 in mutual contact, we could write for temperatures, T, higher than the expected electron gas degeneration temperature, T_d, that the resulting differential thermodynamic power, $\Delta\varepsilon_{1,2}$, is

$$\Delta\varepsilon_{1,2} \approx \frac{k_B}{e_0} \ln \frac{n_1}{n_2}. \tag{8.79}$$

Above formula can be applied only to temperatures $T > T_d$; below this temperature nonclassical, quantum statistical approaches have to be applied.

In above formula k_B is Boltzmann's constant, e_0 is the charge of the electron and $n \approx \rho N_A/M$ denotes the electron concentration in the corresponding two metals, ρ indicating their respective densities and M their atomic weights: N_A indicates Avogadro's number. Applying such a simple model and Eq. (8.79), resulting from it, to the case of the two modifications (vitreous and crystalline) of the same metal, it turns out that for the case of the glass-to-crystal contact of the same metal

$$\Delta\varepsilon_{gl/cryst} \approx \frac{k_B}{e_0} \ln \frac{\rho_{gl}}{\rho_{cryst}} \tag{8.80}$$

holds. For the metal liquid to crystal junction case of the same metal one has to introduce in the framework of the same approximated picture into above formula the ratio $(\rho_{liq}/\rho_{cryst})$. In this way the temperature dependence of the thermoelectric power, $\Delta\varepsilon_{gl/cryst}$, of the glass/metal Seebeck thermal element should be determined by the temperature dependence of the ratio of the respective densities, $\rho_{gl}(T)/\rho_{cryst}(T)$. From the commonly observed temperature dependencies

$$\rho_{cryst}(T) \cong B_1 T + \rho_c(0) \tag{8.81}$$

and

$$\rho_{gl}(T) \cong B_1 T + \Delta\rho_{gl}(0), \tag{8.82}$$

denoting $\Delta\rho(0) = \rho_{cryst}(0) - \rho_{gl}(0)$, one obtains for the usual case $\rho_{gl}(0) < \rho_{cryst}(0)$ that in Eq. (8.80) we should have $\ln(\rho_{gl}(T)/\rho_{cryst}(T)) \approx \Delta\rho(0)/(B_1 T + \rho_c(0))$. Thus,

8.7 Thermoelectric Driving Force at Metallic Glass/Crystal Contacts

we obtain in a first approximation

$$\Delta \varepsilon_{\text{glass/crystal}} \approx \text{const.}/(B_1 T + \rho_c(0)) . \tag{8.83}$$

This result, obtained in the framework of the now generally repudiated classical picture of the electron gas in a solid, is particularly dubious in its behavior at $T \to 0$, where it predicts constant values of $\Delta \varepsilon$. Beside the problems of principal nature (disregard of quantum effects), it has to be also taken into account that the approximation indicated above follows after a truncated expansion of the logarithm at $\Delta \rho(0)/(B_1 T + \rho_c(0)) \ll 1$. Now we have to integrate Eq. (8.83) in accordance with the definition of the thermo-electric EMF [619]

$$\Delta E^{\#} = \int_{T_R}^{T} \Delta \varepsilon_{1,2}(T) dT . \tag{8.84}$$

For the potential determined by the temperature difference between the temperature T of the contact *glass/crystal* and a constant temperature reference contact at temperature T_R, we have thus to expect for $B_1 T \gg \rho_c(0)$

$$\Delta E^{\#}_{\text{gl/cryst}} \approx \frac{\Delta \rho(0)}{B_1 \rho_c(0)} (T_R - T) . \tag{8.85}$$

In order to obtain above linear dependence, here again the expression $\ln(T/T_R)$, resulting from the integration of Eq. (8.84) at $B_1 T \gg \rho_c(0)$ with according to Eq. (8.84), was developed for temperatures $T \ll T_R$ into a truncated power series.

According to the first Thomson relation [619, 620]

$$\Delta \Pi_{1,2} = T \Delta \varepsilon_{1,2} , \tag{8.86}$$

connecting thermoelectric power, $\Delta \varepsilon$, with the Peltier coefficient, $\Delta \Pi_{1,2}$, distinct and simple consequences also follow for the temperature dependence, $\Delta \Pi(T)$, for a metallic glass$_1$/glass$_2$ contact. Equation (8.86) also gives the value of the Peltier heat emitted or consumed in a contact of the discussed type, when an electric current is passed through the circuit with the glass/crystal contact junction. In agreement with the third principle of thermodynamics we should expect in general that

$$\Delta \varepsilon_{1,2}|_{T \to 0} = \left(\frac{d \Delta E^{\#}_{1,2}}{dT} \right) \bigg|_{T \to 0} = 0 , \tag{8.87}$$

as this follows for the thermodynamic functions of any equilibrium system approaching the absolute zero of temperature (see Chapter 9 and also [182, 598]).

From a thermodynamic point of view $\Delta \varepsilon_{1,2}$ has in fact the significance of a thermodynamic function. However, in the case of glasses the frozen-in structure of the glass, as a nonequilibrium system, could require that in the vicinity of $T = 0$ the third principle should not be fulfilled as this is discussed again in more details in Chapter 9 of the present book. Thus, we could expect that (at least for a possible nondegenerate electric charge carrier) in the frozen-in glass $\Delta\varepsilon_{gl/cryst} > 0$ at $T \to 0$, as this is indicated with Eq. (8.87), as this is the case with the frozen-in configurational entropy and enthalpy values, where we have derived $\Delta S_g > 0$, $\Delta H_g > 0$ and experiment has confirmed these expectations. However, in the case of metallic conductors the electricity carriers (the electron gas), below the temperature T_d, have to follow the requirements of quantum statistics, the electron gas has to degenerate as a result (even if we assume its initial Boltzmann-like behavior above T_d). Thus as for any thermodynamic function of a quantum system $\Delta\varepsilon_{g/c} = 0$ has to be expected for $T < T_d$, and thus for temperatures approaching the absolute zero of temperature.

In a more general manner, the aforementioned considerations follow directly from the more realistic quantum models of electron conductivity, for example, in terms of Sommerfeld's quantum electron gas model [620, 621] entirely based on Fermi's theory. According to Sommerfeld we have again to assume that in both contacting metallic electron conducting materials 1, 2 different concentrations n_1, n_2 of electrons exists. In the framework of the Sommerfeld–Fermi's quantum model it follows, however, that now

$$\Delta\varepsilon_{1,2} = \frac{2}{3}\pi^2 \frac{mk_B}{e_0 h^2} \left[\left(\frac{4\pi}{3n_1}\right)^{2/3} - \left(\frac{4\pi}{3n_2}\right)^{2/3}\right] T . \tag{8.88}$$

This is the known expression for the value of $\Delta\varepsilon$ in systems with an electron gas, partly degenerated even above the temperature T_d. For $T \to 0$ one obtains with Eq. (8.88) directly $\Delta\varepsilon_{1,2} = 0$, and thus without problems also

$$\Delta\varepsilon_{glass/crystal} \approx \text{const.} \times T|_{T\to 0} = 0 . \tag{8.89}$$

By means of above dependencies (where Planck's constant, h, appears together with the mass, m, and the charge, e_o, of the electron with already above introduced quantities), here, however, a quadratic $\Delta E_{glass/crystal}$ vs. $(T_R - T)$ dependence has to be expected (after subsequent integration in terms of Eq. (8.84)) instead of the approximation given with Eq. (8.85). Experimental evidence on the course of $\Delta\varepsilon_{1,2}^{\#}(T)$ for various metallic pairs confirms to a great extent this quadratic dependence.

Thus, different are the expectations on the course of the temperature dependence at $T \to 0$ of the thermodynamic function ($\Delta\varepsilon_{glass/crystal}$) and of the respective potential (ΔE) as properties, determined by quantum electric charge carriers (i.e., from the electronic structure) compared with those, which follow for the thermodynamic properties, depending on the bulk constituents of the material vitrified. This follows because $\Delta\varepsilon_{1,2}$ is not a function, determined directly by the configurational structure and the structural changes of the corresponding vitrified solid, as is the case with ΔH_g and ΔS_g. In analyzing the electric properties of the glass, it turns out that differently structured materials (crystal or glass) play the role only of the medium, in which the transport of electricity with the same charge carriers, following quantum statistics, takes place.

In electronic conductivity the properties of the charge carriers, for example, of the electronic gas as a physical state, are decisive. Even when the transport of electrons takes place in a system of frozen-in configurational structure, the electron

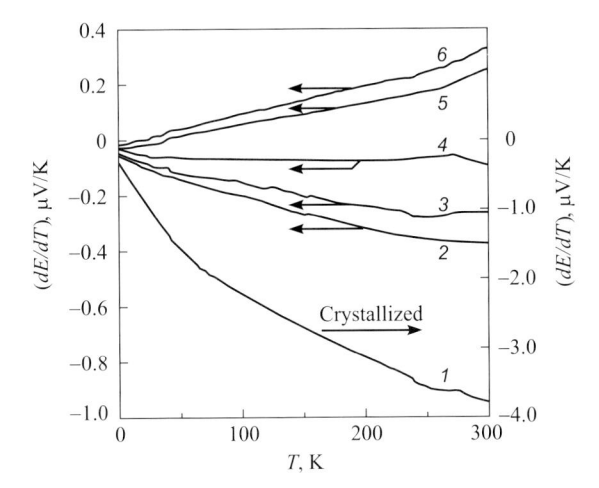

8.7 Thermoelectric Driving Force at Metallic Glass/Crystal Contacts

Figure 8.14 Temperature dependence of the thermoelectric power $\Delta\varepsilon_{1,2} = dE_{1,2}(T,\xi)/dT$ for several metallic glasses of the composition $(Mg_{(1-x)}Zn_x)$ in contact with crystalline Pb: (1) crystallized met-glass $(Mg_{(1-x)}Zn_x)$ with $x = 0.3$; (2) met-glass with $x = 0.2$; (3) met-glass with $x = 0.25$; (4) met-glass with $x = 0.3$ (i.e., with the same composition as the crystalline sample); (5) met-glass with $x = 0.33$; (6) met-glass with $x = 0.35$. Independent of composition, the thermoelectric power of each of the glasses approaches zero at $T \to 0$ either "from above" or "from below" as compared with the crystalline sample. The contact with metallic Pb is used in order to minimize the Thomson effect [619] and, thus, to facilitate the measurements. Data by Basak et al. [622] and by Baibich et al. [623].

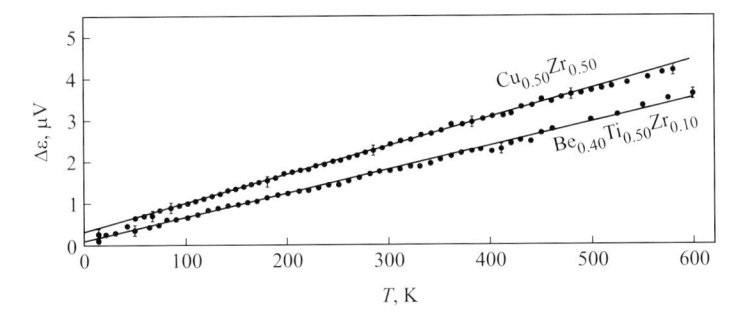

Figure 8.15 Temperature course of the thermoelectric power for two metal alloy glasses ($Cu_{0.5}Zr_{0.5}$ and $Be_{0.4}Ti_{0.5}Zr_{0.1}$) measured against pure Pb according to data by Basak et al. [622]. Note the nearly linear $\Delta\varepsilon$ vs. T course in both cases, as expected from Eq. (8.88).

gas as a whole has to reach its degenerate state at $T \geq 0$, and this determines the direct dependence of $\Delta\varepsilon_{1,2}$ on temperature given in Eq. (8.88). Thus, in general, $\Delta\varepsilon_{\text{glass/crystal}}$ determined by frozen-in systems or even by two frozen-in systems (e.g., $\Delta\varepsilon$ for two contacting alloy glasses) should display a linear $\Delta\varepsilon(T)$ function with $\Delta\varepsilon = 0$ at low temperatures (and even at $T \to 0$) in agreement with Eq. (8.88). Evidence for this conclusion is presented in Figures 8.14 and 8.15.

It also has to be mentioned that from a general thermodynamic point of view thermodynamic potentials (like $\Delta G_g(T,\xi)$) and thermodynamic functions (like the entropy and enthalpy differences, ΔH_g and ΔS_g, of frozen-in, *nonequilibrium* systems as shown in details in both Chapters 3 and 9 and elsewhere [182, 598]) do *not* fulfill the requirements of the third principle in the way it was formulated by Planck for *equilibrium* systems. However, *thermodynamic coefficients* even of nonequilibrium, frozen-in systems (like the specific heats, $C_p(T, \xi)$, or the coefficient of thermal expansion $\alpha(T, \xi)$ or Thomson's coefficient $\alpha_T = d\Delta\varepsilon/dT$) have to approach zero values at $T \to 0$ even for topologically configurational nonequilibrium systems. In the case of vitreous conductors and semiconductors, the fulfillment of the third principle for both thermodynamic coefficients (like Thomson's coefficient $\alpha_T = d\Delta\varepsilon/dT$) and thermodynamic functions (like $\Delta\varepsilon_{1,2}$) or thermodynamic potentials (like $\Delta E_{1,2}$) is guaranteed by the properties of the respective charge carriers, determined by *quantum* statistics. The value of $\Delta\varepsilon$ in both glassy and crystalline materials is determined either by the population of degenerate "quantum" charge carriers (electrons) or (in semiconductors) by charge carriers (holes) that at $T \to 0$ are also degenerate. Thus the third principle is fulfilled at $T \to 0$ for both the thermal electric power, $\Delta\varepsilon_{\text{glass/crystal}}(T)$ and the corresponding thermodynamic coefficient (the Thomson coefficient $d\Delta\varepsilon/dT$): they have to approach zero values at $T \to 0$ as predicted by quantum statistics. However, the configurational properties and thermodynamic functions of glasses, determined in the nonequilibrium, frozen-in state of glasses even at $T \to 0$ by Boltzmann's statistics, do not obey the requirements of both quantum statistics and of the third law of thermodynamics, following from it.

The EMF produced out of glass/crystal contacts is low and could be of significance only as the summaric effect in battery-like arrangements. It has to be taken, however, into account in properly calculating the EMF of the galvanic potential of the *glass* → *crystal* electrolytic cell: for smaller galvanic EMF values, the Seebeck EMF can be comparable with the EMF of the galvanic battery.

Up to here we have given in the present chapter considerations concerning possible new or unexpected applications of glasses. Now follow several observations on the real possibilities in this respect: both in nature and as technical realizations.

8.8
Unusual Methods of Formation of Glasses in Nature and Their Technical Significance

8.8.1
Introductory Remarks

Glasses have been with men since the very beginning of civilization: even primitive societies as we mentioned in Chapter 2 used natural glasses (e.g., obsidian) to cut various tools and even ritual knives out of them. In ancient Egypt and in Mesopotamia we find also the first efforts to produce glass on what was to be considered at these times an industrial scale, to develop it in the Roman Empire and especially in its Mediterranean provinces, for example, in Alexandria, where the first glass factories with real technical significance are found. The history of these developments the interested reader can find in several books [624–626].

Natural glasses, like obsidian, are of volcanic origin, most of them formed in volcanic eruptions and this is the "normal" way of formation of glasses on Earth's surface: they are the remnants of uncrystallized (or partly crystallized) lava and both Mexican obsidian (with its beautiful crystallites) and the nearly crystal free obsidian glasses from the eruptions of the Ararat volcanoes give evidence in this sense: both have been used and are still in use as half noble gemstones. Half of the lithosphere is in fact a magmatic remnant, possessing the features of magmatic crystallization processes.

Nature has given, however, also examples of unusual, even of strange ways of formation of natural glasses. Here we have first to mention the impactites, such as *tektites* which, as it is now assumed with great justification [627], are formed with the impact of extraterrestrial objects (of meteorites) on the earth surface. The Arizona Crater in the USA still provides a fearful example of the energy that can be developed in such catastrophic impacts, which caused millions of molten rock droplets with the strange aerodynamic form of the tektites to fly through the Earth stratosphere and atmosphere to form tektites deposits hundreds of kilometers away. Lunar glass samples, brought by lunar exploration, have the form of ideal spherical droplets: they are also the result of meteoritic impacts. The chemical composition of impact glasses, both on Earth and from the Moon resemble in some respects the composition of the rock or soil on which the meteoric impact took place, however, usually with an extremely reduced percentage of alkaline oxides and other more volatile components: they have evaporated during the extremely high temperatures of the impact. Frozen-in crystalline metallic droplets from the impact meteorite are also found in them. There are also other known cases in which it is assumed that glasses have been formed in the unusual conditions of a thermal shock: thus the Libyan desert glass samples are most probably formed from the impact of lightning in the desert sands.

There are now also other glass samples on Earth which are witnessing the impact of atomic explosions on the Earth's lithosphere: the dreadful *trinitite* or *atomist* glasses, formed on atomic explosion sites, initiated by human will in the last sixty years. They are still radioactive and in a similar way as other impact glasses have a composition, resembling the respective soil (e.g., the *Arkosil* and *Feldspar* sands of the atomic test ground), again with a diminished percentage of the more volatile alkaline oxide components. Thus the remnants of the first experimental atomic blast in Alamogoro can still be found in museum collections as the remainders of the glassy droplets, formed in the fireball of the atomic explosion itself.

Of more significance and a possible global threat are the radioactive remnants of present day atomic fission processes: an industrial problem to which in this chapter several words are devoted below. This is another example, connected with the possible solubility of glasses, in which the waste products of the present day atomic industry are buried as a solute. These glasses will form for thousands of years to come sites on Earth in which the safety of mankind will depend to a great extend on the properly performed prognosis of the solubility of the radioactive waste glasses in geological conditions.

Also connected with the solubility of glasses and the possible formation of vitreous precipitates from supersaturated hydrothermal solutions of silica are the geological problems of precipitation of *hyalite*, an amorphous form of SiO_2, containing a significant percentage of H_2O and its further development into *opals* and the following crystalline minerals of the SiO_2–H_2O sequence. Other problems, connected with the solubility of glasses are discussed below. We would like again to mention the biological significance of the vitreous state and the possibilities to "vitrify" life itself.

8.8.2
Agriglasses, Glasses as Nuclear Waste Forms and Possible Medical Applications of Dissolving Organic Glasses

In Section 8.1, we analyzed several possibilities of the applications of glasses as sources of constant supersaturation in isothermal single crystal synthesis. Here it turned out that even the seemingly insoluble modifications of SiO_2 can be synthesized out of (alkaline) aqueous solutions and thus both cristobalite and quartz can be grown out hydrothermally, using SiO_2 glass as the precursor phase. In several monographs [1, 605, 628] the possibilities and difficulties of these syntheses are discussed in details. In geological conditions, as above said, the solubility of silica in its different modifications explains the observed geological sequences in the formation, growth and development of minerals like hyalite, opal, cristobalite and quartz.

In the middle of the 1950s proposals were advanced to use the solubility of oxide glasses in solving an important agricultural problem: the introduction of microfertilizers into soil. This idea was further on developed in the 1960s and 1970s as a possibility to introduce into soil microfertilizers, such as Se, B, Mo, and so on, by incorporating the corresponding oxides into especially synthesized borate or phosphate glasses with such compositions, as to achieve a constant leaching rate into the aqueous solutions of the humus acids of soil. Thus agricultural experiments, performed according to these ideas, have shown that introducing glass pellets (or glass spheres) into the soil can give optimal results as fertilizer nutrition for perennial plants for several years. A more recent review article of Samouneva *et al.* [616] gives a resume of these efforts with such agriglasses undertaken mainly by a group of scientists in the 1970s. Similar proposals were also made for glass-spheres formed from slightly soluble glasses with constant leaching rate in water, containing bactericide oxides, thus killing amebic populations in tropical wells.

A new trend with great and positive possible consequences in medicine could bring about medicaments or drugs in the form of vitreous organic solutions, or of directly vitrified medicaments: the enhanced solubility of glasses gives remarkable advantages for pharmaceuticals in amorphous forms of administration. Many publications in this respect can be found in the recent international literature and in the proceedings of several conferences in this direction. Here, as evident from the literature, new frontiers are opened of investigation, connected with the analytic, calorimetric and kinetic characteristics of such products and with the mechanism of their dissolution in aqueous biological solutions.

The possibility to immobilize the products of nuclear industry in the form of glassy waste products also depends on the solubility of glasses. A recent monograph [629] gives the necessary details in this respect. Here we would like only to mention two problems in the processes of administration, management and the immobilization of radioactive waste products. First, vitreous solutions have to be found with minimal solubility in water and in the humus acids of soil, in which uranium oxides and the products of uranium fission can be dissolved. Secondly, the solubilization (or better said: the reaction) of the radioactive fission products with the melted glass-forming solvent has to take place at possibly low temperatures (e.g., not exceeding 1000–1200 °C) in order to reduce evaporation of the radioactive waste during the synthesis of the glassy radioactive waste forms. Here experimental investigations include besides solubility, characterization at conditions as nearly as possible to geological realities with perspective of many thousand years. Also, possibilities of crystallization processes and their influence on solubility and security of radioactive waste management have to be analyzed in this respect.

Industrial oxide waste products from the metallurgical industry, from ore deposits, from fly ashes and the incinerator ashes of the industrial metropolis, containing potentially toxic oxides, can be transformed into practically nonsoluble glasses or valuable glass-ceramic materials: an urban problem of no less global significance than the management of radioactive waste materials.

Glasses in every-day applications are usually considered as insoluble and above lines give examples in this respect; soluble glasses, it turns out, however, can be of no less significance than "insoluble" ones. However, in discussing unusual applications of glasses as soluble materials let us also mention the possibility to synthesize ultra thin glassy layers below 1 μ thickness (out of silica or of other silicate glasses) in vacuum-sputtering arrangements in the framework of a method, combining both soluble and insoluble glasses. First on an insoluble glass surface (e.g., of usual window glass or of technical laboratory glass) is evaporated (or vacuum sputtered) a thin layer (approximately 1–2 μ) of a water-soluble glass (e.g., of Graham's glass $NaPO_3$). Then on this soluble glassy layer an insoluble glassy layer with a desired thickness is vacuum-sputtered. Afterwards on this layer a ring is glued out (with some appropriate universal glue) of a material, having a coefficient of thermal expansion similar to this of the insoluble glassy layer. Then the soluble $NaPO_3$-glassy layer is leached-out in water and thus optical glassy windows with thickness ∼1 μ can be obtained. This example illustrates the manifold applications soluble glasses can have in laboratory experimentation.

As another possible medical application of glass, besides the classical bottle or window glass, let us also mention proposals to use microspheres of glasses having constant leaching rate in biological fluids in which clinically essential medicamentation is incorporated (even of radioactive nature). Such glassy microspheres can be entered in the capillary system of the human organism where necessary.

In the introductory chapter of the present book we also mentioned the great problems of vitrification of life, of the possibility thus to realize *absolute anabiosis* and even to revive the ancient ideas of *panspermia*: the spread of life through the universe via microorganisms vitrified in their own biological fluids, sailing on solar winds from star to star and from planet to planet. Here we only would like to repeat that controlled microcrystallization, making out of the body fluid something like a glass-ceramics, and vitrification are in fact the two ways of survival of organisms in the critical sea regions of our planet, where severe cold is followed by temperature rise. At these extreme conditions, it turns out, organisms of insects and small reptiles prepare their organic fluids for both vitrification or crystallization by synthesizing in them substances like glycerol and saccharides, increasing thus the viscosity of the respective biological aqueous solutions.

8.8.3
Glasses as Amorphous Battery Electrodes, as Battery Electrolytes and as Battery Membranes

In Section 8.6 of the present chapter we discussed the possible realizations of *glass* → *crystal* batteries and *glass* → *crystal* rechargeable batteries and accumulators of energy. The necessary thermodynamic formalism in this respect is given there: thus we can determine the electric driving force EMF $\Delta E_g(T)$ of a battery, whose electrodes are constructed from a given substance in its two possible forms: vitreous as the anode and crystalline as the cathode.

There is another possibility which, out of the same principle, could serve to increase the applicability of a battery: to make out of vitrified material one or two of its two initially crystalline electrodes. Thus the initial battery, the electrodes of which were formed out of crystalline materials, may display more or less distinct technical advantages in possible applications. The choice of the most appropriate materials to construct such a device, for example, with a maximal EMF value, has to be made using Eqs. (8.56), (8.74) and (8.77) showing that substances with a maximal value of the product $T_m S_m$ (i.e., with maximal melting point values and highest structural complexity) would give also maximal $\Delta E_g(T)$ values. The *vitreous carbon → graphite* battery gives a good example in this sense.

Let us now consider in some details the "amorphization" of electrodes in order to improve the action of the initially "crystalline" battery. Let us initially have a galvanic cell of the type

$$A_{\text{cryst}} |AB| |CB| C_{\text{cryst}} , \qquad (8.90)$$

in which A_{cryst}, C_{cryst} are electrodes made out of the crystalline material of the metallic substance A and B, AB and CB being the corresponding soluble salts dissolved, for example, in water to form an aqueous electrolyte. Suppose now that we have the possibility to obtain also the metal electrodes A, B as the vitreous form of the respective metallic substances. The question is what would be the EMF of the corresponding galvanic cell

$$A_{\text{glass}} |AB||CB| B_{\text{glass}} \,. \tag{8.91}$$

An analysis shows that the EMF of this new $A_{\text{glass}}/B_{\text{glass}}$ galvanic cell would be

$$\Delta E_{\text{gl}A,B}(T) \cong \Delta E_{\text{cryst}A,B}(T) + \Delta E_{\text{gl}A}(T) - \Delta E_{\text{gl}B}(T)\,, \tag{8.92}$$

where $\Delta E_{\text{gl}A}(T)$ is the EMF of the galvanic cell *glass A* → *crystal A* and $\Delta E_{\text{gl}B}(T)$ corresponds to the EMF of the galvanic cell *glass B* → *crystal B*.

In a previous paper of one of the present authors [177] also the EMF of so-called bipolar galvanic cells is considered, for example, of the type

$$A_{\text{cryst}} |AB_{\text{glass}}||BC_{\text{cryst}}| B_{\text{cryst}} \,. \tag{8.93}$$

The possible application of glassy membranes as in electrolytic cells is widely discussed in literature and needs no special attention here: it is only a further example of the unusual application of glassy materials.

8.8.4
Photoeffects in Amorphous Solids and the Conductivity of Glasses

There are also other unexpected applications of glass, which have to be mentioned. Properties such as the electrical conductivity of glasses (see [630] as an example for an overview of this subject with ionic systems) are also connected with the particular structure of glasses as frozen-in bodies: here defects, inherent in the structure of glasses, determine their principle differences from crystals in both theoretical analysis and possible applications. The usage of glasses as ionic membranes in electrochemistry depends on this their particular property and on the multitude of vitreous compositions and structures possible.

Results with optically irradiated chalcogenide glasses with compositions such as As_2Se_3, As_2S_3, and so on, recently used for information storage show that irradiation brings these amorphous materials to an energetic level even higher than the normally expected "glassy state" values. This is demonstrated by both the increased solubility, C^*_{glass}, and the higher dissolution rates of the optically treated glass-samples. In the previous sections, the Gibbs potential difference, $\Delta G_g(T, \xi)$, existing between the glassy and crystalline states was discussed in terms of Eq. (8.65). It turns out from this equation that differences in the solubility, C_{glass}, of irradiated and nonirradiated amorphous chalcogenide layers could be used in order to determine the change of $\Delta G_g(T, \xi^*)$ upon irradiation and thus to make more or less definitive conclusions on the nature of the irradiation damages in photoactive materials with amorphous structure.

8.9
Some Conclusions and a Discussion of Results and Possibilities

In the present chapter it is shown, how the values of the frozen-in enthalpy, ΔH_g, and of the respective Gibbs potential, $\Delta G_g(T)$, of glasses and of disordered crystals can be evaluated, using simple thermodynamic considerations. The increased value of the configurational disorder in glasses and in defect crystals can and has to be considered. They can most probably be used as a source of increased reactivity and of accumulated energy. A general indication in this respect was given by one of the present authors many years ago [177]. We partly repeat this statement later on in a more recent publication [596]. The simplest way to use this accumulated configurational energy in glasses is to reduce or completely dissemble the kinetic barrier, arresting the disordered configurational structure of glasses below the glass transition temperature, T_g. High melting solids with complicated structures (i.e., with high ΔH_m and ΔS_m values) should be most appropriate as candidates to synthesize (especially at extremely high cooling rates) the most appropriate glassy "accumulators" of energy. The same conclusion applies also to crystals with frozen-in defect structure, discussed here and in Chapter 5.

The easiest way to remove the kinetic barrier, arresting the structure of a glass, or of the glass-like crystal, is to dissolve the defect solid in an appropriate solvent. Thus conditions have been created to use the increased frozen-in free energy, $\Delta G_g(T)$, of glasses, for example, in isothermal crystal growth processes: proposals in this sense were made by classical authors from the 1950s already cited in Section 8.5 and we have employed this possibility in the growth of graphite or diamond single crystals from vitreous carbon using gas transport reactions. This is the first real possibility to free the energy of disorder, accumulated in glasses.

Another promising possibility to employ glasses as a source of accumulated energy is the electrochemical way: to use their increased electrochemical potential and to construct appropriate "physical" galvanic cells (see here Sections 8.6 and 8.8 and our earlier publication [177], where this possibility is discussed in great details and in several variants). In analyzing this problem in such terms, solid electrolyte galvanic cells with glassy constituents in applications in cosmic technologies should be also considered.

Of particular significance could be the enhanced reactivity and higher dissolution rate of glasses (e.g., of organic substances) in such applications as medicine (ultra rapidly dissolving drugs) or in agriculture: in glasses, containing microdopants as fertilizer ions (the so-called agriglasses [616] discussed here earlier in Section 8.8.2). In general, the reaction of a glass with an appropriate gaseous or liquid reagent is the third natural and very real possibility to remove the kinetic barrier, arresting the increased disorder of the system, and thus to free its accumulated energy.

Our analysis shows that, in general (see Eqs. (8.52) and (8.54)), in "normally" quenched glasses (i.e., at $q = (10 \div 10^3)$ K/s) approximately 50% of the enthalpy of melting, $\Delta H_m = T_m \Delta S_m$, and one third of melting entropy, ΔS_m, remains frozen-in as the vitrified enthalpy, ΔH_g, and entropy, ΔS_g. A considerably lower value of the same thermodynamic functions is frozen-in in glass-like crystals (see Eqs. (8.47)–(8.49)).

As mentioned, the removal of the kinetic barrier arresting the structure of glasses can be achieved by appropriate chemical reactions, by dissolution in suitable solvents, or by employing the methods of electrochemistry. Another classical possibility to free the frozen-in enthalpy of a glass, is to increase its temperature above T_g and initiate there its crystallization. In this respect, the introduction of nucleating catalysts into a glass was also considered in Section 8.5. For possible methods of usage in this respect, for example, active crystallization cores could be recommended. In considering glass crystallization as a source of energy, it should be kept in mind that recent estimates show that maybe 99% of the water content of the universe is in a vitreous form (in the comets).

The higher the melting temperature, T_m, and the melting entropy, ΔS_m, of a substance, the more energy can be accumulated at a given cooling rate in its glassy or otherwise defect modifications. At extreme cooling rates ΔH_g values approaching ΔH_m (and even higher) can be accumulated in the vitreous state (see the discussion on the properties of vitreous argon in [1]). One of the advantages of glasses as accumulators of energy is the circumstance that the temperature coefficient of the released Gibbs free energy, $d\Delta G_g(T, \xi)/dT$, is a constant, because the frozen-in entropy of the glass $\Delta S_g = $ const. Electrolytic glass to crystal cells, their potential being determined by the *glass → crystal* reaction can give, according to the formalism we developed in Section 8.6 for metallic alloy systems, approximately $\Delta E_g(T, \xi) = 10$–20 mV at room temperature [177]. However for high melting substances (as in the case with the *vitreous carbon → graphite* galvanic cells, with $T_m \approx 4000$ K for graphite) EMF values $\Delta E = 500$–600 mV have been observed. These results perhaps open new possibilities for carbon-based combustion cells even as technical applications.

In considering the employment of the thermopower of *glass/crystal* or *glass$_1$/glass$_2$* metallic junctions as a possibility to use glasses as sources of constant potentials and hidden energy, it should be noted that the disordered structure of the glassy matrix only influences the drift properties of the electric charge carriers (e.g., the electrons) and in such a way, also influences indirectly the value of the thermoelectric power, $\Delta \varepsilon_{1,2}$. This is the reason why with vitreous conductors and semiconductors, the fulfillment of the third principle for both thermodynamic functions and thermodynamic coefficients, determining the electric properties, is guaranteed in its classical formulations by the very character of the respective charge carriers (e.g., of electrons). Thus the third principle of thermodynamics is fulfilled at $T \to 0$ for the electric properties of glasses and especially for the thermoelectric power, $\Delta \varepsilon_{1,2}(T)$ (see [182] and here Chapter 9 for a general discussion of this topic). At $T = 0$ K the values of the respective thermoelectric power *glass → crystal* $\Delta \varepsilon_{1,2} = d(\Delta E_{1,2}^{\#})/dT = 0$. The EMF of *glass → crystal* metallic thermal contacts is measurable down to relatively very low temperatures. However, with the above mentioned temperature behavior of $\Delta \varepsilon_{1,2}$ the temperature course of the EMF of the *glass → crystal* metallic contact enters absolute zero of temperatures with a course, parallel to the T-axis, in the same way as this is exhibited, in general, by any $\Delta G(T)$ dependence at $T \to 0$. In this way, as with any other substance (or property), no temperature measurement via the EMF of the *glass/crystal* or *glass$_1$/glass$_2$* thermal contact should be possible at absolute zero temperatures.

The main advantage in the possible employment of energy, accumulated and arrested in glass, is that at $T < T_g^-$ it can be preserved for infinitively long times and easily used in appropriate processes of dissolution (at $T < T_g$) or in reaction and crystallization kinetics at any temperature $T > T_g$. Sufficient reaction rates of glassy reagents can be expected to take place even for $T < T_g$ with gaseous transport reactions or with reactions involving liquid components.

Up to now no proper recognition has been given in literature to the possibility to use the accumulated energy of glasses. Besides our papers of 1981, 1986 and 2010 [177, 185, 596] on this subject we can at present mention only a recent publication by Nieuwenhuizen of 2001 [631] where crystallization of glasses at $T > T_g$ is considered as a medium source of freeing accumulated energy in several possible applications, including cosmic ones. We tried to consider here the broader aspects of application of the increased potential and the higher reaction activity of glasses: dissolution, EMF of galvanic cells, chemical reaction kinetics. In essence the chapter of our present book only gives an outline of possibilities, which have yet to be studied in more detail. Many other applications not mentioned here can be expected in further developments and we hope that the chapter and some of the results given here could be of use in analyzing and catalyzing further developments in this interesting field.

In investigating the thermodynamics of glasses, efforts up to now have been mainly concentrated on analyzing glass synthesis versus property dependences for systems, vitrified at nearly equal, "normal" conditions. Here we have also considered mainly the possible applications (and some limitations of these applications) of such "normal" glasses. It may be more interesting to investigate the possibilities to use in the sense, discussed here for glasses, obtained at conditions (e.g., at the methods of hyper-rapid quenching or other methods from the technology on metallic glasses) bringing about systems with considerably higher deviation from equilibrium than usually anticipated. Investigations on the effect of cooling rates [1, 68, 112, 245, 246, 265], or of the rate of pressure increase [383] on vitrification kinetics, glass transition temperature, T_g, and on the frozen-in values of the thermodynamic functions of glasses, show that here significantly new horizons can be opened for new applications of one of the oldest materials known to mankind.

9
Glasses and the Third Law of Thermodynamics
Ivan S. Gutzow and Jürn W. P. Schmelzer

9.1
Introduction

Classical thermodynamics is devoted to the analysis of systems in thermodynamic equilibrium states and to the description of reversible processes in between such states. It is based on three principles. These principles can be formulated either as positive statements (conservation of energy, increase of entropy in irreversible processes and determination of its change in reversible processes, approach of a well-defined value of the entropy at absolute zero temperature) or as negative restrictions prohibiting certain possibilities. Thus, the first two principles of thermodynamics can be expressed in the form of the impossibility to construct a perpetuum mobile of correspondingly the first (a machine which can do work without a supply of an equivalent amount of energy) and second (a cyclic machine which can do work, the only additional effect being the cooling of an energy reservoir) kind, while the third law can be formulated as the law of nonaccessibility of the absolute zero of temperature. While the first two laws of thermodynamics were formulated in the middle of the nineteenth century, the third law was formulated only at the beginning of the twentieth century. In the development of classical thermodynamics of systems in equilibrium in its final stage, the third principle played in this way an exceptional role bringing this science to "the logical end of its own evolution" (expressed by Nernst's own words [632]).

The third law was formulated by Nernst in 1906 and initially denoted as Nernst's heat theorem. Later-on, it was commonly used in a formulation assigned to Planck that the entropy, S, tends to zero when the absolute temperature, T, approaches zero [61, 633]. In the foreword to the third edition of his book [61] in 1910, Planck acknowledged the importance of Nernst's heat theorem and continued:

"Um den wesentlichen Inhalt dieses neuen Theorems ganz rein, in einer für die experimentelle Prüfung möglichst geeigneten Form darstellen zu können, ist es aber nach meiner Meinung notwendig, seine Bedeutung für die atomistische Theorie, die heute noch keineswegs klargestellt ist, einstweilen ganz aus dem Spiel zu lassen ... Andererseits habe ich dem Theorem, um seine Anwendungen so einfach wie möglich zu

gestalten, eine möglichst weitgehende Fassung geben zu sollen geglaubt, und bin dabei, nicht nur in der Form, sondern auch inhaltlich, über die von Nernst selber gegebene noch etwas hinausgegangen. Ich erwähne diesen Punkt auch an dieser Stelle, weil die Möglichkeit im Auge zu behalten ist, dass, wenn sich die weitergehende Fassung nicht bewähren sollte, die ursprüngliche Nernstsche deswegen doch möglicherweise zu Recht bestehen bleiben könnte."

In line with his basic aim to incorporate Nernst's theorem into his course on thermodynamics, Planck decided not to analyze atomistic aspects in his outline. And, as he stressed in the comment, he went even beyond Nernst's formulation in order to allow one its easiest application. He was aware and stated it clearly as well that his generalization may not be correct and once this could be demonstrated, it could be that one has to return to Nernst's original formulation.

Planck originally restricted his formulation of the third law to one-component systems [61, 633] stating:

"Die Entropie eines jeden chemisch homogenen dauernd ungehemmt im inneren Gleichgewicht befindlichen Körpers von endlicher Dichte nähert sich bei bis zum absoluten Nullpunkt abnehmender Temperatur einem bestimmten, von Druck, vom Aggregatzustand usw. sowie von der speziellen chemischen Modifikation unabhängigen Wert. Da die Entropie bisher nur auf eine willkürliche additive Konstante bestimmt ist, können wir unbeschadet der Allgemeinheit diesen Grenzwert gleich Null setzen."

Hereby the notation "chemisch homogen" was earlier defined as a *one-component system*. Note as well the restriction made by Planck excluding systems which are not "dauernd ungehemmt im inneren Gleichgewicht," that is, which are not in equilibrium or not able to reach it due to kinetic reasons. Planck's point of view – the restriction of the validity of Planck's formulation of the third law to one-component systems – was supported initially by Einstein [350] as well. This topic was discussed in detail by Simon in 1937 [51]. Simon noted that if it would be *"postulated that Nernst's theorem should be applied only to pure crystals . . . this would be a very severe restriction, so severe, in fact, that Nernst's theorem could no longer be regarded as a general law at all . . . Summing up, we can state that the present experimental evidence indicates the general validity of Nernst's theorem as a law of thermodynamics. The possibility that some future experiment may not be in agreement with the theorem obviously cannot be excluded, but unless there is some reason from a theoretical point of view to expect such a result, to anticipate it is mere speculation."* In this discussion, he notes as well that the situation is different for frozen-in systems, that is, glasses.

The mentioned restriction in Planck's original formulation is retained even in some modern textbooks on thermodynamics like the excellent textbook written by Kubo [54]. Kubo writes (taking over widely but not fully Planck's already given formulation): "The entropy of a chemically uniform body of finite density approaches a limiting value as the temperature goes to absolute zero regardless of pressure, density or phase. It is therefore convenient to take the state of 0 K as the standard state ... by assuming

$$\lim_{T \to 0} S = S_0 = 0 \text{ ."} \qquad (9.1)$$

So, Kubo, fully follows Planck's original formulation omitting, however, an explicit reference to equilibrium systems and the ability of the systems to approach it at any conditions, however, as it seems, these requirements are implicitly implied also in Kubo's treatment. In other thermodynamic textbooks, like the book of Bazarov [55], the restriction to one-component systems is removed treating the third law as a general theorem valid for systems in thermodynamic equilibrium. Bazarov mentioned, however, in a footnote that he considers the statistical treatment of this law as unsatisfactory.

In the discussed by Bazarov statistical physics approach [62, 63], the third law is commonly treated via the Boltzmann relation

$$S = k_B \ln \Omega \, , \qquad (9.2)$$

(k_B being the Boltzmann constant) connecting the entropy, S, with the number of microscopic configurations, Ω, leading to a given macroscopic state and assuming that Ω is equal to one for $T \to 0$. Fermi [64] discussed the possibility that the ground state of a system also at $T \to 0$ may be degenerate. However, he supposed that – provided this would be the case – the number of microstates should be small so that significant deviations from the third law should not be expected to occur even in this case.

After the formulation of the third law, in fact, as supposed by Nernst, equilibrium thermodynamics as a science was completed and the further significant development of thermodynamics as a science was connected with the analysis of nonequilibrium systems, with applications to irreversible processes, and, what is of particular importance here, with the treatment of kinetically frozen-in, thermodynamically nonequilibrium systems, that is, with glasses. In doing so, usually it is noted [62, 63] that the first principle is applicable in thermodynamics of irreversible processes in its full and most general formulations, and that the second principle is applicable to nonequilibrium only in a restricted way (in its form $dS > 0$ [14, 62, 63]). For the third principle of thermodynamics since Simon's work (see [1]) it is usually stated that it looses its applicability, because, in glasses as nonequilibrium systems, at $T \to 0$ we always have to expect (and it is, in fact, experimentally confirmed; see evidences summarized in [1] and below) that there the entropy $S|_{T \to 0} > 0$ and not $S|_{T \to 0} = 0$ as stated in the formulation assigned to Planck for classical thermodynamics of equilibrium systems [1, 43, 55]. This restriction follows historically from an observation by Einstein from the year 1914 (see [350] and the summary of the subsequent developments in Chapter 5 of our monograph [1] as well as Chapters 2, 5 and 8 of the present monograph) that solids with frozen-in configurational disorder, and in particular glasses, have to be excluded from the third principle.

Simon, with well-known experiments, confirmed the nonzero value of the zero-point entropy ($S|_{T=0} > 0$) for two cases of glass-forming melts (glycerol and vitreous SiO$_2$ [1, 59]). These first results were afterwards verified by measurements on at least 120 other glass-forming melts (see [1] and the details given there). A second source of evidence, showing that, in fact, as predicted by Einstein, a considerable value of configurational entropy exists in defect solids, was given by examining molecular crystals with frozen-in disorder (cf. Chapter 5 of the present book). For such types of matter in the solid state the experiments of Giauque [353], Clusius [352], Eucken [356] and Kaischew [355] confirmed earlier expectations of Schottky [178], Einstein [350] and others [351, 354] that in some molecular crystals $S(0)|_{T=0} > 0$ values are frozen-in, ranging from $R \ln 2$ to $R \ln(3/2)$, in obvious deviation from the third principle of thermodynamics in its classical formulation given by Planck [61, 633]. Later on, the existence of a glass-like transition in defect plastic crystals [237, 377, 634] and in vitrified liquid crystals [347] was also confirmed. A detailed description and analysis of these cases is given in [1, 168] and here in already cited chapters.

However, there is also evidence available that in fact the consequences of the classical formulation of the third principle of thermodynamics assigned to Planck, for example, that the specific heat is equal to zero at temperatures approaching zero is nevertheless obeyed in all experimentally analyzed cases, when the temperature course of the specific heats, $C_p(T)$, of glasses and of defect solids is considered. This thermodynamic coefficient shows the behavior $C_p(T)|_{T \to 0} = 0$. The specific heats of glasses at $T \to 0$ are phononic in their nature and this result is thus an expected finding. There is plenty of theoretical and experimental evidence available, showing that similar consequences from the third principle in the formulation assigned to Planck are, in fact, also obeyed for frozen-in nonequilibrium systems.

Summarizing mentioned experimental findings we conclude: there is a huge amount of experimental evidence accumulated for glasses and similar systems contradicting Planck's classical formulation that entropies have to be nullified at $T \to 0$. On the other hand, $C_p(T) = 0$ holds at $T \to 0$ not only for equilibrium systems, but also for glasses and mentioned other glass-like systems. These results require a more general formulation of the third principle of thermodynamics extending it, in addition to equilibrium systems, also to nonequilibrium bodies. This more general formulation is connected with the impossibility to reach the zero-point of temperature. It can be shown, as demonstrated below, that even with nonequilibrium systems with frozen-in disorder, like glasses and defect crystals, the zero point of temperature cannot be reached. Thus, formulating the third law as the principle of nonaccessibility of absolute zero its range of validity is increased and it becomes – as will be demonstrated below – a general principle valid for both equilibrium and nonequilibrium systems.

9.2
A Brief Historical Recollection

A brief overview on the development of the third law of thermodynamics was given by Einstein in his contribution to *The Scientific Monthly* devoted to Walther

Nernst [635]. Einstein noted that Nernst

> "ascended from Arrhenius, Ostwald and van't Hoff, as the last of a dynasty which based their investigations on thermodynamics, osmotic pressure and ionic theory. Up to 1905 his work was essentially restricted to that range of ideas ... This first productive period is largely concerned with improving the methodology and completing the exploration of a field the principles of which had already been known before Nernst. This work led him gradually to a general problem which is characterized by the question: Is it possible to compute from the known energy ... of a system, the useful work which is to be gained by its transition from one state into another? Nernst realized that a theoretical determination of the transition work, A, from the energy-difference, U, by means of equations of thermodynamics alone is not possible. There could be inferred from thermodynamics that, at absolute zero temperature, the quantities, A and U, must be equal. But one could not derive A from U for any arbitrary temperatures, even if the energy-values or differences in U were known for all conditions. This computation was not possible until there was introduced, with regard to the reaction of these quantities under low temperatures, an assumption which appeared obvious because of its simplicity. This assumption is simply that A becomes temperature-independent under low temperatures. The introduction of this assumption as a hypothesis (third main principle of the theory of heat) is Nernst's greatest contribution to theoretical science. Planck found later a solution which is theoretically more satisfactory; namely, the entropy disappears at absolute zero temperature."

This sketch we would like to follow first in more detail.

When in 1906 Walther Nernst formulated his heat theorem, soon being transformed into the third principle of thermodynamics, he firmly believed that any system, that is, matter in all its forms, has to follow it. Generality and universal applicability is in fact a requirement for any principle of nature. This deep conviction of Nernst can be followed in both his books [60, 632], the first of which, being entirely devoted to the heat theorem, also gives the evidence on the great experimental development, initiated and directed by Nernst in his laboratory to confirm his expectation. There, in both books, one can also find the proud words of Nernst, that *"with the third principle of thermodynamics the whole development of thermodynamics has come to its natural end: nothing of significance should follow"* [632]. Concerning classical thermodynamics, Nernst was in many respects right; the further development of irreversible thermodynamics, to come with De Donder and Prigogine and many others (see [13, 14, 43]), was at the end of the 1920s little or not known even to the greatest representatives of German classical thermodynamics. It is shown here that this new thermodynamic approach was to play a major role in understanding the glass transition.

Nevertheless, soon after Nernst's first publications in the early 1900s, it became evident that gases, by their very nature, do not and cannot follow the third principle. It was again Nernst's deep conviction and great vision, who found in *gas degeneracy* a process, in which matter in gaseous form is changed in such a way that from following Boltzmann's statistics it is transferred to obey quantum laws. Quantum statistics guarantees below the gas degeneracy temperature a peculiar fulfillment of the third principle even for ideal and real gases [62, 63]. Here we have to mention that Nernst foresaw the necessity of gas degeneracy even before quantum statistics was formulated, and that after Planck's first publications, Nernst eagerly accepted the idea [60] that the fulfillment of the third principle is directly connected with quantum mechanics.

Then, in 1914, came a famous lecture by Einstein [350] on the foundations of quantum statistics, in which he brilliantly prophesied that systems with frozen-in statistical disorder, like solid solutions (he also mentioned: "probably also glasses, however, we know – in 1913 – nothing of their structure") should not follow the third principle. Einstein's prediction was developed by Stern [636] for various forms of frozen-in solids. It found its first experimental verification in a series of investigations performed both in German laboratories [352] and then in the work of Giauque [353] and Pauling [354] in the United States on molecular crystals like CO and H_2O. The essence of Einstein's and Stern's objection was that frozen-in states have to display a constant zero-point entropy $S(T)|_{T \to 0} = S(0) > 0$. They should not, as predicted by Nernst [60, 632] (and in a modified form by Planck [61]), display a zero-point entropy $S(0) = 0$, as expected for any homogeneous equilibrium solid. Thus the zero-point entropy of vitreous states, the structure and the possible structural dependence of the zero-point entropy, $S(0)$, of glasses as typical frozen-in structures became of exceptional thermodynamic significance.

Then came the days, when in independent experiments two co-workers of Nernst, Witzel [163] and Simon [21, 58] (and also Giauque [160] in the USA), found on two very different model glasses (glycerol and SiO_2) that both of them do not obey the third principle: they display a considerable frozen-in entropy value $S(0) > 0$ down to $T \to 0$. The interested reader may find in more detail the history of these developments in our monograph [1] and in our articles [182, 191, 634] or in [237, 341]. Additional information on the zero-point entropy, $S(0) > 0$, of glasses for more than one hundred different substances investigated up to now are also given in [1, 191]: they do not all obey the third principle in Planck's sense.

Of great significance for the development of both thermodynamics and of glass science turned out to be Simon's assumption ("glasses are nonequilibrium, nonthermodynamic systems, which have not and cannot fulfill the third principle") and his approximation in calculating $S(0)$. These two basic starting points of the analysis and their consequences in treating thermodynamics of glasses are derived in detail in the monograph [1]. We would like to mention here only the following: Nernst expected (see, e.g., Witzel's paper [163] and [60]) that "something" similar to gas degeneration may also happen with glasses in order that they should nevertheless fulfill the third principle.

In the present chapter we show that with glasses – as with gaseous states – Nernst again was on the right track: in fact as we showed first in [180, 182, 598], glasses follow the third principle, although in a peculiar way. This result we demonstrated using the methods and the principles of the thermodynamics of irreversible processes [13, 14, 43, 184, 637]. Thus, irreversible thermodynamics shows, as classical thermodynamics does, that in fact entropy is frozen-in in glasses. However, it also shows that nevertheless the third principle is followed by glasses – in a way, not recognized by Simon even in his later publications [23, 51]. We will go over to a detailed outline of these results in the following section.

9.3
The Classical Thermodynamic Approach

At the end of the nineteenth century, problems of thermodynamic equilibrium in chemically reacting gas mixtures gained a considerable practical importance. In this connection, the problem arose how one can determine the difference in the Gibbs potential:

$$\Delta G(T) = \Delta H(T) - T\Delta S(T). \tag{9.3}$$

The Delta-sign indicates here the difference of the respective thermodynamic functions in two different states of the systems.

W. Thomson (Lord Kelvin) in 1852 and Berthelot in 1869 tried to solve this problem by attempting to formulate criteria which have to be fulfilled by spontaneously proceeding chemical reactions (cf. [74]). Their result is known as Berthelot's rule, and it can be expressed as $(\Delta H)_{p,T} < 0$. As Nernst already noted in the first edition of his book on thermodynamics [632], Berthelot's rule cannot be considered as a law of nature; anyway, it is fulfilled very often and in particular at low temperatures. So, the stimulus for Nernst's search for an understanding of the behavior of matter at low temperatures was connected with the analysis why Berthelot's rule is fulfilled at low temperatures.

Nernst's leading idea in formulating his heat theorem was that in approaching zero of absolute temperatures, the thermodynamic potential difference $\Delta G(T)$ in the fundamental relation becomes equal to the respective enthalpy difference, $\Delta H(T)$, not only because of $T \to 0$, but also because the entropy $\Delta S(T) \to 0$. Thus, from such consideration it followed not only

$$\Delta H(T)|_{T\to 0} = \Delta G(T)|_{T\to 0} \tag{9.4}$$

but also (and this is the essential point!)

$$\left.\frac{d\Delta H(T)}{dT}\right|_{T\to 0} = \left.\frac{d\Delta G(T)}{dT}\right|_{T\to 0} = 0 \tag{9.5}$$

because of

$$\frac{d\Delta G(T)}{dT} = -\Delta S(T). \tag{9.6}$$

The nullification of $\Delta S(T)|_{T\to 0}$ led to the well-known Nernst construction of the course of the $\Delta G(T)$, $\Delta H(T)$ functions for processes of transitions between two

9 Glasses and the Third Law of Thermodynamics

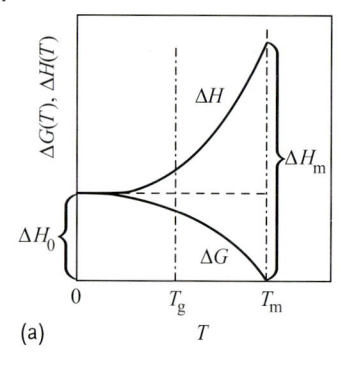

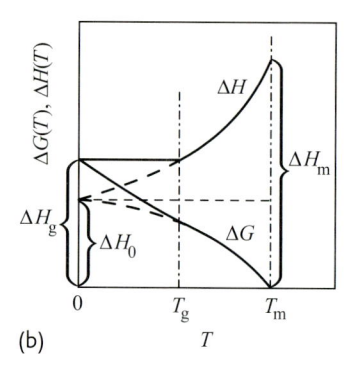

Figure 9.1 Nernst's diagram and Simon's ($\Delta G(T)$, $\Delta H(T)$)-construction [1]: (a) temperature dependence of the enthalpy, $\Delta H(T)$, and of the Gibbs free energy difference, $\Delta G(T)$, melt-crystal for an equilibrium system, obeying the third law of thermodynamics (Nernst's (ΔH, ΔG) diagram); (b) temperature dependence of the enthalpy and the Gibbs free energy for a vitrifying system (solid curves) according to Simon's approximation [23, 58]. The temperature course of $\Delta G(T)$ expected from the third law of thermodynamics is indicated in the construction by a dashed curve. ΔH_0 is the zero-point enthalpy difference of the two different equilibrium states of the same substance (fictive melt, undercooled liquid and crystal), ΔH_g is the zero-point enthalpy difference between the glass and the crystal. Note also that for equilibrium systems the relation $(d\Delta H/dT) = (d\Delta G/dT) = 0$ is obeyed at $T \to 0$, while for a glass at $T \to 0$ the relation $(d\Delta H/dT) = 0$ still holds, however, $(d\Delta G/dT) < 0$.

different states of condensed phases (Figure 9.1). As a consequence it follows that the approach to zero of absolute temperature is accompanied by the approach to zero of the specific heats, $\Delta C_p(T) \to 0$ at $T \to 0$.

In fact, accounting for another well-known thermodynamic dependence

$$\frac{dS(T)}{dT} = \frac{C_p(T)}{T}, \quad (9.7)$$

it becomes evident that Planck's integral (see [61])

$$S(T)|_{T\to 0} = S(0) + \int_0^T \frac{C_p(T)}{T} dT \Big|_{T\to 0} \quad (9.8)$$

has a converging value only if

$$C_p(T)|_{T\to 0} = 0. \quad (9.9)$$

Equation (9.9) is a necessary but not sufficient condition for convergence of the integral. More strictly, $C_p(T)$ should approach $T \to 0$ steeper than $C_p(T) = a_0 T$. And here came into play the Einstein and Debye theories of specific heats of crystalline and amorphous solids [43, 62, 63], considered as an ensemble of quantum oscillators. The quantum theory gave at low temperatures, that is, for $T \ll \theta_D$, where

$$\theta_D = \frac{h}{k_B}\omega_D \quad (9.10)$$

is the characteristic Debye temperature, the dependence

$$C_p(T) = \text{const}_n R \left(\frac{T}{\theta_D}\right)^n \tag{9.11}$$

with $n = 1, 2, 3$ and $\text{const}_1 = \pi^2$, $\text{const}_2 = 43.3$, and $\text{const}_3 = (12/5)\pi^4$ for one-dimensional, two-dimensional and three-dimensional structures, respectively (in Tarassov's classification [270]). Here, as usual, we have indicated with R the universal gas constant, h is Planck's constant and with k_B Boltzmann's constant is denoted. From Nernst's postulate, one obtains thus $C_p \to 0$ and $S \to 0$ for $T \to 0$. Vice versa, the latter condition yields Nernst's postulate, Eq. (9.5).

Using known thermodynamic dependencies, connecting $C_p(T)$ with other thermodynamic coefficients, it can be demonstrated based on Eqs. (9.8) and (9.9) that some of the thermodynamic coefficients (like the coefficient of thermal expansion)

$$\alpha = \frac{1}{V}\left(\frac{\partial V}{\partial T}\right)_p \tag{9.12}$$

are also nullified at $T \to 0$, where others like the compressibility become constant. Equation (9.9) leads also to the consequence that the zero of absolute temperature cannot be reached by any possible process. This result supplies us with the fundamental groundwork for an alternative, and as we will show, the most general formulation of the third principle of thermodynamics, as the principle of unattainability of absolute zero [43, 168, 178, 264].

To reiterate, for an equilibrium system classical thermodynamic theory provides for any system with Nernst's theorem that:

- Its entropy S (more generally, according to Nernst, any entropy difference, ΔS, between different states of the system) approaches zero values at $T \to 0$.
- The specific heats $C_p(T)$ are nullified at $T \to 0$.
- Thus, with $C_p(T) \to 0$ by no means can be the zero point of temperatures be reached.
- All thermodynamic coefficients, which are temperature derivatives of $G(T)$ (like the coefficient of thermal expansion)

$$\alpha = \frac{1}{V}\left(\frac{\partial V}{\partial T}\right)_p = \frac{1}{V}\frac{\partial^2 G}{\partial p \partial T} \tag{9.13}$$

are also nullified at $T \to 0$ [168, 264].
- Thermodynamic coefficients, which are not temperature derivatives of $G(T)$ (like the compressibility)

$$\kappa = -\frac{1}{V}\left(\frac{\partial V}{\partial p}\right)_T = -\frac{1}{V}\frac{\partial^2 G}{\partial p^2}, \tag{9.14}$$

or the electric or magnetic susceptibilities, become constants at $T \to 0$: the dielectric constant behaves as $\varepsilon(T)|_{T\to 0} = \text{const}$. Thus, however, we have [638]:

$$\frac{d}{dT}(\varepsilon)|_{T\to 0} = 0. \tag{9.15}$$

9.4
Nonequilibrium States and Classical Thermodynamic Treatment

The above derivations refer originally to homogeneous equilibrium systems. However, Planck provided via Eq. (9.8) also an example on the thermodynamic treatment of nonequilibrium frozen-in systems [61]. Suppose, he argued in [61] that we have, as predicted by Einstein [350] and Stern [636], originally an equilibrium system (e.g., a solution) in which at some temperature $T = T_E > 0$ has been frozen-in completely or partially its molecular configuration. Let this frozen-in part determine an excess entropy, which can be considered and calculated at $T \geq T_E$ as an ideal solid solution, that is,

$$\Delta S_{ex} \cong -R\left[x_1 \ln x_1 + (1-x_1) \ln(1-x_1)\right]. \tag{9.16}$$

Here x_1 and $x_2 = (1-x_1)$ are the respective molar fractions of the two components 1 and 2 of this solution or of two structural units with nearly equal energy of an otherwise homogeneous body. At $x_1 = x_2 = 0.5$, Eq. (9.16) gives $\Delta S_{ex} \cong R \ln 2$. If the system is frozen-in at T_E, no change will occur down to $T \to 0$ and there in analogy with Eq. (9.8) we have

$$S(T)|_{T \to 0} = S_0 = \lim_{T \to 0} \int_0^T \frac{C_p(T)}{T} dT + \Delta S_{ex}. \tag{9.17}$$

In order to have a converging integral we have again to require that $C_p(T)|_{T\to 0} \to 0$. Thus it turns out that for the considered frozen-in system

$$S(0) \equiv \Delta S_{ex} \approx R \ln 2 \tag{9.18}$$

and this is essential: the phonon specific heats of the frozen-in system below T_E, again given with Eqs. (9.9) – (9.11), allow one to determine $C_p(T)|_{T\to 0} = 0$. This was the argumentation of both Planck and Stern [61, 636] and of Giauque and Pauling ([353, 354, 598], see also [182]) leading to a number of experimental investigations where by the example of several solid solutions and molecular crystals (CO, NO_2, H_2O, etc. [352, 353]) with frozen-in constant disorder in fact $S(0) = R \ln 2$ or $S(0) = R \ln(3/2)$ was found [43, 353, 354, 636]. It is of significance to mention that later on Planck obtained the results, indicated by Eqs. (9.9), (9.16) and (9.17) also by a thorough derivation in terms of the classical Boltzmann statistics [633]. Again reiterating, it is evident that for frozen-in systems classical thermodynamics states:

1. Configurationally determined contributions to the thermodynamic functions (like the entropy S, enthalpy H, free volume V) are frozen-in to constant values and remain so down to $T \to 0$.
2. Despite this fact, thermodynamic coefficients (like $C_p(T)$, $\alpha(T)$) are nullified at $T \to 0$ [61].
3. Thus the zero point of temperatures cannot be reached employing frozen-in nonequilibrium systems: also for them the principle of unattainability of absolute zero remains valid.

In order to reconcile Nernst's heat theorem with the results of experiments, Simon [21, 23, 51] formulated a brilliant general assumption, which survived more than 80 years as one of the fundamentals of glass science. From his and Giauque's results it was evident that glasses are nonequilibrium, kinetically frozen-in thermodynamic systems. So they are in fact nonthermodynamic systems in the classical sense and they do not have to follow the requirements of the third principle, which is reserved for thermodynamic systems only, as argued by Simon [51]. The more technically oriented glass science community accepted Simon's idea with a sort of acclamation: this idea stated that glasses are materials differing from everything else in nature.

In fact, Simon, however, only accepted Planck's treatment of frozen-in crystalline states in its part that a configurational entropy is frozen-in in the glass at the temperature $T_E = T_g$, where T_g indicates Tammann's glass transition temperature, already introduced in glass science at that time. However, Simon in underlining the circumstance that an entropy value can be frozen-in at $T = T_g$, at the same time failed to bring to attention Planck's prediction that Eq. (9.9) guaranteeing fulfillment of the nullification of the specific heats at $T \to 0$ for both equilibrium and nonequilibrium systems, is applicable also to glasses. The fulfillment of Eq. (9.9) guarantees moreover the applicability of the unattainability principle for systems with frozen-in configurational entropy. This fact is of cardinal significance in discussing the applicability of the third law to glasses.

In order to easily calculate the value of frozen-in entropy in terms of classical thermodynamics, Simon reduced in a simple approximation Tammann's glass transition interval (T_g^+, T_g^-) (which in dependence of cooling rate [1] extends over more than (10 ÷ 50) K in real glasses) to the glass transition temperature, T_g, only. Thus he approximated the undercooled liquid from melting point T_m to T_g as a thermodynamic system in full equilibrium, which transformed just at $T = T_g$ to the frozen-in nonthermodynamic, nonergodic system, a glass. In the thus formed glass, according to Simon's approximation, the system retained the thermodynamic parameters and its configurational entropy ΔS_g, corresponding to the temperature $T = T_g$. Thus, according to Simon $\Delta S(0) \equiv \Delta S(T)|_{T \to 0} \cong \Delta S_g$. With this assumption Nernst's diagram on Figure 9.1a is transformed into a picture (see Figure 9.1b) we called in our monograph [1] Simon's diagram: there

$$\frac{d\Delta G(T)}{dT} = \Delta S_g = \text{const.} \quad (9.19)$$

and because of Eq. (9.3) the relation $\Delta G_g(T)|_{T \to 0} \equiv \Delta G_g(0) = \Delta H_g$ holds. Here ΔH_g indicates the frozen-in value of the enthalpy of the glass. The temperature course of the thermodynamic potential $\Delta G_g(T)$ of the glass is according to this approximation a straight line, tangent to the $\Delta G(T)$-course of the undercooled melt at $T = T_g$ [1]. Thus with Simon's approximation the whole thermodynamics of glasses could be easily and in a reproducible way constructed as this is done in great detail in our book [1].

9.5
Zero-Point Entropy of Glasses and Defect Crystals: Calculations and Structural Dependence

In the foregoing sections we have shown that according to both classical thermodynamics and by kinetic approaches, based on nonequilibrium thermodynamics, in the glass-transition region (or approximately at $T = T_g$) is frozen in a configurational entropy ΔS_g, giving at $T \to 0$ rise to a zero-point entropy, $\Delta S(0) > 0$. In a similar way in the mentioned defect molecular crystals (CO, N_2O, H_2O, D_2O, etc., see [354, 634] and Chapter 5) at temperatures $T < T_m$, in the vicinity of the melting point are frozen-in defect crystalline structures, giving at $T \to 0$ the mentioned zero-point entropy $\Delta S(0) > 0$. Experiments performed by Kaischew [355] and by other authors have shown that neither extremely slow crystallization nor prolonged annealing could remove or decrease the zero-point entropy of molecular crystals with frozen-in disorder.

In determining the $\Delta S(0)$ value of defect crystals the vapor pressure $P(T)$ of the samples is measured over a broad temperature interval below T_m and brought to as low temperatures as possible. The results thus obtained are then compared with $P(T)$ values, calculated according to the formula

$$\log P(T) = -\frac{L(T)}{2.3RT} + \frac{1}{2.3}\int_0^T \frac{dT}{T^2}\int_0^T \left[C_p(T)_{vap} - C_p(T)_{cond}\right]dT + i_0, \quad (9.20)$$

where $L(T)$ is the enthalpy of evaporation, $C_p(T)_{vap}$ and $C_p(T)_{cond}$ are the specific heats of the vapor phase and of the liquid or crystalline condensate, and i_0 is the chemical constant [55, 60, 632]

$$i_0 = \frac{1}{2.3R}\left[S(0) - C_p(0)_{vap}\right] \quad (9.21)$$

of the substance under investigation. The values of $C_p(T)_{cond}$ and evaporation and transition heats are experimentally determined. The $C_p(T)_{vap}$ values in the temperature range from $T \to 0$ to T are calculated using band spectral data to construct the partition function of the gaseous phase and to introduce the result into Eq. (9.20), which follows directly (see [55]) from the thermodynamic dependence

$$\ln P = -\frac{\Delta G(T)}{RT} + \text{const.}, \quad (9.22)$$

where $\Delta G(T)$ expressed via Eq. (9.3) gives the Gibbs potential difference defect crystal vapor. At $\Delta S(0) = 0$, Eq. (9.21) would give the i-value of the ideal, defect free crystal according to Eq. (9.8).

Following Simon's approximation, the ΔS_g values of glasses are calculated via the expression

$$\Delta S_g = \Delta S_m - \int_{T_g}^{T_m} \frac{\Delta C_p(T)}{T} dT \approx \Delta S_0, \quad (9.23)$$

where ΔS_m is the molar entropy of melting and $\Delta C_p(T)$ is the experimentally determined specific heat difference liquid (glass)/crystal, which has to be followed to the lowest temperatures. Simon and Lange [21] first reached helium temperatures (4 K) in order to demonstrate the frozen-in entropy of vitreous glycerol and of the SiO_2 glass. With the same data also the frozen-in value of the enthalpy ΔH_g is determined via

$$\Delta H_g = \Delta H_m - \int_{T_g}^{T_m} \Delta C_p(T) dT, \qquad (9.24)$$

where $\Delta H_m = \Delta S_m T_m$. Of particular significance is that ΔH_g can be also directly determined from the difference of dissolution heats of glass and crystal in appropriate solvents or via chemical reactions (in the case of SiO_2 with hydrofluoric acid in platinum calorimeters). At present there are more than 120 glass-forming substances, for which according to Eqs. (9.23) and (9.24) the values of the frozen-in parameters ΔH_g and ΔS_g have been determined (see [1, 237, 341]). The structural and kinetic information, which can be derived from this abundance of data, is discussed in the following section. A survey of typical data and their structural analysis may be traced also in [1, 112].

9.6
Thermodynamic and Kinetic Invariants of the Glass Transition

It was Kauzmann [109] who first observed that in dimensionless coordinates, when the temperature, T, is divided by the temperature of melting, T_m, the course of both kinetic and thermodynamic properties of glass-forming systems with seemingly very different structure changes in a relatively limited range. Thus for the first time a behavior was observed, indicating the possibility T_m to be used as the normalizing factor in a representation of vitrification in terms of the behavior of corresponding states.

Two such important parameters are $(\Delta C_p(T_g)/\Delta S_m)$ and $(\Delta H(T_g)/\Delta H_m)$. They represent important thermodynamic characteristics of the vitrifying system: the first one we called years ago [1, 112, 177] the *thermodynamic structural parameter* of the vitrifying melt, written in a more general way as $a_0 = [\Delta C_p(T)/\Delta S(T)]$. It gives a measure of the increase of structural disorder (ΔS) in a given system, for example, at $T = T_g$, upon unity heat (ΔC_p) introduced into it from outside. The second ratio indicates the energy (as multiples of the melting enthalpy ΔH_m) really frozen-in in the system upon vitrification. The value of a_0 changes in the limits from $a_0 = 1$ (for metallic glasses) to $a_0 = 2$ (in polymers) with a most probable value [1]

$$\frac{\Delta C_p(T_g)}{\Delta S_m} = a_0 = \frac{3}{2} \qquad (9.25)$$

first observed by Wunderlich [173]. The ratio (T_g/T_m) was established as the main kinetic characteristics of vitrification of the form

$$\frac{T_g}{T_m} = \frac{2}{3} \tag{9.26}$$

at "normal" cooling rates ($q \cong 10 \div 10^3$ K/s) for typical glass-formers (in more than a hundred cases, see [1]).

With the mentioned most probable values according to Eqs. (9.25) and (9.26) of the ratio a_0 and (T_g/T_m), we obtain

$$\Delta S_g = \Delta S_m - \int_{T_g}^{T_m} \frac{\Delta C_p(T)}{T} dT \cong \Delta S_m \left[1 - a_0 \ln \left(\frac{T_g}{T_m} \right) \right] \cong 0.33 \Delta S_m, \tag{9.27}$$

$$\Delta H_g = \Delta H_m - \int_{T_g}^{T_m} \Delta C_p(T) dT \cong \Delta H_m \left[1 - a_0 \left(1 - \frac{T_g}{T_m} \right) \right] \cong 0.50 \Delta H_m. \tag{9.28}$$

Thus a simple linear relation has to be observed between ΔS_g and the melting entropy ΔS_m and between ΔH_g and the melting enthalpy ΔH_m of simple (in the sense of one-component) glass-forming systems. Figure 9.2 shows one of these dependencies for a number of systems, which include practically all known glass-forming structures. In considering Eq. (9.27) it should be remembered that ΔS_m includes the main structural characteristics of the melt under investigation. It is seen also from Figure 9.2, that in contrast to Eq. (9.27) in defect crystals not a part of ΔS_m, but a quite independent $\Delta S(0)$ value ($\cong R \ln 2$ or $R \ln(3/2)$) is frozen-in [354, 634].

Let us write now the ideal entropy of mixing (Eq. (9.16)) for a one-component system, in which we consider the free volume $V = N_0 v_0$ as being caused by N_0 holes of molecular dimensions, v_0, introduced into the quasi-crystalline "lattice" of the glass-forming liquid (see the details of this lattice-hole MFA model of liquids in [1] or here in Chapter 3). In this case we have to mix N_A molecules with N_0 holes, the result, leading via Eq. (9.16), to

$$\Delta S_{ex}^0 = -R \ln \left[\ln(1-\xi) + \frac{\xi}{1-\xi} \ln \xi \right], \tag{9.29}$$

if we consider the free-volume MFA model to determine ξ as $\xi = N_0/(N_A + N_0)$ through the relative number of holes. For the usually observed free-volume values in glasses ($\xi = 0.05 - 0.2$) Eq. (9.29) can be approximated by

$$\Delta S_{ex}^0 \approx \Delta S_g = 2.5 R \xi. \tag{9.30}$$

Thus, the frozen-in entropy in simple glasses, where no orientational or configurational entropy contributions have to be expected, should be proportional to the ratio of the molar volumes glass/crystal ($\xi \cong \rho_{glass}/\rho_{cryst}$). This result is in fact observed for the simplest glasses in a remarkable way [1, 191, 206, 207]: SiO_2, GeO_2,

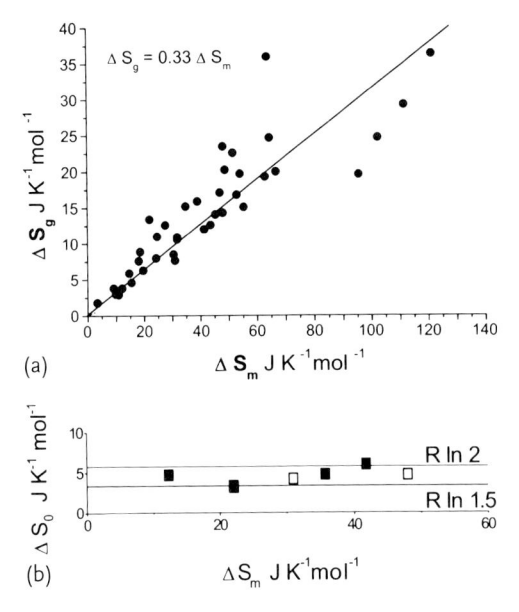

Figure 9.2 Zero-point entropies of glasses and of defect crystals. (a) Dependence of zero-point entropy of 39 typical single-component glasses on melting entropy, ΔS_m. The straight line is drawn according to Eq. (9.25) as $\Delta S_g = (1/3)\Delta S_m$; (b) zero-point entropy of molecular disordered defect crystals (black squares) and of "vitrified" plastic crystals (open squares) vs. the respective ΔS_m-values: no correlation is found here and only $R \ln 2$ to $R \ln(3/2)$-values are observed.

P_2O_5, BeF_2 have nearly identical ξ values, and also identical $\Delta S(0)$ values [1], $\Delta S(0) \approx R/2$ (see [1, 206, 207]).

Thus it follows that in real glass-forming systems as a rule a part of the entropy of melting ΔS_m is frozen-in at $T = T_g$, corresponding roughly to 1/3 of the entropy of melting at $T = T_m$. This result gains an interesting significance if we apply Boltzmann's formula, according to which in general

$$S(T) = R \ln \Omega(T) \tag{9.31}$$

holds, where $\Omega(T)$ indicates in our case the number of thermodynamically significant configurational possibilities of microscopic realization of the system at a given temperature, T. With Eq. (9.27) it follows, that in general in "normally" cooled glasses the number of frozen-in configurations is

$$\Omega(T_g) \cong \sqrt[3]{\Omega(T_m)}. \tag{9.32}$$

9.7
Experimental Verification of the Existence of Frozen-in Entropies

In the previous section we have shown the way frozen-in zero-point entropies are *calculated* out of calorimetric measurements of the specific heats. Here we analyze

Table 9.1 Solubility, C (g per 100 g of the solvent) of glassy and crystalline glucose at 298 K in different organic solvents (according to Parks et al. [610]). Note that the ratio C_{glass}/C_{cryst} does not depend substantially on the solvent, but only on the state of the solute. According to Eq. (9.37) it follows that ΔS_m for glucose has a value of about $\cong 5.5 R$, accounting for the mean value of $\log(\sigma_{glass}/\sigma_{crystal}) = 1.07$ in all four solvents.

Solvent	Glassy solute	Crystalline solute	$\log(C_{glass}/C_{crystal})$
99% ethyl alcohol	4.70	0.44	1.03
Pure ethyl alcohol	1.58	0.22	0.85
Pure isopropyl alcohol	1.07	0.08	1.13
Pure acetone	0.184	0.012	1.12

the methods capable to verify frozen-in entropy values by other experimental methods, considering in particular such experiments in which the frozen-in parameters of the glass are directly "visualized."

The most instructive methods in verifying frozen-in values of entropy could be based on vapor pressure and solubility measurements [107, 185, 255]. Here, however, particular emphasis has first to be given to the very possibility of performing such measurements: they have to be conducted in such a way that the vapor to glass (or the liquid to glass) interface has no time to relax to a more stable configuration. This problem, connected with the vapor pressure measurements of glass, was first posed by Simon [51], although vapor pressure measurements were already used in his times to verify the existence of a frozen-in entropy in defect crystals. First measurements on solubility of a glucose glass (by Parks and Huffman [37]) and crystals in four different solvents are summarized in Table 9.1. Also, quantitative or semi-quantitative data exist on the solubility of several inorganic glasses (Se in CS_2, As_2O_3 and SiO_2 in aqueous solutions and in supercritical water) which are summarized in the monograph [1].

In considering the glass interface relaxation, anticipated by Simon, let us assume that it is governed by a time of molecular relaxation, $\tau(T)_{surf}$, however, the activation energy $U(T)_{surf}$ is only $(1/2 \div 1/3) U(T)_{bulk}$. The rate of this surface change, expressed in terms of a surface structure parameter, ξ_{surf}, should be

$$\left(\frac{d\xi_{surf}}{dt}\right)_T = -\frac{1}{\tau(T)_{surf}} \left[\xi_{surf} - \xi_{eq}(T)_{surf}\right]. \tag{9.33}$$

At a temperature, T, sufficiently below T_g, this rate of change should be sufficiently low to permit the dissolution of every new glass layer to proceed at a constant (or nearly constant) value of the structural parameter, ξ_{surf}. In measuring dissolution rates (which are the only really safe way to perform such experiments, as shown by Grantcharova et al. [185]) we have to compare $\tau(T)_{surf}$ with the characteristic time, $\tau(T)_{sol}$, governing the dissolution process through the concentration gradient in the solute, surrounding the dissolving glass. This time is determined via

$$\frac{d\left[C(t)_{sol} - C(t)_{glass}\right]}{dt} = \frac{1}{\tau(T)_{sol}} \left[C(T)_{glass} - C(T)_{sol}\right] \tag{9.34}$$

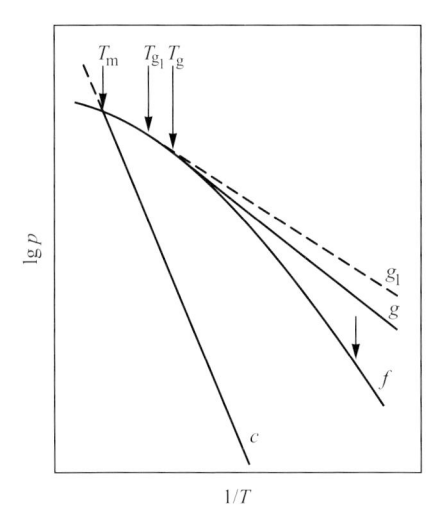

Figure 9.3 Temperature dependence of the vapor pressure of a crystal specified by (c), the undercooled melt (f) and the glass (g), frozen-in at two different temperatures T_g and T_{g1} ($T_{g1} < T_g$). The change of the vapor pressure of the glass upon possible stabilization is indicated by an arrow. Note that the log P_{glass} dependence is in $(1/T)$ representation a straight line, tangent to the log P course of the undercooled melt at $T = T_g$.

from Nernst's dissolution law [185]. At $\tau(T)_{sol} \ll \tau(T)_{surf}$, the true values of solubility, $C(t)_{glass}$, should be obtained as this was performed by Grantcharova and Gutzow and Grantcharova et al. [172, 185] in 1986 with the organic model glass phenolphthalein.

According to Eq. (9.35) (cf. Eq. (2.133))

$$\frac{\Delta G(T)}{T_m \Delta S_m} \cong \left(\frac{1}{2} - \frac{T}{3 T_m}\right) \quad \text{for} \quad T < T_g. \tag{9.35}$$

and Eq. (9.22), we would expect the solubility, σ, of a substance as a glass and as a crystal to be governed by

$$\ln\left(\frac{\sigma_{glass}}{\sigma_{crystal}}\right) = \frac{\Delta G_g}{RT} = \frac{\Delta H_g}{RT} - \frac{\Delta S_g}{R} \tag{9.36}$$

and to have in coordinates $\log \sigma$ vs. $1/T$ the typical course, determined by the constant values of ΔH_g and ΔS_g, as it is illustrated in Figure 9.3 for the case of vapor pressure. Figure 9.3 gives the course of the solubility of both glassy and crystalline phenolphthalein as it was obtained in the mentioned way by dissolution experiments. The values of ΔH_g and ΔS_g calculated according to these results correspond fairly well [172] to the ΔH_g and ΔS_g values obtained for the same substance according to calorimetric $\Delta C_p(T)$ measurements via Eqs. (9.23) and (9.24).

In using the results on the invariants at glass-transition, Eqs. (9.27), (9.28), (9.35), it follows that, in the vicinity of $T = T_g$, the vapor pressure ratio $(P_{gl}/P_{cryst})|_{T=T_g}$ or the solubility ratio $(\sigma_{gl}/\sigma_{cryst})|_{T=T_g}$ in the glass transition interval should be

9 Glasses and the Third Law of Thermodynamics

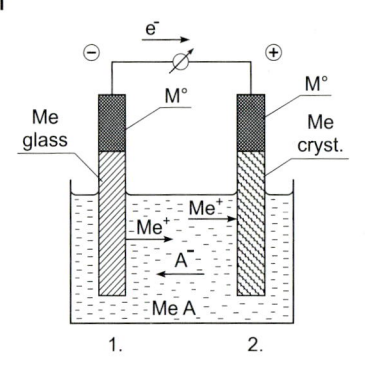

Figure 9.4 A possible realization of the galvanic cell glass/crystal in its normal variant: glass (1) and crystalline (2) metallic electrodes in an aqueous solution of the corresponding metal cation Me.

approximately equal to

$$\log\left[\frac{C_{\text{glass}}}{C_{\text{cryst}}}\right]_{T=T_g} = \frac{\Delta G_{\text{glass}}(T)}{RT}\bigg|_{T=T_g}$$
$$= \frac{\Delta S_m}{2.3R}\left[\frac{1}{2}\left(\frac{T_m}{T_g} - \frac{\Delta S_g}{\Delta S_m}\right)\right] \cong \frac{\Delta S_m}{5.5R}. \quad (9.37)$$

This ratio is in fact observed by Parks and Huffman [37] in the solubility values ($C_{\text{glass}}/C_{\text{crystal}}$) in all four solvents used to dissolve glucose glass and crystal (see Table 9.1) and is roughly followed in all above-mentioned cases of solubility of inorganic and organic glasses. This result gives additional support to the assumption that the difference in the solubility (or vapor pressure) of glasses and crystals is a phenomenon, determined and governed by the frozen-in parameters of the glass, ΔH_g and ΔS_g.

Additional experiments giving also a sound confirmation of the existence of frozen-in parameters in glass are the electric driving force (EMF) of the glass to crystal galvanic element. This possibility was mentioned by Gutzow in [177] and is given by

$$\Delta E_g = -\frac{\Delta G(T)_{\text{glass}}}{zF} = -\frac{1}{zF}\left(\Delta H_g - T\Delta S_g\right). \quad (9.38)$$

In Eq. (9.38), F denotes the Faraday constant and z is the respective change of electric charges in the electrochemical reaction employed. Figure 9.4 shows a possible realization of this effect using a metal glass and its metallic crystalline analog. Figure 9.5 displays the EMF results of the galvanic cell vitreous carbon → graphite case as reported by Das and Hucke [359]. The important feature in Figure 9.5 is the linear character of the ΔE_g dependence on temperature, as it follows for the EMF of any glass → crystal system (cf. Eq. (9.38)).

Such an EMF could be of particular interest in many technical devices as a source of a linearly temperature-dependent potential. In the already cited previ-

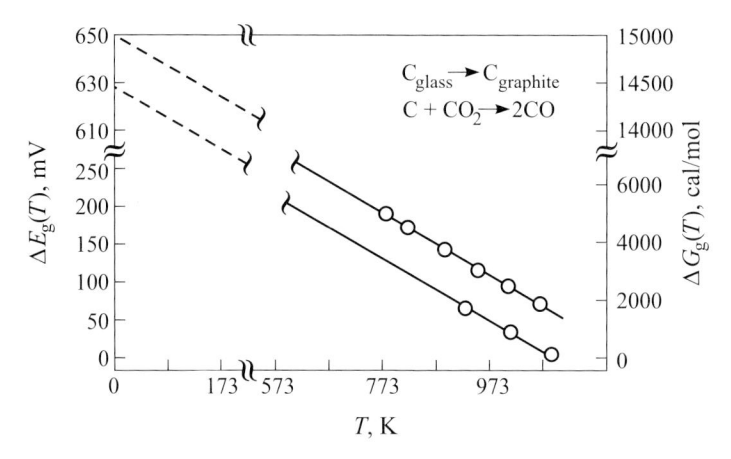

Figure 9.5 Temperature dependence of the EMF of the graphite/vitreous carbon galvanic cell, exploiting Boudouard's reaction $C + CO_2 \rightarrow CO$. Note the linear dependence of ΔE_g on temperature, the slope being determined according to Eq. (9.38) by the constant value of ΔS_g.

ous publication [177] several proposals can be found, giving the possibility to construct, for example, in terms of so-called bipolar cells, convenient devices to use the EMF of the electrochemically performed glass → crystal reaction. These possibilities are inherent in the very nature of glasses and are most easily demonstrated by measuring the electrode potential or even the corrosion potential of glassy metallic alloy electrodes. Here differences of the order of (10–15) mV are to be expected (cf. also [596]). The constancy at $T < T_g$ of both ΔH_g and ΔS_g in Eq. (9.38) determines the straight lines as it is seen both for the carbon glass solubility [601] and the EMF of the vitreous carbon–graphite galvanic element [359, 601].

Another possibility to use the increased potential of the frozen-in glass is the formation of an EMF difference potential, ΔE, stemming from the direct contact of a metallic or semiconductor glass to the respective crystal junctions. There both the Seebeck effect and the Peltier effect, when applied to the frozen-in glassy system, should bring about interesting possibilities (for details see [177, 182, 596, 598]).

Another confirmation of the existence of frozen-in values of glass parameters is connected with volume measurements. A dependence between configurational entropy and free volume, ΔV, follows from every MFA model, as discussed here earlier. In this way we can trace the change of the configurational part, $\Delta S(T)$, of entropy (or its constancy) via direct volume measurements, considering the value of $(V - V_0)$ as the free volume of the system. In this way, using known results on the dependence of V on the rate of cooling, q, we can also establish the real dependence of $\Delta S(T)$ on cooling rates. From existing $V(q)$ data (see [1]) it is obvious, that the greater q, the higher $\Delta S(0)$ values will be frozen-in in the glass, as this is predicted more generally in [112] by Gutzow and Dobreva.

Let us at the end also note that, in terms of viscosity equations connecting directly temperature dependence of viscosity $\eta(T)$ with configurational entropy, for example, according to the Adams–Gibbs relation

$$\eta(T) = \eta_0 \exp\left[\frac{B_0}{T\Delta S(T)}\right], \tag{9.39}$$

it becomes evident that at $T < T_g$ a straight line is to be expected with $\Delta S(T) = \Delta S_g = \Delta S(0)$ at $T < T_g$. This fact additionally confirming the existence of ΔS_g is known from many experiments [1, 180, 598].

9.8
Principle of Thermodynamic Correspondence and Zero-Point Entropy Calculations

Different molecular statistical models have been and can be proposed to calculate the zero-point entropy in glasses. Is there a possibility to prove or disprove the results obtained via existing or still possible models? Where is Ockham's razor in this field of science?

There is indeed a possibility in this respect and it has been formulated many years ago by Leo Szilard [639]: the results of any calculation, made in the framework of molecular, statistical or other models, have to correspond to (and in no way to contradict) any of the three principles of thermodynamics. We have to note, however, that Szilard's principle was originally formulated in accordance with classical thermodynamic (i.e., equilibrium) systems. Further developments and applications of this principle may be found in von Neumann's book [640] (in application to quantum mechanical problems), in Putilov's classical thermodynamics [641] and in [263] concerning the attractive field of imagined automata and molecular engines.

Szilard's principle makes dependent the results of any calculations in the framework of theoretical models on the three principles of thermodynamics. This is because the thermodynamic principles are the generalization of the whole experimental and theoretical knowledge in the field of science. In application to the entropy of glasses and liquids, Szilard's principle of thermodynamic correspondence requires that in approaching absolute zero, the entropy and the specific heats of any real or imagined system considered to be in equilibrium have to approach zero.

Classical examples of fulfillment of this principle give any defect-free crystalline substances, liquid helium and the theoretical models of quantum statistics, according to which the phonon and electronic contributions to the partition function of any system (liquid, glassy or crystalline) approach absolute zero temperatures with both $S(T \to 0) \to 0$ and $C_p(T \to 0) \to 0$ approaching zero values as well. In several recent publications [180, 182, 316] we have given abundant experimental evidence, showing that the electric properties of glasses (e.g., the Seebeck EMF of contacts in metallic glasses) approach in both their entropy and specific heat contributions zero at $T \to 0$: they follow the requirements of Fermi's quantum statistics of electronic systems in fulfilment of the classical formulations of the third principle of thermodynamics.

There are known, several interesting and also severe violations of Szilard's principle inherent in some theoretical speculations even of prominent authors (e.g., Maxwell's demon, Szilard's unimolecular engine, etc., [263]). Flory's model of polymer liquids, constituted of linear polymer chain molecules, leads also, due to a gross miscalculation in the statistical model (see [642–644]), to an entropy $S(0) > 0$ for the fictive equilibrium melt at $T \to 0$. These problems of Flory's model of polymer liquids and glasses were resolved by Milchev [275] (see also [364, 645]) who after pointing out the nonthermodynamic character of Flory's calculation, gave also the proper solution by developing the proof that in fact in Flory's fictive chain polymer liquids model the entropy is nullified at $T \to 0$.

In considering this and similar situations connected with the third law in terms of Szilard's principle and Boltzmann's formula (see Eq. (9.31)), Fermi's words [64] have to be remembered that Boltzmann's approach in statistics "allows some arbitrariness in the choice of the standard zero-point state," that is, the introduction of an additional integration constant in the sense, for example, of

$$\Delta S(0) = k_{\rm B} \ln\left(\frac{\Omega}{\Omega_0}\right) \tag{9.40}$$

However, the necessary introduction of the notions of quantum statistics at $T \to 0$ and of its definition of the phase space and its division defines in such an unambiguous manner Ω_0 in Eq. (9.40) that – as the overwhelming rule – only the value $S(0) = 0$ becomes possible at $T \to 0$. Above statements of masters in quantum statistics like Fermi and Szilard are always to be accounted for when constructing new or reconsidering well-known models in the field of zero-point entropy, in general, and of glasses, in particular.

9.9
A Recapitulation: the Third Principle of Thermodynamics in Nonequilibrium States

The three principles of thermodynamics in their classical formulations either as positive statements or as negative interdictions like the impossibility of different kinds of perpetuum mobile's are well-known to every student of classical thermodynamics. In the basic literature of irreversible processes the first principle is usually formulated as constancy of internal energy, U,

$$U = \text{const.} \tag{9.41}$$

provided it is not supplied from outside via work, heat or addition of matter. To the second principle there is given the more specific form as (cf. Eq. (3.6))

$$dS = d_{\rm e}S + d_{\rm i}S, \quad d_{\rm e}S = \frac{dQ}{T}, \quad d_{\rm i}S \geq 0 \tag{9.42}$$

via the introduction of the entropy production term, $d_{\rm i}S > 0$. In considering all what was said in the present contribution it becomes evident that the statement

$$C_p(T \to 0) = 0, \tag{9.43}$$

which is a form of the third principle directly following from Nernst's heat theorem, may be applied to both systems of classical thermodynamics and to frozen-in systems and, as may be concluded further, to any other types of matter in nonequilibrium states [182]. Equation (9.43) is equivalent to the statement of the principle of unattainability of the absolute zero of temperatures – that is, representing in this way the most general form of the third principle of thermodynamics. Thus the study of glasses enlarges our views on other nonequilibrium and frozen-in systems and on thermodynamics, in general.

A thorough analysis and a recapitulation of new and old developments in the field of the thermodynamics of glass shows that entropy has to be frozen-in and produced in the very process of glass-formation. This result follows from both classical thermodynamic considerations, coming from the times of Nernst, Einstein, Planck and Simon and by the generic non-equilibrium thermodynamic formulations of the kinetics of glass transitions as they are given e.g. in [180, 182, 316]. The entropy, ΔS_g, frozen-in and the entropy produced in vitrification is the typical result of a nonequilibrium process, which is manifested at $T \to 0$ as the zero-point entropy of the vitreous state, $\Delta S(0) > 0$. Present-day kinetics of vitrification lead beyond this statement to definite conclusions concerning the value of the frozen-in entropy $\Delta S(0)$ of the glass and of its dependence on cooling rate and on other factors. The higher the cooling rate, the higher glass transition temperatures result and thus also higher values of ΔS_g and $\Delta S(0)$ are to be expected. The results, concerning the structural dependence of ΔS_0, reported in a general context, give the possibility of generalizations concerning the thermodynamic properties of any other frozen-in nonequilibrium systems. For such nonequilibrium systems the unattainability of the zero point of temperatures is the most appropriate formulation of the third principle of thermodynamics.

Glass transition is a process, in which a metastable system (the undercooled melt) is transferred by a cooling run into a thermodynamically nonequilibrium, frozen-in state, the glass. As a consequence (formulated first by Simon long ago), configurational entropy values $\Delta S(0) > 0$ are frozen-in in glasses. In this way, glasses and mentioned similar systems do not fulfill the third law of thermodynamics in the formulation assigned commonly to Planck. In his treatment, the third law states that at $T \to 0$ the entropy tends to zero. However, thermodynamic coefficients of glasses like the specific heats nevertheless approach zero at $T \to 0$ as classically required as a consequence of the third law in the formulation given above. Consequently, they do not fulfil the third law in the formulation by Planck but fulfill the third law in a modified, enlarged formulation of the principle of nonaccessibility of the absolute zero temperature, as we attempted this in [285]. Replacing the classical formulation by the mentioned principle of nonaccessibility of the absolute zero, it would extend – in line with Nernst's original expectations – the region of applicability of the third law to glasses, other classes of systems with frozen-in disorder: presumably even to any possible systems in nature, in general. In this way, glasses and glass transition give a very serious support for the advantages of a new formulation of one of the greatest principles of Nature.

10
On the Etymology of the Word "Glass" in European Languages and Some Final Remarks

Ivan S. Gutzow

10.1
Introductory Remarks

Several years ago the present author was asked by an outstanding representative of the international glass science community, L. David Pye, to say or even to write something on the etymology of the word "glass" in European languages. This turned out to be a difficult task; here only some results and considerations on this subject are summarized, mostly to provoke further discussion.

Glass even in its first developments was widespread in the Antique world since Egyptian and Mesopotamian times and for hundreds of years glass articles were transported over the seas of the Mediterranean region from place to place, from nation to nation. Glass, and what can be called glass industry of these times, was developed in many places besides Europe and Asia Minor, in India, in China and in Japan, in medieval centuries in the towns of the Muslim nations of Central Asia. Glass windows, mosaics, torches and candelabras decorated cathedrals and churches in medieval Europe, the mosques of the Ottoman Empire and of the Arabic states. Wine glasses and bowls, any sort of glass-plate was a common object in the late days of Roman Rule over the Mediterranean Lands and Europe. More than that glassware of great beauty, the simple little amphoras, the tiny flasks of Egyptian enamel-like glass production can be found in any present-day historic museum all over Europe: in the cities of the former Hellenistic colonies over the Balkan Peninsula, in Greece, in Italy – everywhere. Glass was one of the first products of mankind's creative work and industrial efforts, which really found recognition everywhere. In this way glass and the first products of glass fabrication gave rise to contacts between people and nations. Many books have been written and conferences held which discussed the beginning of glass making, and many of these publications could be recommend to the interested reader. Out of many possibilities here we cite only a few [624–626, 646, 647].

What is of even greater significance than the intriguing history of developing glass production techniques is that glass was one of the earliest products of human civilization. Glass and its products developed connections between nations, and this exchange among people ultimately included words. Thus words, connected with glass and its various forms of usage and production, went all over the world for thousands of years. To follow these words and their now forgotten exact meaning is difficult, but let us first recollect the most significant facts.

10.2
"Sirsu", "Shvistras", "Hyalos","Vitrum", "Glaes", "Staklo", "Cam"

In Assyrian cuneiform stone tablet writings of the eighth century BC, glass (or something maybe resembling glass) is called *sirsu* [626]. In Sanskrit literature of the fourth century BC, the word *kaacha* is used for glass [626]. It is difficult to say now what these words exactly meant and how closely the respective material actually corresponded to present day glass. The colored ancient Egyptian phials and necklace beads are in fact more a glassy frit or an enamel-like opaque glass. However, it seems that the unusual, very particular properties and beauty of the later European glasses of the Roman period, which were shining translucent and thus strikingly different from metals and primitive or even more elaborated ceramics, determined the great variety of words then developed and still in use in many languages to describe this unique material. In this respect glasses deviate strongly from European words used for *crystal*, having the same Greek – Hellenic root (*krystallos*) in all European languages for many hundreds of years. Crystals seemed always to appear in the same constant shape. Glasses were always different.

The word *vitrum* (or *vitrium*) used to indicate glass was introduced at the end of the classical Roman Rule in Europe: at the beginning of the Roman Empire. It was first mentioned, as it seems, by the great Roman orator and statesman Marcus Tullius Cicero in the last century BC [648]. Its significance was explained by Gaius Plinius Secundus (Pliny the Elder) in his well-known *Naturalis Historia*, written in the first century AD. Pliny also told in his treatise his version of the story of the discovery of glass production [625]: by Phoenician sailors who made a fire pyre on a sandy desert beach using *trona* (i.e., natural sodium carbonate) lumps and lime stones to support the kettle. This first glass-production "method," invented or repeated out of older sources in Pliny's story, could not be reproduced in modern times. Nevertheless he was well informed on glass over the world of his century and he called *vitrum graecum* the mosaic glass (sometimes inlaid or covered with gold) to be used in great magnificence in later years in the cathedrals and palaces of the Byzantine Empire [625]. Pliny also mentioned another word for glass: *glaesum*, but it refers more to amber, which as mentioned in Chapter 1 of the present book is in fact vitreous in its nature and is the first organic glass known to mankind. Later on, the word for amber in Latin was *electrum* (from the respective Greek route, leading to "electricity" in later centuries), but *glaesum* remained in the late years of the Roman Empire the word for the glass, produced at these times in the cities along the Rhine.

The Latin word *vitrum* clearly demonstrates the most striking property of silicate glasses and of vitreous materials in general: their translucence, their glitter and their property to "shine." This is because the Latin *vitrum* (or more correctly, its Indo-German root) comes from the Sanskrit word *shvistras*: translucent, light, shiny. It was transformed afterwards to *sviat*, or *zviat* in the Slavonic languages with a similar meaning.

It is not clear what the connection is between *kaacha* and *shvistras*: does the first indicate the first word for glass and does the second give only its most striking property? It seems that the texts in the Assyrian stone writing-tablets about *sirsu* give in fact the first real instruction, and the composition of a glass, something like a prescription of silicate glass production. It seems that this composition essentially persisted for many centuries and that it led also to glass compositions and to glass production techniques, having the shiny, translucent properties to be called real glass in different ways.

Of more significance is that the Latin *vitrum* determined later on the word glass in all Roman languages: *vidrio* (in Spanish), *verre* (in French), *vetro* (in Italian). The Gothic-German and Anglo-Saxon word *glaes* (most probably directly connected with the respective Latin word *glaesum* in Pliny's and later Roman's times, used in fact also as *glesum* in Gothic), was transformed then into *glass* (in English), *Glas* (in German), *gliis* (in Swedish) and *glas* in Irish. It is obviously also determined by one of the striking properties of glass in its Germanic spelling: its *glaze* in modern English (or *Glanz* in German).

In Byzantium times glasses were called *hyalos* in the Greek language of this era. This name comes most probably from the word cold (even icy) in ancient Greek: it is also common with the Hellenistic word for the glassy body of the eye. *Hyalith* is the name of the SiO_2 mineral, discussed already in Chapter 5, formed hydrothermally as a glassy precipitate in nature; it is the first *amorphous* stage in the growth of opals. In fact both quartz crystals (*krystallos*) and Hyalith or opal, because of their unexpectedly good heat conduction, give the impression to be cool, when taken into the hand.

The early medieval old Slavonic word for glass, *staklo*, originated most probably from the Gothic *stehal* – indicating a cup or goblet (for wine). Thus the Slavonic name for glass stresses its most common application: as a material, out of which was formed (a translucent) wine cup in usage with the German neighbors. So it seems that in the first years and centuries of contact between Germanic and Slavonic tribes and nations in Europe a new word was coined. Out of *staklo* the present day names for glass in different Slavonic languages were then directly developed: *steklo* (in present day Russian, *stklo* in old Russian), *stuklo* (in Bulgarian), *szklo* (in Polish), *sklo* (in Czech), *staklo* (in Serbian), and so on. Other nations which were in historic contact with Slavonic countries also took over the above-mentioned root: *stiklas* (in Lithuanian) or *sticlei* (in Rumanian).

The first mentioning of a glass vessel, which can be traced in the Orthodox Church (Bulgarian) Slavonic literature, is the word *stukleniza* from the eleventh century: for a glass phial or flask (as a reliquary of holy water or consecrated oil). In Medieval Russian literature besides *stklo* also a similar word, *sklyainitza*, is used to indicate a glass flask in religious use [626].

In the Turkish language the word *cam* (pronounced as *djam*) is used for glass. Whether another Turkish word for glass, *sirça* (*sircha*), has something to do with the above mentioned Assyrian root *sirsu* is unknown by the present author, but it is a strange coincidence nevertheless. A flask is called in Turkish *şise* (pronounced as shishe) and a glass works in the later years of the Ottoman Empire was accordingly called *shishena*. In the nineteenth century under Ottoman rule in the small Bulgarian city Sopot in the Rose Valley, at the foot of the Balkan Mountains, in such a *shishena* till the year 1848 were produced [649] the small glass phials (*mouskals*) for the rose attar, produced in the nearby rose-distilleries. They were sold all over Europe for prices, equivalent to the weight in gold of such a *mouskal* filled with the precious rose oil.

10.3
"Vitreous", "Glassy" and "Glasartig", "Vitro-crystalline"

The particular richness of the English language with respect to the term glass has to be now mentioned. Following a well-known rule of development of languages (to incorporate foreign words by adopting their general more abstract meanings), the Latin root *vitrum* (developed in English literature to *vitreous*), indicates at present the more general *glass-like* properties of the material and the corresponding physical state itself (*the vitreous state*). In this sense it was chosen also as the title of the former monograph of two of the present authors [1] and of several other well-known similar scientific publications (see [237, 650, 651]). The word *glass* (or *glassy*) should thus refer in English mainly to the material glass, although *glassy state* is also widely used (mostly in American literature) together with *glass science* and *glass transitions* in reference to the *vitreous state*. In this respect it is to be mentioned that in German literature, where there is no direct equivalent of the Latin rooted *vitreous*, the adjective *glasartig* is used.

It is also interesting to note that in another language (in Romanian), which has assimilated both the Slavonic root (*sticlei*) as a noun and the Latin derived pronoun *vitroase*, the combination *starii vitroase* is thus also used to indicate the *vitreous state*. This combination was also the name of a well-known monograph (see [650]) in its first Romanian edition. In Germany, Gustav Tammann coined the word *Glaszustand* [24] as the title of the first monograph in international literature (1933), describing the properties of glass as a particular physical state. In an analogous way *stekloobraznoe sostoyanie* and *steklovanie* are used in Russian scientific literature [70, 652] to indicate (using *steklo* as a root) the *vitreous state* and both *vitrification* and *glass transition* for the respective processes. For the process of secondary crystallization of an already vitrified melt the term *devitrification* stems in English obviously from the already mentioned, more general Latin root, while in German the only available word *Glas* is used to construct the term *Entglasung* for *devitrification*. In Russian the same procedure like in German is used to construct *devitrification*, using again *steklo* as the necessary root (*razsteklovanie*). In other Slavonic languages (like in Bulgarian) vitrification is indicated by the Slavonic root for glass (e.g., *za-*

stuklyavane in Bulgarian) and the Latin word for devitrification (*devitrifikatziya* in Bulgarian) for glass crystallization [649].

The above mentioned more or less internationally accepted scientific terminology refers to processes and states in the framework of *glass science*, a word combination first coined as the title of one of the first books [653] to describe experiments in the field of glass.

It is also of use to refer here to the glass-like transitions in "glassy crystals" we treated in Chapter 5 as particular forms of frozen-in defect states and to the *liquid crystals* discussed in brief there as well. This latter term denotes partly oriented liquids having obtained exceptional applications in present day electronics. The combination *glass ceramics* is also to be recalled in this connection as describing a new material, developed out of glass by its controlled (or even, induced) crystallization. This material opens the wide perspectives of *vitro-crystalline* materials: both as a combination of words and of promising materials, using the properties of two physical states. In this last example again the more general usage of the word *vitreous* becomes evident.

Here maybe is also the place to mention, that in Medieval Venice as *cristallo* were denoted particular glasses with extremely good quality; this word has persisted in the strange combination *crystal glass* up to now for well-known high quality industrial glass products (e.g., to lead oxide crystal glass, etc.). This term, having nothing to do with the real state of the material to which it refers, is still in well-known use and application.

It must also be mentioned, that in the last few years glass science terminology, developed in treating glass transitions, vitrification and devitrification has been used also in other fields of solid state physics and even in cosmic physics and cosmology [654]. It is known that for several years there have been efforts to describe the development of the expanding universe in terms of the general theory of the kinetics of overall crystallization [655], developed many years ago by Kolmogorov and Avrami (see [1, 656]). It is not possible for the present author to decide, whether such efforts have led to results, corresponding to our present day knowledge of the universe. It is only his intention to mention that in a similar way results, notions and ideas on the process of vitrification, on the glass transition and on the thermodynamic properties of glasses as frozen-in systems have been used (employing even the *terminology* developed and specific in glass science) to describe processes of black hole formation and other problems of general cosmological nature. This gives an indication, how general in fact the processes, treated in glass science, only with respect to the change of properties of a given material are. In this sense the theories and notions and ideas of glass science have in fact to be justly considered and reconsidered (and maybe reformulated) as referring generally to any form of condensed matter and to processes possible with any physical state.

10.4
Glasses in Byzantium, in Western Europe, in Venice, in the Balkans and Several Other Issues

Let us now return to our main topic: the history of development of glass and the way it is described in terms of different languages.

Glass was produced in Roman Europe and later on in Byzantium during the Middle Ages not only around Constantinople, but also in Thessalonica, at least, beginning from the ninth century AD till the end of the Byzantine Empire in the fifteenth century [625]. Glass was produced also in these times in the glass kilns of the medieval kingdoms in the European surroundings of the Byzantium Empire. Thus in both Bulgarian medieval capitals not only the glass was produced there, but also, glass fabrication ovens were excavated. Then on the Balkans followed the difficult years of invasion and wars.

The glass, needed in the sixteenth and seventeenth centuries to rebuild Constantinople into the Ottoman capital city of Istanbul was brought mainly from the West: from Venice. Afterwards around Istanbul developed a rich ring of glass-manufacturing enterprises, supplying glass for the blooming Islamic metropolis [657].

It is interesting to note that in the sixteenth century, when a compact Jewish population entered the Ottoman Empire after the persecutions in Spain, the first glass in Thessalonica was produced in a little Jewish held glass-works (a *shishena* in Turkish, as above mentioned). This was perhaps the first glass-kiln in the Balkan Peninsula during Ottoman times. This event can still be traced in the written chronicles of the Jewish diaspora in Thessalonica.

The story of spread and development of glass production all over Europe from the Mediterranean cities, through Alexandria and Venice to Bohemia and Germany is well-known and does not need to be retold here. The interested reader can find information on this subject in the literature [625, 646] and in many classical and present day journal publications. Interesting scientific conferences have been held on this part of the history of glass, on its development through the centuries, on the techniques used, on the invention of the glass pipe, on the island of Murano and Venice, and of the way glass production was brought from there to Bohemia. Particular attention must be paid also to the strange and intriguing story of the development of the *gold ruby glass*, of its secret production in Venice and its re-invention by the German Alchemist Johann Kunckel in the sixteenth century and its production on the Pfauen-Insel near Berlin [658]. Then followed the second reinvention of gold ruby glass by Lomonosov [659] in the eighteenth century in St. Petersburg. Here again are books of interest like [625, 626, 657] and the proceedings of several meetings on this subject, for example, in [646], are to be mentioned.

The history of the word glass is also connected with the development of many other words, connected with glass and its applications in every-day life, in glass-science and in the sciences in general. The *vitreous* and *glassy* states are already discussed above in their general meaning. Let us only recall here the word for a glass window. It is usually stated that from the Latin word *fenestra* comes the present day German *Fenster*, Swedish *fönster*, French *fenetre* for window. The English word for *window* itself is connected, so it seems, to Old Norske's *vindangua* (wind eye) and this is one of the stories of the word for one of the most significant every day applications of glass, connected with nordic fantasy and imagination. Many other interesting examples can be found in the respective literature for the development of words and ideas, connected with glass. It is not the place here to analyze all of these ideas and notions; they are too many and this is a task which demands particular and deep knowledge of the rules of how words change in languages, as they are observed in the long history of mankind.

10.5 Concluding Remarks

We have mentioned and discussed here only the words connected with the most striking properties of glasses in their traditional and even in their possibly unexpected application. We followed in the present book the ways to collect data and results on the properties of glass on their change with temperature and the other external parameters. But there are so many interesting applications of glass from every-day life and industry: in the arts, in architecture [625], and in the religions.

Beginning with ancient Egypt, glass bracelets, necklaces, bowls and plates were found in the graves of many generations, even of nations gone by. The glassy artifacts excavated give an impression on the cultural level, on religious beliefs, on human interrelations: thus glass speaks its language in present-day museums. There are nations, like the now vanished nomadic Turkic people of the Pechenegians, who in the tenth century AD inhabited the lands south of the Danube and west of the Black Sea; from these nomads only the typical blue glass bracelets in their graveyards remained. They indicate now a territory in the also gone Byzantine Empire.

An interesting finding are the ancient Egyptian necklaces (e.g., the one of the queen Hatshepsuth, 1500 BC) which show how glassy (or enamel-like) beads in fact were designated as a falsification of precious or half precious stones. The little enamel-like glass phials from ancient Egypt glass tell in unmistakable words the way glass was produced in this time [625, 658].

The magnificent windows of Medieval European cathedrals open problems not only with the richness of their colors, but also with the possible flow of glass at room temperature: here the theory of glass relaxation still speaks its word, as known from several publications [660–662]. The decorated glass torches in Islamic Mosques have also to be recalled, and in [657] there are brilliant pictures of them. All religions use the extraordinary properties of stained glass and the way it speaks to people.

There are also other problems in this way of thinking. Was (or is, if existing) the Holy Grail a gold plated glass goblet, did the first Christians use and need the golden Church plate of later times? Or was the Holy Grail simply an earthen cup? There was a remarkable lecture [663] on this subject at the International Glass Congress in 1999, and this is in fact only one of many issues which connects the history of glass, main ideas of human development, art, religion and ethical problems.

In above considerations it was mentioned that on the history of glass are written many intriguing books; of particular interest is the analysis of the influence glass technology had on the development of other sciences, arts and technologies. The history of glass is interwoven with the development of European and world civilization. The words, describing glass itself in its different forms in Europe and elsewhere are also part of this development. The change of the word for glass and glassy products in European languages follows and repeats the story of these languages. Glass has been an article produced, changed, interchanged and admired by many European nations in the past centuries. Moreover, glass induced exchange, spread of knowledge between European nations and the nations of the Mediterranean region. Here the story of development of glass and of the words describing it are mostly connected with European technologies, arts and history. Maybe other colleagues from other fields of glass science and history could say additional, and perhaps more essential things on the origin and change of glass and on the words describing glass in other languages: out of Europe, in the long history of Asiatic cultures, in China and in Japan, in India and in the whole development of the Islamic world.

Let us hope it's true that the Slavonic word for glass developed from the first contacts between nations and that it is in fact so that an Indo-German root survived for three thousand years through vitrum to the present day vitreous state as the name for a particular state of matter. This gives hope that the science of glass in all its forms will further survive as a link between people and nations all over the world. It gives also to an conviction that glass and the vitreous state will continue to be words denoting translucence, light, beauty and development of ideas and will never be used in connection with the possibility that human civilization will be bombed into a mass of radioactive waste glass.

The above schematic remarks on the etymology of the words connected with glass were developed in friendly discussion with a well-known Bulgarian author in linguistics, the now late Prof. Dr. Ivan V. Duridanov, member of the Bulgarian Academy of Sciences and one of the founders of the science, called linguistic paleontology. In good remembrance and with gratitude to him, have been written the pages of this chapter.

References

1. Gutzow, I. and Schmelzer, J. (1995) *The Vitreous State: Thermodynamics, Structure, Rheology, and Crystallization*, Springer, Berlin.
2. Mazurin, O.V., Streltsina, M.V., and Shvaiko-Shvaikovskaya, T.P. (1972–1995) *Handbook of Properties of Glasses and Glass-forming Melts*, 9 volumes, Nauka, Leningrad (in Russian).
3. Mazurin, O.V., Streltsina, M.V., and Shvaiko-Shvaikovskaya, T.P. (1980–1990) *Handbook of Glass Data*, 5 volumes, Elsevier, Amsterdam.
4. Angell, C.A. and Rao, K.J. (1972) *J. Chem. Phys.*, **57**, 470.
5. Angell, C.A. (1988) *J. Non-Cryst. Solids*, **102**, 205.
6. Arrhenius, S. (1913) *Das Werden der Welten*, Akademische Verlags-Gesellschaft, Leipzig.
7. Asahina, E. (1966) Freezing and Frost Resistance in Insects, in *Cryobiology* (ed. H.T. Meryman), Academic Press, London, New York, p. 464.
8. Storey, K.B. and Storey, J.M. (1999) *Lifestyles of the Cold and Frozen*, The Sciences, New York Academy Sciences, p. 32.
9. Gutzow, I., Atanassova, S., and Neykov, K. (2002) *Pure Appl. Chem.*, **74**, 1785.
10. Gutzow, I. and Dobreva, A. (1997) Kinetics of Glass Formation, in *Amorphous Insulators and Semiconductors* (ed. M.F. Thorpe and M.I. Mitkova), NATO ASI Series, Kluwer Academic Publishers, p. 21.
11. Gutzow, I., Ilieva, D., Babalievski, F., and Yamakov, V. (2000) *J. Chem. Phys.*, **112**, 10941.
12. Gutzow, I., Yamakov, V., Ilieva, D., Babalievski, F., and Pye, L.D. (2001) *Glass Phys. Chem.*, **27**, 148.
13. De Donder, T. (1936) *Thermodynamic Theory of Affinity: A Book of Principles*, Oxford University Press, Oxford.
14. Prigogine, I. and Defay, R. (1954) *Chemical Thermodynamics*, Longmans, London.
15. Prigogine, I. (1955) *Introduction to Thermodynamics of Irreversible Processes*, Thomas, Springfield.
16. Privalko, V.P. (1980) *J. Phys. Chem.*, **84**, 3307.
17. Gutzow, I., Ilieva, D., Babalievski, P., and Pye, L.D. (1999) Glass Transition: An Analysis in Terms of a Differential Geometry Approach, in *Nucleation Theory and Applications* (eds J.W.P. Schmelzer, G. Röpke, and V.B. Priezzhev), Joint Institute for Nuclear Research Publishing Department, Dubna, Russia, pp. 368–418.
18. Glasstone, S., Laidler, H.J., and Eyring, H. (1941) *The Theory of Rate Processes*, Princeton University, New York, London.
19. Davies, R.O. and Jones, G.O. (1953) *Adv. Phys.*, **2**, 370.
20. Davies, R.O. and Jones, G.O. (1953) *Proc. Royal Soc.* (London) **A217**, 26.
21. Simon, F. and Lange, F. (1926) *Z. Physik*, **38**, 227.
22. Simon, F. (1930) *Ergebnisse der exakten Naturwissenschaften*, **9**, 222.

23 Simon, F. (1956) *The Third Law of Thermodynamics: A Historical Survey*, The 40-th Guthrie Lecture, Yearbook Phys. Society (London) **1**, 1.

24 Tammann, G. (1933) *Der Glaszustand*, Leopold Voss Verlag, Leipzig.

25 Frenkel, Y.I. (1946) *The Kinetic Theory of Liquids*, Oxford University Press, Oxford.

26 Götze, W. (1988) *Z. Phys. Chemie, NF*, **156**, 3.

27 Götze, W., Sjören, L. (1992) *Rep. Prog. Phys.*, **55**, 241.

28 Schilling, R. (1995) Mode Coupling Approach to the Glass Transition, in *Disordered Effects on Relaxational Processes Glasses, Polymers, Proteins* (eds R. Richert and A. Blumen), Springer, Berlin, New York, p. 193.

29 Böhmer, R. and Angell, C.A. (1995) Local and Global Relaxations in Glass Forming Materials, in *Disordered Effects on Relaxational Processes Glasses, Polymers, Proteins* (eds R. Richert and A. Blumen), Springer, Berlin, New York, p. 12.

30 van der Waals, J.D. (1873) Thesis, Leiden University Press, Leiden.

31 van der Waals, J.D. (1899–1900) *Die Kontinuität des gasförmigen und flüssigen Zustandes*, 2nd edition, Johann Ambrosius-Barth Verlag, Leipzig.

32 van der Waals, J.D. and Kohnstamm, P. (1908) *Lehrbuch der Thermodynamik*, Johann Ambrosius-Barth Verlag, Leipzig, Amsterdam.

33 Mayer, J. and Goeppert Mayer, M. (1946) *Statistical Physics*, John Wiley & Sons, New York.

34 Stanley, H.E. (1971) *Introduction to Phase Transitions and Critical Phenomena*, Clarendon Press, Oxford.

35 de Gennes, P.-G. (1974) *The Physics of Liquid Crystals*, Clarendon Press, Oxford.

36 Parks, G.S. (1925) *J. Amer. Chem. Soc.*, **47**, 338.

37 Parks, G.S. and Huffman, F.E. (1927) *Z. Phys. Chemie*, **31**, 1842.

38 Berger, E. (1930) *Glastech. Ber.*, **8**, 339.

39 Blumberg, B.Y. (1939) *Introduction into the Physical Chemistry of Glasses*, Chemistry State Publishers, Leningrad (in Russian).

40 Arcimovich, L.A. (1972) *Elementare Plasmaphysik*, Akademie-Verlag, Berlin.

41 Frank-Kamenetzki, D.A. (1963) *Plasma: The Fourth State of Aggregation of Matter*, Gosatomizdat, Moscow (in Russian).

42 Bernal, J.D. (1957) *Science in History*, C.A. Watts, London.

43 Callen, H.B. (1963) *Thermodynamics: Physical Theories of Equilibrium Thermodynamics and Irreversible Thermodynamics*, Wiley, New York.

44 De Donder, T. and van Rysselberghe, P. (1936) *Thermodynamic Theory of Affinity*, Stanford University Press, Stanford.

45 de Groot, S.R. (1952) *Thermodynamics of Irreversible Processes*, North-Holland Publishing Company, Amsterdam.

46 Denbigh, K.G. (1951) *Thermodynamics of the Steady State*, Butterworth, London, New York.

47 Haase, R. (1963) *Thermodynamik der Irreversiblen Prozesse*, Verlag D. Steinkopff, Darmstadt.

48 Kluge, G. and Neugebauer G. (1994) *Grundlagen der Thermodynamik*, Akademie Verlag, Berlin.

49 Santen, L. and Krauth, W. (2000) *Nature* (L), **6786**, 550.

50 Torquato, S. (2000) *Nature* (L), **6786**, 521.

51 Simon, F. (1937) *Physica*, **4**, 1089.

52 Angell, C.A. (1977) Strong and Fragile Liquids: Glass Transition and Polyamorphous Transitions, in *Amorphous Insulators and Semiconductors* (eds M.F. Thorpe and M.I. Mitkova), NATO-ASI-Series, High Technology, vol. 23, Kluwer Acad. Publishers, p. 1.

53 Angell, C.A. (2000) Glass Formation and the Nature of Glass Transition, in *Insulating and Semiconducting Glasses* (ed. P. Boolchand), World Scientific, Singapore, London, p. 1.

54 Kubo, R. (1968) *Thermodynamics*, North-Holland-Publishing Company, Amsterdam.

55 Bazarov, I.P. (1964) *Thermodynamics*, Higher Education Press, Moscow, 1976 (in Russian); McMillan & Co, New York.

56 Gibbs, J.W. (1875–78) On the Equilibrium of Heterogeneous Substances. *Transact. Connecticut Acad. Sci.*, **3**, 108, 343.

57 Gibbs, J.W. (1928) *The Collected Works*, vol. 1, Thermodynamics, Longmans & Green, New York, London, Toronto.
58 Simon, F. (1926) In: *Handbuch der Physik*, vol. 10 (eds H. Geiger and H. Scheel), Springer, Berlin, pp. 350, 352.
59 Simon, F. (1931) *Z. Anorganische und Allgemeine Chemie*, **203**, 219.
60 Nernst, W. (1918) *Die Theoretischen und Experimentellen Grundlagen des neuen Wärmesatzes*, Verlag W. Knapp, Halle.
61 Planck, M. (1954) *Vorlesungen über Thermodynamik*, 10. Auflage, de Gruyter, Berlin.
62 Landau, L.D. and Lifschitz, E.M. (1969) *Statistische Physik*, Akademie-Verlag, Berlin.
63 Landau, L.D. and Lifschitz, E.M. (1980) *Statistical Physics*, Pergamon, New York.
64 Fermi, E. (1937) *Thermodynamics*, Prentice-Hall Incorp., New York.
65 Einstein, A. (1905) *Annalen der Physik*, **17**, 549.
66 Leontovich, M.A. (1953) *Einführung in die Thermodynamik*, Deutscher Verlag der Wissenschaften, Berlin.
67 Mandelstam, P.I. and Leontovich, M.A. (1937) *J. Exp. Theor. Physics* USSR (JETF), **7**, 439.
68 Gutzow, I. and Dobreva, A. (1992) *Polymer*, **33**, 451.
69 Tool, A.Q. (1946) *J. Amer. Ceram. Society*, **29**, 240.
70 Mazurin, O.V. (1986) *Vitrification*, Nauka, Leningrad (in Russian).
71 Zubarev, D.I. (1992) In *Physical Encyclopaedia* (eds A.M. Prokhorov et al.), **3**, 328.
72 Tammann, G. (1922) *Die Aggregatzustände*, Leopold Voss Verlag, Leipzig.
73 Storonkin, A.V. (1967) *Thermodynamics of Heterogeneous Systems*, Leningrad State University Press, Leningrad (in Russian).
74 Kritchevski, I.R. (1970) *Notions and Basic Concepts of Thermodynamics*, Nauka, Moscow (in Russian).
75 Ehrenfest, P. (1933) *Commun. Leiden Univ.*, **20**(b75), 628.
76 Landau, L.D. and Lifschitz, E.M. (1976) *Elastizitätstheorie*, Akademie-Verlag, Berlin.
77 Justi, E. and von Laue, M. (1934) *Physik Z.*, **35**, 945.
78 Nemilov, S.V. (1988) Thermodynamic Content of the Prigogine–Defay Ratio and the Structural Difference Between Glasses and Liquids, in *The Vitreous State* (ed. E.A. Porai-Koshitz), Leningrad, Nauka, p. 15 (in Russian).
79 Gutzow, I., Petroff, B., Schmelzer, J.W.P., and Pye, L.D. (2002) *Thermodynamic Phase Transitions and the Glass Transition*. Proceedings 14-th Conf. on Glass and Ceramics and II Balkan Glass Conference, Varna, September.
80 Landau, L.D. (1935) *Phys. Z. Sowjetunion*, **8**, 113.
81 Landau, L.D. (1937) *Phys. Z. Sowjetunion*, **11**, 545.
82 Gutzow, I. and Petroff, B. (2003) *The Glass Transition in Terms of Landau's Phenomenological Approach*. Proceedings Parma Conf. Non-Cryst. Solids, Parma, Italy.
83 Landa, L., Landa, K., and Thomsen, S. (2004) *Uncommon Description of Common Glasses, vol. 1: Fundamentals of the Unified Theory of Glass Formation and Glass Transition*, Yanus, St. Petersburg.
84 Onsager, L. (1944) *Phys. Rev.*, **65**, 117.
85 Gebhardt, W. and Krey, U. (1980) *Phasenübergänge und kritische Phänomene*, Vieweg, Braunschweig.
86 Gunton, J.D., San Miguel, M., and Sahni, P.S. (1983) The Dynamics of First-Order Phase Transitions, in *Phase Transitions and Critical Phenomena* (eds C. Domb and J.L. Lebowitz), vol. 8, Academic Press, London, New York.
87 Bartenev, G.M. and Remizova, A.A. (1959) Phase Transitions in Simplest Systems and their Classification, in *Thermodynamics and Structure of Solutions*. Conference Proceedings, Moscow, January 1958, (ed. M.I. Shakhparonov), Sov. Acad. Sci. Publishers, Moscow (in Russian), p. 67/71.
88 Schubert, K. (1968) *Z. Naturforschung*, **23A**, 1276.
89 Tisza, L. (1951) On the General Theory of Phase Transitions, in *Phase Transformations in Solids* (eds R. Smoluchowski, J.E. Mayer, W.A. Weyl), Symposium

Held at Cornell University, 1948, Wiley, New York, p. 1.
90 Üerreiter, K. and Bruns, W. (1964) *Ber. Bunsenges. Phys. Chem.*, **68**, 541.
91 Üerreiter, K. and Bruns, W. (1966) *Ber. Bunsenges. Phys. Chem.*, **70**, 17.
92 Boyer, R.F. and Spencer, R.S. (1944) *J. Appl. Phys.*, **15**, 398.
93 Boyer, R.F. and Spencer, R.S. (1945) *J. Appl. Phys.*, **16**, 594.
94 Boyer, R.F. and Spencer, R.S. (1946) *Advances in Colloid Science*, vol. II, Interscience, New York.
95 Ubbelohde, A.R. (1965) *Melting and Crystal Structure*, Clarendon Press, Oxford.
96 Skapski, A.S. (1956) *Acta Met.*, **4**, 576, 583.
97 Gutzow, I., Razpopov, A., and Kaischew, R. (1970) *Phys. Stat. Solidi*, **a1**, 159.
98 Gutzow, I., Razpopov, A., Pancheva, E., and Kaischew, R. (1972) *Kristall und Technik*, **7**, 769.
99 Volmer, M. (1939) *Kinetik der Phasenbildung*, Th. Steinkopff, Dresden.
100 Ostwald, W. (1896–1901) *Lehrbuch der Allgemeinen Chemie*, Engelmann, Leipzig.
101 Giessen, B.C. and Wagner, C.N.J. (1972) Structure and Properties of Non-Crystalline Metallic Alloys Produced by Rapid Quenching, in *Liquid Metals* (ed. S.Z. Beer), Marcel Decker, New York, p. 633.
102 Turnbull, D. and Fisher, J.C. (1949) *J. Chem. Phys.*, **17**, 71
103 Turnbull, D. (1952) *J. Chem. Phys.*, **20**, 411.
104 Greer, A.L. (1988) *Mater. Sci. Eng.*, **97**, 285.
105 Penkov, I. and Gutzow, I. (1984) *J. Mater. Sci.*, **19**, 233.
106 Gutzow, I., Zlateva, E., Alyakov, S., and Kovatscheva, T. (1977) *J. Mater. Sci.*, **12**, 1190.
107 Grantcharova, E., Avramov, I., and Gutzow, I. (1986) *Thermochim. Acta*, **102**, 249.
108 Mazurin, O.V. (2007) *Glass Phys. Chem.*, **33**, 22.
109 Kauzmann, W. (1948) *Chem. Rev.*, **43**, 219.
110 Gutzow, I. (1972) The Thermodynamic Functions of Super-Cooled Glass-forming Liquids and the Temperature Dependence of their Viscosity, in *Amorphous Materials* (eds R.W. Douglas and B. Ellis), Proc. Third International Conference, Sheffield, 1970. Wiley, London, p. 159.
111 Sakka, S. and Mackenzie, J.D. (1971) *J. Non-Cryst. Solids*, **6**, 145.
112 Gutzow, I. and Dobreva, A. (1991) *J. Non-Cryst. Solids*, **129**, 266.
113 Turnbull, D. and Cohen, M.H. (1960) Crystallization Kinetics and Glass-Formation, in *Modern Apects of the Vitreous State* (ed. J.P. Mackenzie), Butterworths Scientific Publications, p. 38.
114 Flory, P.J. (1936) *J. Amer. Chem. Soc.*, **58**, 1877.
115 Eitel, W., Pirani, M., and Scheel, K. (1932) *Glastechnische Tabellen*, Springer, Berlin.
116 SciGlass 7.5. (2010) *Glass Property Information System*, Institute of Theoretical Chemistry, Shrewsbury, MA, http://www.sciglass.info; cf. Chapter 6 (accessed 28 January 2011)
there exists also a variety of preceding versions.
117 Landoldt-Bönstein, A. (1962) *Zahlenwerte und Funktionen, Physikalische Chemie, Astronomie, Technik*, Bd. II, 2b, 6. Auflage, Springer, Berlin, p. 120, 213.
118 Nemilov, S.V. (1964) *Zh. Prikl. Khimii*, **37**, 1020.
119 Rawson, H. (1967) *Inorganic Glass-forming Systems*, Academic Press, London, New York.
120 Andrade, E.N. (1930) *Nature*, **125**, 309.
121 Andrade, E.N. (1934) *Phil. Magazin*, **17**, 698.
122 Frenkel, Y.I. (1934) *The Theory of Solid and Liquid Bodies*, ONTI, Leningrad (in Russian).
123 Eyring, H. (1936) *J. Chem. Phys.*, **4**, 283.
124 Bezborodov, M.A. (1975) *Viscosity of Silicate Glasses*, Science and Technology Press, Minsk (in Russian).
125 Mackenzie, J.D. (ed.) (1960) *Modern Aspects of the Vitreous State*, Butterworths, London.

126 Kanai, E. and Satoh, T. (1954) *J. Phys. Chem.* (Japan), **9**, 117.
127 Kanai, E. and Satoh, T. (1955) *J. Phys. Chem.* (Japan), **10**, 1002.
128 Vogel, W. (1921) *Phys. Z.*, **22**, 645.
129 Fulcher, G.S. (1925) *J. Amer. Chem. Soc.*, **77**, 3701.
130 Tammann, G. and Hesse, W. (1926) *Z. Anorg. Allg. Chem.*, **156**, 245.
131 Gutzow, I. (1975) *Fiz. i Khim. Stekla*, **1**, 431.
132 Williams, M.L., Landel, R.F., and Ferry, J.D. (1955) *J. Amer. Chem. Soc.*, **77**, 3701.
133 Ferry, J.D. (1961) *Viscoelastic Properties of Polymers*, John Wiley & Sons, New York, London.
134 Donth, E.-J. (1981) *Der Glasübergang*, Wissenschaftliche Taschenbücher Mathematik-Physik, Band 271, Akademie-Verlag, Berlin.
135 Doolittle, H. (1951) *J. Appl. Phys.*, **22**, 1471.
136 Doolittle, H. (1952) *J. Appl. Phys.*, **23**, 236.
137 Bueche, F. (1962) *Physical Properties of Polymers*, Interscience Publishers, New York.
138 Batchinski, J. (1912) *Phys. Z.*, **13**, 1157.
139 Batchinski, J. (1913) *Z. Phys. Chem.*, **84**, 643.
140 Skornyakov, M.M. (1955) *Stroenie Stekol*, Nauka, Moscow, Leningrad, p. 256 (in Russian).
141 Avramov, I. and Milchev, A. (1988) *J. Non-Cryst. Solids*, **104**, 253.
142 Avramov, I. (1990) *Phys. Stat. Solidi*, **a120**, 133.
143 Avramov, I. (1996) *J. Non-Cryst. Solids*, **194**, 122.
144 Avramov, I. (1998) *J. Non-Cryst. Solids*, **238**, 6.
145 Schischakov, N.A. (1954) *Structure of Silicate Glasses*, Academy of Sciences of USSR Publishers, Moscow (in Russian).
146 Sanditov, D.S. and Bartenev, G.M. (1982) *Physical Properties of Disordered Structures*, Nauka, Moscow (in Russian).
147 Macedo, B. and Litovitz, T.A. (1965) *J. Chem. Phys.*, **42**, 245.
148 Adams, G. and Gibbs, J.H. (1965) *J. Chem. Phys.*, **43**, 139.
149 Gutzow, I., Avramov, I., Köstner, K. (1990) *J. Non-Cryst. Solids*, **123**, 97.
150 Dyre, J.C., Hechsher, T., and Niss, K. (2009) *J. Non-Cryst. Solids*, **355**, 624.
151 Richet, P. (2009) *J. Non-Cryst. Solids*, **355**, 628.
152 Dietzel, A. and Brückner, R. (1955) *Glastech. Ber.*, **28**, 455.
153 Dietzel, A. and Brückner, R. (1957) *Glastech. Ber.*, **30**, 73.
154 Hodgdon, J.A. and Stillinger, F.H. (1993) *Phys. Rev.*, **E48**, 207.
155 Dyre, J.C., Christensen, T., and Olsen, N.B. (2006) *J. Non-Cryst. Solids*, **352**, 4635.
156 Dyre, J.C. (2006) *Rev. Mod. Phys.*, **78**, 953.
157 Glotzer, S.C. (2000) *J. Non-Cryst. Solids*, **274**, 342.
158 Roessler, E. (1990) *Phys. Rev. Lett.*, **65**, 1595.
159 Schmelzer, J.W.P., Müller, R., Möller, J., and Gutzow, I. (2003) *J. Non-Cryst. Solids*, **315**, 144.
160 Gibson, G.E. and Giauque, W.F. (1923) *J. Amer. Chem. Soc.*, **45**, 93.
161 Gutzow, I. and Grantcharova, E. (1985) *Commun. Dep. Chem.* (Bulgarian Academy of Sciences), **18**(1), 102.
162 Eitel, W. (1954) *The Physical Properties of the Silicates*, University of Chicago Press, Chicago.
163 Witzel, R. (1921) *Z. Anorg. Allg. Chem.*, **64**, 71.
164 Gutzow, I. (1979) *J. Cryst. Growth*, **48**, 569.
165 Zhurkov, S. and Levin, B. (1950) *Dokl. AN SSSR*, **72**, 269 (in Russian).
166 Kobeko, P.P. (1952) *Amorphous Materials*, Academy of Sciences USSR Press, Moscow, Leningrad (in Russian).
167 Moynihan, C.T., Sasabe, H., and Tucker, J. (1976) Kinetics of the Glass Transition in a Calcium-Potassium Nitrate Melt, in *Molten Salts*, Conference Proceedings, Electrochemical Society Publishing House, New York, p. 182–194.
168 Wilks, J. (1961) *The Third Law of Thermodynamics*, Oxford University Press, Oxford.
169 Bartenev, G.M. (1951) *Doklady Akademii Nauk SSSR*, **76**, 227 (in Russian).

170 Volkenstein, M.V. and Ptizyn, O.B. (1956) *J. Tech. Phys. USSR (JTF)*, **26**, 2204.

171 Volkenstein, M.V. (1959) *Configurational Statistics of Polymer Chains*, Academy of Sciences USSR Publishing House, Moscow, Leningrad (in Russian).

172 Grantcharova, E., Avramov, I., and Gutzow, I. (1986) *Naturwissenschaften*, **73**, 95.

173 Wunderlich, B. (1960) *J. Phys. Chem.*, **64**, 1052.

174 Boltzmann, L. (1896–98) *Vorlesungen über die Kinetische Gastheorie*, Johann-Ambrosius-Barth Verlag, Leipzig.

175 Oblad, A.G. and Newton, R.F. (1937) *J. Amer. Chem. Soc.*, **59**, 2495.

176 Gutzow, I. and Kashchiev, D. (1971) In *Advances in Nucleation and Crystallization of Glasses* (eds L.L. Hench and S.W. Freiman), Amer. Ceram. Society, Columbus Ohio, p. 116.

177 Gutzow, I. (1981) *J. Non-Cryst. Solids*, **45**, 301.

178 Schottky, W., Ulich, H., and Wagner, C. (1929) *Thermodynamik*, Springer, Berlin.

179 Gutzow, I., Schmelzer, J.W.P., and Petroff, B. (2007) *J. Eng. Thermophys.*, **16**, 205.

180 Gutzow, I., Schmelzer, J.W.P., and Petroff, B. (2008) *J. Non-Cryst. Solids*, **354**, 311.

181 Gutzow, I., Schmelzer, J.W.P., and Todorova, S. (2008) *Phys. Chem. Glasses: European J. Glass Sci. Technol.*, **B49**, 136.

182 Gutzow, I. and Schmelzer, J.W.P. (2009) *J. Non-Cryst. Solids*, **355**, 581.

183 Jenckel, E. and Gorke, K. (1952) *Z. Naturforsch.*, **7a**, 630.

184 Haase, R. (1956) *Thermodynamik der Mischphasen*, Springer, Berlin.

185 Grantcharova, E. and Gutzow, I. (1986) *J. Non-Cryst. Solids*, **81**, 99.

186 Schulz, I. and Hinz, W. (1955) *Silikattechnik*, **6**, 235.

187 Simha, R. and Boyer, R.F. (1962) *J. Chem. Phys.*, **37**, 1003.

188 Scott, G.D. (1960) *Nature*, **188**, 908.

189 Bernal, J.D. (1959) *Nature*, **183**, 141.

190 Bernal, J.D. (1964) *Proc. Royal Soc.*, **280**, 299.

191 Gutzow, I. (1962) *Z. Phys. Chem.* (Leipzig) **221**, 153.

192 Wiechert, H. (1989/1992) Zweidimensionale Glasartige Phasen, in *Arbeits- und Ergebnisbericht Sonderforschungsbereich 262: Glaszustand und Glasübergang Metallischer Amorpher Materialien* (eds C. Kolbcher, F. Kremer, and A. Müller), Universität Mainz, Mainz 1992, S. 273.

193 Ziman, J.M. (1979) *Models of Disorder*, Cambridge University Press, London, New York.

194 Ritland, H.N. (1954) *J. Amer. Ceram. Soc.*, **37**, 370.

195 Alfrey, T., Goldfinger, G., and Mark, H. (1943) *J. Appl. Phys.*, **14**, 701.

196 Landau, L.D. and Lifschitz, E.M. (1967) *Elektrodynamik der Kontinua*, Akademie-Verlag, Berlin.

197 Reid, R.C., Prausnitz, J.M., and Sherwood, T.K. (1977) *The Properties of Glasses and Liquids*, McGraw Hill, New York.

198 Volkenstein, M.V. (1959) Optical Properties of Matter in the Vitreous State, in *Investigations in Experimental and Theoretical Physics – a Collection, devoted to the Memory of G.S. Landsberg* (ed. I.L. Fabelinski), Publ. House of Academy of Sciences USSR, Moscow, p. 80/94 (in Russian).

199 Moynihan, C.T. et al. (1976) Structural Relaxation in Vitreous Materials, in *The Glass Transition and the Nature of the Glassy State* (eds M. Goldstein and R. Simha), Annals New York Academy of Sciences **279**, p. 15ff, New York.

200 Jones, G.O. and Simon, F.E. (1949) *Endeavour*, **8**, 174.

201 Johari, G.P. (2000) *J. Chem. Phys.*, **113**, 751.

202 Johari, G.P. (2001) *Chem. Phys.*, **265**, 217.

203 Gonchukova, N.O. (1982) *Fizika i Khimiya Stekla*, **8**, 429 (in Russian).

204 Tammann, G. (1930) *Annal. Phys.*, **5(5)**, 107.

205 Tammann, G. and Jenckel, E. (1930) *Z. Anorg. Allg. Chem.*, **193**, 76.

206 Gutzow, I. (1965) *Relationship between the Structure of a Substance in the Glassy State and the Zero-Point Entropy*. Proceedings of the IV-th All-Union Conference on the Glassy State, Leningrad, 1964 (ed. E.A. Porai-Koshits), Nauka, Moscow, p. 62.

207 Gutzow, I. (1966) *The Structure of Glass* (ed. E.A. Porai-Koshitz), vol. 6, Consultants Bureau, New York, part 1, p. 55.

208 Johari, G.P. (2006) *Philos. Magazine*, **86**, 1567.

209 Oldekop, W. (1957) *Glastech. Ber.*, **30**, 8.

210 Laughlin, W.T. and Uhlmann, D.R. (1972) *J. Phys. Chem.*, **76**, 2317.

211 Thomas, S.B. and Parks, G.S. (1931) *J. Chem. Phys.*, **35**, 2091.

212 Nemilov, S.V. (1976) *Fiz. Khim. Stekla*, **2**, 97.

213 Timura, S., Yokogawa, T., and Niwa, K. (1975) *J. Chem. Thermodyn.*, **7**, 633.

214 Weyl, W.A. and Marboe, E.C. (1962) *The Constitution of Glasses*, Interscience Publishers, New York-London, vol. 1, (1964) vol. II/1, (1967) vol II/2.

215 Smith, G.S. and Rindone, G.E. (1961) *J. Amer. Ceram. Soc.*, **44**, 72.

216 Anderson, C.T. (1937) *J. Amer. Chem. Soc.*, **59**, 1036.

217 Greet, R.J. and Turnbull, D. (1967) *J. Chem. Phys.*, **47**, 2185.

218 Bestul, A. and Chang, S. (1964) *J. Chem. Phys.*, **40**, 3731.

219 Kelley, K.H. (1929) *J. Am. Chem. Soc.*, **51**, 779.

220 Chen, H.S. (1976) *J. Non-Cryst. Solids*, **22**, 135.

221 Chen, H.S. and Turnbull, D. (1967) *J. Appl. Phys.*, **38**, 3646.

222 Dobreva, A. (1992) *Non-Steady State Effects in the Kinetics of Crystallization of Polymer Melts*, PhD thesis, Bulgarian Academy of Sciences, Sofia.

223 Gutzow, I. (1979) Proceedings 1-st International Otto-Schott Kolloquium, *Wiss. Z. Friedrich-Schiller Universitä Jena, Mathematisch-Naturwiss. Reihe*, **28**, 243.

224 Dobreva, A. and Gutzow, I. (1992) *Polymer*, **33**, 451.

225 Eitel, W. (1952) *Thermochemical Methods in Silicate Investigation*, Rutgers University Press, New Brunswick, NJ.

226 Brizke, E.W. and Kapustinski, A.F. (eds) (1949) *Caloric Constants of Inorganic Substances*, Academy of Science Press, Moscow, Leningrad (in Russian).

227 L. Mandelkern (1964) *Crystallization of Polymers*, McGraw-Hill, New York.

228 Dietzel, A. and Poegel, H.J. (1953) *Naturwissenschaften*, **40**, 604.

229 Sternberg, A.A., Mironova, G.S., Zvereva, O.V., and Molovina, M.V. (1989) *Crystal Growth*, vol. 17, Nauka Publishers, Moscow, p. 142 (in Russian).

230 Cargill, G.S. (1975) In *Solid State Physics* (eds H. Ebenreich and F. Seitz), vol. 30, Academic Press, New York, p. 227.

231 Chen, H.S. and Turnbull, D. (1968) *J. Chem. Phys.*, **48**, 2560.

232 Zanotto, E.D. and Müller, E. (1991) *J. Non-Cryst. Solids*, **130**, 220.

233 Gutzow, I., Grigorova, T., Avramov, I., and Schmelzer, J.W.P. (2001) *Phys. Chem. Glasses*, **43C**, 477.

234 Schmelzer, J.W.P. and Gutzow, I. (2006) *J. Chem. Phys.*, **125**, 184511.

235 Nemilov, S.V. (1985) *Fiz. Khim. Stekla*, **11**, 146.

236 Moynihan, C.T. and Gupta, P.K. (1978) *J. Non-Cryst. Solids*, **29**, 143, 158.

237 Nemilov, S.V. (1995) *Thermodynamic and Kinetic Aspects of the Vitreous State*, CRC Press, Boca Raton, London.

238 Schmelzer, J.W.P. and Gutzow, I. (2008) On the Interpretation of the Glass Transition in Terms of Internal Pressure and Fictive Temperature, in *Nucleation Theory and Applications* (eds J.W.P. Schmelzer, G. Röpke, and V.B. Priezzhev), Joint Institute for Nuclear Research Publishing Department, Dubna, Russia, pp. 303–318.

239 Schmelzer, J.W.P. and Gutzow, I. (2009) *J. Non-Cryst. Solids*, **355**, 653.

240 Möller, J., Gutzow, I., and Schmelzer, J.W.P. (2006) *J. Chem. Phys.*, **125**, 094505.

241 Bailey, N.P., Christensen, T., Jakobsen, B., Niss, K., Olsen, M.B., Pedersen, U.R., Schroder, T.B., and Dyre, J.C. (2008) *J. Phys. Condens. Matter*, **20**, 244113.

242 Gupta, P.K. (1988) *J. Non-Cryst. Solids*, **102**, 231.

243 Nemilov, S.V. (2009) *J. Non-Cryst. Solids*, **355**, 607.

244 Bragg, W.I. and Williams, E.J. (1934) *Proc. Royal Soc. Lond.*, **145A**, 699.

245 Tropin, T.V., Schmelzer, J.W.P., and Schick, C. (2011) On the dependence of the properties of glasses on cooling

and heating rates, Part 1: Entropy, entropy production, and glass transition temperature, *J. Non-Cryst. Solids*, **357**, 1291–1302.
246 Tropin, T.V., Schmelzer, J.W.P., and Schick, C. (2011) On the dependence of the properties of glasses on cooling and heating rates, Part 2: Prigogine-Defay ratio, fictive temperature abd fictive pressure, *J. Non-Cryst. Solids*, 1303–1309.
247 Morey, G.W. (1954) *The Properties of Glass*, Reinhold Publ., New York.
248 Gutzow, I. (1972) *Z. Phys. Chem. (N.F.)*, **81**, 195.
249 Kinoshita, A. (1980) *J. Non-Cryst. Solids*, **42**, 447.
250 Schnaus, U.E., Moynihan, C.T. Gammon, R.W., and Macedo, P.B. (1970) *Phys. Chem. Glasses*, **11**, 213.
251 Gutzow, I. and Ilieva, D. (1998) *Glastech. Ber., Glass Sci. Technol.*, **71C**, 109.
252 Gupta, P.K. and Mauro, J.C. (2007) *J. Chem. Phys.*, **126**, 224504.
253 Kivelson, D. and Reiss, H. (1999) *J. Phys. Chem.*, **B103**, 8337.
254 Mauro, J.C., Gupta, P.K., and Loucks, R.J. (2007) *J. Chem. Phys.*, **126**, 184511.
255 Goldstein, M. (2008) *J. Chem. Phys.*, **128**, 154510.
256 Johari, G.P. (2010) *Thermochim. Acta*, **500**, 111.
257 Johari, G.P. (2010) *J. Chem. Phys.*, **132**, 124509.
258 Conradt, R. (2009) *J. Non-Cryst. Solids*, **355**, 636.
259 Gibbs, J.H. and Di Marzio, E.A. (1958) *J. Chem. Phys.*, **28**, 370.
260 Gibbs, J.H. (1960) Nature of the Glass-transition and the Vitreous State, in *Modern Aspects of the Vitreous State* (ed. J.P. Mackenzie), Butterworths, London.
261 Gordon, J.M., Gibbs, J.H., and Fleming, P.D. (1976) *J. Chem. Phys.*, **65**, 2771.
262 Hutchinson, J.M. and Kovasc, A.J. (1977) In *The Structure of Non-Crystalline Materials* (ed. P.H. Gaskell), Symposium Proceedings, Cambridge, p. 167ff.
263 Bennett, C.H. (1987) *Sci. Am.*, **257**, 5.
264 Bennewitz, K. (1926) Der Nernstsche Wärmesatz, in *Handbuch der Physik* (eds H. Geiger and H. Scheel), vol. 9, Springer Verlag, Berlin, p. 141.
265 Tropin, T.V., Schmelzer. J.W.P., and Schick, C. *On the Dependence of the Properties of Glasses on Cooling and Heating Rates: Part 3*, in press.
266 Schmelzer, J.W.P. (2010) On the determination of the kinetic pre-factor in classical nucleation theory. *J. Non-Cryst. Solids*, **356**, 2901–2907.
267 Schmelzer, J.W.P., Müller, R., Möller, J., and Gutzow, I. (2002) *Phys. Chem. Glasses*, **43 C**, 291.
268 Schmelzer, J.W.P., Potapov, O.V., Fokin, V.M., Müller, R., and Reinsch, S. (2004) *J. Non-Cryst. Solids*, **333**, 150.
269 Schmelzer, J.W.P., Zanotto, E.D., Avramov, I., and Fokin, V.M. (2006) *J. Non-Cryst. Solids*, **352**, 434.
270 Tarassov, V.V. (1956) *New Problems of the Physics of Glass*, Gostroiizdat, Moscow (in Russian).
271 McGraw, D.A. (1952) *J. Amer. Ceram. Soc.*, **35**, 22.
272 Bartenev, G.M. (1966) *Structure and Mechanical Properties of Inorganic Glasses*, Building Materials Press, Moscow (in Russian).
273 Reinsch, S. (2001) *Oberflächenkeimbildung von Silikatgläsern der Stoichiometrie des Cordierits und des Diopsids*, PhD thesis, Technische Universität Berlin.
274 Gutzow, I. (1977) Thermodynamical and Model-Statistical Treatment of the Glassy Solidification, in *The Physics of Non-Crystalline Solids*. Proceedings 4-th International Conference Clausthal, 1976 (ed. G. Frischat), Transtech, Aedermansdorf, p. 356.
275 Milchev, A. and Gutzow, I. (1982) *J. Macromol. Sci. – Phys. B*, **21(4)**, 583.
276 Schmelzer, J.W.P., Zanotto, E.D., and Fokin, V.M. (2005) *J. Chem. Phys.*, **122**, 074511.
277 Meixner, J. (1952) *Changements de Phases*, Society de Chimie Physique Publishers, Paris, p. 432.
278 Rehage, G. and Borchardt, W. (1973) The Thermodynamics of the Glassy State, in *The Physics of Glassy Polymers* (ed. R.N. Haward), Appl. Science, London, chapter 1, p. 54.

279 Tool, A.Q. and Hill, E.F. (1925) *J. Soc. Glass Technol.*, **9**, 185.

280 Tool, A.Q. and Eichlin, C.G. (1931) *J. Amer. Ceram. Soc.*, **14**, 276.

281 Landa, L. and Landa, K. (2004) *J. Non-Cryst. Solids*, **348**, 59.

282 Landa, L.M., Thomsen, S., and Hulme, R. (2008) Glass as a Solid Non-Equilibrium Phase under Negative Pressure, in *Nucleation Theory and Applications* (eds J.W.P. Schmelzer, G. Röpke, and V.B. Priezzhev), Joint Institute for Nuclear Research Publishing Department, Dubna, Russia, pp. 295–302.

283 Nieuwenhuizen, T.M. (2000) *J. Phys. Cond. Matter*, **12**, 6543.

284 Nieuwenhuizen, T.M. (2000) *Phys. Rev.*, **E61**, 267.

285 Nieuwenhuizen, T.M. (1997) *Phys. Rev. Lett.*, **79**, 1317.

286 Garden, J.-L., Richard, J., and Guillou, H. (2008) *J. Chem. Phys.*, **129**, 044508, 129901.

287 Rehage, G. and Oels, H.J. (1977) *High Temp.-High Pressures*, **9**, 545.

288 Lillie, H.R. (1933) *J. Amer. Ceram. Soc.*, **16**, 619.

289 Winter, A. (1953) *Verreset Refract.*, **7**, 217.

290 Mazurin, O.V., Startsev, Y.K., and Stolyar, S.V. (1982) *J. Non-Cryst. Solids*, **53**, 105.

291 Mazurin, O.V., Startsev, Y.K., and Pozeluyeva, L.N. (1979) *Fiz. Khim. Stekla*, **5**, 82 (in Russian).

292 Mazurin, O.V., Kluyev, V.P., and Stolyar, S.V. (1983) *Glastech. Ber.*, **56**, 1148.

293 Volkenstein, M.V. (1951) *Molecular Optics*, State Publishers Techn. Teor. Literature, Moscow, pp. 277, 348/350, 463 (in Russian).

294 Berthier, L., Biroli, G., Bouchaud, J.-P., Kob, W., Miyazaki, K., and Reichman, D.R. (2007) *J. Chem. Phys.*, **126**, 184503, 184504.

295 Mueller, H. (1938) *Proc. Royal Soc.* (London) **166**, 425.

296 Vedishcheva, N.M., Wright, A.C., and Porai-Koshits, E.A. (2001) Glass Scientist Extraordinary, in *Proc. Third Conf. on Borate Glasses, Crystals and Melts*, Sofia, 1999: Structure and Applications (eds Y. Dimitriev and A.C. Wright), Society Glass Technology, Sheffield.

297 Lebedev, A.A. (1940) The Structure of Glasses according to *X*-Ray Data and their Optical Properties. *Bull. Acad. Sci. UdSSR (Sect. Phys.)*, **4**, 584.

298 Lebedev, A.A. (1921) *On the Polymorphism and the Annealing of Glass*. Proc. (Trudy) of the State Optical Institute (GOI), **2**, 1 (in Russian).

299 Randall, J.T., Ruxby, H.P., and Cooper, B.S. (1930) *J. Soc. Glass Technol.*, **4**, 219.

300 Tudorovskaya, H.A. (1949) Temperature Dependence of the Coefficient of Refraction of Glasses of the System Na_2O/SiO_2, in *Physico-chemical Properties of the Ternary System Na_2O–PbO–SiO_2* (ed. I.V. Grebenshchikov), Publ. House Academy Sciences USSR, Moscow, p. 186 (in Russian).

301 Evstropyev, K.S. and Skornyakov, M.M. (1949) Dissolution Heats of Glasses and Crystals of the System Na_2O/SiO_2 in H-Acid, in *Physico-chemical Properties of the Ternary System Na_2O-PbO-SiO_2* (ed. I.V. Grebenshchikov), Publ. House Academy Sciences USSR, Moscow, p. 174 (in Russian).

302 Valenkov, N.N. and Porai-Koshits, E.A. (1949) *X*-Ray Investigation of Glass of the System Na_2O/SiO_2, in *Physico-chemical Properties of the Ternary System Na_2O–PbO–SiO_2* (ed. I.V. Grebenshchikov), Publ. House Academy Sciences USSR, Moscow, p. 147 (in Russian).

303 Hägg, G. (1935) *J. Chem. Phys.*, **3**, 10.

304 Porai-Koshits, E.A. (1988) The Structure of Glasses: Geometrical, Structural and Dynamical Aspects, in *The Vitreous State, Proc. VIII All-Union Conference 1986* (ed. E.A. Porai-Koshits), Nauka Publ., Leningrad (in Russian).

305 Schulz, M.M. (1988) Thermodynamics of Melts and Glasses, in *The Vitreous State, Proc. VIII All-Union Conference* (ed. E.A. Porai-Koshits), Nauka, Leningrad, p. 5 (in Russian).

306 Vedishcheva, N.M., Shakhmatkin, B.A., and Wright, A.C. (1996) A Thermodynamic Model for the Calculation of the Physical Properties and Structural

Characteristics of Glasses and Melts, in *Amorphous Insulators and Semiconductors, Proc. NATO Advanced Study Institute, Sozopol* (eds M.F. Thorpe and M.I. Mitkova), Kluwer Academic Publishers, Dordrecht, p. 235.

307 Priven, A.I. (2004) *Glass Technol.*, **45**, 244.

308 Filipovich, V.N. (1965) Relations between Structure of Melts, Glass and Glass-Ceramics, in *Structural Changes in Glasses at Elevated Temperatures* (eds H.A. Toropov and E.A. Porai-Koshits), Nauka, Leningrad, p. 15.

309 Porai-Koshits, E.A. (1965) Glass Structure and the Initial Stages of Glass-Ceramics Formation, in *Structural Changes in Glasses at Elevated Temperatures* (eds H.A. Toropov and E.A. Porai-Koshits), Nauka, Leningrad, p. 1.

310 Andreev, N.S., Mazurin, O.V., and Porai-Koshits, E.A., *et al.* (1972) *The Phenomena of Liquid Phase Separation*, Nauka, Leningrad (in Russian).

311 Golubkov, V.V. (1989) *Fiz. Khim. Stekla*, **15**, 467 (in Russian).

312 Golubkov, V.V. (1997) *The Problem of Inhomogeneous Glass Structure*. Proceedings of the Xth Conference on the Vitreous State, St. Petersburg, Russia.

313 Golubkov, V.V. (1998) *Glass Phys. Chem.*, **24**, 196.

314 Golubkov, V.V, Bogdanov, V.N., and Pakhnin, A.Y. *et al.* (1999) *J. Chem. Phys.*, **110**, 4897.

315 Vasilevskaya, T.I., Golubkov, V.V., and Porai-Koshits, E.A. (1983) An Investigation of the Structure of B_2O_3 with Small Angle X-Ray Diffraction, in *The Vitreous State* (ed. E.A. Porai-Koshits), Nauka, Leningrad, p. 48.

316 Gutzow, I., Pascova, R., and Schmelzer, J.W.P. (2010) *Int. J. Appl. Glass Sci.*, **1**, 221–236.

317 Binder, K. and Kob, W. (2005) *Glassy Materials and Disordered Solids: An Introduction to their Statistical Mechanics*, World Scientific, Singapore.

318 Gibbs, J.W. (1902) *Elementary Principles in Statistical Mechanics Developed with especial reference to the Rational Foundation of Thermodynamics*, Charles Scribner's Sons, New York; Edward Arnold, London.

319 Gutzow, I. (2000) The Generic Phenomenology of Glass Formation, in *Insulating and Semiconducting Glasses* (ed. P. Boolchand), World Scientific, Singapore, London, pp. 65–93.

320 Gutzow, I. and Babalievski, F. (1999) *Compt. Rend. Bulg. Acad. Sci.*, **52**, 61.

321 Angell, C.A. (1968) *J. Am. Ceram. Soc.*, **51**, 117.

322 Simon, S.L. and McKenna, G.B. (2009) *J. Non-Cryst. Solids*, **355**, 672.

323 Feistel, R. and Wagner, W. (2006) *J. Phys. Chem. Ref. Data*, **35**, 1.

324 Feistel, R. and Feistel, S. (2006) Die Ostsee als thermodynamisches System, in *Irreversible Prozesse und Selbstorganisation* (eds L. Schimansky-Geier, H. Malchow, and T. Pöschel), Logos-Verlag, Berlin.

325 Bottinga, Y. and Richet, P. (1995) *Geochim. Cosmochim. Acta*, **59**, 2725.

326 Wolf, G.H. and McMillan, P.F. (1995) Pressure effect on silicate melt structure and properties. *Rev. Mineral.*, **32**, 507.

327 Gupta, P.K. (1987) *J. Amer. Ceram. Soc.*, **70**, C-152.

328 Skripov, V.P. and Faizullin, M.Z. (2006) *Crystal-Liquid-Gas Phase Transitions and Thermodynamic Similarity*, Fizmatlit Publishers, Moscow, 2003; English translation: Wiley-VCH Verlag GmbH, Berlin, Weinheim.

329 Skripov, V.P. and Faizullin, M.Z. (2005) Solid-Liquid and Liquid-Vapor Phase Transitions: Similarities and Differences, in *Nucleation Theory and Applications* (ed. J.W.P. Schmelzer), Wiley-VCH Verlag GmbH, Berlin, Weinheim, pp. 4–38.

330 Feistel, R. and Hagen, E. (1995) *Prog. Oceanogr.*, **36**, 249.
In addition, the PC-program *Seawater.exe* was employed (available via http://www.io-warnemuende.de, accessed January 28, 2011) supplied by courtesy of R. Feistel, Leibniz Institut für Ostseeforschung, Rostock-Warnemünde, Dr. R. Feistel.

331 Sharma, S.K., Virgo, D., and Kushiro, I. (1979) *J. Non-Cryst. Solids*, **33**, 235.

332 Stebbins, J.F. and McMillan, P.F. (1989) *Amer. Mineralogist*, **74**, 965.
333 Waff, H.S. (1975) Geophysical Research **2**, 193.
334 Woodcock, L.V., Angell, C.A., and Cheeseman, P. (1976) *J. Chem. Phys.*, **65**, 1565.
335 Xue, X., Stebbins, J.F., Kanzaki, M., McMillan, P.F., and Poe, B. (1991) *Am. Mineral.*, **76**, 8.
336 Nielsen, A.I., Christensen, T., Jakobson, B., Niss, K., Olsen, N.B., Richert, R., and Dyre, J.C. (2009) *J. Chem. Phys.*, **130**, 154508.
337 Dobreva, A. and Gutzow, I. (1997) *J. Non-Cryst. Solids*, **220**, 235.
338 Moorgani, K. and Coey, J.M.D. (1984) *Magnetic Glasses*, Elsevier Publ., Amsterdam, New York.
339 Binder, K. and Young, A.P. (1986) *Rev. Mod. Phys.*, **58**, 801.
340 Giauque, W.F. and Stout, J.W. (1936) *J. Amer. Chem. Soc.*, **58**, 1144.
341 Donth, E. (2001) *The Glass Transition: Relaxation Dynamics in Liquids and Disordered Materials*, Springer, Berlin.
342 Battezzati, L. and Garone, E. (1984) *Z. Metallkunde*, **75**, 305.
343 Jäckle, J. (1981) *Phil. Mag.*, **B44**, 533.
344 Johari, G.P. (1980) *Phil. Mag.*, **B41**, 41.
345 Johari, G.P. (2000) *J. Chem. Phys.*, **112**, 7518, 8958.
346 Grebovicz, J. and Wunderlich, B. (1981) *Mol. Cryst. Liq. Cryst.*, **76**, 287.
347 Sorai, M. and Seki, S. (1973) *Mol. Cryst. Liq. Cryst.*, **23**, 299.
348 Sorai, M., Tani, K., Suga, H., and Yoshioka, H. (1983) *Mol. Cryst. Liq. Cryst.*, **97**, 365.
349 Sorai, M., Tani, K., Suga, H., and Yoshioka, H. (1983) *Mol. Cryst. Liq. Cryst.*, **95**, 11.
350 Einstein, A. (1914) *Verh. Dtsch. Phys. Ges.*, **16**, 820.
351 Pauling, L. and Tolman, R.C. (1925) *J. Amer. Chem. Soc.*, **47**, 2148.
352 Clusius, K. and Teske, W. (1938) *Z. Phys. Chem.*, **6**, 135.
353 Clayton, J.O. and Giauque, W.F. (1932) *J. Amer. Chem. Soc.*, **54**, 2610.
354 Pauling, L. (1935) *J. Amer. Chem. Soc.*, **57**, 2680.
355 Kaischew, R. (1938) *Z. Phys. Chem.*, **B40**, 273.
356 Eucken, A. and Veith, H. (1937) *Z. phys. Chem.*, **B35**, 463.
357 Schulz, A.K. (1954) *J. Chem. Phys. Biol.*, **51**, 324.
358 Schulz, A.K. (1954) *Kolloid Z.*, **138**, 75.
359 Das, S.K. and Hucke, E.E. (1975) *Carbon*, **13**, 33.
360 Blachnick, R. and Hoppe, A. (1979) *J. Non-Cryst. Solids*, **34**, 191.
361 Debenedetti, P.G. (1996) *Metastable Liquids*, Princeton Univ. Press.
362 Wong, J. and Angell, A. (1976) *Glass Structure by Spectroscopy*, Marcel Decker, New York.
363 Zubarev, D. N., Morozov, V. G., Röpke, G. (2002) *Statistical Mechanics of Non-Equilibrium Processes* vol. 1, Fizmatlit, Moscow, 66–68, (in Russian).
364 Petroff, B., Milchev, A., and Gutzow, I. (1996) *J. Macromol. Sci.-Phys.*, **B 35**, 763.
365 Gupta, P.K. and Moynihan, C.T. (1976) *J. Chem. Phys.*, **65**, 4136.
366 Varshneya, A.K. (1993) *Fundamentals of Inorganic Glasses*, Academic Press, Boston, New York.
367 Winter-Klein, A. (1943) *J. Amer. Ceram. Soc.*, **26**, 189.
368 Reiner, M. (1964) *Phys. Today*, **17**, 62.
369 Stevels, M. (1971) *J. Non-Cryst. Solids*, **6**, 307.
370 Gutzow, I. and Babalievski, F. (1998) *A new approach in the thermodynamics and kinetics of vitrification*. Proceedings 18th Intern. Congress on Glass, San Francisco, E–4, pp. 37–42.
371 Bartenev, G.M. and Lukyanov, I.A. (1955) *J. Phys. Chem.* (USSR), **29**, 1486 (in Russian).
372 Avramov, I. (1998) *Glastech. Ber.*, **71C**, 198.
373 Avramov, I., Vassilev, T., and Penkov, I. (2005) *J. Non-Cryst. Solids*, **351**, 472.
374 Beaman, R. (1952) *J. Polymer Science*, **9**, 470.
375 Kelker, H. and Hatz, R. (1980) *Handbook of Liquid Crystals*, Verl. Chemie, Weinheim, Deerfield Beach, Florida, Basel, p. 340.

376 Yoshioka, H., Sorai, M., and Suga, H. (1983) *Mol. Cryst. Liq. Cryst.*, **95**, 11.

377 Suga, H. (1986) Thermodynamic Aspects of Glassy Crystals, in *Dynamic Aspects of Structural Changes in Liquids and Glasses* (eds C.A. Angell and M. Goldstein), Ann. New York Acad. Sci., **484**, 249.

378 Eisenberg, D. and Kauzmann, W. (1993) *Structure and Properties of Water*, 1986, p.77/80, 93/99 of the Russian translation: Gidrometeoizdat, Leningrad.

379 Jäckle, J. (1984) *Physica*, **B127**, 79.

380 Bernal, J.D. (1961) *Sci. Am.*, **8**, 908.

381 Woodcock, L.V. (1976) *J. Chem. Soc. Farad. Trans.*, **II72**, 1661.

382 Beal, R.J. and Dean, R.P. (1968) *Phys. Chem. Glasses*, **9**, 125.

383 Gutzow, I., Dobreva, A., Ruessel, C., and Durschang, B. (1997) *J. Non-Cryst. Solids*, **219**, 313.

384 Gutzow, I., Durschang, B., and Ruessel, C. (1997) *J. Mater. Sci.*, **32**, 5389, 5405.

385 Winkelmann, A. and Schott, O. (1894) *Ann. Phys. Chem.*, **51**, 697.

386 Rabinovich, S.G. (2005) *Measurement Errors and Uncertainties: Theory and Practice*, 3rd edn., Springer, Berlin.

387 Mazurin, O.V. and Gankin, Y.V. (2004) *J. Non-Cryst. Solids*, **342**, 166.

388 Mazurin, O.V. and Gankin, Y.V. (2004) In: Proc. of XX Congress on Glass, Kyoto University Press, Kyoto.

389 Mazurin, O.V. (2007) *Glass Technol.: Eur. J. Glass Sci. Technol.*, **A48**, 297.

390 Kutub, A.A. (1995) *J. Mater. Science*, **30**, 724.

391 Inaba, S., Oda, S., and Morinaga, K. (2002) *J. Non-Cryst. Solids*, **306**, 42.

392 Takahashi, K., Osaka, A., and Furuno, R. (1983) *J. Ceram. Soc. Japan*, **91**, 199.

393 Lenoir, M., Grandjean, A., Linard, Y., Cochain, B., and Neuville, D.E. (2008) *Chem. Geol.*, **256**, 316.

394 Inaba, S., Fujino, S., and Morinaga, K. (1999) *J. Amer. Ceram. Soc.*, **82**, 3501.

395 Huang, W.C., Jain, H., Kamitsos, E.I., and Patsis, A.P. (1993) *J. Non-Cryst. Solids*, **162**, 107.

396 Mohamed, A.A., El-Sayed, K., and Zayan, M. (1975) *El-Mously, M.K., Egypt. J. Phys.*, **6**, 37.

397 Bokov, N.A. and Golubkov, V.V. (2008) *Fiz. Khim. Stekla*, **34**, 685.

398 Abd El Latif, L. (2005) *J. Pure Appl. Ultrason.*, **27**, 80.

399 Frischat, G.H., Krause, W., and Hubenthal, H. (1984) *J. Amer. Ceram. Soc.*, **67**, C10.

400 Milberg, M.E., Belitz, R.K., and Silver, A.H. (1960) *Phys. Chem. Glasses*, **1**, 155.

401 Mazurin, O.V. and Gankin, Y.V. (2007) Proc. XXI Congress on Glass, Strasbourg.

402 El-Damrawi, G. (2001) *Phys. Chem. Glasses*, **42**, 56.

403 Shabanova, E.B. (1967) *Trudy Gorkij Politekhnical Institute*, **23**, 38.

404 Masuda, H., Nishida, H., and Morinaga, K. (1998) *J. Japan Inst. Metals*, **62**, 444.

405 El Batal, F.H. (2007) *Nuclear Instr. Methods Phys. Res.*, **B254**, 243.

406 Field, M.B. (1969) *J. Appl. Phys.*, **40**, 2628.

407 Farley, J.M. and Saunders, G.A. (1975) *Phys. Status Solidi*, **A28**, 199.

408 Drake, C.F., James, B.W., Kheyrandish, H., and Yates, B. (1983) *J. Non-Cryst. Solids*, **57**, 305.

409 Chopra, N., Mansingh, A., and Mathur, P. (1992) *J. Non-Cryst. Solids*, **146**, 261.

410 Sidkey, M.A., El Mallawany, R., Nakhla, R.I., and Abd El-Moneim, A. (1997) *J. Non-Cryst. Solids*, **215**, 75.

411 Yoshida, S., Aono, S., Matsuoka, J., and Soga, N. (2001) *J. Ceram. Soc. Japan*, **109**, 753.

412 Redman, M.J. and Chen, J.H. (1967) *J. Amer. Ceram. Soc.*, **50**, 523.

413 Burger, H., Vogel, W., Kozhukharov, V., and Marinov, M. (1984) *J. Mater. Sci.*, **19**, 403.

414 Lin, F.C. (1963) *Glass Ind.*, **44**, 19, 87, 102.

415 Dutreilh-Colas, M., Thomas, P., Champarnaud-Mesjard, J.C., and Fargin, E. (2003) *Phys. Chem. Glasses*, **44**, 349.

416 El-Damrawi, G., and Doweidar, H. (2001) *Phys. Chem. Glasses*, **42**, 116.

417 Doweidar, H., Moustafa, Y.M., Abd El-Maksoud, S., El-Damrawi, G., and Mansour, E. (2001) *Phys. Chem. Glasses*, **42**, 333.

418 El-Damrawi, G. (1994) *J. Non-Cryst. Solids*, **176**, 91.

419 Doweidar, H., El-Igili, K., and Abd El-Maksoud, S. (2000) *J. Phys. D: Appl. Phys.*, **33**, 2532.

420 Doweidar, H., Megahed, A.A., Abd Al-Maksoud, S., El-Shahawi, M.S., and El-Fol, Y. (1994) *Phys. Chem. Glasses*, **35**, 187.

421 Doweidar, H., El-Damrawi, G., Moustafa, Y.M., and Hassan, A.K. (1999) *Phys. Chem. Glasses*, **40**, 252.

422 Volf, M.B. (1988) *Mathematical approach to glass*, Elsevier, Amsterdam, Oxford, New York, Tokyo.

423 SciGlass (Glass Property Information System) (1996–2010) ITC, Inc., http://www.sciglass.info.

424 Appen, A.A. (1974) *Khimiya Stekla* (Chemistry of Glass), Nauka Publ. House, Leningrad (in Russian).

425 Bacon, C.R. (1977) *Amer. J. Sci.*, **277**, 109.

426 Balmakov, M.D. and Bratova, I.V. (1982) *Fiz. Khim. Stekla*, **8**, 675 (in Russian).

427 Balmakov, M.D. (1977) *Fiz. Khim. Stekla*, **3**, 255 (in Russian).

428 Belousov, Y.L. and Akulova, M.V. (1992) *Fiz. Khim. Stekla*, **18**, 94 (in Russian).

429 Belousov, Y.L. and Firsov, V.A. (1991) *Fiz. Khim. Stekla*, **17**, 411 (in Russian).

430 Boni, R.E. and Derge, G. (1956) *J. Metals*, **206**, 53.

431 Borisova, Z.U. (1983) *Chalcogenide Semiconducting Glasses*, Leningrad Gos. Univ., Leningrad (in Russian).

432 Bottinga, Y., Richet, P., and Weill, D.F. (1983) *Bull. Mineral.*, **106**, 129.

433 Bottinga, Y. and Weill, D.F. (1970) *J. Amer. Sci.*, **269**, 169.

434 Bottinga, Y. and Weill, D.F. (1972) *J. Amer. Sci.*, **272**, 438.

435 Bottinga, Y., Weill, D.F., and Richet, P. (1982) *Geochim. Cosmochim. Acta*, **46**, 909.

436 Braginsky, K.I. (1973) *Steklo Keram.*, **7**, 10 (in Russian).

437 Clarke, J.R. and Turner, W.E.S. (1919) *J. Soc. Glass Technol.*, **3**, 260.

438 Demkina, L.I. (1989) *Fiz. Khim. Stekla*, **15**, 717 (in Russian).

439 Demkina, L.I. (1991) *Fiz. Khim. Stekla*, **17**, 899 (in Russian).

440 Demkina, L.I. (1976) Density and Optical Constants of Glasses, in *Fiziko-Khimicheskie Osnovy Proizvodstva Opticheskogo Stekla (Physico-Chemical Background of Production of Optical Glasses)* (ed. L.I. Demkina), Nauka Publ. House, Leningrad, p. 78 (in Russian).

441 Dietzel, A. (1942) *Sprechsaal*, **75**, 82.

442 Flom, Z.G. and Kofman, A.G. (1985) *Steklo Keram.*, **7**, 10 (in Russian).

443 Fluegel, A. (2007) *Glass Technol., Europ. J. Glass Sci. Technol. A*, **48**, 13.

444 Fuxi, G. (1995) *J. Non-Cryst. Solids*, **184**, 9.

445 Gan, F.-S. (1963) *Scientia Sin.*, **12**, 1365 (in Russian).

446 Gan F.-S. (1974) *Scientia Sin.*, **17**, 534 (in Russian).

447 Gilard, P. and Dubrul, L. (1928) *Verre, Silic. Ind.*, **9**, 25.

448 Gilard, P. and Dubrul, L. (1937) *J. Soc. Glass Technol.*, **21**, 476.

449 Gilard, P. and Dubrul, L. (1937) *Les bases physico-chimiques de l'industrie du verre*, France Thone, Paris.

450 Giordano, D., Dingwell, D.B., and Romano, C. (2000) *J. Volcanol. Geotherm. Res.*, **103**, 239.

451 Giordano, D. and Dingwell, D.B. (2003) *J. Phys.: Condens. Matter.*, **15**, S945.

452 Goleus, V.I., Belyi, A.Y., Sardak, E.M., and Belyi, Y.I. (1996) *Steklo Keram.*, **8**, 6.

453 Goleus, V.I., Belyi, Y.I., Marchenko, A.V., and Shashek, L. (1989) *Fiz. Khim. Stekla*, **15**, 283 (in Russian).

454 Goleus, V.I., Belyi, Y.I., Nosenko, A.V., and Kozyreva, T.I. (1991) *Fiz. Khim. Stekla*, **17**, 361 (in Russian).

455 Goleus, V.I., Belyi, Y.I., Nosenko, A.V., Kozyreva, T.I., and Maltseva, V.V. (1991) *Fiz. Khim. Stekla*, **17**, 200 (in Russian).

456 Goto, A., Oshima, H., and Nishida, Y. (1997) *J. Volcanol. Geotherm. Res.*, **76**, 19.

457 Gudovich, O.D. and Primenko, V.I. (1985) *Fiz. Khim. Stekla*, **11**, 149.

458 Hormadaly, J. (1986) *J. Non-Cryst. Solids*, **79**, 311.

459 Hrma, P., Piepel, G.F., Vienna, J.D., Cooley, S.K., Kim, D.S., and Russell, R.L. (2001) *Database and Interim Glass Property Models for Hanford HLW Glasses*, Pac. Northwest Nat. Lab., PNNL-13573. http://www.pnl.gov/main/publications/external/technical_reports/pnnl-13573.pdf, (accessed 28 January 2011).

460 Hrma, P., Piepel, G.F., Redgate, P.E., Smith, D.E., Schweider, M.J., Vienna, J.D, and Kim, D.S. (1995) *Ceram. Trans. Environ. Issues Waste Manag. Technol.*, **61**, 505.

461 Hrma, P. and Robertus, R.J. (1993) *Ceram. Eng. Sci. Proc.*, **14**, 187.

462 Hrma, P., Smith, D.E., Matyas, J., Yeager, J.D., Jones, J.V., and Boulos, E.N. (2006) *Glass Technol., Europ. J. Glass Sci. Technol. A*, **47**, 78.

463 Huggins, M.H. and Sun, K.H. (1943) *J. Amer. Ceram. Soc.*, **26**, 4.

464 Huggins, M.H. and Sun, K.H. (1947) *J. Phys. Colloid Chem.*, **51**, 438.

465 Khalimovskaya-Churkina, S.A. and Priven, A.I. (2000) *Glass Phys. Chem. (Engl. Transl. of Russian Fiz. Khim. Stekla)*, **26**, 531.

466 Kharyuzov, V.A. and Zorin, A.P. (1967) *Elektronnaya Tekhnika Ser. XIV*, **3**, 65 (in Russian).

467 Kokorina, V.F. (1996) *Glasses for Infrared Optics*, CRC, Boca Raton.

468 Kopp, H. (1865) *Philos. Trans. Royal Soc. London*, **155**, 71.

469 Kozyukov, V.M. and Mazurin, O.V. (1994) *Glass Phys. Chem. (Engl. Transl. of Russian Fiz. Khim. Stekla)* **20**, 302.

470 Kucuk, A., Clare, A.G., and Jones, L. (1999) *Glass Technol.*, **40**, 149.

471 Lakatos, T. and Johansson, L.G. (1978) *Glastekn. Tidskr.*, **33**, 55.

472 Lakatos, T., Johansson, L.G., and Simmingskold, B. (1972) *Glass Technol.*, **13**, 88.

473 Lakatos, T., Johansson, L.G., and Simmingskold, B. (1976) *Glastekn. Tidskr.*, **31**, 31.

474 Lakatos, T., Johansson, L.G., and Simmingskold, B. (1975) *Glastekn. Tidskr.*, **30**, 7.

475 Lakatos, T., Johansson, L.G., and Simmingskold, B. (1979) *Glastekn. Tidskr.*, **34**, 9.

476 Lakatos, T., Johansson, L.G., and Simmingskold, B. (1979) *Glastekn. Tidskr.*, **34**, 61.

477 Lebedev, A.A. (1921) *Trudy GOI*, **10**, 1 (in Russian).

478 Leko, V.K. (1980) *Fiz. Khim. Stekla*, **6**, 553 (in Russian).

479 Leko, V.K. and Mazurin, O.V. (2000) *Glass Phys. Chem. (Engl. Transl. of Russian Fiz. Khim. Stekla)*, **26**, 226.

480 Leko, V.K. and Mazurin, O.V. (2003) *Glass Phys. Chem. (Engl. Transl. of Russian Fiz. Khim. Stekla)*, **29**, 16.

481 Lyon, K.C. (1944) *J. Amer. Ceram. Soc.*, **27**, 186.

482 Lyon, K.C. (1974) *J. Res. Nat. Bur. Stand.*, **78A**, 497.

483 Makishima, A. and Mackenzie, J.D. (1973) *J. Non-Cryst. Solids*, **12**, 35.

484 Mazurin, O.V. (1962) *Elektricheskie Svoistva Stekla (Electrical Properties of Glass)*, Inostrannaya Literatura, Leningrad.

485 Mazurin, O.V., Tretiakova, N.I., and Shvaiko-Shvaikovskaya, T.P. (1969) *Metod rascheta viazkosti silikatnykh stekol (Method of calculation of viscosity of the silicate glasses)*, Deposited in VINITI, No. DEP1091-69 (in Russian).

486 Mazurin, O.V., Nikolina, G.P., and Petrovskaya, M.L. (1988) *Raschet vyazkosti stekol, (Calculation of viscosity of glasses)*, LTI Im. Lensoveta, Leningrad (in Russian).

487 Moore, J. and Sharp, D.E. (1958) *J. Amer. Ceram. Soc.*, **41**, 35.

488 Myuller, R.L. (1960) *Stekloobraznoe Sostoyanie (Glassy State)*, Nauka, Moscow, p. 61 (in Russian).

489 Myuller, R.L. (1954) *J. Fiz. Khim.*, **28**, 2189 (in Russian).

490 Novopashin, A.A. and Seregin, N.N. (1979) *Fiz. Khim. Stekla*, **5**, 431 (in Russian).

491 Okhotin, M.V. (1954) *Steklo Keram.*, **1**, 7 (in Russian).

492 Okhotin, M.V. (1947) *Stekolnaya i Keramicheskaya Promyshlennost*, **3**, 12 (in Russian).

493 Philips, C.J. (1964) *Glass Technol.*, **5**, 216.

494 Primenko, V.I. and Galyant, V.I. (1989) *Stroitelnye Materialy i Konstruktsii*, **2**, 38 (in Russian).

495 Priven, A.I. (2002) *Fundamentals of the calculation of concentration-temperature-time dependencies of properties of oxide glass-forming substances in wide composition area and temperature range*, Dr. Sci. Thesis, St. Petersburg (in Russian).

496 Priven, A.I. (1997) *Glass Phys. Chem.* (Engl. Transl. of Russian Fiz. Khim. Stekla), **23**, 333.

497 Priven, A.I. (1997) *Glass Phys. Chem.* (Engl. Transl. of Russian Fiz. Khim. Stekla), **23**, 344.

498 Priven, A.I. (1997) *Glass Phys. Chem.* (Engl. Transl. of Russian Fiz. Khim. Stekla), **23**, 416.

499 Priven, A.I. (1998) *Glass Phys. Chem.* (Engl. Transl. of Russian Fiz. Khim. Stekla), **24**, 19.

500 Priven, A.I. (1998) *Glass Phys. Chem.* (Engl. Transl. of Russian Fiz. Khim. Stekla), **24**, 31.

501 Priven, A.I. (1998) *Glass Phys. Chem.* (Engl. Transl. of Russian Fiz. Khim. Stekla), **24**, 67.

502 Priven, A.I. (1999) *Glass Phys. Chem.* (Engl. Transl. of Russian Fiz. Khim. Stekla), **25**, 491.

503 Priven, A.I. (2000) *Glass Phys. Chem.* (Engl. Transl. of Russian Fiz. Khim. Stekla), **26**, 441.

504 Priven, A.I. (2000) *Glass Phys. Chem.* (Engl. Transl. of Russian Fiz. Khim. Stekla), **26**, 455.

505 Priven, A.I. (2000) *Glass Phys. Chem.* (Engl. Transl. of Russian Fiz. Khim. Stekla), **26**, 541.

506 Priven, A.I. (2001) *Glass Phys. Chem.* (Engl. Transl. of Russian Fiz. Khim. Stekla), **27**, 360.

507 Priven, A.I. (2001) *Glass Phys. Chem.* (Engl. Transl. of Russian Fiz. Khim. Stekla), **27**, 435.

508 Priven, A.I. (2001) *Glass Phys. Chem.* (Engl. Transl. of Russian Fiz. Khim. Stekla), **27**, 527.

509 Priven, A.I. (2003) *Glass Phys. Chem.* (Engl. Transl. of Russian Fiz. Khim. Stekla), **29**, 60.

510 Priven, A.I. (2005) *Glass Technol.*, **46**, 263.

511 Priven, A.I. (1988) *Fiz. Khim. Stekla*, **14**, 589 (in Russian).

512 Rao, Q., Piepel, G.F., Hrma, P., and Crum, J.V. (1997) *J. Non-Cryst. Solids*, **220**, 17.

513 Ratcliffe, E.H. (1963) *Glass Technol.*, **4**, 113.

514 Rubenstein, C. (1964) *Glass Technol.*, **5**, 36.

515 Sasek, L. (1972) *Silikaty*, **16**, 209 (in Czech).

516 Schwiete, H.E. and Ziegler, G. (1955) *Glastechn. Ber.*, **28**, 137.

517 Sergeev, O.A. and Shashkov, A.G. (1983) *Teplofizika Opticheskikh Sred (Thermophysics of Optical Media)*, Nauka i Tekhnika, Minsk.

518 Sharp, D.E. and Ginther, L.B. (1951) *J. Amer. Ceram. Soc.*, **34**, 260.

519 Shaw, H.R. (1965) *Amer. J. Sci.*, **263**, 120.

520 Shaw, H.R. (1972) *Amer. J. Sci.*, **272**, 870.

521 Shchavelev, O.S. and Alekseenko, M.P. (1976) Thermal and Thermo-optical Properties of Glasses, in *Fiziko-Khimicheskie Osnovy Proizvodstva Opticheskogo Stekla (Physico-Chemical Background of Production of Optical Glasses)* (ed. L.I. Demkina), Nauka Publ. House, Leningrad, p. 117 (in Russian).

522 Shchavelev, O.S., Mokin, N.K., Babkina, V.A., and Plutalova, N.Y. (1989) *Fiz. Khim. Stekla*, **15**, 614.

523 Shchavelev, O.S., Babkina, V.A., and Golovin, A.I. (1993) *Fiz. Khim. Stekla*, **19**, 800.

524 Shchavelev, O.S., Polukhin, V.N., Murashev, S.V., Amosova, L.P., Yakobson, N.A., and Shchavelev, K.O. (1996) *Glass Phys. Chem.* (Engl. Transl. of Russian Fiz. Khim. Stekla), **22**, 418.

525 Stebbins, J.F., Carmichael, I.S.E., and Moret, L.K. (1984) *Contrib. Mineral. Petrol.*, 131.

526 Sun, K.-H. (1943) *J. Amer. Ceram. Soc.*, **30**, 282.

527 Takahashi, K. (1953) *J. Soc. Glass Technol.*, **37**, 3N.

528 Winkelmann, A. (1893) *Ann. Phys. Chem.*, **49**, 401.

529 Winkelmann, A. and Schott, O. (1894) *Ann. Phys. Chem.*, **51**, 730.

530 Winkelmann, A. (1897) *Ann. Phys. Chem.*, **61**, 105.

531 Young, J.C. and Finn, A.F. (1940) *J. Res. Nat. Bur. Stand.*, **25**, 759.

532 Loschmidt, J. (1865) Sitzungsberichte der kaiserlichen Akademie der Wissenschaften Wien, *Proc. Acad. Sci. Vienna*, **52**, 395–413.

533 Biscoe, J. and Warren, B.E. (1938) *J. Amer. Ceram. Soc.*, **21**, 287.

534 Bray, P.J. and O'Keefe, J.G. (1963) *J. Phys. Chem.*, **65**, 37.

535 Dell, W.J., Bray, P.J., and Xiao, S.Z. (1983) *J. Non-Cryst. Solids*, **58**, 1.

536 Green, R.I. (1942) *J. Amer. Ceram. Soc.*, **25**, 83.

537 Jellison, G.E., Feller, S.A., and Bray, P.J. (1978) *Phys. Chem. Glasses*, **19**, 52.

538 Loshagin, A.V., Nepomiluyev, A.M., Buler, P.I., and Sosnin, Y.P. (1990) *Fiz. Khim. Stekla*, **16**, 9 (in Russian).

539 Loshagin, A.V. and Sosnin, Y.P. (1994) *Glass Phys. Chem.* (Engl. Transl. of Russian Fiz. Khim. Stekla), **20**, 250.

540 Scheerer, J., Muller-Warmuth, W., and Dutz, H. (1973) *Glastechn. Ber.*, **46**, 109.

541 Silver, A.H. and Bray, P.J. (1958) *J. Chem. Phys.*, **29**, 984.

542 Svanson, S.E., Forslind, E., and Krogh-Moe, J. (1962) *J. Phys. Chem.*, **66**, 174.

543 Yun, Y.H. and Bray, P.J. (1978) *J. Non-Cryst. Solids*, **27**, 363.

544 Zhdanov, S.P. and Schmidel, G. (1975) *Fiz. Khim. Stekla*, **1**, 452 (in Russian).

545 Gehlhoff, G. and Thomas, M. (1926) *Z. Tech. Phys.*, **7**, 260.

546 Fluegel, A. (2007) *Modeling of glass liquidus temperatures using disconnected peak functions*. Presentation at ACerS Glass and Optical Materials Division Meeting, Rochester, NY, USA.

547 Fluegel, A. (2007) *Statistical Calculation and Development of Glass Properties*, http://glassproperties.com (accessed 28 January 2011).

548 Weinberg, M.C., Uhlmann, D.R., and Poisl, W.H. (1997) *The boron oxide anomaly revisited*. Proc. Second Conf. on Borate Glasses, Crystals and Melts (eds A.C. Wright, S.A. Feller, and A.C. Hannon), Society Glass Technology, Sheffield, Abingdon, p. 63.

549 Budhwani, K. and Feller, S. (1995) *Phys. Chem. Glasses*, **36**, 183.

550 Feller, S., Boekenhauer, R., and Zhang, H. et al. (1994) *Chimika Chron. New Ser.*, **23**, 315.

551 Esteves, M.J.C., Cardoso, M.J.E.M., and Barcia, O.E. (2001) *Indian Eng. Chem. Res.*, **40**, 5021.

552 Mendeleev, D.I. (1932) *Fundamentals of Chemistry*, Khimiya Publ. House, Moscow and Leningrad (in Russian).

553 Shultz, M.M. (1984) *Fiz. Khim. Stekla*, **10**, 129.

554 Yesin, O. (1947) *J. Fiz. Khim.*, **21**, 479.

555 Zachariasen, W.H. (1932) *J. Amer. Ceram. Soc.*, **54**, 3841.

556 Schtschukarew, S.A. and Müller, R.L. (1930) *Z. Phys. Chem.*, **5**, 5.

557 Yakhkind, A.K. (1965) Proc. IV Conf. Stekloobraznoe Sostoyanie (Glassy State), Moscow, Leningrad, p. 79 (in Russian).

558 Yakhkind, A.K. (1988) *Fiz. Khim. Stekla*, **14**, 723 (in Russian).

559 Yakhkind, A.K. (1989) *Fiz. Khim. Stekla*, **15**, 494 (in Russian).

560 Yakhkind, A.K. and Myasnikova, E.A. (1989) *Fiz. Khim. Stekla*, **15**, 726 (in Russian).

561 Yakhkind, A.K. (1990) *Fiz. Khim. Stekla*, **16**, 192 (in Russian).

562 Appen, A.A. (1949) *DAN SSSR*, **69**, 841 (in Russian).

563 Demkina, L.I. (1947) *DAN SSSR*, **58**, 807 (in Russian).

564 Dyshlova, T.A. and Sheludyakov, L.I. (1980) *On the relationship between chemical composition, structure and thermal expansion of silicate and aluminosilicate glasses*, Comm. No. 3, VINITI, Alma-Ata, N 5007-80 (in Russian).

565 Giordano, D. and Dingwell, D.B. (2003) *Earth Planet. Sci. Lett.*, **208**, 337.

566 Giordano, D. and Dingwell, D.B. (2004) *Earth Planet. Sci. Lett.*, **221**, 449.

567 Mysen, B.O. (1999) *Geochim. Cosmochim. Acta*, **63**, 95.

568 Mysen, B.O. and Frantz, J.D. (1993) *Eur. J. Mineral.*, **5**, 393.

569 Mysen, B.O. and Virgo, D. (1978) *Am. J. Sci.*, **2**, 1307.

570 Abbas, A., Delaye, J.-M., Ghaleb, D., and Calas, G. (2003) *J. Non-Cryst. Solids*, **315**, 187.
571 Delaye, J.-M. and Ghaleb, D. (1996) *J. Non-Cryst. Solids*, **195**, 239.
572 Fábián, M., Sváb, E., Proffen, T., and Veress. E. (2008) *J. Non-Cryst. Solids*, **354**, 3299.
573 Inone, H., Aoki, N., and Yasui, I. (1987) *J. Amer. Ceram. Soc.*, **70**, 622.
574 Soppe, W. and Hartog, H.W. (1989) *J. Non-Cryst. Solids*, **108**, 260.
575 Varsamis, C.-P.E., Vegiri, A., and Kamitsos, E.I. (2002) *Phys. Rev. B*, **65**, 104203
576 Shakhmatkin, B.A., Vedishcheva, N.M., Shultz, M.M., and Wright, A.C. (1994) *J. Non-Cryst. Solids*, **177**, 249.
577 Shakhmatkin, B.A., Vedishcheva, N.M., and Wright, A.C. (1997) Proc. Second Conf. on Borate Glasses, Crystals and Melts (eds A.C. Wright, S.A. Feller, and A.C. Hannon), Society Glass Technology, Abingdon, Sheffield, p. 189.
578 Priven, A.I. and Mazurin, O.V. (2003) *Glass Technol.*, **44**, 156.
579 Pushkareva, M.V. (1993) *Principles and Method of Calculation of Viscosity of Glasses in Wide Range of Compositions and Temperatures*, Ph. D. Thesis, Belgorod.
580 Murase, T. (1962) *Hokkaido Univ. Fac. Sci. J. Ser. 3*, **1**, 487.
581 Urbain, G., Cambier, F., Deletter, M., and Anseau, M.R. (1981) *J. Amer. Ceram. Soc.*, **80**, 139.
582 INTERGLAD (International Glass Database System) (1991–2010) New Glass Forum, http://www.newglass.jp/interglad_n/gaiyo/info_e.html
583 Backman, R., Karlsson, K.H., Cable, M., and Pennington, N.P. (1997) *Phys. Chem. Glasses*, **38**, 103.
584 Baret, G., Madar, R., and Bernard, C. (1991) *J. Electrochem. Soc.*, **138**, 2830.
585 Doerner, P., Gauckler, L.J., Krieg, H., Lukas, H.L., Petzow, G., and Weiss, J. (1979) *CALPHAD Comput. Coupling Phase Diagr. Thermochem.*, **3**, 241.
586 Rockett, T.J. and Foster, W.R. (1965) *J. Amer. Ceram. Soc.*, **48**, 75.
587 Hanni, J.B., Pressly, E., Crum, J.V., Minister, K.B.C., Tran, D., Hrma, P., and Vienna, J.D. (2005) *J. Mater. Res.*, **20**, 3346.
588 Dreyfus, C. and Dreyfus, G. (2003) *J. Non-Cryst. Solids*, **318**, 63.
589 Hunold, K. and Bruckner, R. (1980) *Glastech. Ber.*, **53**, 149.
590 Taylor, T.D. and Rindone, G.E. (1970) *J. Amer. Ceram. Soc.*, **53**, 692.
591 Gehlhoff, G. and Thomas, M. (1925) *Z. Techn. Phys.*, **10**, 544.
592 Gehlhoff, G. and Thomas, M. (1926) *Z. Techn. Phys.*, **3**, 105.
593 Priven, A.I. and Fluegel, A. (2007) *Evaluation of the most probable values of physical properties of materials and automated procedure of expertise of experimental data about properties of materials*, US Provisional Patent Application No. 60986951, EFS ID 2451705, US PTO.
594 Esbensen, K.H. (2000) *Multivariative data analysis – in practice*, CAMO ASA, p. 600.
595 Dyakonov, T.K. (1956) *Problems of the Theory of Similarity in Physico-Chemical Processes*, Publ. House Soviet Academy Sciences, Moscow.
596 Gutzow, I. and Todorova, S. (2010) *Phys. Chem. Glasses: Eur. J. Glass Sci. Technol.*, **B51**, 83.
597 Gutzow, I., Pye, L.D., and Dobreva, A. (1995) *J. Non-Cryst Solids*, **180**, 107.
598 Gutzow, I., Petroff, B., Möller, J., and Schmelzer, J.W.P. (2007) *Phys. Chem. Glasses: Eur. J. Glass Sci. Technol.*, **B48**, 168.
599 Kouchi, A. (1984) *Nature*, **330**, 550.
600 Kouchi, A. and Kuroda, T. (1990) *Jpn. J. Appl. Phys.*, **29**, 807.
601 Gutzow, I. et al. (2003) *J. Mater. Sci.*, **38**, 3747.
602 Pascova, R. and Gutzow, I. (1986) *Phys. Chem. Glasses*, **27**, 140.
603 Guencheva, V., Stoyanov, E., Gutzow, I., Guenter, C., and Ruessel, C. (2004) *Glass Sci. Technol.*, **5**, 244.
604 Guencheva, V., Ruessel, C., Fokin, V.M., and Gutzow, I. (2007) *Phys. Chem. Glasses: Eur. J. Glass Sci. Technol.*, **B48**, 256.
605 Iler, R. (1979) *The Chemistry of Silica*, Wiley, New York.
606 Mitcherlich, E. (1855) *J. Prakt. Chem.*, **66**, 257.

607 Ringer, A. (1912) *Z. Anorg. Chem.*, **32**, 212.
608 Gmelin, L. (1952) *Gmelins Handbuch der anorganischen Chemie*, 8th edition, No. 10, p. 257.
609 Gmelin, L. (1952) *Gmelins Handbuch der anorganischen Chemie*, 8th edition, No. 17, p. 278.
610 Parks, G.S., Huffman, F.E., and Cattoir, F.R. (1928) *J. Phys. Chem.*, **32**, 1366.
611 Nacken, R. (1945) *Hydrothermale Synthese von Quarzkristallen aus Quarzglass*, FIAT Review, German Science Reports No. 601.
612 Nacken, R. (1950) *Chem. Z.*, **74**, 745.
613 Thomas, L.A., Wooster, N., and Wooster, W.A. (1949) *Discuss. Faraday Soc. (London)*, **5**, 341.
614 Kolb, E. (1968) *Solubility of Selenium and Tellurium*. Proc. International Symposium, Montreal, October 1967, (ed. W.C. Cooper), Plenum Press, London.
615 Gutzow, I., Todorova, S., Kostadinov, L., Stoyanov, E., Guenter, C., Dunken, H., and Rüssel, C. (2005) Diamonds by Transport Reactions with Vitreous Carbon and from the Plasma Torch, in *Nucleation Theory and Applications* (ed. J.W.P. Schmelzer), Wiley-VCH Verlag GmbH, Berlin, Weinheim, pp. 256–308.
616 Samouneva, B., Bozadjiev, P., Djambaski, P., and Rangelova, N. (2000) *Glass Technol.*, **41**, 206.
617 Heusler, K.E. and Huerto, D.J. (1989) *Electrochem. Soc.*, **136**, 65.
618 Zaprianova, V., Raicheff, R., and Dimitrov, V. (1994) *J. Mater. Sci. Lett.*, **13**, 927.
619 Laski, G. (1928) Thermoelektrizität, in *Handbuch der Physik*, vol. 13 (eds H. Geigerand and R. Seheel), Springer Verlag, p. 183.
620 Zemansky, N.W. (1968) *Heat and Thermodynamics*, McGraw-Hill Book Company, New York, London.
621 Sommerfeld, A. (1962) *Vorlesungen über statistische Physik, Band V: Thermodynamik und Statistik*, Dieterich Verlag, Wiesbaden.
622 Basak, S., Nagel, R., and Giessen, B.C. (1980) *Phys. Rev.*, **B21**, 4049.
623 Baibich, N.N., Muir, W.B., Altounian, Z., and Guo-Hua, T. (1982) *Phys. Rev.*, **B 26**, 2963.
624 Kisa, A. (1908) *Das Glas im Altertum*, Verlag W. Hiersemann, Leipzig.
625 Kachalov, N. (1959) *Glass*, Publ. House Soviet Acad. Sciences, Moscow (in Russian).
626 Bezborodov, M.A. (1956) *Glass-Fabrication in Ancient Russia*, Publishing House of the Academy of Sciences of Belorussia, Minsk, (in Russian).
627 Pye, L.D., O'Keefe, J.A., and Frechette, V.D. (eds) (1984) *Natural Glasses*, North-Holland Physics Publishing, Amsterdam.
628 Smakula, A. (1962) *Einkristalle*, Springer, Berlin.
629 Ojovan, M.I. and Lee, W.E. (2005) *An Introduction to Nuclear Waste Immobolization*, Elsevier, Amsterdam.
630 Huner, C.C. and Ingram, M.D. (1986) *Phys. Chem. Glasses*, **27**, 51.
631 Nieuwenhuizen, T.M. (2001) ArXiv: cond-mat/0102528, vol. 128.
632 Nernst, W. (1926) *Theoretische Chemie*, 16. Auflage, F. Enke Verlag.
633 Planck, M. (1930) *Einführung in die Theorie der Wärmelehre*, S. Hirzel Verlag, Leipzig.
634 Gutzow, I., Petroff, B., Todorova, S., and Schmelzer, J.W.P. (2005) An Empirical Approach in the Thermodynamics of Amorphous Solids and Disordered Crystals, in *Nucleation Theory and Applications* (eds J.W.P. Schmelzer, G. Röpke, and V.B. Priezzhev), Joint Institute for Nuclear Research Publishing Department, Dubna, Russia, p. 146ff.
635 Einstein, A. (1942) The work and personality of Walther Nernst. *Sci. Mon.*, **54**, 195.
636 Stern, O. (1916) *Ann. Phys.*, **49**, 823.
637 Kammer, H.-W. and Schwabe, K. (1984) *Einführung in die Thermodynamik Irreversibler Prozesse*, Akademie Verlag, Berlin.
638 Sychev, V.V. (1973) *Complex Thermodynamic Systems*, Consultants Bureau, New York, London.
639 Szilard, L. (1925) *Z. Phys.*, **2**, 32.

640 von Neumann, J. (1932) *Mathematische Grundlagen der Quantenmechanik*, Springer Verlag, Berlin.

641 Putilov, K.A. (1971) *Thermodynamics*, Nauka, Moscow.

642 Fast, J.D. (1962) *Entropy: the Significance of Entropy and its Applications in Science and Technology*, McGraw Hill Book Company, New York.

643 Flory, P.J. (1953) *Principles of Polymer Chemistry*, Cornell University Press, Ithaca, New York.

644 Flory, P.J. (1956) *Proc. Roy. Soc.* (London), **A234**, 60.

645 Milchev, A. (1983) *Comptes Rend. Bulgarian Acad. Sci.*, **36**, 1415.

646 Kordas, G. (ed.) (2002) *Hyalos, Vitrum, Glass: History, Technology and Conservation of Glass and Vitreous Materials in the Helenic World*. Proc. 1st Int. Conf. Athens 2002, Glassnet Publ., Athens.

647 Schulz, H. (1928) *Die Geschichte der Glaserzeugung*, Akademische Verlagsgesellschaft, Leipzig.

648 Engels, S. and Stolz, R. *et al.* (eds) (1989) *ABC Geschichte der Chemie*, VEB Deutscher Verlag für Grundstoffindustrie, Leipzig.

649 Gutzow, S. (1964) *Technology of Glass*, State Publ. House, Sofia (in Bulgarian).

650 Balta, P. and Balta, E. (1976) *Introduction to the Physical Chemistry of the Vitreous State*, Editura Academiei, Bukarest, Abacus Press, Kent, England.

651 Zarzycki, K. (1991) *Glasses and the Vitreous State*, Cambridge Univ. Press, Cambridge.

652 Mazurin, O.V. (1978) *Vitrification and Stabilization of Inorganic Glasses*, Nauka Publishers, Leningrad.

653 Pye, L.D., Stevens, H.J., and La Course, W.C. (1972) *Introduction to Glass Science*, Plenum Press, New York.

654 Nieuwenhuizen, T.M. (1998) *Phys. Rev.*, **81**, 2201.

655 Kämpfer, B., Lukacs, R., and Paal, G. (1994) *Cosmic Phase Transitions*, Teubner, Stuttgart, Leipzig.

656 Avramova, K. (2002) *Cryst. Res. Technol.*, **37**, 491.

657 Kücük, E. (1999) *The Turkisch Glass Industry and Sisecam*, Istanbul.

658 Siedler, W.J. (1996) *Auf der Pfaueninsel*, Bertelsmann-Verlag, Berlin.

659 Kudryavtsev, M.V. (1954) *The Life and Work of M.A. Lomonosov*, Foreign Language Publ. House, Moscow.

660 Avramov, I. and Avramova, K. (2007) *J. Non-Cryst. Solids*, **353**, 218.

661 Zanotto, E.D. and Gupta, P.K. (1998) *Am. J. Physics*, **66**, 392.

662 Zanotto, E.D. and Gupta, P.K. (1999) *Am. J. Physics*, **67**, 260.

663 Pye, L.D. and Montenero, A. (2001) Plenary Lecture held at the XIXth International Congress on Glass, Edinburgh.

Index

a
Amorphous solids   1, 11, 179, 195, 311, 364

b
Bartenev–Ritland equation   61, 77, 112

c
Collection of glass data   6, 224, 253, 309

d
Defect crystals   313, 328, 353, 368, 370
Devitrification   1, 85, 92, 95, 100, 383
Disordered solids   181

e
Ehrenfest ratio   126
Entropy production   93, 102, 113, 115, 161, 316

f
Fictive pressure   144, 146
Fictive temperature   26, 143, 144, 146, 208
Free volume   9, 54, 76, 105, 150, 165, 167, 171, 189
Frenkel–Kobeko rule   93, 111, 162, 205

g
Generic theory of vitrification   119, 120
Glass formation   1, 7, 25, 36, 231
Glass stabilization   7
Glass transition   1, 40, 56, 63, 81, 82, 88, 93, 97, 110, 112, 119, 120, 125, 128, 140, 142, 152, 159, 161, 162, 186, 188, 195, 197, 206, 216, 307, 313, 317, 319, 354, 383
Glasses   1, 5, 8, 11, 13, 20, 36, 47, 69, 74, 82, 83, 92, 122, 150, 152, 153, 158, 159, 162, 173, 179, 182, 191, 215, 216, 220, 223, 232, 246, 252, 253, 255, 257, 264, 267, 271, 279, 286, 296, 301, 307, 311, 312, 329, 332, 348, 353, 356, 358, 362, 372, 377, 379, 383

Glasses at low temperatures   6, 67, 123, 193, 346, 355
Glassy polymers   3, 4

h
Heat capacity   19, 79, 127, 258

k
Kauzmann paradox   88, 216
Kinetic invariants   206, 217, 328

l
Liquid crystals   1, 11, 181, 183, 208, 217, 219, 312, 327, 360

m
Metallic glasses   3, 4, 217, 347, 369, 376

p
Phase transition   26, 33–35, 125, 126, 207
Phases   12, 26, 27, 31
Prediction of glass data   223, 255, 265, 267, 268, 279, 280, 285, 296, 301, 306
Prigogine–Defay ratio   125, 126, 130, 131, 133, 135, 137, 143

r
Reactivity of glasses   329, 353
Relaxation   13, 82, 91, 97, 104, 113, 160, 165, 174, 176, 177, 196, 307, 314, 333, 372, 385
Reliability of glass data   224, 229, 232, 244, 249

s
SciGlass   224, 225, 227, 230, 242, 253, 257, 276, 286, 297, 300
Silicate glasses   4, 51, 56, 154, 156, 184, 192, 233, 235, 238, 239, 258, 274, 312, 380
Simon's approximation   67, 69, 158, 185, 193, 194, 213, 318, 334, 336, 364, 367

Solubility of glasses   312, 329, 332, 334, 338, 349, 350, 372, 374
Spin glasses   1, 183, 219
Structural order parameter   97, 99, 105, 108, 123, 131

t

Thermodynamic coefficients   31, 76, 127, 128, 131, 135, 142, 166, 348, 378
Thermodynamic fluctuations   50
Thermodynamic invariants   218
Thermodynamics of irreversible processes   13, 89, 95, 101, 160, 179, 195, 314, 359, 363
Third law of thermodynamics   21, 82, 192, 364

v

Vapor pressure of glasses   120, 332, 333, 335, 339, 368, 372, 374
Viscosity   10, 46, 48, 50, 51, 53–55, 57, 84, 149, 150, 165–167, 169, 171, 188, 196, 220, 231, 256, 260, 278, 280, 282–285, 293, 321, 375
Vitreous state   1, 2, 7, 11, 39, 68, 82, 83, 332, 342, 355, 386
Vitrification   1, 17, 40, 43, 63, 81, 82, 91, 95, 117, 126, 142, 158, 179, 184, 191, 215, 219, 311, 315, 318, 327, 378, 383

z

Zero-point entropy   73, 155, 181, 182, 211, 213, 362, 368, 376, 378